Conservation Drones

Conservation Drones

Mapping and Monitoring Biodiversity

Serge A. Wich, PhD
Liverpool John Moores University, UK, and University of Amsterdam, The Netherlands

Lian Pin Koh, PhD
University of Adelaide, Australia, and Conservation International, USA

Great Clarendon Street, Oxford, OX2 6DP,
United Kingdom

Oxford University Press is a department of the University of Oxford.
It furthers the University's objective of excellence in research, scholarship,
and education by publishing worldwide. Oxford is a registered trade mark of
Oxford University Press in the UK and in certain other countries

First Edition published in 2018
Impression: 1

Published in the United States of America by Oxford University press
198 Madison Avenue, New York, NY 10016, United States of America

British Library Cataloguing in Publication Data
Data available

Library of Congress Control Number: 2018933616

ISBN 978–0–19–878761–7

DOI:10.1093/oso/9780198787617.001.0001

Printed and bound by
CPI Group (UK) Ltd, Croydon, CR0 4YY

Preface

Serendipity plays a large part in science and was also crucial in how we started our drone work. Both of us had been working in Southeast Asia's forests to study its rich biodiversity for many years when we first met in May 2011, over a coffee in the aptly named Greencafe of the Swiss Federal Institute for Science in Zurich. We planned to discuss the ongoing impact of oil palm plantation development on the species we studied and particularly the orangutans. Soon enough we were discussing how costly and time intensive it is to obtain data at regular intervals for species such as orangutans over large areas and how wonderful it would be if we could monitor the nests that orangutans build from the air. Lian Pin had just started to fly remote control fixed wing aircraft so we decided to see if we could fit a camera under a fixed wing, and fit an open source flight controller to it as well so that we could fly pre-programmed missions. A few months later, after some tests in Switzerland, we found ourselves in Sumatra flying a Bixler fixed wing with an APM 1 and camera over the forests. In a few days we flew 36 missions and collected a wealth of data that made us very enthusiastic about this technology. We shared our ideas and findings with colleagues and because of their positive feedback started the conservationdrones.org website and coined the term conservation drones. Since then we have had the opportunity to work in a large number of locations around the world and posted these on our website in the hope that this would inspire others to explore the use of drones for their work. We hope that we have stimulated a few people to start using drones through the website and our presentations and papers. The last seven years have seen an enormous growth in the use of drones for conservation work and a large amount of exciting work has been done by researchers around the world. With this book we aim to share some of our own work, but fortunately much more the work of others, in the field of drones, conservation, and ecological research. We hope that this will further contribute to excite people about the opportunities that this field offers for conservation.

Acknowledgements

We owe a great amount of gratitude to many colleagues and friends who contributed to this book in numerous ways. We apologize for those who we have inadvertently forgotten to mention. Some contributed by reviewing chapters, some by providing discussion and ideas about certain topics, some by providing images and figures, some by collecting data, some by stimulating us to write this book, some by helping us to build drones, some by kindly sharing their drone expertise with us, some by providing us with hardware, some by allowing us time at work to focus on drones when many people thought this was 'just a hobby', some by providing a supportive work environment, and many by being on various data collection missions with us in the field and being a crucial part of the experience we have collected over the years. In alphabetical order we thank Christiaan Adams, Marc Ancrenaz, Sander van Andel, Chris Anderson, Brandon Basso, Frederic Bezombes, Noémie Bonnin, Hidde Boonstra, Claire Burke, David Burton, Rhett Butler, Ahimsa Campos-Arceiz, Mike Checkley, Adrienne Chitayata, Juanita Choo, Kenneth Clarke, Chris Collins, David Dellatore, Brenden Duffy, Regina Frey, Andy Goodwin, Benoit Goossens, Molly Hennenkam, Christiaan van der Hoeven, Jeff Kerby, Adam Kilpatrick, Francis Lilley, Wee Siong Lim, Steve Longmore, Tony Lynam, David Mobach, Matt Nowak, Keeyen Pang, Ryan Pang, Alex Piel, Lilian Pintea, Maisie Rashman, Ramesh Raja Segaran, Carel van Schaik, Ahmed Al-Shamma'a, Ian Singleton, Danica Stark, Andy Tattersall, Gokarna Thapa, Graham Usher, Jaap van der Waarde, Peter Wheeler, Simon Wunderlin, and the large number of field staff that have supported our various missions.

We specifically would like to express our gratitude to Andy Goodwin, Dominique Chabot, Jeff Kerby, and Owen McAree for kindly reviewing most or all of the chapters. We appreciate Noémie Bonnin's tremendous help with going through some of the literature for Chapter 3. Perry van Duijnhoven has voluntarily designed many of the figures in this book and been very patient with the many 'small' amendments we asked for.

Finally SW is grateful for the love and support from Tine, Amara, and Lenn during the writing of this book and being patient with me being away during the many fieldtrips 'droning' around the world. Their patience has been enormous during the weekends and evenings that were absorbed by writing. LPK dedicates this book to his lovely wife, Juanita, for holding the fort at home while he travelled around the world, 'working'.

During writing LPK was supported by the Australian Research Council, and Conservation International. We thank the following organizations in no systematic order for supporting our work financially or in other ways. National Geographic, Waitt Foundation, Mongabay, US Fish and Wildlife Services, Denver Zoo, Philadelphia Zoo, Conservation International, Wildlife Conservation Society, Chester Zoo, Ugalla Primate Project, RemoteInsights, Google, 3DRobotics, WWF Netherlands and several other WWF offices, Orangutan Information Centre, ISTAT Foundation Humanitarian Grant, Orangutan Conservancy, Swiss Federal Institute of Technology, PanEco, Jane Goodall Institute, University of Zurich, Liverpool John Moores University, HUTAN, Indianapolis Zoo, Margot Marsh, Science and Technology Facilities Council (STFC), and the Sumatran Orangutan Conservation Programme.

Contents

CHAPTER 1

Deciding to use a drone

1.0 Introduction

The recent proliferation of non-military applications of drones in many aspects of our society, from filming to asset management and farming, has also spread into conservation (Koh and Wich 2012, Chabot and Bird 2015, Floreano and Wood 2015, Wich 2015, Christie, Gilbert et al. 2016, Crutsinger, Short et al. 2016). Within the conservation community drones have mainly been used for three aspects of conservation efforts: 1) to determine the distribution and density of animal species; 2) for mapping of land-cover and land-cover change; 3) anti-poaching activities. These aspects will be dealt with in more detail in later chapters of this volume along with related aspects of spatial ecology that are relevant to conservation science and management. In this chapter we will provide a brief introduction to these three main aspects and then focus on which considerations are important to take into account when deciding to start applying drones for conservation or not. We will start, however, with providing an overview of alternative aerial data-collection technologies. Deciding which technology to use for data gathering is not always straightforward and will likely depend on multiple factors (Morgan, Gergel et al. 2010, Toth and Jóźków 2016). We will give an overview of some aspects that are important to consider when starting the process of thinking about data collection. Then we will discuss which equipment and infrastructural needs are important to consider when using drones and we will end by discussing some general considerations related to the usage of drones for conservation.

1.1 Alternative technologies to acquire aerial data

Conservationists have a long history of using aerial and space platforms to obtain imagery for land-use mapping and monitoring, as well as for the study of animal distribution and density. Here we discuss the most common ones: manned aircraft, satellites, kites, and balloons. All of these come with a set of advantages and disadvantages that are useful to think through when considering collecting aerial and/or space data for conservation projects (see Table 1.1).

1.1.1 Manned planes and helicopters

Aerial photography from manned planes started with one of the pioneers of flying, Wilbur Wright, in 1908 and seriously took off during World War I to provide imagery of the frontlines (Professional Aerial Photographers Association 2016). After World War I, mapping of areas for non-military purposes started in earnest (Professional Aerial Photographers Association 2016). It has been and still is an important aspect of natural resource management (Paine and Kiser 2012). Manned aircraft (planes or helicopters, Figure 1.1) have been used widely by conservationists to map land-cover and to examine land-cover change (Al-Bakri, Taylor et al. 2001, Plieninger 2006, Gerard, Petit et al. 2010). Similarly, aerial census work to obtain data on animal distribution and density has been extensively used (Caughley 1977, Buckland, Anderson et al. 2001, Buckland, Anderson et al.

Conservation Drones: Mapping and Monitoring Biodiversity. Serge A. Wich & Lian Pin Koh. Oxford University Press (2018).
 DOI:10.1093/oso/9780198787617.001.0001

Table 1.1 Aerial and space data collection platforms and their constraints

	Area covered	Resolution	Regulations*	Wind	Flight duration	Costs	Other constraints
Satellites	Large	Relatively low	NA**	NA	Multiple years	High for high resolution	Cloud cover
Manned aircraft	Large	Medium	Aviation authorities	Light-strong	Hours	High	
Kites	Small	High	Aviation authorities	Light-strong	As long as there is wind	Low	
Balloons	Small	High	Aviation authorities	Light	Multiple days	Low	Hydrogen or helium might not be locally available or be difficult to transport to the site
Drones	Medium	High	Aviation authorities	Light-moderate	Hours or parts of hours	Medium	

* Regulations for flying aerial platforms vary between countries, often depend on the weight of the platform, and the purpose of the flight (e.g. hobby, research, or commercial). ** NA = not applicable.

Figure 1.1 Researchers climbing into a hovering helicopter above the Batang Toru river in Sumatra, Indonesia, and researchers preparing for a flight with a Cessna fitted with gliders to find orangutan nests in the peat swamp forests of Indonesian Borneo. © Serge Wich.

2004). In some cases, such as that of the Cape fur seals (*Arctocephalus pusillus pusillus*) in southern Africa aerial census work has allowed for almost four decades of population monitoring (Kirkman, Yemane et al. 2013). For instance aerial counts of animals in Africa have been conducted since the mid-1950s (Jachmann 2002), but in contrast to surveys with drones the detection of animals is mainly done by human observers and not from photos after the flight. There have been some concerns that aerial counts of large herbivore species are considerably lower than ground counts (Jachmann 2002) and can be influenced by factors such as the low probability of spotting single animals, small groups of animals, and animals that were not visible because they were behind or under vegetation. In addition, animals that react to the aircraft and move are more likely to be recorded

(Jachmann 2002). Furthermore, there are potential biases introduced by factors relating to humans such as visual acuity, concentration, and visual attention (Fleming and Tracey 2008). Important, but not often considered, is how the characteristics of different survey platforms (helicopters and fixed wing planes) can affect the detection by observers on a manned aircraft (Noyes, Johnson et al. 2000, Fleming and Tracey 2008).

Although using manned aircraft has been widely applied, there are also three challenges to using these. First, aerial surveys with manned aircraft often face high costs. These costs are highly variable for different parts of the world but can be substantial and often prohibitive for conservationists, particularly when regular flights are required. Second, availability of manned aircraft can be an issue in certain parts of the world. For instance, in parts of Indonesia where we have worked it is sometimes not possible to rent small planes to fly low over forest areas or remote areas can be too far away for a small plane to reach. In some areas with large river systems planes that can land on water can be an option. Third, there is substantial risk associated with aerial surveys using manned aircraft. Crashes are a leading cause of death among wildlife biologists (Sasse 2003). Fourth, flying manned aircraft for anti-poaching efforts is extremely risky and can lead to pilots being killed.[1]

In many instances the census method applied is substantially different from census work using drones because much of the census work using manned aircraft is conducted directly by the observers on board the aircraft (Buckland, Anderson et al. 2001, Buckland, Anderson et al. 2004), whereas with drones the most common method is to have a camera that faces down to take images on which animals or their signs are counted later on (Chabot and Bird 2012, Koh and Wich 2012, Chabot and Bird 2015, Wich 2015). Direct counts from a manned aircraft are time efficient because at the end of the flight the data is ready to be analysed or close to being in the right format for analyses. This stands in contrast to data collection with a camera on a drone where after a drone flight there are potentially thousands of photos that need to be examined to detect animals or their signs. Although inspecting images is conducted mostly manually at the moment, there are promising computer vision methods being developed to (semi) automate this process (e.g. Chen, Shioi et al. 2014, van Gemert, Verschoor et al. 2014). An important advantage of collecting data with a camera on board a drone is that there is a record of the data that can be re-analysed and shared if needed.

[1] https://www.theguardian.com/world/2016/jan/30/british-pilot-killed-by-elephant-poachers-tanzania (accessed 7 November 2017)

1.1.2 Satellites

Classifying land-cover types and changes in land-cover types over time is usually conducted with the aid of satellite images (Horning, Robinson et al. 2010) and, particularly at the global scale, it will continue to be conducted in this way (Hansen, Potapov et al. 2013). Since 1972 satellites have been taking images of the Earth at various resolutions (see Table 1.1), at various temporal intervals, and with various sensors (Table 1.1; for extensive lists of satellites and their specifications see Horning, Robinson et al. 2010, Toth and Jóźków 2016). Analyses using these images have given us a wealth of information about global land-cover and change as well as detailed assessments of smaller areas.

There are, however, four challenges with using satellite images. First, in the humid tropics, but also in the Arctic, areas can be covered by clouds for prolonged periods of time, which renders satellite image acquisition of an area impossible or significantly delays acquisition (Hansen, Roy et al. 2008, Mulaca, Storvoldb et al. 2011). In arctic regions passive sensors also do not provide information during the prolonged dark periods and therefore do not provide scientists with the data they require in a region that is experiencing rapid climate and ecological change (Post, Forchhammer et al. 2009, Serreze, Barrett et al. 2009, Post, Bhatt et al. 2013). Second, conservationists would often prefer to use high resolution imagery and, even though these high resolution images are available, these are costly and therefore often not within the budgets of conservation workers (Table 1.2). Third, even though several satellites acquire imagery with short

Table 1.2 Selection of satellites

Satellite	Pixel resolution (m)*	Repeat frequency (days)	Estimated costs (US $) km^2**
Worldview-3	1.24	<1	14.50–17.50
GeoEye-I	1.65	2–8	14.50–17.50
Spot 4/HRVIR	20	2–3	0.8
Landsat/TM & ETM+	30	16	Free
Sentinel 2	10, 20, 60	5	Free
MODIS	250, 500, 1000	1–2	Free

Table based on Horning, Robinson et al. 2010 with additional information. * All pixel resolutions are for the multispectral images. ** Price estimates for archived images and depends on exact bands required (http://www.landinfo.com/satellite-imagery-pricing.html). Even though prices are per km^2 the minimum area that needs to be purchased is larger. For example for Spot 4 the minimum area is 1,000 km^2. See http://www.wmo-sat.info/oscar/ for a complete list of Earth observation satellites that have been launched or are being planned for launch in the near future.

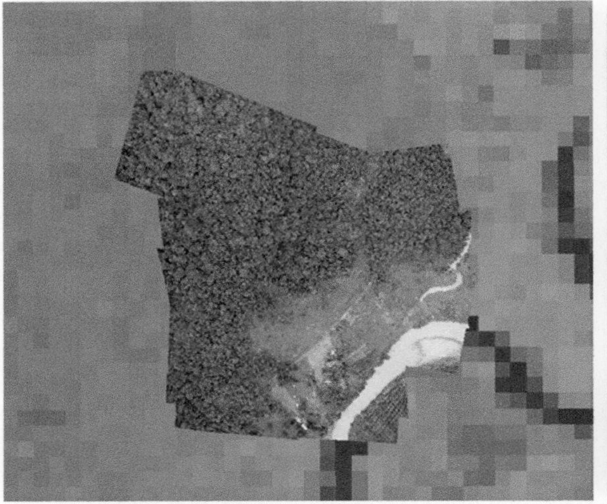

Figure 1.2 Left panel: Drone georeferenced orthomosaic superimposed on a Landsat satellite image from an area in Sumatra. Landsat pixel resolution is 30m per pixel, drone orthomosaic pixel resolution is 3cm per pixel. The orthomosaic was produced with Pix4Dmapper (www.pix4d.com). Right panel: Single image from drone for the same area as the orthomosaic. © Serge Wich.

refresh rates (Table 1.2), the almost persistent cloud cover in some tropical areas and the high costs of high resolution images make repeat data for land-cover analyses prohibitively expensive for many conservationists. Fourth, in cases where immediate data are necessary it is possible to task satellites to obtain data within a few days, but this comes at a premium price. Drones can therefore be a useful addition to the use of satellite images because they allow for the acquisition of very high resolution imagery and allow for a large flexibility in the timing of such data acquisition (Figure 1.2).

Recently a number of companies (e.g. Planet Labs[2]) have been deploying small satellites (cube-sats, nanosats) into orbit with the aim of providing high resolution images (~1 m) at a much reduced price and at daily refresh rates once the full constellations of satellites are up.

[2] www.planet.com (accessed 7 November 2017)

The World Meteorological Organization's Observing Systems Capability Analysis and Review tool (OSCAR)[3] is a complete list with current status of the very many Earth observation satellites that have been launched or are being planned in the near future.

Although satellites have mainly been used by conservationists to obtain land-cover classifications and analyses of land-cover change, the high resolution satellite images have recently also been used to detect animals or their signs. An early comparison of aerial counts from planes and counts of several geese species using very high resolution satellite images showed promising results (Laliberte and Ripple 2003). A pioneering study published in 2012 managed to obtain a population estimate of emperor penguins (*Aptenodytes forsteri*) from space using a combination of medium and very high resolution satellite images (Fretwell, LaRue et al. 2012). Subsequent studies have used satellite images (GeoEye-1) to obtain accurate counts of Burchell's zebras (*Equus quagga burchellii*) and blue wildebeest (*Connochaetes taurinus*) in the Serengeti National Park, in Tanzania (Yang, Wang et al. 2014). Studies of marine mammals such as elephant seals (*Mirounga leonina*) (McMahon, Howe et al. 2014) and Southern right whales (*Eubalaena australis*) (Fretwell, Staniland et al. 2014) have been conducted as well. Although these approaches are very promising in terms of their accuracy compared to other census methods, they have several challenges such as cloud cover, difficulty or impossibility of detecting animals under vegetation, and the necessity of complex analyses (Yang, Wang et al. 2014).

1.1.3 Kites

In 1882 the English meteorologist E. D. Archibald claimed to have taken the first aerial photos from a kite, but this imagery no longer exists (Verhoeven 2009). The earliest surviving images are from Arthur Batut in 1888 (Batut 1890). Due to the weight of cameras being used a series of kites had to be used to lift the cameras. This method provided the world with some of the first stunning images from the air such as those of the devastation in the aftermath of the San Francisco earthquake (Professional Aerial Photographers Association 2016). Nowadays there are opportunities to 3D print aerial camera mount kits that pivot[4] so that the camera remains in a stable position under relatively small kites, or to purchase them online.[5] Kite photography provides a good opportunity for a very low-cost aerial photography platform that allows for very high resolution mapping. Although kites are probably best suited for small-scale area mapping, area coverage can be substantial if large areas can be covered, for example by foot along a coastline or by boats in an estuary or along rivers. This method has been used in the aftermath of the BP Deepwater Horizon disaster off the coast of Louisiana to map coastal areas.[6] Kite photography is less suitable for areas that are forested because only the edge of the forest can be photographed. An additional benefit of kite photography is that relatively high wind speeds do still allow for operations, whereas for lighter drones high winds can mean that the wind speed is higher than the speed at which the drone is safe to operate. Varying wind speeds have been argued to be less suitable for kite photography operations (Verhoeven 2009), but the development of kites with self-stabilizing cameras (Murray, Neal et al. 2013) will probably allow for the usage of kites in more varying wind speeds as well because photos under nadir can still be obtained.

Although kites have been used extensively in archaeology to map excavation sites and cultural heritage sites (Verhoeven 2009) the use of kites for conservation research has been limited so far, but recently researchers have been arguing that kites are suitable for environmental research as a low-cost option that might have less constraints on usage by regulations, are logistically easier in remote areas, and are useful in areas where wind speeds are often not conducive for flying of smaller drones (Anderson 2016, Anderson, Griffiths

[3] http://www.wmo-sat.info/oscar/ (accessed 7 November 2017)

[4] Fotr example, http://www.thingiverse.com/search/page:1?q=picavet (accessed 7 November 2017)

[5] https://publiclab.myshopify.com/collections/mapping (accessed 7 November 2017)

[6] https://publiclab.org/archive (accessed 7 November 2017)

et al. 2016, Duffy and Anderson 2016). An early study assessed the possibility of using kites for shoreline mapping (Scoffin 1982), and subsequent studies have continued to use kites for mapping of vegetation, ice networks, and intertidal systems, and making orthomosaics to map vegetation types (Boike and Yoshikawa 2003, Smith, Chandler et al. 2009, Bogacki, Giersz et al. 2010, Bryson, Johnson-Roberson et al. 2013). Organizations such as Public Lab[7] have been promoting kite and balloon aerial photography and provide details on the kit that can be used on their website.

Studies that have used kites for aerial census of animals are sparse. Studies have applied kite photography to count Adelie penguins (*Pygoscelis adeliae*) (Fraser, Carlson et al. 1999) and New Zealand sea lions (*Phocarctos hookeri*) (Cawthorn 1993), and more recently beluga whales (*Delphinapterus leucas*).[8]

1.1.4 Balloons and blimps

The use of balloons to take aerial images has a long history that started in 1858 in Paris when the French photographer Gaspard-Félix Tournachon flew with a balloon over Paris (Newhall 1982). Unmanned balloons in archaeology have been in use since 1932 using hot air, hydrogen or helium to provide lift (Guy 1932, Verhoeven 2009). Balloons lack a system to power them through the air and non-rigid airship systems (i.e. systems that completely rely on the internal pressure to maintain their shape) that do have a propulsion mechanism are called blimps. Balloons and blimps of various kinds have been used to map vegetation and analyses of such images have been applied to distinguish various plant species assemblages (Miyamoto, Yoshino et al. 2004), mapping of agricultural fields (Thornton, Fawcett et al. 1990, Inoue, Morinaga et al. 2000, Jensen, Apan et al. 2007), plant communities (Pitt, Glover et al. 1996), vegetation mapping (Miyamoto, Yoshino et al. 2005), river corridors (Vericat, Brasington et al. 2009), and geomorphology and soils (see references in Verhoeven 2009). The development of advanced tethered balloons has recently also occurred for the police and military.[9]

Balloons can be very useful aerial platforms as they can stay aloft for prolonged periods of time (days rather than hours). They can either be fixed in one location by tethering them to the ground or they can be moved around. The fact that they can be tethered has the advantage that the sensors can be powered from the ground which alleviates issues surrounding the batteries that are needed to power sensors over prolonged periods of time. Tethering also allows for a direct download of the data to a ground station.

1.2 To drone or not to drone

There are several aspects to consider before deciding whether a drone is the right platform. A key aspect in thinking about using a drone or not is to determine exactly what the data are that one would like to collect and whether these can be collected by an alternative aerial or space platform. A simple decision tree can guide this process. The first question in this tree can be whether the data that one requires are collected in other ways or are not collected at all at present. If the data are not being collected at the moment and there are no alternative technologies (see section 1.1 on alternative technologies) to obtain the data then it is worth investigating whether using drones would be possible. If the data are being collected by other methods or can potentially be collected by other methods then it is important to evaluate whether the costs, both money and time, of those methods are greater or smaller than using drones. If the costs of current methods or non-drone alternatives are higher than using drones then the use of a drone is worth examining further. At present there is unfortunately little information available about the costs of different data collection methods (for a rare, but not all-inclusive comparison of elephant surveys see Vermeulen, Lejeune et al. 2013; see also Groves, Alcorn et al. 2016 for an example on counting salmon nesting sites). Both these studies indicated that drone surveys were more costly than the manned aircraft

[7] www.publiclab.org (accessed 7 November 2017)

[8] http://www.ifaw.org/united-states/news/how-carefully-count-beluga-herd-including-every-baby-beluga (accessed 7 November 2017)

[9] http://www.military.com/NewContent/0,13190, Defense watch_072105_Helms,00.html (accessed 7 November 2017)

alternatives, but both do not include all costs, such as those from data analysts. Although it is often known what the costs of an actual survey are in terms of renting a plane and car, obtaining a satellite image, ground teams, food, travel, accommodation, etc., there is little information available about costs of the data analyses in terms of software, hardware, and labour. If these were similar between the various data collection options comparing these would be relatively straightforward, but unfortunately they are not. Although it is beyond the scope of this book to try to attempt to estimate the costs of the various methods in detail an example might help to understand the complexity of such comparisons (Table 1.3).

To determine the distribution and density of orangutans (*Pongo* spp.) researchers have traditionally relied on conducting ground surveys during which the sleeping platforms (here referred to as 'nests') that orangutans build are counted along a straight line transect that is walked by a minimum of two observers (van Schaik, Priatna et al. 1995). Recently, researchers have been using drones to fly over the rainforests where orangutans occur and count their nests on photos taken from the drone. Both methods can yield density estimates (Hennekam 2015, Wich, Dellatore et al. 2016). For this simple comparison we will assume that the transect can be reached in a day walking from the basecamp which is on the edge of the forest. In addition, we assume that permissions to collect both the ground and aerial data have equal costs.

The ground survey method requires two trained researchers to walk for a day to reach the transect, spend one day on conducting the data collection, and one day to walk back. Training researchers to collect such data usually takes 2–3 days. During a transect data are recorded on a datasheet with a pen. Equipment needed during the survey includes a compass, global positioning system (GPS), a measuring tape (25 or 50 m), a machete, and a pair of

Table 1.3 Comparison of ground and drone orangutan surveys

Item	Ground survey	Drone survey
Training data collection and data analyses	One week	Two weeks
Time to collect data	3 days	0.5 days
Data processing software costs	Free	US$549 and up
Computer	Modest processing power	High processing power
Survey equipment costs	~1500 US$*	~6000 US$*
Time to reach density estimate	4 days	2 days

* For ground surveys two pairs of binoculars, one GPS, a laptop with modest processing power, and the other equipment was estimated to be 1500 US$. For the drone survey we estimated the cost by using a basic drone from www.hornbillsurveys.com (assessed on 11 February 2016: US$4200, costs of shipping this to Indonesia, and a laptop with high processing power for data analyses (~1500 US$)). The drone images can be processed into a georeferenced orthomosaic with several software packages (see Chapter 7 for details). Although non-commercial options exist, these can be complicated to use for non-experts and therefore a commercial user-friendly package is advisable. The most affordable desktop commercial option for educational users (we assume here that the researchers are affiliated to a university) at the moment is Agisoft Photoscan (www.agisoft.com, assessed on 11 February 2016). Several cloud options are also available (details in Chapter 7).

binoculars. In addition, equipment to camp out in the forest is needed as well as food for three days. Once the forest team returns to camp, the data can be entered on a computer in a free software package (Thomas, Buckland et al. 2009) and processed in a few hours to yield a density estimate. Thus in four days a density estimate for that particular transect can be achieved.

For the drone surveys a team of two researchers would need to be trained for two weeks or so (depending on the system) before they would be sufficiently knowledgeable and confident to pilot the drone. They would need a drone with a camera, their survey equipment, and a laptop to act as a ground control station. The data collection for the transect will involve some careful planning of the mission including preparation of the flight, the flight itself, and some post-flight checking and packing of the system. All combined, that could take as little as three hours. The researchers have chosen to fly in a lawnmower pattern over the area where the transect is located so that all the images can be processed into a georeferenced orthomosaic on which the nests can be counted, and the surface area is known, which is needed to calculate density. After the flight the images will be downloaded to the computer and processed by the photogrammetry software to make the orthomosaic. The images contain a geotag (containing latitude, longitude, and altitude) that comes from the GPS in camera or from the GPS that is used by the autopilot (more details in Chapter 7). The time needed to process the orthomosaic will depend on the specification of the computer, but will be a few hours at least. So realistically after day one the drone team has completed the data collection and has a georeferenced orthomosaic, while the ground team just arrived at the transect location. On day two the drone team will spend the first part of the day on carefully looking for nests on the images. The camera on the drone is usually set at a short interval, so typically one hour of flight will have yielded 1800 photos. A trained observer can check an image for nests in 1 minute, which would result in 30 hours of checking photos for nests if all the images need to be assessed. In this case only a subset for the transect area would need to be assessed and one morning for two observers would likely suffice. Once the number of nests is known the density can be calculated by using a previously established relationship between the density established from the ground surveys and aerial surveys (Hennekam 2015). Thus after two days the drone team has a density estimate.

The above comparison is obviously somewhat simplistic, but provides some of the aspects that are relevant when comparing methods. The example is based on a single survey with one line transect. In this case the ground team would be less costly to deploy than the drone team. This is due to the high upfront costs of the drone equipment, a high processing power computer, and costly data-processing software. In reality surveys are almost never restricted to a single line transect in a single area. Usually large-scale survey designs are used over large areas. As scale increases the data collection with the drone system will become less costly per incremental area surveyed than the ground surveys because the time needed for the ground surveys quickly becomes very large (Wich, Singleton et al. 2016). There is a clear need for careful and all-inclusive comparisons between survey methods so that conservationists can make decisions about which data collection method to use based on careful costs estimates for the various methods.

Generally when considering whether to use drones there are several aspects to consider. There is the choice of drone system (e.g. multirotor or fixed wing) and the ground control hardware and software associated with drones, but there are several other aspects too (Table 1.4). There is a wide variety of drone options on the market (see Chapter 2) for a variety of price brackets. Then there is a variety of sensors that can be fitted to the drone depending on what the requirements are for the user's data collection needs (see Chapter 3). Most of the sensors collect some sort of image and very quickly these can run into many terabytes so storage and back-up can become a non-negligible cost. There is also the cost of batteries, although this is not high if the batteries are carefully maintained (see Chapter 2). Finally, there are costs of flight training and, depending on the regulations, costs of the qualification to fly as a pilot (see Chapter 4).

In addition to the cost evaluation there are a number of other general considerations when planning

Table 1.4 Aspects to consider when considering using a drone

System	Flight parameters	Sensors	Telemetry/ Transmission	Maintenance	Logistics
Ease of software user interface	Flight duration	Sensor compatibility of platform	Telemetry distance	General user support	Ease of transportation
System complexity	Flight velocity (cruise)		Legal aspects of telemetry transmission power	Repair (self/service)	Can the batteries be transported internationally?
System material	Flight velocity (maximum)		Photo/video transmission distance	Can spare parts be easily purchased?	
System propulsion	Maximum wind speed		Remote control range		
Set up complexity	Required size of take-off area				

Based on information from Patrick Ribeiro from https://openforests.com/

Table 1.5 Drone planning and operating aspects in different environments

Operating environment/ specific challenge	Safety & regulations	Societal considerations	Wind	Fine particles	Solar effects (glint, shadows, albedo)	Spatial constraints (GNSS difficulties)	Telemetry issues	Topography issues
Coastal	X	X	X	X	X			X
Dry land	X	X	X	X				X
Polar	X	X	X					X
Dense forest	X	X	X			X	X	X
High altitude	X	X	X					X

Based on (Duffy, Cunliffe et al. 2017).

to use drones and some of those vary on location (Duffy, Cunliffe et al. 2017). The Duffy et al. study compiled the experiences of a group of researchers that have been flying drones in a number of key environments to provide advice to the environmental drone community. They describe a set of aspects to consider when flying drones in different environments (Table 1.5). These relate to important considerations to make pre- and during flight operations in terms of safety and regulations, factors influencing safe operations, and successful data acquisition.

A few key considerations are highlighted here and in Chapter 4.

1.2.1 Terrain

An important aspect of flying drones is the terrain. Drones are very capable of flying level above flat terrain and flight is efficient then in terms of energy usage because the motor can turn at a relatively constant speed. Most drones can be programmed to follow the terrain through the use of a digital elevation model or elevations that a user programs in, but climbing does consume more battery power so terrain-following flights will consume more energy than level flights. Although high resolution digital elevation models exist, the digital elevation models most often used in ground control software are often of relatively low resolution unless a user provides higher resolution data. For instance the standard global 30 m resolution SRTM (Shuttle Radar Topography Mission)[10] has the advantage of having global coverage but the disadvantage of relatively low resolution for areas that have very undulating terrain. Such low resolution means that certain peaks or valleys might not be indicated by the digital elevation model and that the drone therefore needs to fly at a higher elevation above ground to reduce the risk of crashes.

1.2.2 Weather

An important aspect of drone flying is to be aware of the potential risks of adverse weather conditions. There are a growing number of specific apps that help plan flights in relation to predicted and actual weather conditions (e.g. UAV Forecast). Adverse weather can impact on drone flight itself and also have potential implications for data collection. Most small drone systems on the market do not handle rain well. There are exceptions, and some systems can withstand some rainfall and can even land in water if needed. Irrespective of the capability of the drone to fly in rain an important question is whether the sensor on the drone will still be able to collect meaningful data during rain. Most data collection by sensors would be negatively impacted by rain. Visual sensors either would get wet lenses or simply images would be hampered by rain between the drone and the land surface. Acoustic sensors could also be negatively influenced by rain either in terms of the sensor becoming humid and yielding different responses or by the rain influencing the transmission of sound that the sensor is trying to record. Flying in the rain is therefore not advisable.

Wind is another factor that needs to be carefully assessed. Drone manufacturers usually provide an upper limit for wind speed that a particular drone system can be flown in and these should always be adhered to. Measuring the wind on the ground is standard best practice and anemometers are widely available and affordable. It is important to be aware, though, that wind speed is usually much higher in the air than at ground level. So caution is needed. But even below the maximum limit one should consider if there is a risk of quickly changing weather or whether wind and in particular gusts could negatively influence data collection. For mapping missions some wind is not necessarily negative because a fixed wing drone can maintain a stable crab angle during stable wind speeds. This can actually improve the data collection as long as the mission is carefully planned perpendicular to the wind direction.

Drones have been widely used under very different temperatures from tropical regions to polar conditions (Koh and Wich 2012, Nilssen, Storvold et al. 2014, Ratcliffe, Guihen et al. 2015). Both can pose challenges. The tropics are usually very humid and it is well known that prolonged humidity can be damaging to electrical circuits and can also damage camera equipment by moisture getting trapped between the lenses. Researchers in humid areas often keep large air-tight boxes to store electronic equipment in, with silica gel to absorb the moisture. Therefore under such conditions drones and their sensors need careful storage to prevent the damaging influence of moisture on the equipment. Also in very hot areas where equipment is exposed to the sun for prolonged periods of time it is advisable to use lightly coloured systems as much as possible because they absorb less heat. For instance black foam on fixed wing systems can become very hot and lose its rigid structural properties and adhesives can delaminate. Altitude estimates from the barometer in autopilot systems can also be influenced by the aircraft being in the full sun due to air temperatures rising in the aircraft. Although flight time seems not to be affected much by the impact of tropical temperatures on batteries, this is not the case in very cold conditions. In arctic conditions flight time is reduced due to shorter battery life. Some drone manufacturers supply battery warmers to get the batteries up to 15 °C and also provide upper limits for the usage of their drones. A number of multirotor aircraft have also

[10] https://earthexplorer.usgs.gov/

started to build heaters into the vehicle so that batteries remain at a constant temperature during flight. Some autopilot systems (e.g. Pixhawk2) have a heating element in the data processor to allow it to work in colder regions. Under very cold conditions there is a chance that equipment becomes brittle and breaks so care needs to be taken to avoid that. But generally flying under cold conditions is possible.

1.2.3 High altitude flying

The last aspect to consider is that when flying at high altitudes above sea level flight time is reduced. At high altitudes the air is less dense so more power is used to keep the aircraft airborne and this reduces flight time. Flying in such areas can often be improved by using a different set-up than used at standard altitudes, but even standard set-ups for multirotors have been successfully flown above ground elevations as high as 5000 m.

1.2.4 Landing area

An important factor influencing the possibility of using drones is the availability of landing areas. Both fixed wing drones and multirotor drones need an open area to land in and these can be limited in certain environments such as forests (Duffy, Cunliffe et al. 2017). The size of the area needed is larger for a fixed wing system than a multirotor system because fixed wing systems generally cannot perform a vertical take-off and landing (VTOL). As a result, fixed wing systems need to glide down which often needs a considerable horizontal flight path over which the flight altitude decreases at a relatively shallow angle against the wind. Particularly when there is tall vegetation surrounding the landing area the size of the landing area itself needs to be quite large to enable a safe landing. There are also fixed wings (specifically delta-shaped wings) that can circle (loiter) down to the ground or close to the ground and then do a horizontal glide down over a small distance. But even those systems need considerably more space than multirotors that can perform a true VTOL. Recent hybrid models that combine horizontal propellers for vertical take-off and vertical propellers for horizontal flight on a fixed wing are attempts to overcome this and reduce the take-off and landing area size constraints for fixed wing systems.

However, surveys over large forested areas where there are very few open areas will continue to be a challenge, particularly when these are not crossed by roads that can be used for take-off and landing, but only by river systems. Although it is possible to take off and land a multirotor on boats, this is generally not possible with a fixed wing unless there is a net in which the drone can be caught. Drone surveys on seas have used nets or hooks from cranes to catch drones during the landing phase (Koski, Abgrall et al. 2010). This requires considerable flying skills though and is not without risk of damaging the drone or having it fall into the water. Particularly when flying over the ocean the salt water can quickly lead to corrosion and malfunction of electrical systems. There are, however, fixed wing systems that have been designed to be able to land in fresh and saltwater and recovered without damage to the system.[11] There are also multirotor systems that claim to be waterproof and therefore be able to take off and land in water, but the usage of these systems is not widespread and waves are likely a risk to such systems in addition to leaks. Salt air can also damage the ground control station and the transmitter. So particularly when flying from small boats this can be a challenge because waves splash over the boat. An additional challenge is getting the autopilot to initialize on a moving boat, especially a small boat that moves substantially. Gimbal systems that stabilize the take-off area by maintaining it horizontally could facilitate this. It is also important to realize that without a firm anchor to maintain the position of the boat, the boat will move and therefore any automated return to launch flight mode will not land the drone on the boat, but in the water due to the shifted position of the boat.

1.2.5 Regulations

When considering using a drone for survey work it is important to determine whether the flights would be allowed within the regulatory framework of the country the flight is planned to occur in (see Chapter

[11] http://www.questuav.com (accessed 7 November 2017)

4, Duffy, Cunliffe et al. 2017). For instance, in many countries the regulations stipulate that flights are only allowed within visual line of sight (VLOS, 500 m) or extended visual line of sight (750 m). Similarly there often is a ceiling to how high it is permitted to fly above the ground (400 ft). For surveys over large areas the VLOS requirements are often limiting unless there are multiple take-off and landing areas that one can use. In this situation a drone survey might become cost prohibitive compared to using alternative methods due to the need to move from one area to another.

CHAPTER 2

Typology and anatomy of drones

2.1 Multirotor

The most popular type of consumer-grade drones is a multirotor, which as the name implies, relies on one or more rotors that generate the lift required for flight (Figure 2.1). The aircraft is able to move forwards, backwards, left, and right, by adjusting the relative speeds at which its rotors are spinning. The most common type of multirotor drone is the quadcopter (four rotors). But there are hexacopters (six rotors), octacopters (eight rotors), and other configurations as well, each of which is used for slightly different applications. For example, hexacopters may be more appropriate than quadcopters in situations where greater lifting power is needed. Having a greater number of rotors also confers redundancy to the aircraft such that there is a higher chance of the aircraft making a successful landing even if any one of its rotors should fail. This book does not discuss the technical details of aerodynamics as there are many existing resources that cover that topic (for example, Marques and Ronch 2017).

2.2 Fixed wing aircraft

This type of drone looks and functions just like a manned aeroplane (Figure 2.2). It usually has one or two horizontally oriented rotors that generate thrust to propel the drone forwards. The movement of air over and under its wings generates the lift that allows a fixed wing drone to take off and maintain straight and level flight (Marques and Ronch 2017). This aircraft changes the direction of its flight by the, movement of its control surfaces (ailerons, rudder and elevator). For example, the ailerons that are located near the tips of the main wings allow the aircraft to 'roll' left or right. Similarly, the rudder that is located on the 'fin' (or vertical stabilizer) of the tail assembly allows the aircraft to 'yaw' left or right, whereas the elevator located on the tailplane (or horizontal stabilizer) allows the aircraft to 'pitch' up or down (Marques and Ronch 2017).

2.3 Hybrid VTOL

A hybrid VTOL is an aircraft design that combines the features of the multirotor and fixed wing aircraft (Table 2.1). The two most common types of VTOL are the 'tail-sitter' and 'tilt-rotor'. As the name suggests, a tail-sitter is usually a fixed wing aircraft that rests on its tail when it is on the ground (Figure 2.3). At take-off, the vertically oriented rotors provide the downward thrust to lift the aircraft off the ground. The sophisticated flight controller carefully maintains the stability of the aircraft through constant adjustments to the differential thrust of the two rotors as well as the control surfaces on its wings. When the aircraft has reached a safe altitude, the operator can command it to transition to forward flight as the aircraft assumes a more conventional horizontal fixed wing profile.

The tilt-rotor is the second type of VTOL aircraft that is also based on a fixed wing airframe (Figure 2.4). However, in this case the aircraft takes off from the ground in its typical conventional fixed wing profile with vertically oriented rotors that provide downward thrust for lift. Once safely off the

Conservation Drones: Mapping and Monitoring Biodiversity. Serge A. Wich & Lian Pin Koh. Oxford University Press (2018).
 DOI:10.1093/oso/9780198787617.001.0001

Figure 2.1 A multirotor aircraft on a surveying mission. © Lian Pin Koh.

ground, the aircraft transitions into forward flight by tilting its rotors 90 degrees to produce horizontal thrust.

2.4 Features of multirotor vs fixed wing vs VTOL drones

Multirotor and fixed wing drones have very different capabilities that need to be carefully considered. The appropriate type of aircraft should be chosen for each specific ecology or conservation application. The following table summarizes the most important pros and cons of each aircraft type (Table 2.1).

2.5 Essential components

In addition to the actual airframe and regardless of whether it is a multirotor or fixed wing type, almost all drones comprise a few key essential components. These are invariably a power source, a flight controller, and a ground control station.

2.5.1 Power source

Most drones are powered by rechargeable lithium-ion (Li-ion) or lithium-ion polymer (LiPo) batteries. These are the same batteries commonly found in computer laptops and other mobile devices. By packing large amounts of energy into a small amount of material, Li-ion or LiPo batteries offer excellent energy to weight ratios that make them a great energy source for drones (Figure 2.5).

However, the downside of Li-ion or LiPo batteries is that they are rather unstable (Lisbona and Snee 2011). Catastrophic failure can result from overcharging, over-discharging, over-temperature, short circuit, crushing, or penetration. The outcomes from mishandling or improper care can range from unreliable performance to vigorous battery fires.

It is important to always closely adhere to the battery charging guidelines for each type of drone model and battery charger. Most importantly, one should never leave a charging battery unattended.

Figure 2.2 A fixed wing drone used for forest mapping in Malaysia. © Lian Pin Koh.

Table 2.1 Pros and cons of the three common types of conservation drones

	Multirotor	**Fixed wing**	**Hybrid VTOL**
Launch locations	Small take-off and landing area required; can launch in a wide range of environmental conditions.	Requires large, open area for launch and landing.	Small take-off and landing area required; can launch in a wide range of environmental conditions.
Payload	Can support heavier payload (e.g. several kilograms).	Can support lighter payload (e.g. less than a kilogram), although gasoline powered fixed wings may carry heavy payloads.	Can support lighter payload (e.g. less than a kilogram).
Flight time	Shorter flight time (e.g. less than an hour).	Longer flight time (e.g. more than an hour).	Intermediate flight time (e.g. 30–60 minutes).
Pilot experience	Can be flown with minimal or no training and experience.	Substantial experience and training required to effectively pilot, especially for take-off and landing.	Intermediate pilot experience.

Figure 2.3 A tail-sitter hybrid VTOL aircraft developed in Switzerland. © Wingtra.com.

Figure 2.4 A tilt-rotor hybrid VTOL aircraft developed in the United States. © BirdsEyeView Aerobatics.

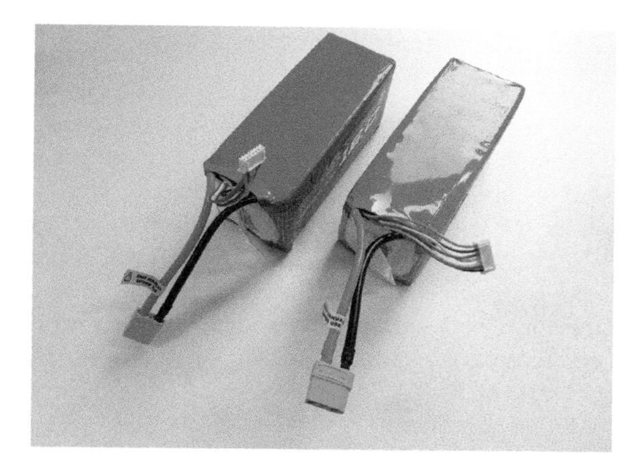

Figure 2.5 A pair of LiPo batteries used in conservation drones. © Lian Pin Koh.

2.5.2 Flight controller

There are a variety of different flight controllers available on the market today that also range in sophistication. The simplest flight controller is the same system found in all remote-controlled model aircraft. This would be a radio receiver that converts command signals transmitted from a ground operator into proportional movements of the aircraft's control surfaces to direct its flight path.

The more typical flight controller used in drones today is comprised of an array of electronic components including the radio receiver, a space-based radionavigation system such as the Global Positioning System (GPS) or Global Navigation Satellite System (GLONASS), and an Inertial Measurement Unit (IMU). The IMU itself usually is an integrated set of sensors that may include

Figure 2.6 An early version of the ArduPilotMega – an Open Source autopilot platform commonly used in conservation drones. © Lian Pin Koh.

an accelerometer, gyroscope, magnetometer, and airspeed sensor. The modern flight controller also includes a microprocessor capable of processing data received from the IMU and other peripheral sensors to maintain straight and level flight of the aircraft without operator input (Figure 2.6). In more advanced systems, the flight controller is also able to command the aircraft to fly a mission automatically to pre-programmed geographical coordinates. It is this feature that makes the modern drone an extremely versatile tool for collecting remotely sensed data over a landscape or seascape. In fact, the cost of commercially available ready-to-fly drones has dropped so dramatically in the past five years that few drone operators actually need to understand how a flight controller works anymore.

2.5.3 Ground control station

A ground control station (GCS) is a ground-based system that allows the operational control or programming of a drone mission. In the context of drones used in ecology and conservation applications, the GCS is most commonly a computer laptop running dedicated software. This GCS would also include a radio telemetry system, usually in the form of a serial data modem, that transmits and receives flight data and commands between the aircraft and the GCS.

Other functionalities of the GCS may include the programming of a mission by specifying target areas to survey, waypoints, and other essential instructions such as flying height and speed (Figure 2.7).

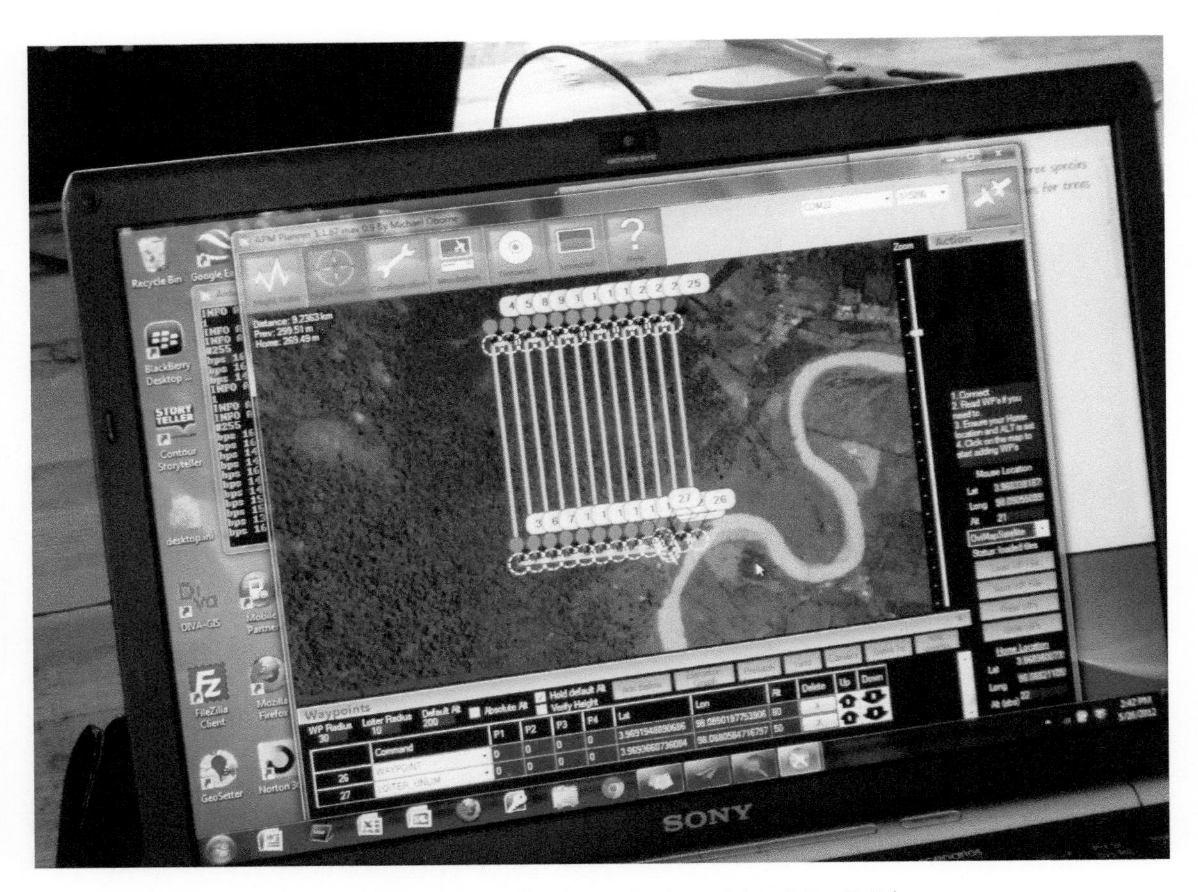

Figure 2.7 The Mission Planner Ground Control Station used for programming drone missions. © Lian Pin Koh.

However, GCS are not restricted to the laptop form, as other mobile devices such as tablets and mobile phones can also run GCS applications to perform similar functions. In fact, in some cases, such as when the operation is to be conducted in remote locations and without the ability to establish a proper basecamp, having a more mobile GCS would help to increase the chances of success. Recent advances in mobile GCS capabilities are increasing their versatility in conservation related applications, including the ability to 'live' stream video data to the ground operator.

2.6 Summary

- The most popular aircraft types include multirotor, fixed wing, and 'vertical-take-off-and-landing' (VTOL) craft.
- The essential components of a drone are a power source, flight controller, and ground control station.
- Each type of drone has its pros and cons. It is important to use the appropriate type based on the requirements and operating conditions of the mission.

CHAPTER 3

Sensors

3.0 Introduction

Data collection for conservation purposes such as animal density and distribution or land-cover change needs consideration of the various sensor types that are available and their specifications in relation to the data that are needed. For many animal surveys a simple RGB (Red Green Blue) photo camera might suffice, but for crepuscular and nocturnal animals a thermal sensor might be more applicable. Land-cover classification might be possible with RGB images, but could be more accurate or allow for the classification of more land-cover types when using more bands from the electromagnetic spectrum that capture information outside of visible wavelengths (Figure 3.1). Modified RGB cameras that allow data capture in the near infrared or multispectral cameras are often applied for the collection of such data. For an ecologist interested in vegetation structure it might be sufficient to derive measurements from the DSM (Digital Surface Model) and DTM (Digital Terrain Model) produced with photogrammetry software, but in dense vegetation types such as tropical rainforests using LiDAR (Light Detection and Ranging) would provide data that would allow for more accurate results because there would be more ground hits. Determining which sensor to use is not always clear-cut and is often constrained by budget. In this chapter an overview of several sensors will be given as well as specific applications that they have been used for. Because most conservation researchers will start with a specific question and then subsequently will start exploring which sensor or set of sensors will be suitable for their data collection, we will approach the sensor issue from the application end instead of vice versa.

3.1 Sensor types

The number of sensors that can be fitted to drones and/or have been specifically designed to be fitted under drones is expanding rapidly (van Blyenburgh 2013). The compilation by van Blyenburgh (2013) contains 406 sensors that range from the visual range to the microwave range. Several recent excellent overviews also provide lists of sensors used under drones, their specifications, and their specific applications (Everaerts 2008, Colomina and Molina 2014, Salamí, Barrado et al. 2014, Shahbazi, Théau et al. 2014, Linchant, Lisein et al. 2015, Pajares 2015, Christie, Gilbert et al. 2016, Toth and Jóźków 2016). Here we will restrict ourselves to those that have specific conservation applications, although several studies that focused on agriculture will also be mentioned due to their potential application in conservation. Due to the very rapid development in specific sensor models it is inevitable that many of the specific sensor brands and models mentioned in this chapter will be superseded shortly after publication. But the information will still help the reader to find further specifications on these sensors that are useful for planning their own work. For such specific details readers are advised to consult the specific sources that have been used in the chapter, the above-mentioned overview papers, and online search engines.

Conservation Drones: Mapping and Monitoring Biodiversity. Serge A. Wich & Lian Pin Koh. Oxford University Press (2018).
 DOI:10.1093/oso/9780198787617.001.0001

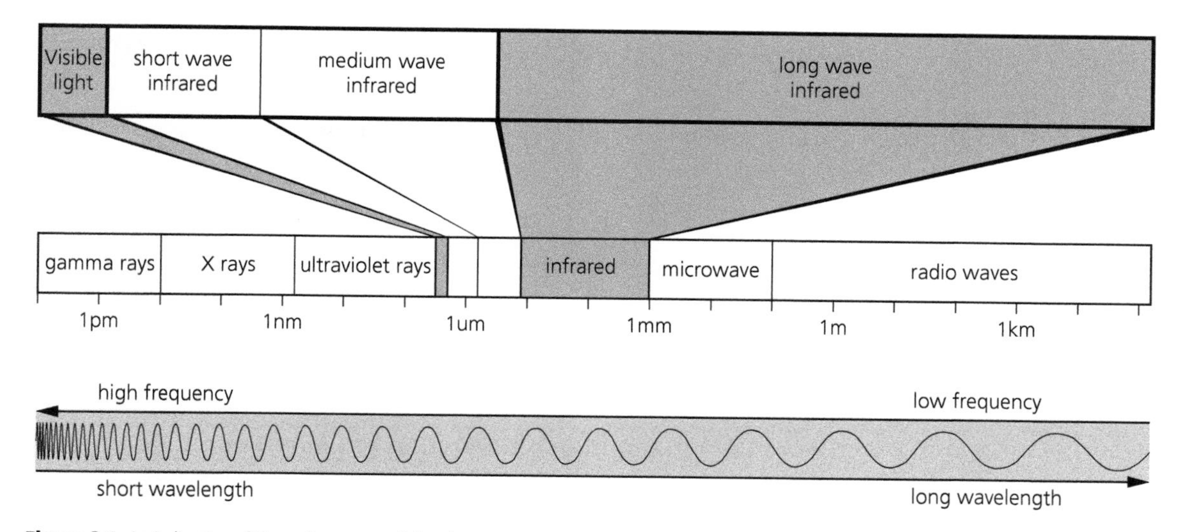

Figure 3.1 An indication of the various parts of the electromagnetic spectrum and their wavelengths. © Perry van Duijnhoven.

Table 3.1 Ground sample distance and footprint for several sensors

Camera	Canon S110 (RGB)	Sony A6000 (RGB)	Flir Vue (thermal)
Focal length (mm)	5.2	20	13
Number of pixels	4000 × 3000	6000 × 4000	640 × 512
Sensor size (mm)	7.6 × 5.7	23.5 × 15.6	9.0 × 6.8
Ground sample distance (GSD, cm)*	3.65	1.95	10.22
Footprint (m)*	146.2 × 109.6	117.5 × 78.0	69.2 × 52.3

* Above ground level (AGL) was 100 m for all calculations.

Table 3.2 Common sensors used in conservation research

Sensor type	Brand examples	Wavelength range*
RGB (Red, Green, Blue)	Canon, Nikon, Sony	390–700 nm
Multispectral	Tetracam, Micasense, Parrot Sequoia, Slantrange, converted cameras	300–1100 nm
Hyperspectral	Rikola Ltd, Headwall photonics, Resonon	380–2500 nm
Thermal	FLIR, Xenics, Optris, TeAx-Technology	7–13.5 μm
LiDAR	Riegl, Velodyne, YellowScan	

* = The range is not for one camera specifically but indicates the total range that is generally available. Specific cameras might only use part of that range.

Because the focus of this chapter is on applications and not on details of the various sensors we will only give a brief overview of common sensors that have been used in conservation work (Tables 3.1, 3.2, Figure 3.2). Details on specific sensors will be given in the section on each sensor, but some common issues to think about will be discussed here. An important consideration when using drones to collect data is to determine whether the system will need one sensor or multiple sensors during the same flight, or whether data with multiple sensors can be collected over multiple flights. The latter is, for instance, likely to be an option when flights are conducted for land-cover

Figure 3.2 Various sensors used under drones. Top left: GoPro © Andy Goodwin; top right: Canon RGB camera © Andy Goodwin; middle left: Multispectral Parrot Sequoia © Serge Wich; middle right: Thermal sensor © Maisie Rashman; Lower left: A Vulcan Y6 multirotor with a Velodyne HDL-32E LiDAR used in the DigiArt project. The DigiArt project has received funding from the European Union's Horizon 2020 research and innovation programme under grant agreement No 665066 © Frederic Bezombes; Lower right: Telemetry system tracking Noisy Miner birds (inset) © Oliver Cliff.

classification because the objects of interest are motionless, but this approach is likely not applicable when studying animals that move. In that case a system with multiple sensors to collect data at the same time is likely necessary. There are not many systems on the market that are adapted for carrying multiple sensors at the same time. Some others allow for multiple interchangeable sensors, but with only one sensor on board during a flight.[1] Although such off-the-shelf systems might work well for certain applications, there might be situation in which researchers will need to build their own systems or order customized systems in which researchers can choose sensors themselves and design the system around a sensor or set of sensors. Another important aspect is how images will be georeferenced and, in the case of multiple sensors, how images from the different sensors will be co-registered. Although some sensors come with a built-in GPS unit and IMU (see Chapter 2 for details), many do not. There are several workflows possible in which the images are either automatically geotagged or geotagged with the aid of post-flight software on the ground control station or other software packages.[2] Some of these workflows also allow for the insertion of attitude data from the autopilot (e.g. through Pix4D[3]: details in Chapter 7).

A key aspect of any mission to collect data should be whether the sensor and flying height combination will yield images of sufficient resolution to discern the object of interest, or in the case of land-cover mapping, vegetation types of interest. Ground control software such as Mission Planner[4] automatically calculates the ground resolution (often called GSD (Ground Sample Distance) which is the distance between the centre of two adjacent pixels in an image) once the correct data for the camera have been entered or a sample image has been uploaded. However, it is straightforward to calculate this without software with the following formula:

$$GSD = Pixel\,size * \frac{AGL}{Focal\,length}$$

AGL is the above ground level in metres. For instance for a Canon S110 the focal length is 5.2 mm, the sensor resolution is 4000 × 3000 pixels, and the sensor size is 7.6 mm × 5.7 mm. Dividing either 7.6 by 4000 or 5.7 by 3000 gives the pixel size 0.0019 mm. GSD when flying at 100 m (AGL) is then:

$$GSD = 0.0019 * \frac{100}{5.2} = 0.0365$$

The image footprint on the ground can then be calculated by multiplying the number of pixels by the size of each pixel. In this case the width of the image is 146 m (4000 × 0.0365) and the height of the image is 109.5 (3000 × 0.0365).

This assumes that the camera is placed top facing forward. This is the usual camera position in drones because it allows the distance between the flight lines to be as wide as possible for a given side overlap between the images. If the camera would be side facing forward flight lines would have to be closer to each other for a similar side overlap between images. The two reasons to avoid this are that a) a fixed wing system might not be able to make tight enough turns and building in overshoot to allow the drone to line up with the flight lines is not using flight time effectively, and b) it means flight distance is larger to cover the same surface area on the ground which again is not effective. When considering ground sample distance of images it is important to keep in mind that this differs widely among sensors (Table 3.1) and that, for instance, thermal sensors generally have a much lower number of pixels than RGB cameras.

Examining the relationship between ground resolution and flying altitude is important so that a researcher can assure that a specific sensor and flying altitude combination will provide the ground resolution needed. Obviously, smaller objects need a higher ground resolution, which can be obtained by flying lower, but whether that is an option depends on a range of factors such as vegetation height, the terrain, and how low or high one needs to fly to cover sufficient area for a survey. In addition, when flying low with a fast flying drone there is a chance for blurred images if the shutter speed

[1] eBee: https://www.sensefly.com/drones/ebee-ag.html
[2] http://www.geosetter.de.
[3] https://pix4d.com/ (accessed 7 November 2017)
[4] http://ardupilot.org/planner/docs/mission-planner-overview.html (accessed 7 November 2017)

is too slow compared to the speed at which the drone is flying at. A suggested rule of thumb to avoid motion blur in images due to the speed of the drone is that the distance covered during the time the shutter is open to take the image is smaller than half the size of the pixel (Bosak 2013). Using this rule might still lead to images that are not sharp if there are sudden motions of the drone during the image capture. Therefore a simpler option is to always use a very fast shutter speed and use the 'Shutter Priority' mode with the ISO setting at auto so that the exposure can be compensated when the aperture reaches its maximum size. Generally a shutter speed of 1/1000 sec works well during conditions when there is relatively low light (early and late in the day and in cloudy and overcast conditions). A faster shutter speed can be used during brighter conditions. If flying lower is not an option it is sometimes possible to use the optical zoom on a sensor, but using this might come at the cost of more vibration of the camera and hence blurred images. Examining published papers and reports, and checking the online DIY drone communities,[5] to determine what previous researchers have been using, often helps with deciding which specifications are required to obtain sufficient ground detail. Related to this is the question of whether the sensor will need to be pointed straight down or at an angle. Whereas for mapping a sensor that faces straight down is required, this might not be the case for spotting animals that are often found under trees in which case a camera at an oblique angle might be more beneficial. Experience has also taught us that it is important to set the camera focus to infinity and not on auto-focus because that will often lead to blurred images. To further improve exposure it is beneficial to use the matrix (or pattern) metering mode when this is available so that the camera is using the full extent of its field to calculate the exposure setting.

3.1.1 Visible spectrum sensors (RGB)

Due to the, initially, high costs and relatively large weight of multispectral and hyperspectral cameras the most common sensors have been standard consumer cameras that obtain imagery in the visible part of the spectrum (Table 3.3; Lebourgeois, Bégué et al. 2008, Rump, Zinke et al. 2011). There have been a large number of brands that researchers have fitted under drones, but common ones include Canon, Sony, and Nikon. Although there are a large range of cameras that have been used on drones it is important to carefully determine whether the sensor size and number of pixels will be sufficient for the particular research aim, because these factors in combination with the flying height and focal length are important in determining the resolution of an image (see section 7.2). Another important consideration when a drone is built and not purchased off-the-shelf is by which mechanism the camera can capture images at a pre-defined interval. This does not only apply to RGB cameras, but also to multispectral and hyperspectral cameras. There are several common options for this depending on the camera. First, cameras might have built-in time interval settings, but these often are restricted to a few intervals. Cameras with a GNSS often also allow for the interval to be set at a flight distance interval (e.g. Parrot Sequoia). Second, some cameras allow for the installation of certain apps on the camera that can be used to set the time interval that images will be taken. Third, specifically for Canon cameras there is the opportunity to use CHDK (Canon Hack Development Kit).[6] This allows for putting new firmware on the SD card in the camera so that self-written scripts can be loaded that allow for images to be taken at a set interval. It is not known whether this will remain available for future Canon cameras. Fourth, several autopilot systems (e.g. the 3DRobotics Pixhawk) can trigger cameras at set intervals or set flight distances with the ground control software.[7] Depending on the particular camera brand, different types of specific cables and camera triggers will need to be used (e.g.[8]). Fifth, there are infrared triggers that can be connected to the autopilot and that can trigger the camera through its infrared sensor (e.g. stratosnapper[9]). Alternatively some drones have triggers connected to a servo that physically presses the shutter button. Once the images have been obtained these can be used for several purposes,

[5] www.diydrones.com

[6] http://chdk.wikia.com (accessed 7 November 2017)

[7] Mission Planner: http://ardupilot.org/planner/docs/mission-planner-overview.html

[8] http://www.seagulluav.com/ (accessed 7 November 2017)

[9] http://littlesmartthings.com/stratosnapper2/ (accessed 7 November 2017)

Table 3.3 Applications, sensor types, and drone systems

Application	Sensor(s)	Species/topic	Aim	Drone (model if known)*	Detection method	Source
Animals (birds)						
Animal detection/birds	RGB (Sony Nex5 camera)	Black kites (*Milvus migrans*)	Conspicuousness of bird nests	MR (Hexacopter)	Manual	(Canal, Mulero-Pázmány et al. 2016)
Bird detection	RGB (GoPro Hero 4)	Tristan Albatross (*Diomedea dabbenena*)	Population count	MR (DJI Phantom 2)	Manual	(McClelland, Bond et al. 2016)
Bird detection and disturbance	RGB (Sony RX-1, mvBlueCOUGAR-X, Sony A7-R, Phase 1, GoPro Hero)	Various waterfowl species	Determining disturbance on birds with various drone models	FW (UAVER Avian-P, Skylark II, Drone Metrex Topodrone-100), MR (DJI Phantom, FoxTech Kraken-130)	Manual	(McEvoy, Hall et al. 2016)
Animal detection/birds	RGB (Canon EOS M)	Lesser frigatebird (*Fregata ariel*), crested tern (*Thalasseus bergii*), royal penguin (*Eudyptes schlegeli*)	Counting animals	MR (X8 by 3D Robotics) and FW (FX79 airframe)	Manual	(Hodgson, Baylis et al. 2016)
Animal detection/birds	RGB (GoPro Hero 1 and 3)	Osprey (*Pandion haliaetus*), bald eagle (*Haliaeetus leucocephalus*), ferruginous hawk (*Buteo regalis*), red-tailed hawk (*B. jamaicensis*)	Determining disturbance of ground and aerial surveys	MR (Draganflyer X-4)	Manual	(Junda, Greene et al. 2016)
Animal detection/birds	RGB (GoPro Hero 1 and 3)	Osprey (*Pandion haliaetus*), bald eagle (*Haliaeetus leucocephalus*), ferruginous hawk (*Buteo regalis*), red-tailed hawk (*B. jamaicensis*)	Surveying raptor nests	Multirotor (Draganflyer X-4)	Manual	(Junda, Greene et al. 2015)
Bird count	RGB (Canon Powershot S90)	Common tern (*Sterna hirundo*)	Colony count	FW (AI-Multi UAs)	Automatic	(Chabot, Craik et al. 2015)
Animal detection/birds	RGB (DJI Phantom 2 built-in camera)	Hooded crow (*Corvus cornix*)	Check nest contents	MR (DJI Phantom 2 Vision)	Manual	(Weissensteiner, Poelstra et al. 2015)
Animal detection/birds	RGB (Panasonic LUMIX G5)	Adélie penguin (*Pygoscelis adeliae*)	Measuring reaction to drones	MR (MK ARF Okto XL)	Manual	(Rümmler, Mustafa et al. 2015)
Animal detection/birds	RGB (Canon EOS M)	Gentoo penguin (*Pygoscelis papua*)	Assessing detectability	MR (Cinestar 6)	Manual	(Goebel, Perryman et al. 2015, Ratcliffe, Guihen et al. 2015)
Animal detection/birds	RGB (Sony NEX-5, Canon EOS-M and the Olympus E-P1)	Gentoo penguin (*Pygoscelis papua*), chinstrap penguin (*Pygoscelis antarctica*)	Basic detection, wildlife reaction, effectiveness comparison of 3 aircraft and 3 cameras	3 MRs (MK-OktoXL, APQ-18 and APH-22)	Manual	(Goebel, Perryman et al. 2015)

(*continued*)

Table 3.3 Continued

Application	Sensor(s)	Species/topic	Aim	Drone (model if known)*	Detection method	Source
Animal reaction/birds	RGB (GoPro Hero 3)	Mallard (*Anas platyrhynchos*), greater flamingo (*Phoenicopterus roseus*), common greenshank (*Tringa nebularia*)	Assessing behavioral reactions to a multirotor drone	MR (Phantom, Cyleone, France)	Manual	(Vas, Lescroël et al. 2015)
Animal detection/birds	Electro-optical/infrared (EO/IR) sensor	Various seabird species, turtles, seals, and cetaceans	Assessing detectability	2 FWs (MQ-9 Predator B and Puma AE)	Manual	(Brooke, Graham et al. 2015)
Animal detection/birds	RGB (Canon S100 and GoPro Hero3)	Various waterbird species	Assessing detectability and behavioral reaction to UAS	Petrol-powered VTOL (Honeywell RQ-16 T-Hawk) and FW (AeroVironment RQ-11A)	Manual	(Dulava, Bean et al. 2015)
Animal detection/birds	RGB (GoPro Hero 2 and Panasonic LX3)	White stork (*Ciconia ciconia*)	Assess power lines risk on birds	FW (Easy fly St-330)	Manual	(Mulero-Pázmány, Negro et al. 2014)
Bird detection	EO/thermal sensor and RGB (GoPro Hero 3 black edition)	Greater sage-grouse (*Centrocercus urophasianus*)	Counting and monitoring	FW (Raven RQ-11A)	Automatic	(Hanson, Holmquist-Johnson et al. 2014)
Animal detection/birds	RGB (GoPro 2)	Steller's sea eagle (*Haliaeetus pelagicus*)	Nest check and birds reaction	MRs (APM and Naza-M based)	Manual	(Potapov, Utekhina et al. 2013)
Animal detection/birds	RGB (Panasonic Lumix FT-1)	Black-headed gull (*Chroicocephalus ridibundus*)	Detect changes in breeding populations	FW (Multiplex Twin Star II, no autopilot)	Manual	(Sardà-Palomera, Bota et al. 2012)
Animal detection/birds	RGB (Pentax Optio A20)	Canada geese (*Branta canadensis*) and snow geese (*Chen caerulescens*)	Comparing photographic counts from repeated flights to repeated visual ground counts	FW (CropCam)	Manual and automatic	(Chabot and Bird 2012)
Animal detection/birds	Thermal	Sandhill crane (*Grus canadensis*)	Assessing detectability	FW (Raven RQ11)	Manual and automatic	(Hutt 2011)
Animal detection/birds	RGB (Evoltmodel 420)	Red knot (*Calidris canatus*), egret species (*Ardea alba, Bubulcus ibis, Egretta* sp.), pelican (*Pelecanus* sp.), wood stork (*Mycteria americana*)	Studying bird detection	MR (Nova 2)	Manual	(Brush and Watts 2008)
Animal detection/birds	RGB (Canon Elura 2) and near infrared	White ibis (*Eudocimus albus*), Egret (*Egretta* sp.), wood stork (*Mycteria americana*)	Basic detection	FW (FoldBat)	Manual	(Jones, Pearlstine et al. 2006)

Animals (aquatic)						
Animal behaviour	RGB (proprietary DJI camera)	Green turtle (*Caretta caretta*)	Investigating green turtle and fish cleaning	MR (DJI Phantom 3)	Manual	(Schofield, Papafitsoros et al. 2017)
Animal sex ratio determination	RGB (proprietary DJI camera)	Green turtle (*Caretta caretta*)	Determining the sex ratio of green turtles	MR (DJI Phantom 3)	Manual	(Schofield, Katselidis et al. in press)
Animal density/marine	RGB (GoPro Hero 3+)	Blacktip reef sharks (*Carcharhinus melanopterus*) and pink whiprays (*Pateobatisfai*)	Density assessment	MR (DJI Phantom 2)	Manual	(Kiszka, Mourier et al. 2016)
Animal detection/fish	RGB (GoPro 3+)	Chinook salmon (*Oncorhynchus tshawytscha*)	Counting salmon spawning sites	MRs (Aeryon Scout and Mikrokopter)	Automatic	(Groves, Alcorn et al. 2016)
Animal detection/herptiles	RGB (Canon S100)	Estuarine crocodiles (*Crocodylus porosus*)	Basic detection	FW (Bormatec-MAJA)	Manual	(Evans, Jones et al. 2015)
Animal detection/herptiles	Electro-optical/infrared (EO/IR) camera	Sea turtles (*Caretta caretta*)	Basic detection	FWs (MQ-9 Predator B and Puma AE)	Manual	(Brooke, Graham et al. 2015)
Animal detection	RGB (GoPro Hero 3 Black Edition)	Kemp's ridley sea turtle (*Lepidochelys kempii*)	Abundance and movements of turtles	MRs (DJI Phantom 1 and 2)	Manual	(Bevan, Wibbels et al. 2015)
Animal detection/marine mammals	RGB (Sony NEX-5, Canon EOS-M and the Olympus E-P1)	Antarctic fur seal (*Arctocephalus gazella*) and leopard seals (*Hydrurga leptonyx*)	Basic detection, wildlife reaction, effectiveness comparison of 3 UAVs and 3 cameras	MRs (MK-OktoXL, APQ-18 and APH-22)	Manual	(Goebel, Perryman et al. 2015)
Detection and disturbance	RGB (Nikon D300)	Spotted (*Phoca largha*) and ribbon seals (*Histriophoca fasciata*)	Basic detection and disturbance evaluation	FW (ScanEagle)	Manual	(Moreland, Cameron et al. 2015)
Animal detection/marine mammals	RGB (Sony HDR-CX760 and PJ650)	Grey seals (*Halichoerus grypus*) and harbour seals (*Phoca vitulina*)	Basic detection, photo-ID and measurement, animal reaction, comparison of 3 UAVs	MRs (DJI 450, Cinestar 6, Vulcan 8)	Manual	(Pomeroy, O'Connor et al. 2015)
Animal detection/marine mammals	Electro-optical/infrared (EO/IR) camera	Hawaiian monk seals (*Neomonachus schauinslandi*)	Basic detection	FWs (MQ-9 Predator B and Puma AE)	Manual	(Brooke, Graham et al. 2015)
Animal detection/marine mammals	RGB (Olympus E-PM2)	Killer whales (*Orcinus orca*)	Basic detection	MR (APH-22)	Manual	(Durban, Fearnbach et al. 2015)
Animal detection/marine mammals	RGB (GoPro cameras)	Bowhead whales (*Balaena mysticetus*)	Identification of bowhead whales and behavioral reaction to UAVs	FW (Brican TD100E)	Manual	(Koski, Gamage et al. 2015)

(*continued*)

Table 3.3 Continued

Application	Sensor(s)	Species/topic	Aim	Drone (model if known)*	Detection method	Source
Animal detection/marine mammals	RGB (Nikon® D90)	Dugongs (*Dugong dugon*)	Basic detection	FW (ScanEagle)	Manual	(Hodgson, Kelly et al. 2013, Brooke, Graham et al. 2015)
Animal detection/herptiles	RGB (Nikon® D90)	Sea turtles (*Caretta caretta*)	Basic detection	FW (ScanEagle)	Manual	(Hodgson, Kelly et al. 2013)
Animal detection/fish	RGB (Canon EOS Kiss X)	Wild pacific chum salmon (*Oncorhynchus keta*)	Aerial census of adult chum salmon using	Helicopter: gasoline-engine (Voyager GSR260)	Manual	(Kudo, Koshino et al. 2012)
Animal detection/herptiles	RGB (IXUS 120)	Green turtles (*Chelonia mydas*)	Assess turtles' spatial distribution	FW (Swinglet CAM)	Manual	(Winter 2012)
Animal detection/herptiles	RGB (EVolt camera)	Alligator (*Alligator mississippiensis*)	Basic detection	FW (University of Florida's Nova 2)	Manual	(Watts, Perry et al. 2010)
Animal detection/marine mammals	RGB (Video camera)	Whale-like targets, harbour seals (*Phoca vitulina*), California sealions (*Zalophus californianus*)	Basic detection and factors influencing detection rate	FW (Insight A-20)	Manual	(Koski, Allen et al. 2009, Koski, Abgrall et al. 2010, Koski, Thomas et al. 2013, Koski, Gamage et al. 2015)
Animal detection/herptiles	RGB (CMOS camera and Canon Elura 2) and near-infrared sensor	Alligator (*Alligator mississippiensis*)	Basic detection	FW (MLB FoldBat)	Manual	(Jones, Pearlstine et al. 2006, Watts, Perry et al. 2010)
Animal detection/marine mammals	RGB (CMOS camera and Canon Elura 2) and near-infrared sensor	Florida manatees (*Trichechus manatus latirostris*)	Basic detection	FW (MLB FoldBat)	Manual	(Jones, Pearlstine et al. 2006, Martin, Edwards et al. 2012)
Animals (terrestrial mammals)						
Animal detection/terrestrial mammals	RGB (Nikon D3X)	Caribou (*Rangifer tarandus caribou*)	Assessing detectability using caribou targets	FW (TD100E)	Manual	(Patterson, Koski et al. 2016)
Animal detection/primate	RGB (Canon S100)	Sumatran orangutans (*Pongo abelii*)	Orangutans nest detection	FW (Skywalker 2013)	Manual	(Wich, Dellatore et al. 2016)
Animal reaction/terrestrial mammals	X	American black bear (*Ursus americanus*)	Behavioral and physiological reaction to drone	MR (Iris, 3DR)	X	(Ditmer, Vincent et al. 2015)
Animal impact on habitat	RGB (Canon S100)	Eurasian beaver (*Castor fiber*)	Impact of beavers on ecosystem	MR (Y6, 3DR)	Manual	(Puttock, Cunliffe et al. 2015)

Animal detection/primate	RGB (Canon Powershot SX230 HS)	Chimpanzees (*Pan troglodytes*)	Chimpanzees nest detection	FW (Maja, Bormatec)	Manual	(van Andel, Wich et al. 2015)
Animal detection/terrestrial mammals	Thermal (Thermoteknix Micro CAM)	Giraffes (*Giraffa camelopardalis*)	Rhinoceros anti-poaching	FW (Easy Fly St-330)	Manual	(Mulero-Pázmány, Stolper et al. 2014)
Animal detection/terrestrial mammals	RGB (Panasonic Lumix LX-3, GoPro Hero2), Thermal (Thermoteknix Micro CAM)	Rhinoceros (*Diceros bicornis* and *Ceratotherium simum*)	Rhinoceros anti-poaching	FW (Easy Fly St-330)	Manual	(Mulero-Pázmány, Stolper et al. 2014)
Animal detection/terrestrial mammals	RGB (Ricoh GR3)	African elephant (*Loxodonta africana*)	Basic detection and animals reaction	FW (Gatewing X100)	Manual	(Vermeulen, Lejeune et al. 2013)
Animal detection/primate	RGB (Canon IXUS 220 HS and Pentax Optio WG-1 GPS)	Sumatran orangutans (*Pongo abelii*)	Assess the potential of drone for conservation	FW (Bixler with APM)	Manual	(Koh and Wich 2012)
Animal detection/terrestrial mammals	RGB (Canon IXUS 220 HS; Pentax Optio WG-1 GPS)	Sumatran elephants (*Elephas maximus sumatranus*)	Demonstrate the potential of UAVs for environmental and conservation applications	FW (Bixler with APM)	Manual	(Koh and Wich 2012, Wich, Dellatore et al. 2016)
Animal detection/terrestrial mammals	RGB (Pentax Opio A10 and A40)	Black bear (*Ursus americanus*), woodland caribou (*Rangifer tarandus caribou*), white-tailed deer (*Odocoileus virginianus*), grey wolf (*Canis lupus*), North American beaver (*Castor canadensis*) lodges	Basic detection	FW (CropCam)	Manual	(Chabot 2009)
Animal detection/terrestrial mammals	RGB	Bison (*Bison bison*)	Population survey of bison	FW (Nova 2)	Manual	(Watts, Bowman et al. 2008, Watts, Perry et al. 2010)
Animal detection/terrestrial mammals	RGB (Evoltmodel 420)	Bison (*Bison bison*)	Design of georeferencing techniques in the national bison range	FW (Nova 2)	Manual	(Wilkinson 2007, Wilkinson, Dewitt et al. 2009)
Automated detection						
Automated detection	Thermal (SenseFly LLC, Thermomapper) and RGB (Canon S110)	Grey seal (*Halichoerus grypus*)	Wildlife surveys	FW (SenseFly eBee)	Automatic	(Seymour, Dale et al. 2017)
Automated detection	Thermal (FLIR Tau 640)	Cows (*Bos taurus*) and humans (*Homo sapiens*)	Wildlife survey and anti-poaching	MR (Y6)	Automatic	(Longmore, Collins et al. 2017)

(continued)

Table 3.3 Continued

Application	Sensor(s)	Species/topic	Aim	Drone (model if known)*	Detection method	Source
Automatic animal detection/terrestrial mammals	Thermal (FLIR Tau 2-640)	Koala (*Phascolarctos cinereus*)	Wildlife automated detection	MR (S800 EVO)	Automatic	(Gonzalez, Montes et al. 2016)
Automatic animal detection/dugong	X	Dugong (*Dugong dugon*)	Using deep convolutional neural networks for detection	X	Automatic	(Maire, Alvarez et al. 2015)
Automatic animal detection/terrestrial mammals	Thermal (Flir Tau640) and RGB (Nikon D7000)	American bison (*Bison bison*), red fallow deer (*Dama dama*),grey wolves (*Canis lupus*), white-tailed deer (*Odocoileus virginianus*) and elks (*Cervuscanadensis*)	Species detection	Helicopter (Responder INGRobotic Aviation, Sherbrooke, QC, Canada)	Automatic	(Chrétien, Théau et al. 2015, Chrétien, Théau et al. 2016)
Automatic animal detection/terrestrial mammals	Thermal (Tamarisk 640)	Common hippopotamus (*Hippopotamus amphibius*)	Species detection	FW (Falcon)	Automatic	(Lhoest, Linchant et al. 2015)
Automatic animal detection/birds	RGB (Sony Nex-5)	Black-faced spoonbill (*Platalea minor*)	Automating spoonbill detection to facilitate surveys	FW (CropCam)	Automatic	(Liu, Chen et al. 2015)
Automatic animal detection/terrestrial mammals	RGB (Olympus PEN Mini E-PM1)	Pocket gopher (*Thomomys talpoides*)	Pocket gopher mound feature detection	FW (RQ-84Z Aerohawk)	Automatic	(Whitehead, Hugenholtz et al. 2014)
Automatic animal detection/birds	RGB (Sony Nex-5 and Olympus Pen E2)	Common gull (*Larus canus*)	Automatic bird count, determination of canopy height and DSM	MRs (Falcon 8 and MD4-1000)	Automatic	(Grenzdörffer 2013)
Automatic animal detection/marine mammals	RGB with polarizing filter (Nikon 12 megapixel)	Dugongs	Automatically identify marine mammals	FW (ScanEagle)	Automatic	(Maire, Mejias et al. 2013)
Automatic animal detection/terrestrial mammals	Thermal (FLIR Tau 640)	Roe deer (*Capreolus capreolus*)	Rescue fawns from mowing machines	MR (Falcon 8)	Automatic	(Israel 2011)
Automatic animal detection/marine mammals	RGB (Firefly MV USB)	Whales	Algorithms for object identification and tracking	MR	Automatic	(Selby, Corke et al. 2011)

Automatic animal detection/marine mammals	Multispectral (MANTIS-3)	Belugas and other species of baleen whales	Spectral detection of marine mammals	FW	Automatic	(Schoonmaker, Wells et al. 2008)
Automatic animal detection/birds	RGB (Canon Elura 2 scan video recorder and CMOS analog video chip)	Various bird species and decoys	Assess accuracy of automatic identification	FW (no autopilot)	Manual	(Abd-Elrahman 2005)
Telemetry						
Automatic animal detection/birds	BeagleBone VHF receiver	Bicknell's thrush (*Catharus bicknelli*) and Swainson's thrush (*C. ustulatus*)	Detecting VHF tagged birds with a receiver on the drone	MR (Sky Hero Spyder X8)	Automatic	(Tremblay, Desrochers et al. 2017)
Automatic animal detection/birds	Low-power RF tag localization	Noisy miners (*Manorina melanocephala*)	Real-time autonomous localization	MR (Falcon 8)	Automatic	(Cliff, Fitch et al. 2015)
Automatic animal detection	Tag localization	Animals in general	Radio tag simulation evaluation	X	Automatic	(Posch and Sukkarieh 2009)
Fire detection						
Fire detection	Multispectral (AMS-Wildfire sensor)	X	Fire detection and monitoring	FW (Ikhana)	Automatic	(Ambrosia, Wegener et al. 2011)
Fire detection	Hyperspectral/thermal	X	Development of a system for fire detection and monitoring	FW (based on Dornier DO27)	Automatic	(Rufino and Moccia 2005, Martínez-de-Dios, Merino et al. 2007, Merino, Caballero et al. 2012)
Fire detection	RGB/hyperspectral/fire sensor	X	Fire detection and monitoring	Helicopter and airship Karma	Automatic	(Rufino and Moccia 2005, Martínez-de-Dios, Merino et al. 2007, Merino, Caballero et al. 2012)
Land-cover studies						
Land-cover classification	RGB (Canon S100)	X	Vegetation structure mapping	MR (Y6, 3DR)	Automatic	(Cunliffe, Brazier et al. 2016)
Land-cover classification	Hyperspectral (PIKA II)	X	Vegetation classification	FW (Arcturus T-16)	Automatic	(Mitchell, Glenn et al. 2016)
Land-cover classification	RGB (GoPro Hero 3)	X	Grassland fraction measurements	MR (DJI Phantom 2)	Automatic	(Chen, Yi et al. 2016)
Land-cover classification	RGB (Sony a6000)	X	Mapping and monitoring of tundra vegetation	MR (Spyder PX8Plus 1000)	Automatic	(Fraser, Olthof et al. 2016)

(*continued*)

Table 3.3 Continued

Application	Sensor(s)	Species/topic	Aim	Drone (model if known)*	Detection method	Source
Land-cover features	RGB (Canon S110)	X	Top-of-canopy height and above ground carbon density	FW (Kestrel, Linn Aerospace)	Automatic	(Messinger, Asner et al. 2016)
Land-cover classification	Side-looking synthetic aperture radar (SAR)	X	Trees measurement	MR (DJI Phantom 2)	Automatic	(Li and Ling 2016)
Land-cover classification	RGB (Sony NEX-5)	X	Forest monitoring (canopy height and closure)	MR (MD4-1000)	Automatic	(Zhang, Hu et al. 2016)
Land-cover classification	Modified RGB (NIR) Canon PowerShot SD780 IS	Various peatland species	Vegetation classification and CH4 fluxes	MR (MicrodroneMD4–200)	Semi-automatic	(Lehmann, Münchberger et al. 2016)
Land-cover features	RGB and modified RGB multispectral	Various plant species	Invasive plant species detection and health	FW (Gatewing X100)	Automatic	(Michez, Piégay et al. 2016)
Land-cover classification	RGB (Mobius and GoPro Hero 3)	x	Map fish nursery areas	MR (propriety prototype quadcopter)	semi-automatic	(Ventura, Bruno et al. 2016)
Land-cover features	Modified RGB camera (Panasonic Lumix DMC-GF1)	Olive trees	Canopy reconstruction	FW (mX-SIGHT)	Automatic	(Díaz-Varela, de la Rosa et al. 2015)
Land-cover classification	RGB and modified RGB	X	Tree species identification	FW (Gatewing X100)	Automatic	(Lisein, Michez et al. 2015)
Land-cover features	Modified RGB multispectral (Canon S110 near infra-red)	X	Forest characteristics	FW (SenseFly eBee)	Automatic	(Puliti, Ørka et al. 2015)
Land-cover classification	RGB (GoPro) and thermal (Optris PI)	X	Polar region mapping	MR (Tarot FY690)	Automatic	(Stuchlík, Stachoň et al. 2015)
Land-cover features	RGB (Canon ELPH 520 HS)	X	Measure forest canopy structure	MR	Automatic	(Zahawi, Dandois et al. 2015)
Land-cover classification	RGB (GOPRO 3+)	X	3D models of trees	MR (Iris, 3DR)	Automatic	(Gatziolis, Lienard et al. 2015)
Land-cover features	RGB	X	Map canopy gap	FW (Carolo P200)	Automatic	(Getzin, Nuske et al. 2014)
Land-cover classification	RGB and modified RGB multispectral	X	Vegetation classification	MR (MD4-200)	Automatic	(Gini, Passoni et al. 2014)

Land-cover features	Modified RGB camera (Panasonic Lumix DMC-GF1)	Orchards	Tree height	FW (mX-SIGHT)	Automatic	(Zarco-Tejada, Diaz-Varela et al. 2014)
Land-cover classification	RGB (Canon S90)	X	Classify wetland vegetation types to investigate relationships between habitat and breeding least bitterns	FW (AI-multi)	Automatic	(Chabot and Bird 2013, Chabot, Carignan et al. 2014)
Land-cover classification	LiDAR	X	Tree DBH	MR	Automatic	(Chisholm, Cui et al. 2013)
Land-cover classification	Hyperspectral	X	Land surface reflectance	Not stated	Automatic	(Duan, Li et al. 2013)
Land-cover classification	Modified RGB camera (Canon IXUS 400 and Panasonic LUMIX LX-3)	Waterlogged bare peat, *Sphagnum* spp., *Eriophorum vaginatum* and *Betula pubescens*	Peat bog vegetation	MR (Ifgicopter and MD 4-1000)	Automatic	(Knoth, Klein et al. 2013)
Land-cover features	RGB	X	Assess forest understory floristic biodiversity	FW (Carolo P200)	Automatic	(Getzin, Wiegand et al. 2012)
Land-cover classification	RGB (Pentax Optio A40) and Modified RGB (Sigma DP1)	X	Clustering tree species	Multirotor (MD4-200)	Automatic	(Gini, Passoni et al. 2012)
Land-cover classification	RGB (Canon PowerShot A650)	Tree species	Automatically identify tree genera/species	Remote controlled helicopter (no autopilot)	Manual	(Peck, Mariscal et al. 2012)
Land-cover mapping	RGB (Panasonic Lumix LX-3)	Lesser kestrel (*Falco naumanni*)	Mapping habitat under kestrel flight paths	FW (ST-model Easy Fly)	Automatic	(Rodríguez, Negro et al. 2012)
Land-cover classification	Multispectral (Condor-1000 MS5 5 CCD multi-channel camera)	X	Land-cover classification	Helicopter (CAMCOPTER,S-100 UAV)	Automatic	(De Biasio, Arnold et al. 2010)
Land-cover classification	RGB (Olympus stylus)	X	Measuring bare areas	FW (RnR APV3) and helicopter (Raptor Model 30)	Automatic	(Breckenridge and Dakins 2011)
Land-cover features	Modified RGB multispectral	*Phragmites australis*	Invasive plant species detection	FW (AggieAir)	Automatic	(Zaman, Jensen et al. 2011)
Land-cover classification	RGB	Various plant species	Feature extraction	FW (based on J3 Cub)	Automatic	(Bryson, Reid et al. 2010)

(*continued*)

Table 3.3 Continued

Application	Sensor(s)	Species/topic	Aim	Drone (model if known)*	Detection method	Source
Land-cover classification	RGB (Canon SD 550 and Canon SD 900)	Grass, shrub, bare	Species level identification/ vegetation classification	FW (MLB BAT 3)	Automatic	(Laliberte and Rango 2009, Rango, Laliberte et al. 2009, Laliberte, Goforth et al. 2011)
Land-cover classification	Thermal	olive orchards	Canopy conductance and crop water stress index	Helicopter (Benzin Acrobatic, Vario, Germany)	Automatic	(Berni, Zarco-Tejada et al. 2009a)
Land-cover classification	RGB (Canon PowershotG5 and Canon EOS 5D)	X	Classify Riparian vegetation types	Radio controlled paraglider ('Pixy' drone)	Automatic	(Dunford, Michel et al. 2009)
Land-cover classification	Multispectral (MCA-6), thermal camera (Thermovision A40M)	X	Vegetation monitoring (LAI, chlorophyll content, and water stress detection)	Helicopter (frame from Benzin Acrobatic, Vario, Germany)	Automatic	(Berni, Zarco-Tejada et al. 2009b)
Land-cover classification	RGB (A101fc 1280x1024 Bayer) and hyperspectral camera	Vineyards	Creating vigour map of vineyards	FW (RCATS/APV-3)	Automatic	(Johnson, Herwitz et al. 2003)
Miscellaneous						
Songbird recordings	Acoustic sensor (Zoom H1)	Several song bird species in the US	Determining the feasibility of songbird recordings with a drone	MR (DJI Phantom 2)	Manual	(Wilson, Barr et al. 2017)
Vegetation monitoring	Hyperspectral	X	Vegetation index	MR (MK-OktoXL)	Automatic	(Aasen, Burkart et al. 2015)
Water sampling	Water sampling equipment	X	Testing feasibility of aerial water sampling	MR (AscTec Firefly)	Manual	(Ore, Elbaum et al. 2015)
Miscellaneous/precision agriculture	RGB (Panasonic Lumix DMC-GF1)	X	Crop biomass estimation (NGRDI and LAI)	MR (Flightcopter)	Automatic	(Jannoura, Brinkmann et al. 2015)
Vegetation monitoring	RBG GoPro Hero3 and VNIR hyperspectral sensor	Vineyard	Leaf area index	MR (OnyxStar BAT-F8, Altigator, Belgium)	Automatic	(Kalisperakis, Stentoumis et al. 2015)
Vegetation monitoring	Hyperspectral	Maize, potato, and sunflower	Leaf area index			(Duan, Li et al. 2014)
Miscellaneous	RGB (GoPro Hero 3)	Green algae (*Cladophora glomerata*)	Algal cover mapping	MR	Automatic	(Flynn and Chapra 2014)
Vegetation monitoring	Multispectral (ADC Air digital camera)	X	Assessment of green leaf cover (NDVI)	Helicopter (Predator Gasser)	Automatic	(McGwire, Weltz et al. 2013)

Vegetation monitoring	Hyperspectral and multispectral	Vineyard	Estimating leaf carotenoid content in vineyards	FW (mX-SIGHT and Viewer)	Automatic	(Zarco-Tejada, Guillén-Climent et al. 2013)
Vegetation monitoring	Multispectral (Tetracam ADC-lite)	Vineyard	Vineyard management by NDVI	MR (VIPtero, a modified Mikrokopter Hexa-II)	Automatic	(Primicerio, Di Gennaro et al. 2012)
Vegetation monitoring	SAR	X	Development of SAR for environmental monitoring	FW (Aludra MK1)	Automatic	(Koo, Chan et al. 2012)
Miscellaneous/precision agriculture	Multispectral	X	Crop biomass estimation	Helicopter (based on Mikado Logo 600)	Automatic	(Honkavaara, Saari et al. 2013)
Collection of blow samples	Blow sample sensor	Various marine mammal species	Obtaining blow samples	Helicopter	Manual	(Acevedo-Whitehouse, Rocha-Gosselin et al. 2010)
Miscellaneous	RGB, modified RGB and thermal camera	Antarctic moss	mapping moss beds in Antarctica	MR	Automatic	(Lucieer, Robinson et al. 2010, Lucieer, Robinson et al. 2012, Turner, Lucieer et al. 2012, Lucieer, Malenovský et al. 2014, Lucieer, Turner et al. 2014, Turner, Lucieer et al. 2014)
Vegetation monitoring	Multispectral (Condor-1000 MS5 5 CCD)	Various types of vegetation	Classify different land types and calculate vegetation indices (NDVI)	Helicopter (CAMCOPTER S100)	Automatic	(De Biasio, Arnold et al. 2010)
Vegetation monitoring	Modified RGB camera	X	Crop Monitoring (GNDVI and LAI)	FW (Vector-P)	Automatic	(Hunt, Hively et al. 2010)
Vegetation monitoring	Modified RGB cameras	Spring barley	Environmental monitoring (NDVI map)	FW (CropCam)	Automatic	(Gay, Stewart et al. 2009)
Miscellaneous/precision agriculture	Thermal	Cotton (*Gossypium hirsutum*) and winter wheat (*Triticum aestivum*)	Plant water stress	FW	Automatic	(Sullivan, Fulton et al. 2007)
Vegetation monitoring	Multispectral	Maize field	Vegetation monitoring (LAI, NDVI)	Helicopter	Automatic	(Sugiura, Noguchi et al. 2005)

* FW = fixed wing drone, MR = multirotor drone.

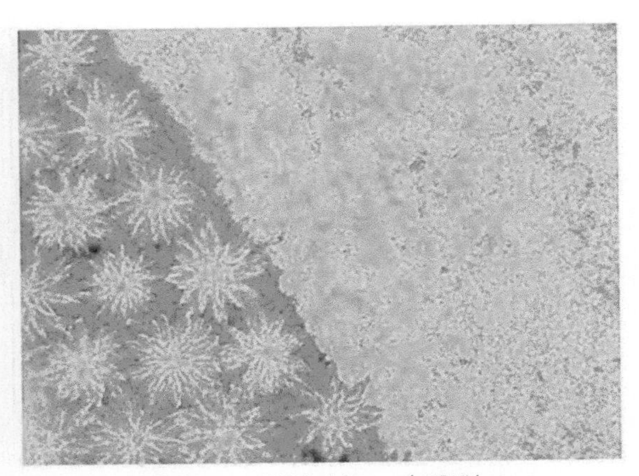

Figure 3.3 Left panel: NIR image of boundary of oil-palm plantation and forest from modified Canon SX230, right panel: NDVI images as generated with ImageJ script from (http://www.maxmax.com/). © conservationdrones.org.

such as animal monitoring or land-cover mapping (see Chapters 6 and 7).

3.1.2 Multispectral cameras

Modified RGB cameras

Due to the cost of multispectral cameras many users have resorted to converted standard RGB cameras as an alternative to true multispectral cameras (Lebourgeois, Bégué et al. 2008, Berra, Gibson-Poole et al. 2015). In standard RGB cameras the sensor is sensitive to part of the non-visual spectrum from 700–1250 nm. However, to restrict the camera to only the visible part of the spectrum the 700–1250 nm region is usually blocked by a hot mirror filter. Theoretically, once the hot mirror filter is removed the camera's RGB channels provide counts of R+NIR, G+NIR, B+NIR. If then, for instance, a longpass filter (a filter that only allows for longer wavelengths to be passed through) is placed in front of the lens so that the wavelengths corresponding to blue are blocked, the sensor only receives the NIR part of the spectrum on the blue channel. Once this NIR component is then subtracted from the digital counts of the other two channels (where R+NIR and G+NIR have been recorded) one is left with the parts of the spectrum that correspond to R, G, and NIR (Rabatel, Gorretta et al. 2014, Berra, Gibson-Poole et al. 2015). From these, measures related to plant photosynthetic activity such as NDVI (Normalized Difference Vegetation Index, Figure 3.3) or related measures can be calculated depending on which filter is being used (Hunt, Hively et al. 2010, Berra, Gibson-Poole et al. 2015). The NIR band in combination with the R and G bands (NIR-R-G) might also allow for more accurate land-cover classification than RGB and is therefore not only useful for studies requiring vegetation indices, but also for researchers interested in land-cover.

There are two options to obtain a modified camera. The first is the do-it-yourself option of removing the hot mirror filter and adding a different filter. This is not necessarily a complicated procedure and can thus be done at home or in the lab with proper instructions or online tutorial videos (e.g.[10]), but needs some care and work in a clean, dust-free environment. Second, there are also off-the-shelf options to purchase modified cameras from vendors such as MaxMax that can convert a multitude of camera brands and types from low-cost digital cameras to high-end cameras.[11] There are also companies such as Mapir[12] that sell one camera with various filter modifications that can either be used on their own or in a pack of four cameras under a multirotor. These cameras are also part of a user-friendly workflow from the

[10] https://publiclab.org/ (accessed 7 November 2017)

[11] https://www.maxmax.com/ (accessed 7 November 2017)

[12] https://www.mapir.camera/ (accessed 7 November 2017)

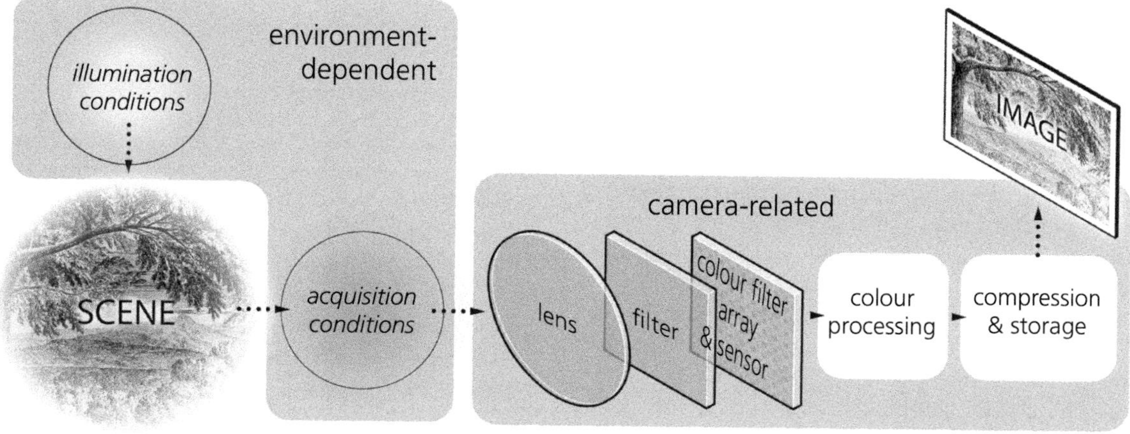

Figure 3.4 Aspects of image acquisition (based on Lebourgeois, Bégué et al. 2008). © Perry van Duijnhoven.

acquisition of the images to software that can yield the images that are ready for interpretation.

These converted cameras do come with certain challenges though. Many of these also apply to non-converted consumer-grade digital cameras. The main limitations are geometric and radiometric (Wackrow, Chandler et al. 2007, Honkavaara, Arbiol et al. 2009). The extent to which these need to be corrected for depends to a large extent on the aim of the study. For a simple count of animals there is likely no correction needed, but for very accurate orthomosaics needed for land-cover change analyses there is likely a need to make careful geometric and radiometric corrections (Wackrow, Chandler et al. 2007, Lebourgeois, Bégué et al. 2008, Honkavaara, Arbiol et al. 2009). When working on projects it is therefore important to consider the various factors that can influence the conversion from the object and the digital recording in the camera (Figure 3.4). There is a large body of highly specialized literature on these topics (Wackrow, Chandler et al. 2007, Lebourgeois, Bégué et al. 2008, Honkavaara, Arbiol et al. 2009), but also some excellent overviews that are less technical (e.g. Whitehead and Hugenholtz 2014). A final, but important, point is that the wavelength bands (spectral resolution) of the modified cameras are often wide and overlapping which leads to limited spectral precision, which can lead to less accurate results.

True multispectral cameras

Recently, several true multispectral cameras (i.e. cameras with different sensors for the different bands, also called camera arrays) have been developed specifically for drones. One important advantage of these cameras above the modified cameras is that there is no spectral contamination between the various bands because they are obtained with independent sensors. The number of bands, spectral range, and resolution depends on the specific camera. The MicaSense RedEdge[13] camera for instance has five bands (Blue, Green, Red, Red Edge, and Near infrared), whereas the MicaSense/Parrot Sequoia has four bands (Green, Red, Red Edge, and Near infrared) and a standard RGB camera.[14] Similar options and some with even more bands are offered by companies like Tetracam[15] and Slantrange.[16] These cameras also have the benefit that they sometimes come with a built-in downwelling irradiance sensor so that images can be radiometrically calibrated in real time based on sky conditions. Nevertheless, it is

[13] https://www.micasense.com/ (accessed 7 November 2017)

[14] https://www.parrot.com/us/Business-solutions/parrot-sequoia (accessed 7 November 2017)

[15] http://www.tetracam.com/ (accessed 7 November 2017)

[16] http://www.slantrange.com (accessed 8 November 2017)

often recommended that where possible ground-based calibration panels are used that are placed in the area that is being mapped, or that before and after the flight an image of a calibration panel is captured allowing for more analytical options. In addition, some newer models also come with a built-in GNSS, IMU (inertial measurement unit), and magnetometer, and still only weigh 72 grams (e.g. Sequoia). Some of these are well integrated with photogrammetry software such as Pix4D mapper[17] to produce NDVI or other outputs.

But even with these specifically designed cameras there can be issues relating to the different perspective centres, lens distortion effects, and viewing angles of the different lenses, which can lead to misalignment of the registration for the images from the various lenses. Some companies provide software packages to correct for the misalignment of the bands, but these do not always fully resolve the precise band-to-band registration challenges and further post-processing might be needed (Jhan, Rau et al. 2016).

3.1.3 Hyperspectral cameras

Hyperspectral cameras aim to acquire the full spectral range for each pixel in a scene. This leads to a large number (often 100–300) of narrow bands (2–20 nm bandwidth) for a continuous spectral range (Jhan, Rau et al. 2016). This distinguishes these sensors from multispectral sensors where a small number of spectral bands are sampled at a lower resolution. The miniaturization of hyperspectral cameras is fairly recent and the sensors are still costly and thus less widespread in their usage than RGB and multispectral cameras (Colomina and Molina 2014, Whitehead and Hugenholtz 2014, Toth and Jóźków 2016). The 'push broom' line scanning process that hyperspectral cameras use to obtain images is different than that of RGB and multispectral cameras that use rolling or total shutters. As with other sensors radiometric and geometric issues can arise with these sensors and must be carefully calibrated before conducting quantitative analyses (Hruska, Mitchell et al. 2012, Mitchell, Glenn et al. 2012, Aasen, Burkart et al. 2015, Mitchell, Glenn et al. 2016).

3.1.4 Thermal imaging cameras

Thermal sensors detect the thermal radiation from objects. For conservation these have been used to detect animals, poachers, vegetation, or fire. Until recently thermal sensors were usually accompanied by large, heavy, and costly cooling systems. The usage of new materials for the development of a new generation of thermal sensors has led to cameras that can operate at ambient temperatures. These cameras have become sufficiently small and affordable for widespread usage under drones. Thermal imaging cameras have been used in combination with RGB and multispectral cameras, which makes co-registration of the bands from the various cameras necessary (Berni, Zarco-Tejada et al. 2009b, Baluja, Diago et al. 2012, Chrétien, Théau et al. 2015, Chrétien, Théau et al. 2016).

There are several companies producing thermal imaging cameras specifically for drones (e.g. FLIR, Optris, Xenics). From a user point of view it is important to consider whether it is only temperature differences or also absolute temperature per pixel that are needed. For the latter, one needs to use a radiometric thermal camera; these are normally more costly than the non-radiometric thermal cameras that can be used to assess relative temperature differences in an image. From a practical point of view it is important to consider whether there is a need for a camera with a built-in recording unit or not. Having a recording unit built-in reduces extra wires on the drone, which is generally desirable because it means fewer parts that can break. FLIR is the main company that now produce a range of (non)radiometric thermal sensors with data capture options built-in. TeAx-Technology also sells modified FLIR thermal cores with recording backpacks and software to analyse the video output. Several other companies such as Optris and Xenics sell thermal cameras that need an external recording unit.

3.1.5 Light Detection and Ranging (LiDAR)

All the above mentioned sensors are passive sensors that detect naturally reflected or emitted

[17] https://pix4d.com/ (accessed 7 November 2017)

electromagnetic radiation. LiDAR sensors are active sensors (they emit their own radiation to illuminate targets) that emit very rapid laser pulses that are used to measure the distance to an object. These sensors have been used extensively on manned airplanes, satellites, and on the ground for a variety of conservation questions (Davies and Asner 2014). Only recently have the systems become sufficiently small and lightweight to be fitted on drones (Jaakkola, Hyyppä et al. 2010, Wallace, Lucieer et al. 2012a,b). Despite the fact that LiDAR systems have been reduced in price and size, they are still relatively large and costly and their usage in conservation has been very limited so far. A particular challenge in lightweight and relatively unstable drones is that LiDAR requires extremely high geometric precision from the GNSS and IMU (Whitehead and Hugenholtz 2014).

3.1.6 Synthetic Aperture Radar (SAR)

SAR, another active sensor technique, has traditionally only been used under manned aircraft and satellites, but due to reductions in size can now be used on large (> 20 kg) drone systems (Essen, Johannes et al. 2012, Koo, Chan et al. 2012, Remy, de Macedo et al. 2012). The development to further miniaturization is occurring (Suomalainen, Franke et al. 2014, Li and Ling 2015, Li and Ling 2016), but due to their high costs and relatively large size the usage in conservation of SAR under drones is in its infancy.

3.1.7 Telemetry

Globally there are a huge number of projects in which animals have been fitted with some sort of tagging device from which the signal can be received by hand-held radio-receivers, manned aircraft with receivers, cell-phone towers, satellites, and others (Thomas, Holland et al. 2012). Despite there being enormous opportunities for tracking animals with drone trackers there has to our knowledge been no off-the-shelf system developed for this purpose. As a result the very few studies that have used drones to locate animals have fitted standard radio-receivers underneath drones or have developed purpose-built systems (Cliff, Fitch et al. 2015, Tremblay, Desrochers et al. 2017).

3.1.8 Miscellaneous sensors

Sensor development for drones is a rapidly expanding field and this will lead to new opportunities for conservation monitoring. The development of new gas sensors will allow the monitoring of various gases such as carbon dioxide, methane, and water vapour (Baer, Gupta et al. 2012). Other sensors that are being developed for drones and that could have applications for conservation are sonar, sensors that allow drones to be mobile communication systems that will allow for data transmission in remote areas, and packages in which multiple sensors are integrated (Pajares 2015). There has been very little research conducted with acoustic sensors on drones (Wilson, Barr et al. 2017). The potential for acoustic surveys of large areas with drones is immense. The main drawback at the moment is the noise made by the motors and the potential for the frequency range of the sounds made by the drone to overlap with the frequency of the calls of animals. Although this is certainly an issue of concern there are surprisingly few studies that have tested whether animal calls can be recorded with directional microphones. Preliminary research has been conducted with bat detectors (Koh, pers. comm.), because it was thought that the high frequency calls of the bats would not be in the same frequency band as the sounds from the drone. Researchers have also started to use drones in very creative ways to collect samples from animals, as highlighted by a study on various marine mammals to collect blow samples (Acevedo-Whitehouse, Rocha-Gosselin et al. 2010, Bennett, Barrett et al. 2015).

3.1.9 Live transmission

An important aspect for data collection is whether data need to be transmitted back in real time to a user on the ground or whether data can be stored onboard for download later. Where live transmission of data is required it is important to consider whether the data that is transmitted live can be of lower resolution than the data to be stored onboard. Transmission of lower resolution data, in the case of RGB video, can often be achieved at lower costs and longer distances than transmission in the native resolution, so part of the decision can be due to budget. But lower resolution data might not give the user the information needed so careful testing is advisable so that the system that is purchased or built provides the user with the resolution needed for the specific use case. One conservation use case for which live transmission of data is needed is poaching. To intercept poachers it is important to obtain data in real time so that interventions can be made as swiftly as possible.

3.2 Applications of sensor types

3.2.1 Animal studies

The usage of drones to detect animals or their signs for a whole array of questions has mainly been conducted with RGB cameras (Table 3.3), but thermal sensors have also been used (Table 3.3, Figure 3.5). Studies have used them to detect animals in terrestrial and marine environments (Table 3.3) and RGB cameras have generally performed well for the detection of species ranging from birds (Chabot, Carignan et al. 2014, Chabot, Craik et al. 2015) to large mammals in terrestrial (Vermeulen, Lejeune et al. 2013, Mulero-Pázmány, Stolper et al. 2014) and marine settings (Koski, Allen et al. 2009, Hodgson, Kelly et al. 2013, Koski, Gamage et al. 2015). For some species detecting indirect signs has been successful, for example great ape nests (van Andel, Wich et al. 2015, Wich, Dellatore et al. 2016) and gopher mounds (Whitehead, Hugenholtz et al. 2014). The use of thermal sensors has been limited thus far, but they have been used for animals such as rhinoceros in Africa, koalas in Australia, roe deer in Germany, cows and people in the UK, grey seals in Canada, and rabbits and chickens in Denmark (Israel 2011, Christiansen, Steen et al. 2014, Mulero-Pázmány, Stolper et al. 2014, Chrétien, Théau et al. 2015, Lhoest, Linchant et al. 2015, Chrétien, Théau et al. 2016, Gonzalez, Montes et al. 2016, Longmore, Collins et al. 2017, Seymour, Dale et al. 2017). These are particularly useful when detection needs to occur at night or dawn/dusk, but should also be able to provide data during the day for animals that stand out against the background due to their temperature.

3.2.2 Locating radio-tagged animals

An exciting prospect for drone usage in animal studies is to use drones to help locate animals with radio tags. At present locating tagged animals is often done on foot, which is time consuming and not always feasible in rugged terrain or remote areas that are too far away from base camps. In addition manned aircraft and satellites have been used to locate animals with various tag systems, but this comes at high costs (Kenward 2001, Thomas, Holland et al. 2012). There are a number of technologies that researchers have been using to tag animals and obtain information on their movement patterns such as VHF collars or GPS collars that communicate with satellites, phone networks, or GPS loggers that communicate with GPS loggers on other individuals when they are in the vicinity and exchange data (Cagnacci, Boitani et al. 2010). Researchers have recently begun exploring opportunities for the usage of drones to locate animals that have VHF or other types of tags, in an effort to reduce the costs of locating tagged animals. Most of these have explored the feasibility of this by examining various concepts that can be applied in combination with their mathematical solutions and some basic simulations and by ground testing that did not involve animals (Posch and Sukkarieh 2009, Soriano, Caballero et al. 2009, Korner, Speck et al. 2010, Santos, Barnes et al. 2014). Several studies have explored the concept, models, simulations, and some tests without animals, for fish in particular (Jensen and Chen 2013, Leonardo, Jensen et al. 2013, Jensen, Geller et al. 2014). There are, however, very few studies that have actually used drones in the field to detect animals with tags (Cliff, Fitch et al. 2015, Tremblay, Desrochers et al. 2017). Both these studies show promising results and were able to show that birds could be detected and, in the case of the Tremblay, Desrochers et al. (2017) study, showed that over 50 per cent of the tagged birds could be detected and that signal strength received on the drone was significantly stronger than that on the ground receiver. Tremblay et al. also report no significant interference from the drone electronics on the receiver.

Figure 3.5 Examples of animal images with RGB and thermal sensors. Top left: Asian elephant in Sumatra, lower left: Indian rhinoceros in Nepal, top right: baboons and ankole cattle in Knowsley Safari, UK, lower right: cows in the Netherlands. © conservationdrones.org.

3.2.3 Land-cover classification

Land-cover classification is of major importance in conservation and a relatively large number of studies have used drones to obtain very high resolution imagery for such classifications (Table 3.2). Many of these studies have used RGB cameras for classification of vegetation types in areas ranging from forests to wetlands (Dunford, Michel et al. 2009, Rango, Laliberte et al. 2009, Chabot and Bird 2013, Wich, Koh et al. in press). Researchers have also used converted and multispectral cameras or used combinations of RGB and multispectral cameras for land-cover classification to benefit from the added information of the multiple spectral bands (Table 3.2). A converted RGB camera was used in a study to classify vegetation types in a peat bog in lower Saxony in Germany (Knoth, Klein et al. 2013). Vegetation in grasslands in the USA has been classified using combinations of RGB and multispectral cameras and automated workflows have been developed for the various steps needed for this (Laliberte, Goforth et al. 2011). Such studies have not been restricted to terrestrial areas but have also focused on classifying the habitats of juvenile salmon in which RGB, converted RGB multispectral, and thermal sensors were used (Woll, Prakash et al. 2011). A hyperspectral camera under a large fixed wing drone has been used to distinguish grasses from other vegetation in shrubland areas (Mitchell, Glenn et al. 2016).

3.2.4 Land-cover change detection

Land-cover change monitoring has usually been conducted with satellites (Broich, Hansen et al. 2011, Hansen, Potapov, Utekhina et al. 2013), but more recently drones have been used for this at a small spatial scale and with a much higher ground resolution. Forest change detection with a LiDAR system was used in a eucalypt plantation in Tasmania, Australia (Wallace, Lucieer et al. 2012a).

3.2.5 Land-cover features

Estimating tree height for forests is usually conducted from the ground or by using airborne LiDAR. Both are costly and LiDAR is not generally available around the globe so in many countries tree height estimates are very restricted both spatially and temporally. Researchers have started to use drones with various sensors to estimate tree height using structure from motion techniques (see Chapter 7). LiDAR sensors have been used for forest inventories and tree height and crown width estimates (Wallace, Lucieer et al. 2012b). Drones with a LiDAR have also been used to determine tree diameter at breast height (DBH) in forest stands (Chisholm, Cui et al. 2013). In relatively open olive orchards a fixed wing drone using a modified consumer-grade camera from which the infrared filer was removed was used to determine the height of the olive trees (Zarco-Tejada, Diaz-Varela et al. 2014). Tree heights, above ground biomass, and other forest characteristic comparisons between point clouds, derived from images obtained with RGB cameras or modified RGB multispectral cameras and point clouds obtained with LiDAR carried by drones, have also been made in forest stands by various research groups with promising results (Dandois and Ellis 2010, Dandois and Ellis 2013, Lisein, Pierrot-Deseilligny et al. 2013, Dandois, Olano et al. 2015, Puliti, Ørka et al. 2015, Thiel and Schmullius 2016a, Thiel and Schmullius 2016b). Point clouds from drone RGB images have also been used to compare measures calculated from the point clouds with ground-based measures such as above ground woody biomass, canopy height, and openness in forests of different phases of recovery (Zahawi, Dandois et al. 2015). In addition to comparisons of forest and tree characteristics from drone-derived point clouds with LiDAR, there have also been comparisons of tree stem reconstructions in open forest stands from point clouds made with an RGB camera with those a terrestrial laser scanner (TLS; Fritz, Kattenborn et al. 2013). Studies have also used the orthomosaics derived from RGB images to count fallen stems in a broadleaved forest in Japan (Inoue, Nagai et al. 2014). Researchers have also used a drone with an RGB camera to map canopy cover and the sizes of gaps in the canopy and compare these to measurements from manned airborne LiDAR (Getzin, Nuske et al. 2014). The same research group has also used a drone to obtain RGB aerial images of canopy gaps to assess forest understory floristic biodiversity (Getzin, Wiegand et al. 2012). Studies have also been able to detect several tree species and determine their health in riparian forests in Belgium using a combination of RGB and modified RGB multispectral cameras (Michez, Piégay et al. 2016b). Other studies have relied solely on RGB sensors to distinguish tree species in deciduous forests (Lisein, Michez et al. 2015). Invasive plants in riparian areas in Belgium and wetlands in the USA have also been detected using RGB and/or modified RGB multispectral sensors (Zaman, Jensen et al. 2011, Michez, Piégay et al. 2016a). Estimating biomass of crops has been conducted using both multispectral and RGB sensors (Honkavaara, Saari et al. 2013, Jannoura, Brinkmann et al. 2015).

3.2.6 Water content/stress

Monitoring water content and stress in plants has also been achieved with thermal sensors under drones, and this approach has been particularly used in the agricultural sector where vineyards and orchards have been used to test these approaches (Sullivan, Fulton et al. 2007, Berni, Zarco-Tejada

Figure 3.6 Forest fires in the Sabangau peatswamp forests. © Bernat Ripoll Capilla/Borneo Nature Foundation.

et al. 2009a, Gonzalez-Dugo, Zarco-Tejada et al. 2013, Gago, Douthe et al. 2014). Similar research has been conducted with multispectral sensors for citrus fruits (Stagakis, González-Dugo et al. 2012). Researchers have also used a combination of hyperspectral and thermal sensors to measure water stress in plants due to diseases (Calderón, Navas-Cortés et al. 2013).

3.2.7 Leaf area index

A important measure in studies that attempt to determine the primary productivity of plants and the biochemical cycles of plants is to determine the leaf area index (LAI)(Bréda 2003). Measures of this index are usually achieved from the ground (Bréda 2003) or from satellites (Chen, Pavlic et al. 2002, Nguy-Robertson and Gitelson 2015). More recently drones with RGB (Córcoles, Ortega et al. 2013) or hyperspectral sensors (Duan, Li et al. 2014) have been used to map the LAI. Canopy reconstruction using structure-for-motion (SfM) techniques have also been used to estimate LAI in vineyards (Mathews and Jensen 2013).

3.2.8 Fires

Fire detection is an important aspect of land management in many parts of the world. Thermal sensors and mixtures of RGB, hyperspectral, and thermal sensors have been used in fire detection (Rufino and Moccia 2005, Martínez-de-Dios, Merino et al. 2007, Merino, Caballero et al. 2012, Yuan, Zhang et al. 2015, Twidwell, Allen et al. 2016) and the development of georeferenced orthomosaics from thermal data for extraction of wildfire data (Zhou, Li et al. 2005). NASA and the US Forest Service have been using large drones with sensors that can register 16 spectral bands for fire detection (Ambrosia, Wegener et al. 2011). A smaller drone system with a thermal imaging camera has also been used by the US Forest Service and NASA for fire management (Hinkley and Zajkowski 2011). In Indonesia the regular outbreaks of fires have huge implications for climate change mitigation, health, tourism, and the economy (van der Werf, Dempewolf et al. 2008, Gaveau, Salim et al. 2014). Recently, off-the-shelf multirotor systems have been used by organizations in the field to assist them to obtain real-time data on the extent of these fires (Figure 3.6).

3.2.9 Miscellaneous applications

Drones have recently been used or have been suggested to collect data that are useful for epidemiological studies (Barasona, Mulero-Pázmány et al. 2014, Fornace, Drakeley et al. 2014, Jones 2014). So far drones have been collecting high-resolution land-cover images with RGB cameras that provide information that is relevant for animal species that can carry diseases (Barasona, Mulero-Pázmány et al. 2014, Fornace, Drakeley et al. 2014). Drones with RGB cameras have also been used to monitor green algae (*Cladophora glomerata*) distributions in the Clark Fork River in western Montana (Flynn and Chapra 2014). In this study, image orthomosaics were processed on which algae could be automatically identified. Surface water contamination has recently also been investigated in Campania (Italy) with the aid of two types of blimps that carried a thermal imaging sensor (Lega and Napoli 2010). Contaminated water has a different thermal signature than the surrounding water, which allows for the detection of contaminated discharges into streams, rivers, and seas (Lega and Napoli 2010). Drones have also been used to study various aspects, such as extent and health of Antarctic moss beds, which are important indicators of regional climate change and thus important to study in this sensitive ecosystem (Lucieer, Robinson et al. 2012, Turner, Lucieer et al. 2012, Lucieer, Malenovský et al. 2014, Lucieer, Turner et al. 2014, Turner, Lucieer et al. 2014). These studies have used RGB, multispectral, hyperspectral, and thermal sensors to study various aspects of moss beds. Another interesting application of drones is the study of impact that animals, such as beavers, have on their habitat (Puttock, Cunliffe et al. 2015). Using structure-from-motion to prepare very high resolution orthomosaics allowed the researchers to rapidly (one day of data collection) obtain insight into the impact of beavers on their habitat such as gnawed-through trees, the beaver dam, changes to the watercourses, and channel morphology (Puttock, Cunliffe et al. 2015). This approach allows for detailed monitoring of how animals change their habitat and can thus be an important tool to increase our understanding of ecosystem engineering by animals (Wright and Jones 2006).

Although the potential applications for recordings of animal calls from drones are numerous there have been virtually no studies conducted. The only published study that we are aware of (Wilson, Barr et al. 2017) used a commercially available multirotor with a low-cost acoustic recorder attached to an eight-metre long fishing line. The multirotor was hovered at three different altitudes (28, 48, and 68 m) to record birdsongs played back from a speaker on the ground, with calls from five species. Additionally the drone was used to fly at 50 m over areas in which ground point counts were also made. This allowed for a comparison of ground observations (audible and visual) and the data acquired with the recorder under the drone for a number of bird species. Before analyses the recordings were passed through a high pass filter to reduce the noise from the drone. The results showed that the detection rate in the playback experiments did not depend on flying altitude and ranged between 41.2 per cent and 75.8 per cent depending on the loudness of the bird species. The comparison of ground (using audio data only so that data are comparable to the drone recordings) and drone observations in the point counts showed that fewer species were recorded by the drone (37 vs 32 species) and that species richness and total count were significantly lower in the drone data. The authors conclude that their results are only valid for their specific set up and that other drone/acoustic recorder combinations will likely lead to different results, but that further research is needed to fully explore the potential of using drones for acoustic surveys. A preliminary study has investigated whether calls of the grey-headed flying-fox (*Pteropus poliocephalus*) could be recorded with a bat detector that was fixed with a long cable under a multirotor drone (Koh, pers. comm.). Analyses of the frequency of the bat calls and the noise produced by the drone showed that most of the noise from the drone was around 4 kHz, whereas the bat calls were around 39 kHz and thus well separated from the drone noises (Figure 3.7).

Much more research should be conducted in this field so that drone noise is minimized and acoustic sensors are directional so that the potential

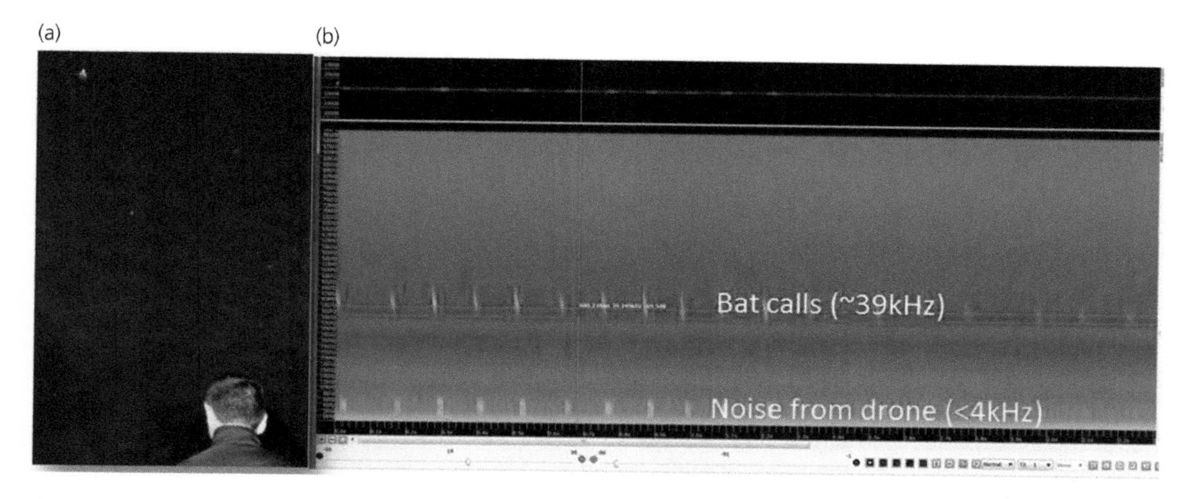

Figure 3.7 a) Lian Pin Koh flying a Y6 multirotor with a bat detector underneath. b) Sonogram showing time on the x-axis and frequency on the y-axis. The bat calls are visible around 39kHz, whereas the noise from the motors and propellers of the drone is around 4kHz. © Lian Pin Koh.

for recording of calls is optimized. Alternatively, or in addition, sound absorbing material could be placed between the recording unit and the motor/propellers on the drone, but all this is very fertile ground for further experimenting. The potential for acoustic surveys with drones is large. Particularly promising is the prediction from Wilson, Barr et al. (2017) that there will be a time when drones will be flying transects of more than 20 km and instantaneously identifying bird calls and geotagging the recordings. Such development could be useful for a number of animal groups that are vocal such as birds and primates.

Although much of the data gathering with drones has focused on sensing the electromagnetic spectrum there are vast opportunities beyond such sensors to use drones for environmental applications. Particularly promising are the usage of drones to sample air quality and water (Ore, Elbaum et al. 2015, Villa, Gonzalez et al. 2016, Kersnovski, Gonzalez et al. 2017). A recent paper discussed the development of a multirotor system to conduct water sampling and showed very promising field tests of the system (Ore, Elbaum et al. 2015). Air quality systems have been proposed, of particular usefulness in the case of forest fires where safety of personnel is of importance as well as obtaining air quality information (Twidwell, Allen et al. 2016). In forest fire situations drones can be used to obtain data on air quality, provide visual information on potential areas that can be used for fire breaks, facilitate searching for water sources that can be used for supplying fire fighting vehicles with water, provide visual information about homes and other infrastructure, and potentially function as relay systems for cellular or radio coverage in areas where this is needed (Twidwell, Allen et al. 2016). In the biodiversity-rich peat forests of Indonesia drones have already been used by the Orangutan Tropical Peatland Project (OuTrop) during the massive forest fires in 2015.[18] Drones enabled the fire-fighting teams to obtain information about the extent and location of fires, which increased the safety of the staff involved and their efficiency in extinguishing fires or placing fire breaks. A particularly exciting application of drones is to use them to obtain blow samples from marine mammals, which allows for analyses of stress hormones, DNA, viruses, and bacteria, and thus allows for a non-invasive method to obtain a wealth of data on whales (Acevedo-Whitehouse, Rocha-Gosselin et al. 2010, Bennett, Barrett et al. 2015).

[18] https://www.orangutan-appeal.org.uk/about-us/projects/article/outrop-project-update (accessed 8 November 2017)

CHAPTER 4

Surveillance

4.1 Terrestrial

One of the most pressing conservation challenges in recent decades is the illegal trade and trafficking of wildlife and associated products across international borders. The seemingly insatiable global demand for wildlife products drives the poaching of a wide range of imperiled species, including elephants for ivory and tigers for skins and bones. The value of the global illegal wildlife trade is estimated at up to $20 billion per year (Duffy 2016a). There are also growing concerns that this hugely lucrative industry is funding terrorist and criminal groups around the world (Duffy 2016b).

Global interest in the use of drones as a tool for combatting illegal wildlife trade arguably began in 2012 with a $5 million grant to WWF from technology giant Google to support the deployment of conservation drones in Asia and Africa (Ogden 2013). The authors of this book were involved in helping WWF Nepal and their local government partners in a series of pilot projects to trial this technology in Chitwan National Park and Bardiya National Park (Figure 4.1).

Other organizations soon jumped on the conservation drone bandwagon, including the Lindbergh Foundation that funded the creation of the Air Shepherd Initiative.[1] Notably, Air Shepherd uses drones in combination with predictive computer algorithms to stop poachers before they can reach target animals in southern Africa (also see Chapter 6).

Elsewhere in the world, drones have also been used as tools for the surveillance of other kinds of illegal activities and their impacts. Several non-governmental organizations in Southeast Asia, including the Sumatran Orangutan Conservation Programme[2] and Greenpeace, have been flying drones over peatlands to monitor the spread of forest fires (Figure 4.2). These fires are often started and used by farmers as the cheapest way to clear lands in preparation for the next growing season (Page and Hooijer 2016). Given the remoteness and inaccessibility of burnt areas, the drone footage collected by non-governmental organizations is often the only available evidence for these activities. The impact of these fires extends beyond borders, most evidently resulting in a persistent and economically important transboundary haze across parts of Southeast Asia for several months each year (Lee, Jaafar et al. 2016).

4.2 Marine

The biggest conservation threat in the marine environment arguably is illegal, unreported, and unregulated fishing (Arias and Pressey 2016). These activities are contributing to growing and unsustainable demands on global fish stocks. They also represent lost revenues to the legitimate fishing industry and government departments, resulting in significant economic impacts.

Addressing the problem of illegal fishing, in particular, would involve developing and imple-

[1] http://airshepherd.org/ (accessed 8 November 2017)

[2] https://sumatranorangutan.org/ (accessed 8 November 2017)

Conservation Drones: Mapping and Monitoring Biodiversity. Serge A. Wich & Lian Pin Koh. Oxford University Press (2018).
 DOI:10.1093/oso/9780198787617.001.0001

Figure 4.1 A park ranger launching a conservation drone to patrol a national park in Nepal. © Lian Pin Koh.

menting effective monitoring regimes (Humber, Andriamahaino et al. 2017). In parts of the world where large oceanscapes are under protection, substantial investments in funds, manpower, and time would be required to support regular monitoring for the effective enforcement of regulations.

As a result, some government departments and non-governmental organizations have turned to drones as a tool to help reduce the cost of marine conservation. Together with the Wildlife Conservation Society and the government of Belize, the authors of this book were involved in a pilot project to trial the use of drones for patrolling marine protected areas within the Glover's Reef Marine Reserve (Figure 4.3).

In Australia, the coastal habitats of the Great Barrier Reef are an ecologically important and sensitive ecosystem that is under pressure from human activities and climate change (Hedge, Molloy et al. 2017). A team of indigenous rangers are using drones to patrol the Great Barrier Reef for the purposes of monitoring habitat health, wildlife, and human activities (GBRMPA 2017). The ability of drones to provide a vantage point 100 m over inaccessible coastal areas allows the rangers to perform their tasks much more cost effectively.

4.3 Key issues

Although drone technology is a promising and empowering tool for environmental surveillance, it does face several technical, ethical, and regulatory challenges.

4.3.1 Technical challenges

Flight time: The biggest technical hurdle to the effective use of drones for habitat and wildlife surveillance is the limited flight time. As we discussed in Chapter 2, fixed wing drones generally offer a longer flight time than multirotor drones, and would be more appropriate for surveillance

Figure 4.2 Image of a burning peat swamp forest captured by a conservation drone in Sumatra, Indonesia. © conservationdrones.

applications. Even so, most affordable and operationally ready fixed wing drones in the market today would not be able to operate for much more than an hour or two. This typically translates to a total flight distance of up to 100 km, or areal coverage of a few hundred hectares, in the most ideal weather conditions. Although this amount of flight time might be sufficient for the surveillance of small patches of forest and short stretches of coastline, it would need to be significantly longer for this technology to be a scalable and cost effective alternative to traditional methods of monitoring. Recent advances in solar powered fixed wing drones to increase flight time might provide a solution in the not too distant future (see also Chapter 8).

Landing strip and piloting skill: One of the most compelling features of drones is their apparent ease of use, especially in terms of the ability to preprogram survey missions as discussed in Chapter 2. However, in the case of fixed wing drones, which would be the drone type more appropriate for surveillance missions, there is a requirement to locate a large enough landing strip to allow for landing. This is typically a clearing in the forest or coast of at least 50 × 100 m. Furthermore, depending on the presence of potential obstacles adjacent to this landing space, the pilot would also need to be sufficiently skilled to land the aircraft. In the near future, fixed wing drones are likely to have more reliable capabilities for automated landing that take account of weather, obstacles, and payload.

Telemetry: A key tactical requirement of most surveillance operations is the real-time communication of data between the drone and the ground

Figure 4.3 A fixed wing drone used for patrolling marine protected areas in Belize. © conservationdrones.

operator to enable rapid response to a developing situation (Figure 4.4). This data telemetry is important in terms of both receiving a 'live' stream of flight details (e.g. drone location) and images from the drone to the operator, and transmitting new commands from the operator to the drone to effect an appropriate action, such as changing the mission to focus on the object of interest. Unfortunately, most non-military grade radio telemetry systems used in commercially available drones are limited in their transceiving range. The most sophisticated of such long-range radio systems typically do not work beyond several tens of kilometres and may be subjected to increased regulatory restrictions.

4.3.2 Ethics and privacy

Unknown impacts on wildlife: Although drones are increasingly being used for wildlife applications, we still do not have a good understanding of the potential physiological responses of wildlife to drones that are operating in close proximity (Hodgson and Koh 2016) (also see Chapter 6). Different animal populations can respond idiosyncratically to drones depending on the animals' physiology, previous exposure to similar disturbances, and the prevailing environmental conditions (Ditmer, Vincent et al. 2015, Pomeroy, O'Connor et al. 2015, Vas, Lescroël et al. 2015). Past research on other types of stressors on animals have demonstrated significant impacts that include abandonment of nests and chicks during breeding seasons, and increased susceptibility to predation (Coetzee and Chown 2015).

Privacy: Another issue that introduces complications is concerns about the privacy rights of people who are under surveillance during drone missions (Humle, Duffy et al. 2014, Sandbrook 2015, Finn and Wright 2016, Wich, Scott et al. 2016, Winter, Rice et al. 2016, Duffy, Cunliffe et al. 2017). Although the individuals detected and observed within protected areas would be the intended targets of surveillance,

Figure 4.4 A long range data telemetry system used to maintain communications with a conservation drone during a surveillance mission in Belize. © conservationdrones.

they still would retain basic rights to their privacy in most societies (Sandbrook 2015, Wich, Scott et al. 2016, Duffy, Cunliffe et al. 2017). There have been recent discussions in Australia and the United States on the need to clarify such privacy issues, as well as conflicts over drone applications and surveillance laws in countries such as Sweden.

4.3.3 Regulations

Regulations for the operation of drones differ between countries and it is important that operators always consult these before operating and where needed apply for the necessary permits. In case of doubt about regulations it is sensible to contact the civil aviation authority of the country of operation. Although no complete global database of drone regulations exists, the droneregulations site[3] is a good starting point for a search on regulations. A main limitation for drone operation at present is that it is generally not permitted to operate beyond visual line of sight, which often is 500 m horizontal distance and 400 ft vertical distance. For drones to be used as a cost effective surveillance tool, they would need to be scalable and cover significant extents of the land- or seascape. Even if the technical challenges to the operation of drones over large distances and areas are overcome, there still exists the legislative issue of operating a drone beyond visual line of sight (BVLOS) (Chabot and Bird 2015). In countries such as Australia, the regulations applicable to BVLOS operations remain nascent as different segments of society debate the various concerns about allowing drone operators to do so. In many other countries, there are not even regulations regarding this kind of operation.

[3] https://www.droneregulations.info/ (accessed 8 November 2017)

CHAPTER 5

Mapping

5.1 Land-cover mapping

One of the most common applications for drones is mapping. The simplest way to map is to fly a drone at the required altitude and acquire a georeferenced aerial photograph of the immediate surroundings below the drone (Figure 5.1). This often works well if a very high resolution image is required and the area of interest is small enough to be captured in a single photograph (Duffy, Cunliffe et al. 2017).

However, biologists often also need to map a larger area for research in landscape ecology or to develop conservation strategies at that scale. A common approach to achieving this is to take a series of overlapping aerial photographs as the drone flies over a landscape (Nex and Remondino 2014) (see also Chapter 7). This generates a dataset that can be processed, using methods from photogrammetry, such as Multi-View Stereo Structure from Motion (MVS-SfM), to produce a map of the area of interest, known as an orthomosaic (Figure 5.2).

Additionally, the drone is often programmed by the operator to fly to defined waypoints (or coordinates) in a lawnmower type pattern to collect the required data in the most efficient way. This is commonly known as mission planning.

Due to their flight duration and regulations often restricting flights to only visual line of sight, the spatial scale at which drones are most cost effective as a data collection platform is in the range of tens to hundreds of hectares, beyond which it would be more cost effective to collect the data using other platforms such as manned aircraft and satellites. If in the future flight duration and, particularly, regulations will allow for beyond visual line of sight flying, the area that can be mapped with drones will increase.

Another less common drone application related to mapping is 'drone truthing'. Satellite-based remotely sensed data are often used by biologists for land-cover classification at the scale of hundreds to thousands of square kilometres. A main challenge of this technique is the high cost of conducting ground surveys to collect reliable data for training the classification algorithm and validating classified areas (also known as 'ground truthing'). Instead of sending biologists to remote sites to 'ground truth' any particular land-cover, they can now use drone-derived maps as training data (Szantoi, Smith et al. 2017). In this way, drones are allowing biologists to 'drone truth' the development of satellite-derived maps more cost effectively.

5.2 Monitoring of vegetation condition

Biologists are interested in monitoring the health of a forest or other natural habitats for various reasons. For example, the dieback of certain tree species could be a harbinger of a more serious and widespread outcome for an entire landscape. The ability to detect changing vegetation conditions would enable biologists to address the underlying causes of the problem in time.

The use of drones for vegetation monitoring is in fact just an extension of their use in mapping. The advantage that a drone has over more traditional platforms of data collection is that, if the target site

Conservation Drones: Mapping and Monitoring Biodiversity. Serge A. Wich & Lian Pin Koh. Oxford University Press (2018).
 DOI:10.1093/oso/9780198787617.001.0001

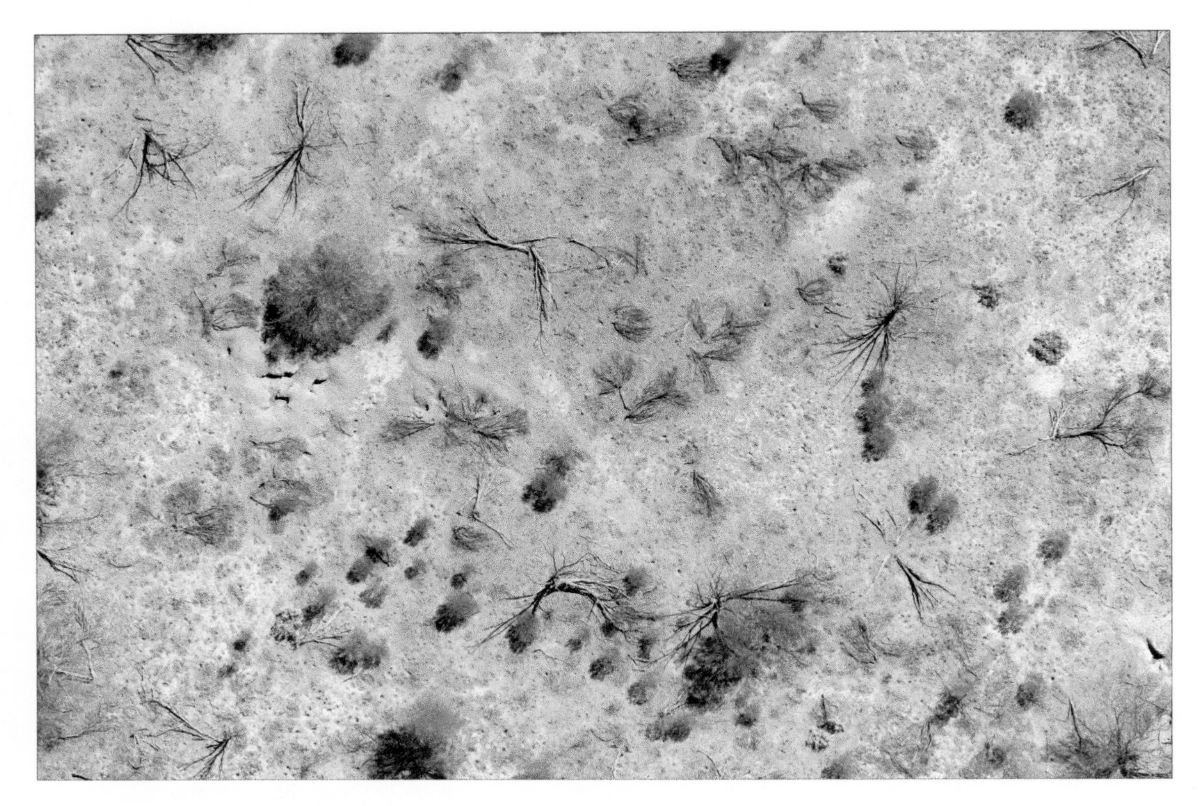

Figure 5.1 Single image taken by a multirotor drone in a study of the habitat of the burrowing bettong. © Lian Pin Koh.

is of an appropriate scale for this technology, drones can be deployed as often as necessary to collect a time series of maps for monitoring (Hird, Montaghi et al. 2017), whereas satellite data collection is often at fixed intervals and can be hampered by cloud cover. Additionally, in the specific use case of monitoring vegetation condition, drones can also be used to collect appropriate spectral data through a variety of sensors, including multispectral, hyperspectral, near infrared, and infrared cameras, allowing for more direct comparisons with satellite datasets (see also Chapter 3). Some data collected by these sensors can be further analysed to give us an indication of underlying causes of decline such as disease, high salinity, and water stress (Soubry, Patias et al. 2017).

5.3 Biomass estimation

Drones are also being used to produce Digital Surface Models, which are geographically accurate representations of the vegetation and other structures within a virtual landscape on a computer. There are many uses for DSM in conservation, the most important of which arguably is to get an estimate of the vegetation volume of a particular forest, and from that, an estimate of biomass (Bendig, Bolten et al. 2014, Messinger, Asner et al. 2016). The biomass of a forest is a key indicator of the carbon stock contained within the forest, which has important implications for informing climate change mitigation strategies (Rodríguez-Veiga, Wheeler et al. 2017).

5.4 Key considerations

5.4.1 Mission planning

All else being equal, the spatial footprint of a drone-acquired image is primarily determined by the height at which the photograph is taken. For example, a multirotor drone hovering at 400

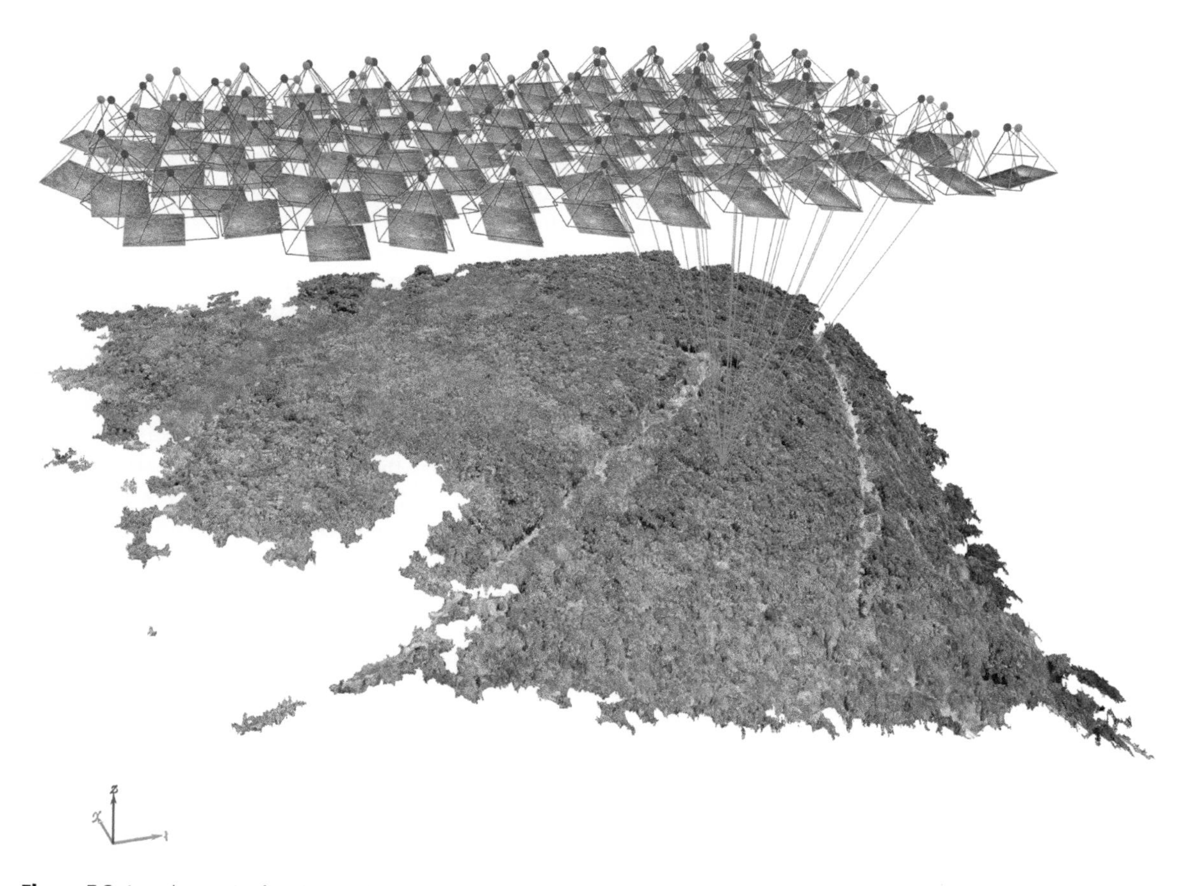

Figure 5.2 An orthomosaic of a 1-hectare forest plot in New Caledonia mapped by a multirotor drone. Multiple small images above the orthomosaic represent individual raw images captured by the drone that were used for the generation of the orthomosaic. © Lian Pin Koh.

feet would be able to capture an image with a larger footprint than one flying at a lower altitude. The practical implication of this for a mapping mission is that the horizontal distance that a drone has to travel in the mission is inversely proportional to its height above the ground (or forest canopy). In other words, one can fly higher to map quicker.

However, it is important to note that the trade-off of capturing data at high altitudes is the loss in image quality (i.e. spatial resolution), simply due to the subject being physically farther away from the camera. This can be an acceptable trade-off in some use cases, such as when the map is to be used for land-cover classification in a landscape with distinctly different land-cover types (e.g. intact forests versus monocultures). In other applications, such as when the goal is to count orangutan nests in the forest canopy, the mapping mission would usually need to be carried out at a much lower altitude (Figure 5.3).

Of course, there are other factors that could affect the spatial resolution of a drone-acquired image, including the sensor size and focal length of the camera. A camera with a longer lens would produce an image of higher spatial resolution than one with a shorter lens. Therefore, in theory one could potentially get a high quality image even at high altitudes, simply by using an appropriate camera system. However, apart from the cost consideration of the camera itself, the drone would also need to be big enough to handle a bigger payload.

Figure 5.3 A very high resolution image of two orangutan nests acquired during a conservation drone mission over a rainforest in Sumatra, Indonesia. © conservationdrones.

5.4.2 Data processing software

There is an increasing range of data processing software available in the market that could be useful for producing the intended products of a mapping mission (see also Chapter 7). Most of these software tools are based on a computer vision technique known as Multi View Stereo—Structure from Motion, a workflow that brings together multiple computer vision techniques to yield the desired output, which may include point clouds, orthomosaics (two-dimensional maps) and Digital Surface Models (three-dimensional models).

5.4.3 Environmental factors

The environment can have a big influence on how a mapping mission needs to be planned and executed. For example, the ideal time of day to conduct a mapping mission for the purpose of land-cover classification is usually between 1000 h and 1400 h during the day. The images captured during this time would have the least amount of shadows as the Sun would be largely directly overhead. Shadows are to be avoided as they are artefacts that could affect data processing.

The other important consideration is wind, especially when a fixed wing drone is used for mapping. If a mapping mission is programmed such that the transects of the mission run parallel to the prevailing wind direction, the drone would be experiencing a headwind when traveling in one direction and a tailwind in the other. In this scenario, there is a risk that the drone would be flying faster than its intended speed with the tailwind, resulting in insufficient overlaps between consecutive images captured in this part of the mission (and too much overlap when facing a headwind, which is less of an issue). If a mapping mission has to be carried out in windy conditions, it is more desirable to orient the transects such that they are perpendicular to the prevailing wind. Another undesirable effect of wind is that it may result in the movement of vegetation, resulting in slightly inconsistent locations of objectives in the images acquired by the drone, which may ultimately interfere with data processing.

Of course there are several other environmental considerations that may be important to mapping that are not discussed here, including incoming solar radiation, solar flare activities, and human generated magnetic interference.

CHAPTER 6

Animal detection

6.1 Animal distribution and density data collection

A key task for conservation science is to collect accurate and precise data on animal distribution and density so that population sizes can be estimated. A multitude of methods have been used to try to achieve that goal, but all are costly and time consuming (Buckland, Anderson et al. 2001, Buckland, Anderson et al. 2004, MacKenzie 2006, Kühl, Maisels et al. 2008, Campos Cerqueira and Aide 2016). A commonly used method to obtain such data is to use line transects or point samples to collect data on either the animals or their indirect signs (e.g. nests or dung) and then use well-developed analytical methods to analyse the data (Buckland, Anderson et al. 2001, Buckland, Anderson et al. 2004). Such data are often collected on foot, driving, flying in manned planes or helicopters, and from ships. These methods generally yield accurate data, but are often time consuming and costly. For instance, renting ships to conduct surveys to determine the distribution and density of marine mammals is often costly, whereas depending on vessels of opportunity means that the timing of the survey and the data collection design will not be systematic even though costs will be reduced. Aerial surveys with manned aircraft often also face high costs for plane or helicopter rental and in many locations availability is limited or absent. In addition, there is substantial risk associated with such aerial surveys: crashes are a leading cause of death among wildlife biologists (Sasse 2003) and fatalities in general aviation were found not to be declining (Wiegmann and Taneja 2003). Ground surveys on foot often have the disadvantage that in large areas there are considerable amounts of time and costs associated with reaching survey locations. In our own work with Sumatran orangutans some sample locations take five days on foot to reach (and five days back) due to the hilly terrain and large tracts of rainforest (see Chapter 1). Thus with the actual data collection one sample area can be 12 days' effort which leads to high costs. Driving in more open areas is used for surveys as well, but usually only roads (paved or unpaved) can be used, which can lead to biased survey results. In addition, several of these methods require either high rent or purchase costs plus maintenance costs. It is therefore important to explore other methods to obtain data on animal distribution and density.

6.1.1 Alternative methods

Conservation scientists have been exploring technological options (camera traps and acoustic sensors) with the aim to reduce the costs and time investment for such data collection and collect data on animals that are elusive and otherwise difficult to study (Griffiths and Van Schaik 1993, Karanth 1995, Silveira, Jácomo et al. 2003, Fretwell, LaRue et al. 2012, Fretwell, Staniland et al. 2014, McMahon, Howe et al. 2014, Yang, Wang et al. 2014, Spillmann, van Noordwijk et al. 2015, Campos Cerqueira and Aide 2016). Even though these methods provide good data they are often not easy to deploy and generally have high costs due to the costs of camera traps, acoustic sensors, and the manual labour needed

Conservation Drones: Mapping and Monitoring Biodiversity. Serge A. Wich & Lian Pin Koh. Oxford University Press (2018).
 DOI:10.1093/oso/9780198787617.001.0001

to empty out the memory cards. Conservationists have been using satellite upload links for data from camera traps networks that are linked to a central satellite upload node; this is not commonly used due to the costs.[1] Researchers have also used passive acoustic sensor networks with radio antennas to link these to a computer at a permanent site where recordings can then be monitored online (Kalan, Piel et al. 2016). But this is not an out-of-the-box solution and therefore not commonly used. Although the opportunity exists to place camera or microphone networks in remote locations, the fact that data need to be retrieved manually or with costly satellite links sustains interest in other methods to obtain data at lower costs. Of course the feasibility of this depends on the species that is being studied.

Due to the increasing resolution of sensors on satellites there have been a few recent studies that have used these to determine the distribution of terrestrial and marine mammals (Fretwell, Staniland et al. 2014, McMahon, Howe et al. 2014, Yang, Wang et al. 2014). This is a very promising method, but there are several limitations. First, due to the limited resolution only large species in open areas can be detected. Second, the complexity of the analyses puts constraints on this type of work for organizations who lack that technical skill. Third, the costs of very high resolution satellite images are high and, particularly when frequent images are needed, this puts this method beyond most conservation organizations and researchers at academic institutions. Without doubt the costs of very high resolution satellite images will decrease as well as the complexities of the analyses and as such this method will likely be much more common in the future. Specifically, with the rise of small cube/nano-satellites that will increase sample rate and decrease costs there will be more scope for usage of this method.

More recently drones have been added to that toolbox. It is, therefore, important to determine whether drones can obtain the required data in a more cost-effective way than alternatives (see Chapter 1 for an example). In addition, there are several other considerations. First of all it needs to be considered whether the animals or their signs can be recorded with the sensor used on the drone (for details on sensors see Chapter 3). Whether this is feasible depends to a large extent on several factors that all relate to the main question of whether the animal or its sign can be detected from the air. Factors that influence detection probability are size of the animal (this influences flying altitude), activity pattern (nocturnal animals might be difficult to detect during the day because they could be in tree holes, holes in the ground, or under very tight canopy), camouflage, whether it occurs under dense forest canopy, and whether it is startled by the drone. Animals that will flee when they detect the drone will be difficult to count in a reliable way; there can also be ethical concerns about disturbing animals and guidelines for this have been suggested (Hodgson and Koh 2016).

Additionally it needs to be determined whether the area that needs to be covered for the survey work can realistically be reached and covered with a drone. Although flying times for drones can be substantial, the increased flying time often comes at increased costs (Chapter 2). Therefore careful consideration of different models with different capabilities is needed before deciding whether a drone can be used to detect animals.

Another important consideration is (as indicated in Chapter 1) the analyses of the data. It is important that the complete workflow from capturing the data until the results that are required are carefully thought through before starting to acquire a drone and using one (Table 6.1). There are many steps to consider here, from whether the animal species of interest can be detected, whether the site can be reached, weather conditions, and flying regulations (Chapter 8), to potential societal issues, logistical issues such as having electricity to charge batteries for planes, cameras, and ground control stations, and others (Duffy, Cunliffe et al. 2017).

6.2 Drone usage for animal distribution and density

Manned aircraft surveys have been used extensively for studies on animal distribution and density (Buckland, Anderson et al. 2001, Buckland, Ander-

[1] https://www.zsl.org/conservation-initiatives/conservation-technology/instant-detect (accessed 8 November 2017)

Table 6.1 Workflow considerations

Workflow part	Factors to consider
Detection probability	Size of animal, sensor resolution, flying height, camouflage, vegetation cover.
Survey site location	Take-off and landing (TOL) areas, distance from where animals need to be surveyed to nearest TOL area, flight duration and distance covered by drone. Permission from landowners to conduct flight. Wind and temperature. Potential location influence on GNSS and compass.
Regulations	Check local and national regulations.
Privacy issues	Will data of people or objects be collected that need approval?
Social issues	Do people in the vicinity of the flight need to be informed or asked for permission or otherwise engaged with?
Logistics	Will there be electricity to charge batteries for drones, cameras, ground stations, computers, GPS, and other electronics?
Storage	Assure that sufficient storage space is available for all the data that will be collected.
Spare parts	Make sure spare parts of essential items are available in case equipment fails.
Analyses	What type of analyses will be conducted and what software and hardware are needed to achieve those? Is in-house capacity and knowledge available or do parts of the analyses need to be outsourced (consider budget)?

son et al. 2004), but the usage of drones is still in its infancy for such studies. Although drones have been used to study a wide number of species there are very few studies that explicitly try to derive animal distribution or density from the data (Hodgson, Kelly et al. 2013, Vermeulen, Lejeune et al. 2013, Chabot, Craik et al. 2015, Wich, Dellatore et al. 2016), and even fewer that subsequently derive estimates of population size from those (Hodgson, Kelly et al. 2013, Vermeulen, Lejeune et al. 2013, Hodgson, Baylis et al. 2016). Understandably, many studies have focused on whether particular species can be detected at all and which factors influence detection (Linchant, Lisein et al. 2015), so we will start with examining a few examples.

6.2.1 Comparing traditional surveys to drone surveys

For drones to become widely used in wildlife studies it is crucial that they provide counts of animals or their signs that are sufficiently accurate compared to terrestrial or marine counts. Although there remains a large amount of work to be conducted in this field both in terms of field tests and the development of statistical methods to account for detection variability of animals or their signs in images, there have been several studies that compared the number of animals or their signs that were found on aerial images to those found during terrestrial or marine counts. Some studies also assessed systematically how various factors such as flying height, vegetation, time of day, size of waves on sea, and others could influence detectability. It needs to be remembered that ground counts often miss animals or their signs, and these then have to be corrected for by models that fit detection functions to the observed data (Buckland, Anderson et al. 2001, Buckland, Anderson et al. 2004). This means that ideally studies that compare ground and aerial surveys would know the exact number of animals or signs that are there so that the detectability for both methods compared to the true number can be assessed. Without attempting to provide an exhaustive overview of studies that compared aerial and ground counts, some examples are discussed here of studies that compared ground and aerial data gathered with drones.

A study on black-headed gull (*Chroicocephalus ridibundus*) nests in Spain found that there was relatively little difference between aerial and terrestrial counts (Sardà-Palomera, Bota et al. 2012). A comparison between direct counting and drone data was made twice, and although in both cases the aerial data yielded higher nests counts, the differences were small—6.1 per cent (15 nests) and 0.8 per cent (2 nests), respectively. A study that surveyed flocks of Canada geese (*Branta canadensis*) and snow geese (*Chen caerulescens*) compared results from manual counts of the geese on aerial images to counts made by an observer on the ground using a spotting scope and handheld counter (Chabot and Bird 2012). Perhaps not surprisingly, this study found that the low contrast of the Canada geese with the ground resulted in generally lower counts from the air than from the ground while the aerial surveys had higher counts than the ground survey counts for the snow geese which have a higher contrast with the ground. The Canada geese results also showed a higher coefficient of variation for the aerial counts (11–106 per cent) compared to the snow geese (1–6 per cent). The coefficients of variation for the ground counts were 1–6 per cent for the Canada geese and 11 per cent for the snow geese, indicating that the coefficient of variation from aerial counts for the snow geese is lower than that for the ground counts. A recent study used drones to very precisely count lesser frigatebirds (*Fregata ariel*), crested terns (*Thalasseus bergii*), and royal penguins (*Eudyptes schlegeli*) on Australian islands (Hodgson, Baylis et al. 2016). This study showed that the counts derived from drone data were higher than those made from the ground and that the counts from the drones also had a lower variance. The last study on birds to highlight here focused on the common tern (*Sterna hirundo*) and found that plot and full colony aerial counts correlated significantly and that full colony aerial counts lead to population estimates that ranged from 93 to 96 per cent of the ground counts (Chabot, Craik et al. 2015). The authors suggest that the discrepancy between ground and aerial counts could be due to three not mutually exclusive factors. First, the image analysts could have missed some of the terns during the aerial counts because in some cases contrast between the terns and background was limited. Second, because during ground counts terns are flushed off the nests the ground counts included active and inactive nests. During aerial counts only active nests are counted (i.e. those with terns on them) and this could have led to a difference between the two methods. Third, there is a possibility that nests were sometimes left unattended.

Studies have also examined detectability in large mammals. A study on caribou examined the question of variables influencing caribou detection by placing surrogate caribou consisting of unpainted fir plywood boards in six different habitats and assessing the effects of a variety of factors on whether a surrogate caribou was manually detected or not on the images. They found that 77.5 per cent of the caribou were detected, but that habitat was an important factor. Detection was higher in open habitat than burned habitat and forest. The contrast of the surrogate caribou against the landscape was also of influence on detection. Detection was also somewhat higher during evening flights (87 per cent) than morning flights (75 per cent) (Patterson, Koski et al. 2016). The Patterson, Koski et al. (2016) study is unique in that it incorporated as a variable the distance of the surrogate caribou from the centreline of the image (in most cases the centreline of the flight). This is a very relevant variable because it allows for an estimate of how that distance influences detection and thus how this should be corrected for by fitting detection functions (Buckland, Anderson et al. 2001, Buckland, Anderson et al. 2004). Interestingly, in the caribou study it was not one of the main variables explaining detection even though it was included in one of the top three statistical models that were evaluated.

For marine mammals surrogate targets have also been used to investigate whether weather conditions, the width of the search, and flying altitude can influence detectability (Koski, Allen et al. 2009). In this study fully or partially inflated kayaks (3m long) of three different colours were used at sea to simulate two whale species: bowhead whales (*Balaena mysticetus*) and grey whales (*Eschrichtius robustus*). The study showed that the best model for detection included Beaufort wind force, target colour, and target inflation. The study further found that detection rates during the drone study were similar to those reported in the literature for manned aerial surveys of marine mammals. Another study on marine mammals assessed whether flying altitude, sun glitter, sea state, and turbidity influenced dugong (*Dugong dugon*) sightings on RGB images (Hodgson, Kelly et al. 2013). This study made several assumptions to get around the fact that no alternative aerial or boat survey was conducted simultaneously to which the drone data could be compared.

It was assumed that there were no trends in the distribution of the dugongs in the surveyed area, that the number of dugongs that are available to be photographed is constant during one flight, but can differ between flights, and that the covariates do not influence the availability of dugongs that can be photographed. The results from the study showed that only turbidity influenced the sighting rate of dugongs in images. These results are promising because the drone data seem less limited in how sea state influences dugong sightings than manned aerial surveys, which thus opens opportunities for such surveys to be conducted by drones instead of manned aircraft.

It is essential though that direct comparisons between drone and manned aerial surveys are conducted. Several studies have also started to examine whether ground and aerial surveys from drones find similar numbers of great ape sleeping platforms (called nests) and which factors influence detection from the air (van Andel, Wich et al. 2015, Wich, Dellatore et al. 2016). A study on chimpanzees (*Pan troglodytes*) in Gabon used a fixed wing system with an RGB consumer-grade camera to determine which proportion of nests found during ground surveys could be detected in the aerial data and whether canopy structure, and other variables such as the height at which a nest was built, could explain detection variability. The most important variable explaining nest detection from the air was canopy openness. In closed canopy inland forest only 8 per cent of individual nests (33 per cent of nest groups) that were found during the ground surveys could be found on aerial images. In the more open coastal forest this percentage rose to 48 per cent of the individual nests (68 per cent of nest groups). Preliminary results on nests (Figure 6.1) of the Sumatran orangutan (*Pongo abelii*) showed that, similar to the study on chimpanzees, significantly fewer nests were found during the drone survey than the ground survey (Wich, Dellatore et al. 2016).

However, the number of nests found per distance covered during the ground and aerial surveys showed significant positive correlation. This indicates that despite the fewer nests found during the aerial surveys the relative density is similar for ground and aerial surveys (Figure 6.2). Although

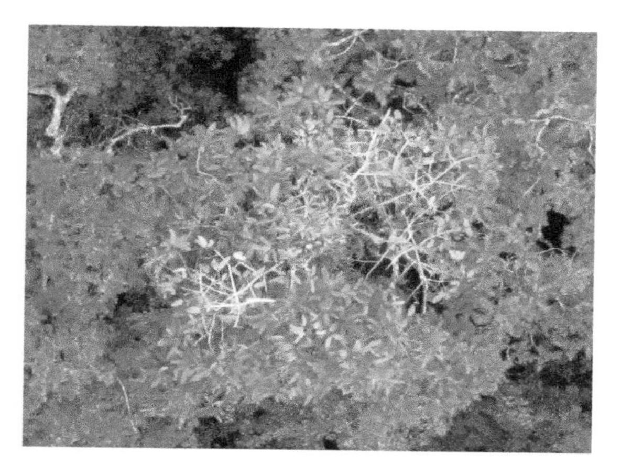

Figure 6.1 Orangutan nests on photos acquired with a RGB camera on a drone. © conservationdrones.

no statistical evaluation of the variables influencing nest detection were performed, the mean tree DBH (diameter at breast height) and height of the nests in a tree appeared to be similar for nests only observed on the ground and those observed from the air. Nests that were observed from the air did seem to occur more in pioneer trees from the genera *Macaranga* and *Mallotus*, which might indicate that nests in certain tree species have a higher detectability that those in other species.

In riverine systems there have been promising studies counting salmon (Kudo, Koshino et al. 2012, Groves, Alcorn et al. 2016). One of these studies used a petrol-engined remote-control helicopter with a first person view (FPV) camera to guide the flying in a study to count adult chum salmon in Japan (Kudo, Koshino et al. 2012). The study used counts from catching the salmon with dragnets in the river during the same sample period as a comparison to the number of salmon found on the images acquired from the remotely controlled helicopter. The number of salmon found through an automated procedure and those counted from the net catch showed a significant positive correlation and indicates that counts from drones can be used as an alternative to net counts. The Groves, Alcorn et al. (2016) study used multirotors with RGB cameras to obtain images of salmon nesting sites (redds) and found that the nest counts from the drone images were more accurate than those made from manned helicopter counts.

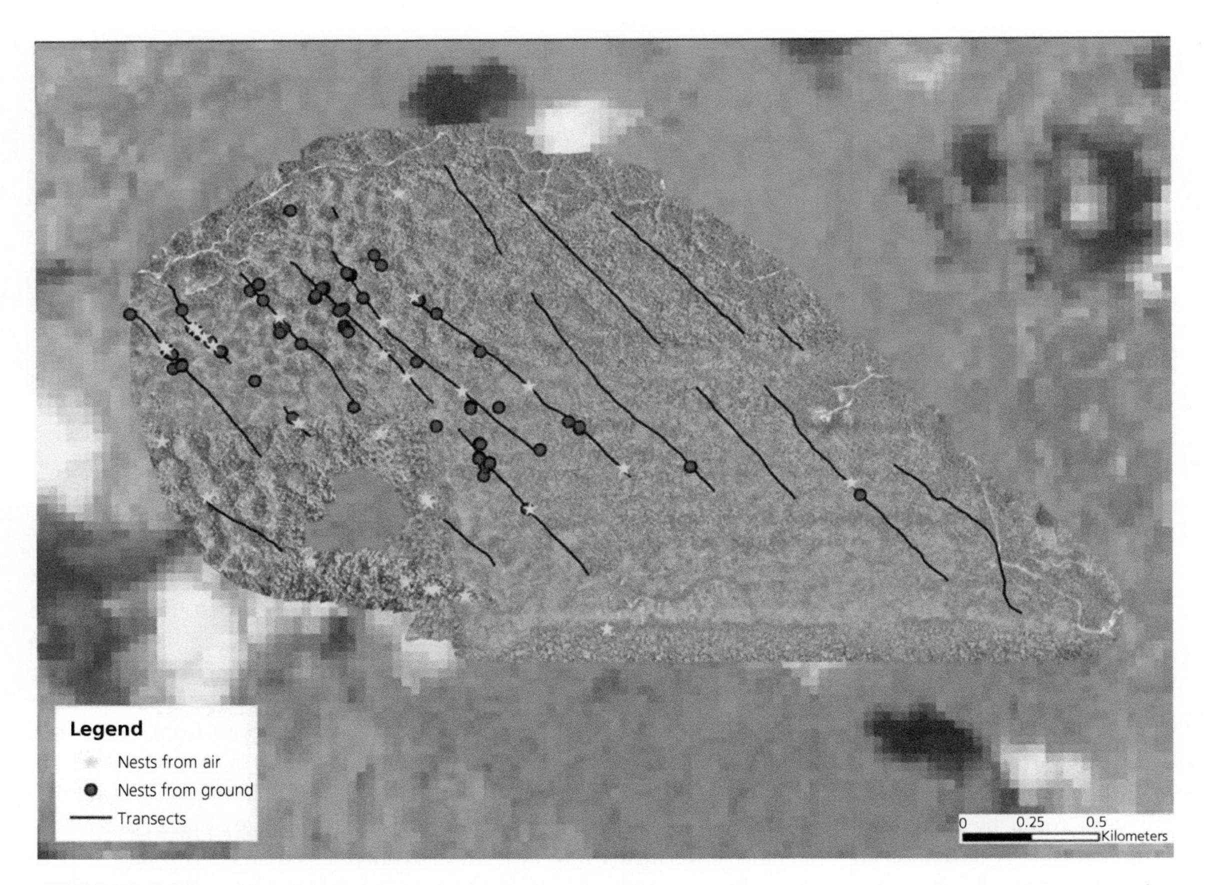

Figure 6.2 The map shows an orthomosaic based on 2238 images with a ground resolution of 5.22 cm/pixel side and a total area of 5.22 km^2 as generated in Pix4Dmapper. The orthomosaic is placed over a Landsat 7 satellite image with a 30 m/pixel side resolution. The orangutan nests that were observed during the ground surveys are indicated as red dots and those observed on the images acquired from the drone as yellow stars. For details see Wich, Dellatore et al. (2016).

6.2.2 Estimating density and population size

To estimate the true animal density from drone surveys it is important that the area sampled by the aerial system is measured, the animals or their indirect signs therein are counted without there being any double counting due to animal movements, and corrections are applied if detection is incomplete. Thus surveys with drones need to solve the same issues as other survey methods face when detection is not complete (Buckland, Anderson et al. 2001, Buckland, Anderson et al. 2004). That means that in addition to comparisons to ground data there is a need to determine whether detection is influenced by the distance from the midline in the photo (the line of flight if the aircraft does not have a crab angle, which should be adjusted for, e.g. Patterson, Koski et al. 2016). In a study on elephants, transect flights were used to estimate elephant density for transects (Vermeulen, Lejeune et al. 2013). Some simultaneous ground and aerial counting was conducted, which indicated that on images all animals that were observed from the ground could be seen and that therefore aerial counts can be used to estimate density because the strip width covered by the images from the drone were known as well. Even though this was the case it was noted that single observers that look for elephants on images tend to miss quite a large proportion (14.7 per cent) of the actual elephants on the photos. Doubling the number of people checking the images reduced this percentage to 7.8 per cent, which is still high and indicates that similar tests should be conducted for smaller animals as well to determine what the percentages are for those. Although several studies have used multiple researchers to count animals, signs, or surrogates on images the percentages are not always provided which makes it difficult to assess how commonly one analyst misses objects and how much this improves by adding analysts. Animal density studies have also been used in a marine setting. A study at Moora island, which is part of French Polynesia, studied density differences in two shallow sites for blacktip reef sharks (*Carcharhinus melanopterus*) and pink whiprays (*Pateobatis fai*) (Kiszka, Mourier et al. 2016). Due to the potential influence of glare and wind on detection, flights were only conducted when there was no wind and glare. The authors report that there was no difference in detection between the various image analysts and reported density differences between the two blocks they studied; densities were higher in the areas where provisioning occurs. Due to the low flight altitude it is likely that there were no detection differences with distance to the midline, but the authors do not specifically address this issue. It is promising that drones can be used for density assessments in marine environments and there are huge opportunities for more studies because the authors report that several other species were observed on the images.

6.3 Anti-poaching efforts

Poaching is one of the largest threats to wildlife in general (Fa and Brown 2009) and the largest threat to some highly valuable animals such as rhinoceroses, elephants, and tigers that are experiencing major population decreases in many areas (Biggs, Courchamp et al. 2013, Wittemyer, Northrup et al. 2014, Naidoo, Fisher et al. 2016, Poulsen, Koerner et al. 2017). The areas in which such species need to be protected are often large while at the same time resources (staff, equipment, and operational finances) are often limited (Olivares-Mendez, Bissyandé et al. 2013). This means that areas are often not as well patrolled as they could be. Drones offer a potential means to change this due to their relatively low costs. An additional advantage is that they make less noise than manned aircraft and thus there is less chance that the aircraft will be detected by poachers (Olivares-Mendez, Bissyandé et al. 2013). As a result several conservation organizations have been using drones for anti-poaching efforts (WWF, Africa Parks, International Anti-Poaching Foundation, and Peace Parks Foundation) and several organizations focus specifically on the usage of drones for anti-poaching.[2] The media (online and in printed magazines) have also reported widely on the potential and actual uses of drones in anti-poaching efforts.[3]

Even though drones have been used to detect poachers (in one case even resulting in the death of a poacher[4]) there is little published scientific study on the usage and success of drones in anti-poaching efforts (Linchant, Lisein et al. 2015). Perhaps this is not surprising as sensitive information about detection success might deliberately not be published or studied, but the lack of such information makes it very difficult to determine the potential usage of drones for anti-poaching efforts and the success so far.

Although there is a risk of oversimplifying matters, drones can facilitate anti-poaching efforts by: a) finding wildlife that might be at risk of being poached; and b) detecting poachers themselves either before or after they have killed an animal. Several studies have been conducted on detecting wildlife and people with drones (see Chapter 3, and the examples given in this chapter). These have shown that rhinos, elephants, and people can be detected with RGB and thermal imaging cameras, with successful detection and identification depending on the time of day and flying height (Vermeulen, Lejeune et al. 2013, Mulero-Pázmány, Stolper et al. 2014). Vermeulen, Lejeune

[2] http://www.spots.org.za/Site/index.php, http://www.shadowview.org/, http://airshepherd.org/, http://bat-hawkrecon.com/ (accessed 8 November 2017)

[3] http://www.huffingtonpost.com/2015/03/23/drones-rhino-poaching_n_6922804.html, http://www.bbc.com/news/business-28132521, https://www.scientificamerican.arcom/ticle/drones-bring-fight-and-flight-to-battle-against-poachers/ (accessed 8 November 2017)

[4] http://www.shadowview.org/news/poachers-caught-shadowview-drones/ (accessed 8 November 2017)

et al. (2013) showed that elephants in an open area remain visible on images acquired with a fixed wing drone flying at 300 m above ground level (AGL) and the specific camera they used (Ricoh GR3, 28 mm). The Mulero-Pázmány, Stolper et al. (2014) study found that on RGB stills (Panasonic Lumix LX-3) rhinos remained detectable in grassland and more closed forest habitats up to 239 m AGL, but that people only remain detectable until 158 m AGL in the same habitats. On the RGB stills rangers that were wearing camouflaged clothing were more difficult to distinguish. This study also found that for RGB images midday showed the best result due to the sun being at or close to its zenith. A GoPro 2 camera used on the same fixed wing only allowed for detection of animals and people when flying below 50 m AGL. The thermal imaging camera (Thermoteknix Micro CAM 640×480 pixels) allowed for detection of animals and people up to 155 m, but generally did not allow for sufficient resolution to identify species or to distinguish between animals and people. This indicates that RGB stills were most successful, which is not surprising due to their higher resolution. Thermal imaging provided the best results during early morning and night. The authors conclude that flying between 100 and 180 m is likely the best compromise between high detectibility of animals and people with low detectibility of the drone itself by poachers on the ground. Of course these recommendations and results are based on the specific camera and lens combinations used in these studies and cannot therefore necessarily be extrapolated to other camera/lens combinations.

Another main challenge is the vast areas that need to be patrolled. Compared to the large size of some of these areas (e.g. Kruger National Park covers 19,480 km^2; Olivares-Mendez, Bissyandé et al. 2013), the area that can be patrolled during one drone flight is extremely small. A drone flying at a cruise speed of 45 km/h at 100 m AGL with a FLIR Tau 640 with a standard 13 mm lens will only have a maximum footprint width of 82.84 m resulting in a total of 3.7 km^2 being covered during a one hour flight, which is equivalent to 0.019 per cent of the whole park. This indicates that trying to completely cover a park the size of Kruger is not realistic at all. As a result of this discrepancy between the size of the area that needs to be covered and that covered during a single drone flight researchers have been examing how to optimize flights to intercept poachers[5] (Snitch 2013, Park, Serra et al. 2015a, Park, Serra et al. 2015b, Shaffer and Bishop 2016, Haas and Ferreira 2017). Somewhat similar models have also been developed for ground-based patrols (Fang, Nguyen et al. 2016, Kar, Ford et al. 2017). Such approaches have also been used in policing of other crimes where they have often been called 'interdiction patrolling' (or 'hot spots policing'). These approaches rely on data-based analytical models that predict where a crime scene (a poaching event in this case) might occur. In the case of poaching this is based on the assumption that poaching is a deterministic process that shows spatial clustering, as was shown for elephant poaching events in Tsavo National Park in Kenya (Shaffer and Bishop 2016). This study showed that most poaching events occurred in areas that are largely open with low shrubs (60 per cent of all poaching events), are within 4 km of a water source (62 per cent), and occur within 4 km of a road (89 per cent) (Figure 6.3). Such clustering indicates that anti-poaching efforts can likely be focused on areas showing such features. Based on the size of a high-density poaching area of 23.2 km^2 in the centre of Tsavo NP, the authors calculate that it would need a drone equipped with a FLIR Tau 2 640 with a 7.5 mm wide angle lens (resulting in a horizontal field of view of 187 m), flying at 91 m AGL, and cruising at 59.5 km/h about 2 hours and 34 minutes to cover the whole area. This shows that predictive models can greatly reduce the time (effort) needed to monitor key areas with drones as compared to ground-based patrols.

Other models have been specifically modelled with data from rhinos. Park, Serra et al. (2015a) developed a model for Kruger National Park that used data on the location of rhinos (data from 20 GPS tagged animals for 24 months) and poaching incidents (17 months of data) to predict where future events would likely occur, and used this as a basis to develop flight plans for drones in which these would optimize the number of predicted rhinos

[5] http://www.npr.org/2013/06/11/188638982/to-crack-down-on-rhino-poaching-authorities-turn-to-drones (accessed 8 November 2017)

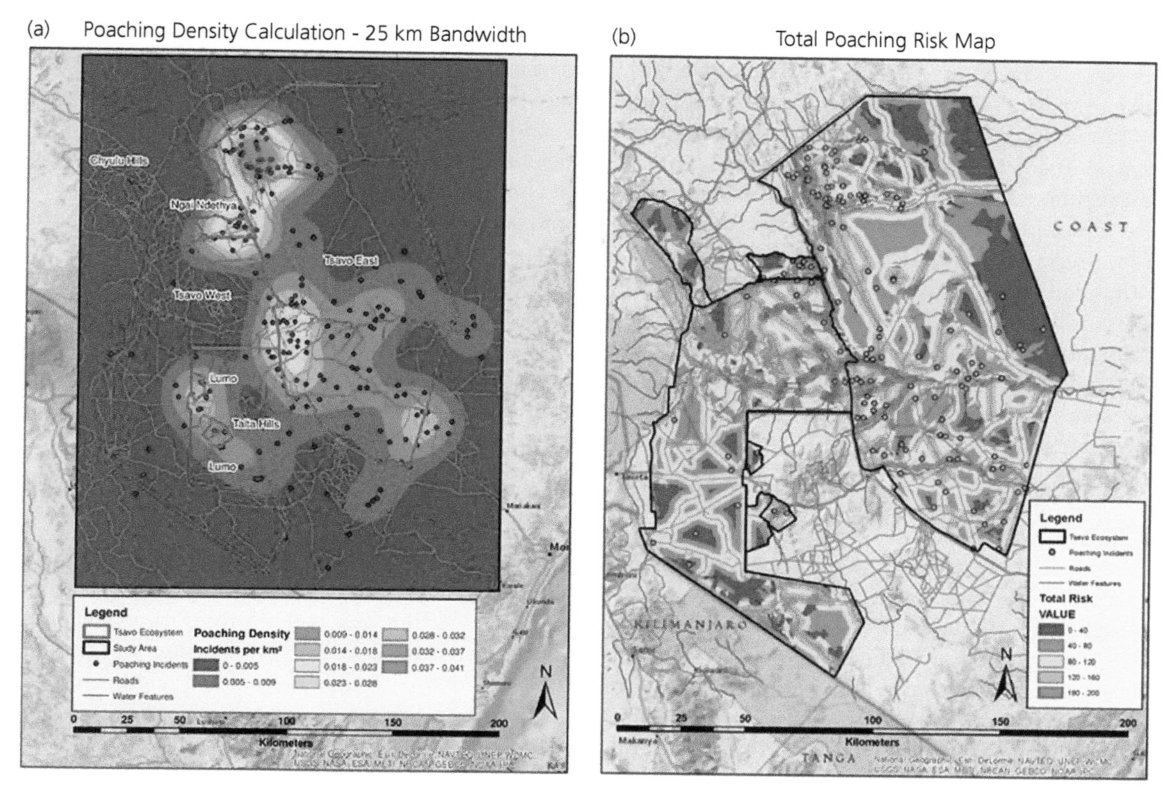

Figure 6.3 a) Map of Tsavo National Park showing elephant poaching density. b) Map showing total elephant poaching risk in Tsavo National Park. From Shaffer & Bishop (2016). © M. J. Shaffer.

covered during a flight. Another anti-poaching model that has recently been published differs from the models used by Park et al. in that it also includes the interactions between rhinos, poachers, and the anti-poaching units (Haas and Ferreira 2017). This approach models poachers as being maximizing agents that have knowledge about the previous routes walked by patrol units. The model then produces patrol routes that are a random selection from candidate routes which makes them harder to predict by the poachers (Haas and Ferreira 2017).

Another option to maximize the efficiency of drones' flying time would be to use tracking devices on highly prized animals and then have drones fly specifically over areas where these animals occur for patrolling or only fly to such animals if certain distress signals have been transmitted (although this might then be too late; O'Donoghue and Rutz 2016). A key aspect for the latter is that response time is fast so that rangers can get to a potential poaching site rapidly by helicopter, or in cases where those are not available use a drone to go there to collect data (RGB/thermal) to be transmitted back to the rangers as they approach the site over the ground.

In addition to such predictive models to use drones more effectively, efforts are underway to at least in theory develop completely automated missions that can be used in anti-poaching efforts (Olivares-Mendez, Bissyandé et al. 2013, Olivares-Mendez, Fu et al. 2015). The idea here is to develop missions in which drones can automatically take off from and land on mobile and fixed recharge stations while flying missions during which drones use onboard processing to detect and track animals, people, or cars, while also detecting and storing information on poachers' faces, and sending pings with location coordinates back to rangers if suspicious activities are detected (Rodríguez-Canosa, Thomas et al. 2012, Olivares-Mendez, Bissyandé et al. 2013, Olivares-Mendez, Fu et al. 2015, Ward, Hensler et al. 2016). Although it is hard to predict when such systems will actually be operational, the development of tracking and face detection algorithms from the perspective of a drone is continuing fast and experiments with landing on recharge stations are occurring for multirotors (Olivares-Mendez, Fu et al. 2015). Although recharging stations for fixed wing systems might be technically difficult due to landing and take-off, they might be achievable due to the development of hybrid drones that take off as a multirotor and continue to fly as a fixed wing. Such hybrid systems in combination with recharge stations could potentially lead to networks of self-charging drones in conservation areas. One potentially complicated aspect here is how such recharge sites will be protected from poachers who would likely attempt to either intercept the drone or assure that the recharge sites are not functioning properly. Another promising development which could be combined with hybrid systems would be the incorporation of photovoltaic cells on drones to increase flight duration. As such work is also focusing on systems with small wingspans (~2 m) this is promising for the near future (Morton and Papanikolopoulos 2016).

An important consideration in using drones and other technology that is potentially logging information about poachers or highly prized animals is that third parties are interested in such data as well. These could be members of the public such as photographers, but also poachers themselves (Cooke, Nguyen et al. 2017). In Banff National Park (Canada) VHF radio receivers used by researchers and managers to track animals were banned from the park for the public because photographers used these to find animals that were tagged.[6] It is very likely that poachers could use such VHF receivers as well to find animals and in India attempts were made by poachers to hack into an email account containing GPS information for an endangered Bengal tiger which was fitted with an Iridium GPS satellite collar.[7] It is therefore important that the conservation community, when using such technology, is proactive in determining how to mitigate such threats. For the use of drones this will likely need encryption of transmitted data and encryption of control systems so that drones cannot be hacked into and taken over by third parties.

The issues surrounding data security in an anti-poaching setting will likely only grow in the near future because of the development of sensor networks for conservation that will integrate the various emerging technologies being developed (Figure 6.4; Pimm, Alibhai et al. 2015, Marvin, Koh et al. 2016). Such sensor networks will probably consist of ground sensors for a variety of data—acoustic, visual (RGB and thermal), vibration (to detect cars and walking people), and odour (to detect air molecules that could be related to dead animals)—in combination with swarms of drones collecting data with a variety of sensors, but also functioning as relay stations for the ground sensors so that data will be quickly transmitted to rangers. Collared animals (either VHF or GPS) will likely be part of this network as well. All such information can then be integrated in an updated version of software such as the Spatial Monitoring and Reporting Tool (SMART[8]) in which all such sensor information will be integrated with data from ground patrol data on travel routes, snare locations, killed animal locations, and more. Fast analyses can then provide information to rangers on where to go next.

It is important to realize that, although technology can play a major role in detecting poachers before or after an animal has been killed, technology by itself will not reduce poaching. Reduction of poaching is likely to be much more dependent on factors such as effective law enforcement and working with local communities. For instance in India and Nepal the rhino population has increased 17-fold since the late 19th century, when there were only approximately 200 individuals left (Talukdar, Emslie et al. 2008).

[6] http://www.pc.gc.ca/apps/scond/Cond_E.asp?oID=24602&oPark=100092 (accessed 8 November 2017)

[7] http://timesofindia.indiatimes.com/city/bhopal/Hacking-tiger-collar-Cyber-poachers-fox-wildlife-brass-MP-forest-officials/articleshow/22826103.cms?referral=PM (accessed 8 November 2017)

[8] http://smartconservationtools.org/ (accessed 8 November 2017)

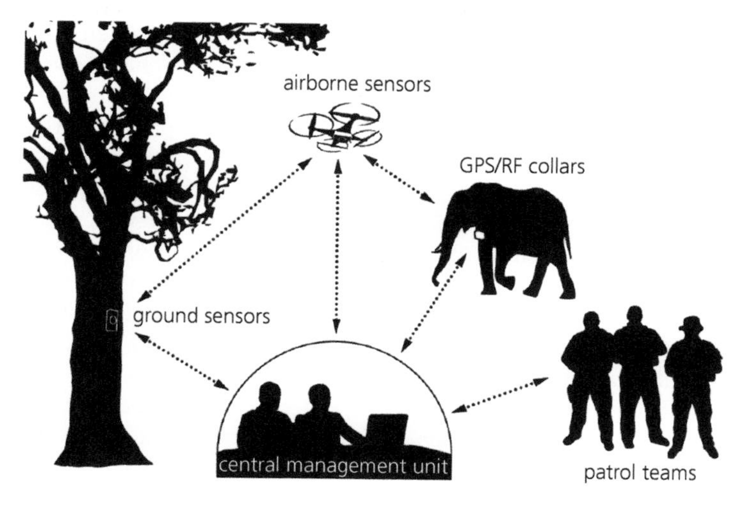

Figure 6.4 Network between several sensors and humans. © Perry van Duijnhoven.

In the case of Nepal there has been zero poaching of rhinos since 2011. The two main factors that have been attributed to this remarkable success are the effective involvement of local communities and strict enforcement of the laws that have been facilitated by institutional and legislative changes (Aryal, Acharya et al. 2017). Therefore drones and other technologies should not be seen as a silver bullet, but more as part of a set of tools that can facilitate the reduction of poaching. As important is to ensure that drones used as a conservation tool are not perceived negatively by local communities due to the military connotations that many see with drones. If such negative perceptions arise there is a chance that it could backfire on the usage of drones in a conservation setting, so it is important to work closely with local communities when drones are being used (Sandbrook 2015). In addition, it is important to carefully assess the effectiveness of drones as a tool in conservation compared to other aspects needed for conservation such as park staff, their resources, training, education, awareness, and improving the judicial system to name but a few (Humle, Duffy et al. 2014).

6.4 Disturbance

A general concern with any type of animal surveys should be the potential disturbance the survey might cause to animals or habitats. Disturbance will vary between survey methods, animal species surveyed, and habitat, but ground and aerial surveys will likely lead to some form of behavioral or physiological reaction, such as flushing of birds during ground surveys (MacArthur, Geist et al. 1982, Stankowich 2008, Chabot, Craik et al. 2015). Even camera trap methods, which are generally considered non-intrusive, are documented to have some influence on wildlife (Wegge, Pokheral et al. 2004, Meek, Ballard et al. 2014, Meek, Ballard et al. 2015, Meek, Ballard et al. 2016). Minimizing the disturbance of surveys on animals and habitat should be something that wildlife biologists take into consideration and try to quantify so that choices among survey methods can be based on comparative research (e.g. Potapov, Utekhina et al. 2013, Le Maho, Whittington et al. 2014, Weissensteiner, Poelstra et al. 2015, Junda, Greene et al. 2016). An example of such a comparative study that used humans or a remotely controlled small vehicle (rover) to obtain radio-frequency identifications of king penguins (*Aptenodytes patagonicus*) found that penguins showed significantly lower and shorter stress responses when approached by the rover than by humans (Le Maho, Whittington et al. 2014). The disturbance of manned aerial surveys on various animal species is relatively well documented and can occur from large distances (Andersen, Rongstad et al. 1989, Wilson, Culik et al. 1991, Bleich, Bowyer et al. 1994, Born, Riget et al. 1999, Delaney, Grubb et al. 1999, Giese and Riddle 1999, Patenaude, Richardson et al. 2002, Frid 2003). The negative impact can be large, as shown by a study on Adélie penguins (*Pygoscelis adeliae*) which found that manned aircraft already started to visually disturb birds at distances greater than 1000 m, and multi-day exposure to a helicopter even led to birds not returning to their nests, causing the colony to decline 15 per cent (Wilson, Culik et al. 1991).

One of the reasons to use drones to survey animals is that these would lead to less disturbance because of their reduced noise compared to manned aircraft, and hence be preferable. A comparison of noise levels between manned helicopters, manned fixed wings, and various sizes of drone systems indeed showed that when standardized to a similar altitude the manned aircraft had higher sound levels than the unmanned systems (Christie, Gilbert et al. 2016) and, more generally, suggestions for standardizing methods to quantify noises that might disturb animals have been made (Pater, Grubb et al. 2009). A small fixed wing drone produced 50dBA (Decibel A-weighting, a sound level measuring unit) on the ground when flying at 100 m altitude, whereas a manned fixed wing aircraft produced 75dBA at the same altitude. A relatively large quadcopter produces approximately 55dBA, whereas a manned helicopter produces a noise level of approximately 95dBA at similar altitude (Christie, Gilbert et al. 2016). As a consequence the disturbance levels of drones are lower than that of manned aircraft when operated at similar altitude (Sleno and Mansfield 1978, Acevedo-Whitehouse, Rocha-Gosselin et al. 2010, Mulaca, Storvoldb et al. 2011, Moreland, Cameron et al. 2015). A number of studies have been specifically examining or reporting whether animals show behavioral, or in fewer cases physiological, responses to drones flying over or near them (see Table 6.2) and several review studies on the disturbance of drones on wildlife have recently emerged (Smith, Sykora-Bodie et al. 2016, Borrelle and Fletcher 2017, Mulero-Pázmány, Jenni-Eiermann et al. 2017).

We highlight a few studies here as examples before providing some of the key results of the review studies. In a study that aimed to assess the differences in breeding populations by counting nests of the black-headed gull (*Chroicocephalus ridibundus*) there was no disturbance reported even though the fixed wing drone flew at 30–40 m above ground level (Sardà-Palomera, Bota et al. 2012). Disturbance was estimated by examining whether the birds were flying on the images. In only 1.3 per cent of the images were birds flying, which was interpreted as no reaction to the overflight of the drone (Sardà-Palomera, Bota et al. 2012). During tests to specifically examine the potential reaction of animals to a low flying Gatewing X100 fixed wing, the drone was flown at 100 m above elephants, Buffon kob, and baboons during two test flights in which 10 flight lines each were flown above a waterhole. No reaction of any of the three species was detected by an observer on the ground who was close to the waterhole (Vermeulen, Lejeune et al. 2013). Reactions do occur though in some species and studies in which multirotors have been used to assess the nests of birds—Steller's sea eagle (Potapov, Utekhina et al. 2013) and hooded crow (Weissensteiner, Poelstra et al. 2015)—to count bird

Table 6.2 Animal responses to drones

Species	Drone	Behavioral response	Source
Birds			
Several bird species	Multirotor (DJI Phantom 2)	None to minimal	(Wilson, Barr et al. 2017)
Four raptor species: osprey, bald eagle, ferruginoushawk, and red-tailed hawk	Multirotor (Draganflyer X-4)	Varies between species, but relatively limited	(Junda, Greene et al. 2016)
Tristan albatross	Multirotor (DJI Phantom 2)	No response observed	(McClelland, Bond et al. 2016)
Lesser frigatebird, crested tern, royal penguin	Multirotor (X8 by 3D Robotics) and fixed wing (FX79 airframe)	No group startle response observed	(Hodgson, Baylis et al. 2016)

Various waterfowl species	Fixed wings (UAVER Avian-P, Skylark II, Drone Metrex Topodrone-100), multirotor (DJI Phantom, FoxTech Kraken-130)	Response overall minimal, but dependent on drone system and flying height	(McEvoy, Hall et al. 2016)
Mallards, greater flamingos, common greenshanks	Multirotor (Phantom)	Minimal response	(Vas, Lescroël et al. 2015)
Common tern	Fixed wing (AI-Multi UAs)	Minimal	(Chabot, Craik et al. 2015)
Several bird species	VTOL* (Honeywell RQ-16 T-Hawk)	Minimal, potentially only < 30 m	(Dulava, Bean et al. 2015)
Gentoo penguins and chinstrap penguins	3 Multirotors (MK-OktoXL, APQ-18, and APH-22)	No behavioral responses observed (lowest flying height 23m)	(Goebel, Perryman et al. 2015)
Adélie penguin	Multirotor (MK ARF Okto XL)	Minimal to high depending on flying altitude	(Rümmler, Mustafa et al. 2015)
Hooded crow	Multirotor (DJI Phantom 2 Vision)	Moderate (lower than when human climbs tree)	(Weissensteiner, Poelstra et al. 2015)
Greater sage-grouse	Fixed wing (Raven RQ-11A)	None to minimal response	(Hanson, Holmquist-Johnson et al. 2014)
Steller's sea eagle	Multirotors (APM and Naza-M based)	None to minimal	(Potapov, Utekhina et al. 2013)
Canada geese and snow geese	Fixed wing (CropCam)	No disturbance of birds. Flying at 183m	(Chabot and Bird 2012)
Black-headed gull	Radio controlled fixed wing aircraft (Multiplex Twin Star II)	Drone flew at 30–40m high, but no disturbance of birds	(Sardà-Palomera, Bota et al. 2012)
Sandhill crane	Fixed wing (Raven RQ11)	None to minimal response	(Hutt 2011)
Marine mammals			
Antarctic fur seal and leopard seals	3 Multirotors (MK-OktoXL, APQ-18, and APH-22)	No response observed	(Goebel, Perryman et al. 2015)
Bowhead whales	Fixed wing (Brican TD100E)	No response observed	(Koski, Gamage et al. 2015)
Spotted and ribbon seals	Fixed wing (ScanEagle UAS)	Reduction of response compared to manned aircraft, but response present	(Moreland, Cameron et al. 2015)
Grey seals and harbour seals	3 Multirotors (DJI 450, Cinestar 6, Vulcan 8)	Moderate behavioral response varied with flying height	(Pomeroy, O'Connor et al. 2015)
Killer whales	Multirotor (APH-22)	No response observed	(Durban, Fearnbach et al. 2015)
Various marine mammals	RC helicopter	No response observed	(Acevedo-Whitehouse, Rocha-Gosselin et al. 2010)
Florida manatee	Fixed wing (FoldBat)	No response observed	(Jones, Pearlstine et al. 2006)
Terrestrial mammals			
American black bear	Multirotor	Moderate to high	(Ditmer, Vincent et al. 2015)
Giraffes	Fixed wing (Easy Fly St-330)	No response observed	(Mulero-Pázmány, Stolper et al. 2014)
Rhinoceros	Fixed wing (Easy Fly St-330)	No response observed	(Mulero-Pázmány, Stolper et al. 2014)
African elephant	Fixed wing (Gatewing X100)	No response observed	(Vermeulen, Lejeune et al. 2013)
Reptiles			
Kemp's ridley sea turtle	Multirotor (DJI Phantom 1 and 2)	No response observed	(Bevan, Wibbels et al. 2015)

*VTOL = Vertical take-off and landing

chicks have reported that sometimes the parent birds fly away or produce alarm vocalizations, or in some cases fly close towards the drone, but in both these studies disturbance from the drone was equal to or lower than that of other survey methods such as climbing a tree or a camera mounted on a telescope pole. A study to systematically investigate the potential disturbance of a multirotor drone on the behaviour of Adélie penguins (*Pygoscelis adeliae*) was conducted by examining their behavioral responses to various flight paths (Rümmler, Mustafa et al. 2016). This study indicated that taking off from 50 m away from the penguins did seem to cause behavioral changes during the take-off phase, remained elevated during the various flight paths above the penguins, and even increased during lower vertical flying above the penguins (Figure 6.5).

A thorough study on three bird species (semi-captive mallards, *Anas platyrhynchos*, wild flamingos, *Phoenicopterus roseus*, and wild common greenshanks, *Tringa nebularia*) used 204 flights to assess the influence of drone colour, drone approach speed, repeated flights, and direction of approach on the behaviour of these species (Figure 6.6; Vas, Lescroël et al. 2015).

Their results indicated that the semi-captive or wild setting did not influence the behavioral response and that responses were similar for the three species. Approach speed, drone colour, and repeated flights all had no significant influence on the birds' reaction. Approach angle, however, did have an impact on birds' behaviour, with angles of 20°, 30°, and 60° not leading to any reaction, but angles of 90° leading to a response. The authors suggest that the 90° angle might be associated with an attack from a predator by the birds and that future studies with multirotors and fixed wings need to assess this. Overall, in 80 per cent of the tests a drone could be flown to within 4 m of a

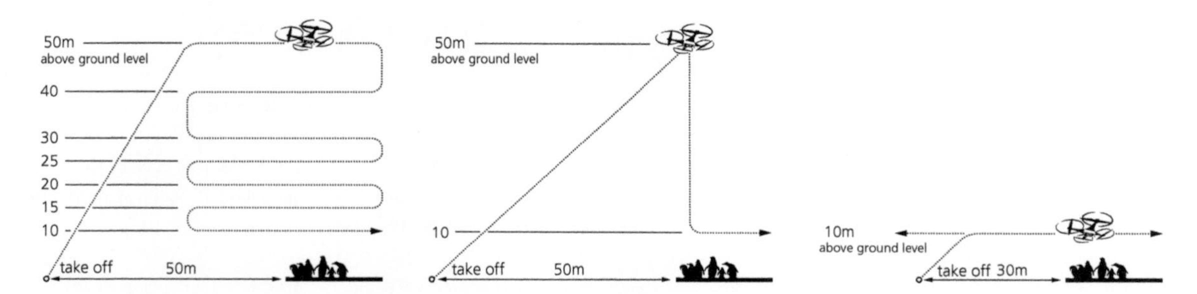

Figure 6.5 Flight paths during the study on penguins (based on Rümmler, Mustafa et al. 2016). © Perry van Duijnhoven.

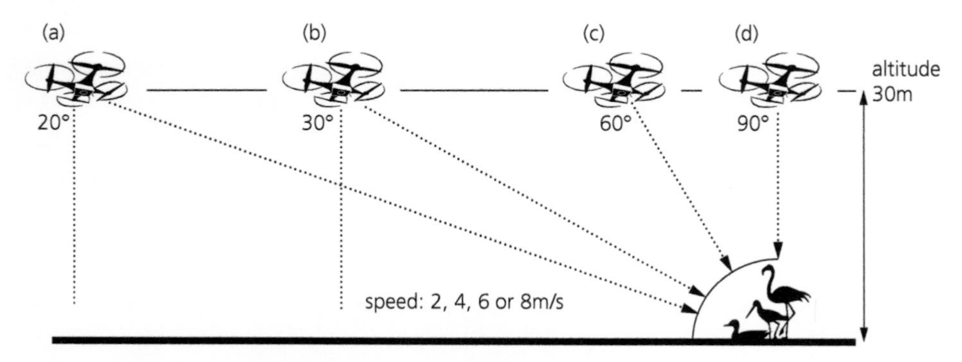

Figure 6.6 Factors used in flight tests with birds. The drone (in three different colours: white, black, blue) was launched vertically up to 30m, then descended towards the birds at different speeds (2, 4, 6, or 8 m/s) and different angles ($\alpha = 20°, 30°, 60°$, or $90°$) (based on Vas, Lescroël et al. 2015). © Perry van Duijnhoven.

bird without leading to any visual impact on their behaviour (Vas, Lescroël et al. 2015). Another study that focused on waterfowl and the disturbance of several fixed wing and multirotor drones found that the level of disturbance was generally low (McEvoy, Hall et al. 2016).

A review of the impact of drones on marine mammals concluded that generally there is a paucity of data on the behavioral responses of marine mammals towards drones and that the available information is largely anecdotal (Smith, Sykora-Bodie et al. 2016). The review indicates that disturbance does occur, but is likely a function of flying altitude and that the data do not allow for distinguishing the impact of either noise or visual cues (the actual drone or its shadow) or their combination. There appears to be a general consensus that the usage of drones elicits fewer behavioral responses than usage of manned aircraft, but that more research on this topic is needed (Smith, Sykora-Bodie et al. 2016). Such research should not only focus on behavioral responses, but also examine potential physiological responses that could indicate stress, such as heart rate (e.g. Ditmer, Vincent et al. 2015 for black bears). Smith, Sykora-Bodie et al. (2016) provide two specific recommendations for the research community. First, they recommend studying the behavioral responses of different taxa to different types of drones and measuring the threshold distance at which reactions occur. Second, examine factors that could potentially evoke or change the behavioral responses such as acoustic characteristics of the study site, age–sex class studied, presence of anthropogenic activities, approach angle, and visual and acoustic properties of the drone (Smith, Sykora-Bodie et al. 2016).

To measure the response of animals studies have almost exclusively relied on behavioral responses that could be detected visually, but several authors of such studies (Vas, Lescroël et al. 2015) have cautioned against purely relying on visual assessments because studies have shown that there can be non-visual physiological responses to disturbance from manned aircraft such as an increased heart rate (Wilson, Culik et al. 1991). It is important, therefore, that similar studies on the impact of drones assess physiological responses (e.g. heart rate and changes in hormone levels) in addition to behaviours that can be visually studied. Unfortunately these studies are still rare. As far as we are aware the first such study was conducted on black bears (*Ursus americanus*; Ditmer, Vincent et al. 2015). In this study black bears were fitted with a GPS collar to monitor location and implanted cardiac biologgers to monitor heart rate (Figure 6.7). Data from the GPS collars were retrieved when the bears were hibernating in the winter den whereas the data from the biologgers were retrieved via a transcutaneous telemetry link. The results of this study indicate that all four bears showed a steep increase in their heart rate, but that in only one out of the 17 flights did one of the bears increase its rate of movement in response to the drone. Although heart rate went up steeply it also returned to its pre-flight level quickly after the flight, but this relatively fast return could be due to these bears living in a landscape that has been heavily impacted by humans. Therefore studies on bears in less human-impacted landscapes need to be conducted in the future.

A largely understudied area is whether the noise from drones impacts calling behaviour of animals. It is well appreciated that anthropogenic noise (such as urban noise) influences the pitch and other variables of bird calls (Slabbekoorn and Peet 2003, Slabbekoorn and den Boer-Visser 2006, Job, Kohler et al. 2016). It would therefore be interesting to examine whether transient noise from drones impacts bird call behaviour. As part of a larger study on the usage of drones to record bird calls, researchers compared whether bird song output as recorded by an autonomous recording unit was different before, during, and after the drone hovering above point count stations (Wilson, Barr et al. 2017). The authors conclude that there may have been a small decrease in bird song output during the period that the drone was hovering, but that it was not possible to separate potential changes in song output from masking effects of the noise from the drone. It would be very useful to conduct similar studies with different species and drone/recorder combinations to gain a better understanding of whether drone noise influences animal calling behaviour.

Largely missing from the current literature as well are studies on the noise that drones produce and

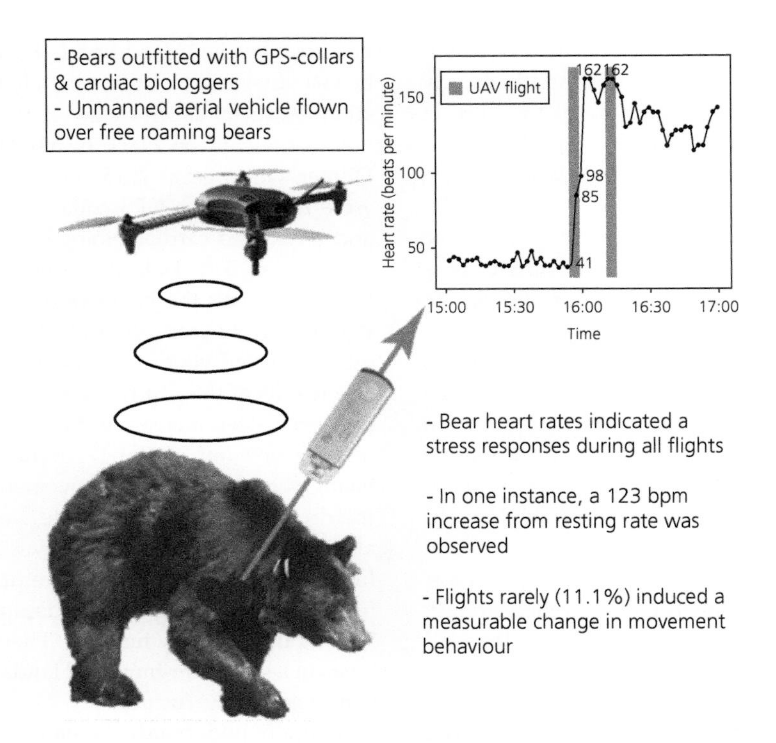

Figure 6.7 Overview of a study that assessed the disturbance of a drone on black bears (from Ditmer, Vincent et al. 2015). Reprinted with permission from Biology Letters.

the auditory sensitivity of animals' hearing. Studies in this field could facilitate researchers' decisions on how far away from animals to take off and land, and how high to fly over animals without animals potentially hearing the noise from the drone. A promising study measured the sound levels of two commercially available drone systems (a multirotor, Skyranger from Aeryon Labs, Canada, and a fixed wing, eBee by Sensefly, Switzerland) at 1.5 m above the ground, and then modelled the sound levels at various flying heights (Figure 6.8; Scobie and Hugenholtz 2016). The modelling results were subsequently linked to known hearing thresholds of various species (white-tailed deer, domestic dog, domestic cat, northern bobwhite, and mallard between 0.25 and 20 kHz; Scobie and Hugenholtz 2016). The results indicated that the eBee sound pressure measure levels were variable in the lower frequencies and lower above 12.5 kHz, but similar between 1.25 kHz and 12.5 kHz. On average for a particular setting of temperature (20°C) and humidity (80 per cent) the distances at which the various species can hear the Skyranger is somewhat larger than the eBee. Although sound attenuation is nonlinear for temperature and humidity variation, the results in general show that to avoid the drones being heard by the animals the Skyranger should be flown at greater distances from them than the eBee. In addition the higher hearing sensitivity of cats means that both systems must be flown further away from cats than the other species to avoid audible detection. As our knowledge of hearing of animals progresses (e.g. Heffner 2004), more studies into the reaction of animals, based on sound pressure level measures in the field coupled to behavioral responses and knowledge of the actual hearing sensitivity of animals, can help in understanding how drones affect animal behaviour and can facilitate minimizing such impacts. Although it might be expected that petrol-powered aircraft are generally more noisy than electric aircraft, it is interesting that a study on several waterbird species found that there may be an increase in flushing behaviour, but only below a flying height of 30 m

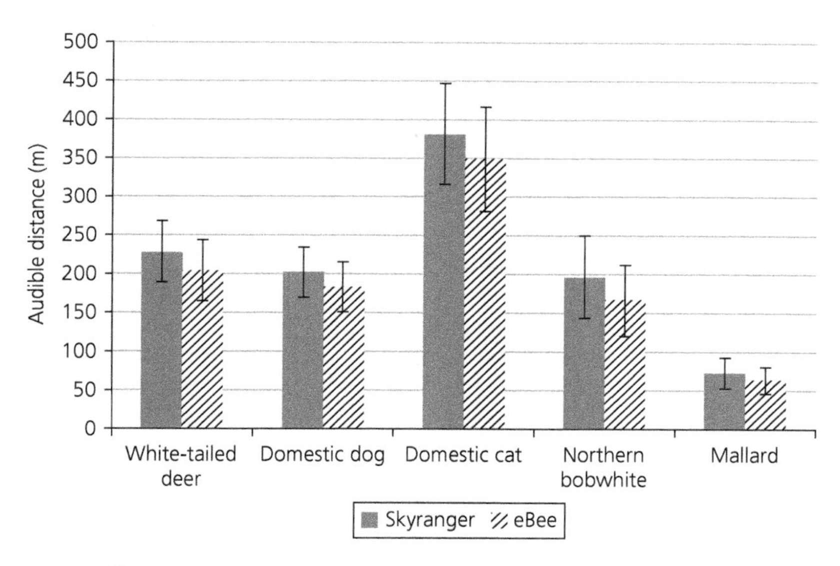

Figure 6.8 Distance at which different species can still hear two drones with means and SE indicated in the figure. Based on Scobie & Hugenholtz (2016).

above the birds (Dulava, Bean et al. 2015). Especially considering that at 15 m the noise of the petrol-powered drone in the study is between 81 and 90 dBA (Dulava, Bean et al. 2015).

As mentioned earlier it is important to assess the impact of drone surveys relative to other survey methods. Unfortunately there are very few studies that assess this aspect. Junda, Greene et al. (2016) compared vocal and movement behavioral responses of four raptor species to humans on foot approaching a nest without a drone flying above the nests to humans approaching the nest and then a drone flying above the nest. The results indicate that there is considerable variation in the reaction of the four raptor species, with ospreys not showing a larger reaction to the condition with drones compared to the human-only condition, but bald eagles showing a significantly stronger reaction to the condition with the drone. Overall, the authors conclude that the reactions of the birds to the drone would not discourage them to use drones for raptor nest surveys. They further indicate that more research needs to be conducted in which a drone-only condition is compared to a human climbing towards a nest to obtain nest data or using a long pole with a mirror.

At present there is a limited, but growing, number of studies that have examined the impact of drones on wildlife disturbance, and a recent review study was able to find important patterns in these studies (Mulero-Pázmány, Jenni-Eiermann et al. 2017). First, the flight pattern made a significant difference to disturbance, with animal reactions being stronger to flight patterns that were targeted towards the animals compared to lawnmower flight patterns. This can be interpreted as that flights in which a drone 'simulates' a direct approach (attack) to an animal will lead to higher disturbance levels than flights in which the animals are not directly approached. Second, as expected by other studies that have examined the general influence of noise on animals (Shannon, McKenna et al. 2016), noisier fuel-powered drones lead to more disturbance than electric-powered systems. Third, different animal types respond differently to drones with birds being more disturbed than terrestrial mammals and aquatic species showing the least response. Aquatic species likely react less because the water probably insulates them from the visual and audio stimuli from the drone. The difference between birds and terrestrial mammals is explained by the authors as influenced by the natural higher threat levels that birds have from aerial predators compared to large terrestrial mammals. Fourth, breeding animals generally react less than non-breeding animals, an effect likely caused by breeding animals being less likely to leave their nest. Fifth, larger aggregations of animals show a stronger response than smaller ones. Although the above patterns will facilitate making decisions about using drones, there are still not sufficient data to make reliable predictions about how a given animal species in a specific setting will react to a specific drone system. Therefore it has been argued that the precautionary principle should be applied when planning to use a drone for wildlife studies and that there is a need for the development of a code of practice with specific guidelines to mitigate against potential disturbance of drone flights to wildlife (Hodgson and Koh 2016, Mulero-Pázmány, Jenni-Eiermann et al. 2017). A preliminary list of guidelines based on these studies is presented here.

1) Use the precautionary principle in lieu of evidence.
2) Use (where available) an institutional animal ethics process to provide oversight to drone derived animal observations and experiments.
3) Follow the relevant civil aviation rules.
4) Carefully select appropriate drone and sensor equipment. Generally use the smallest and least noisy system possible.
5) Use ground and aerial procedures that minimize the disturbance on animals. Considerations here are: placement of ground control station away from the area where the animals are; minimize the flight time; maximize flying height; avoid drone manoeuvres above the animals; use lawnmower flight patterns; minimize flights during the breeding period or other sensitive periods; minimize resemblance between drone shape and an aerial predator when studying species that have aerial predators; use indirect shallow angle approaches instead of direct ones that are vertically downwards; monitor animals before, during, and after the flight and cease operations when disturbance is too high (where possible determine non-acceptable disturbance in advance); for nest inspections, fly at times in which eggs/chicks are out of risk or carefully monitor reaction and cease when reaction is non-acceptable; for flights around aggressive raptor's territories perform these at day times when the temperature is low and birds are less prone to fly.
6) Publish details to increase the body of literature on animal disturbance.

It deserves repeating that the disturbance of drones should ideally be compared to the disturbance of other survey methods and there is little comparative research on this topic (e.g. Junda, Greene et al. 2016). It is well established that other survey methods can lead to disturbance as well (Götmark 1992) and even though drones might lead to disturbance it might be lower than that of other methods (Junda, Greene et al. 2016, Borrelle and Fletcher 2017) and therefore a drone survey might still be acceptable if its disturbance is relatively low compared to other survey methods.

CHAPTER 7

Data post processing

7.0 Introduction

The wide spectrum of data collection needs that drones can facilitate has led to an explosion of usage in conservation research and management. As detailed in the previous chapters for conservation there are two main non-real time monitoring applications: mapping (see Chapters 3–5) and animal monitoring (Chapters 3, 5, and 6), and a third application, surveillance, for which real-time monitoring is more relevant (Chapter 4). For all these applications the most important aspect is to derive the relevant data from the images. The amount of data increases rapidly into tens of thousands of individual images, large numbers of orthomosaics, and if video is used hundreds of hours of video. Because in the vast majority of cases data are analysed manually (see Table 3.2) this places profound constraints on the speed at which data can be analysed and therefore the speeds at which conservation managers can react. Of course for many applications the speed of analyses is not of the essence, but even in such cases the costs associated with manual analyses are a constraint and, as shown in Chapter 1, the amount of manual labour for analyses increases quickly. Thus while the usage of drones leads to a reduction in costs at the acquisition side, manual analyses create an increase in costs on the analyses side. Due to this there is a strong demand for automated analyses that are ideally free or affordable so that NGOs, local communities, researchers, and others can use this. There are a variety of analyses that have been conducted to automate land-cover classification, features of the landscape, and the detection and identification of animals (Table 3.2). In this chapter we will give an overview of these, but we will also provide a very basic introduction to the software process used to derive point clouds and orthomosaics from the images collected on such missions.

7.1 Photogrammetry basics

Photogrammetry has been defined as 'the art, science, and technology of obtaining reliable information about physical objects and the environment through processes of recording, measuring, and interpreting photographic images and patterns of recorded radiant electromagnetic energy and other phenomena' (Wolf, Dewitt et al. 2014). The usage of consumer grade cameras and drones poses several challenges to conventional photogrammetry methods and therefore makes these difficult and time consuming to implement (Westoby, Brasington et al. 2012, Gross and Heumann 2016). Among these are the variability in the camera pose (attitude) and illumination between images, the difference in the percentage of overlap between images, perspective distortion due to low flights, limited accuracy of GPS and IMU on board the drone, perspective distortions due to the lower flight altitude of drones compared to manned aircraft, and lens distortions (Laliberte, Rango et al. 2007, Barazzetti, Remondino et al. 2010, Turner, Lucieer et al. 2012, Westoby, Brasington et al. 2012, Gross and Heumann 2016).

Conservation Drones: Mapping and Monitoring Biodiversity. Serge A. Wich & Lian Pin Koh. Oxford University Press (2018).
 DOI:10.1093/oso/9780198787617.001.0001

The development of SfM (Structure-from-Motion) has made it possible to derive high-quality outputs (point clouds and orthomosaics) despite the above-mentioned issues, although there is a need for detailed comparisons between the outputs from traditional photogrammetry methods with the newer drone-derived outputs (Colomina and Molina 2014). This method has its roots in the software development that dealt with feature extraction and the advances in computer vision (Westoby, Brasington et al. 2012) and has relatively recently become available through the development of several user-friendly software packages that are specifically developed for images acquired from drones. The main differences between this method and traditional photogrammetry are that there is no need to have a number of features on the images for which the 3D location is known and that even without information on camera parameters the reconstruction can occur. In SfM the reconstruction of the geometry of the scene, and the position and orientation of the camera, are resolved automatically through a procedure that uses features extracted from a number of overlapping images and without the need for camera details such as focal length, number of pixels, sensor size, and lens distortion metrics (Snavely, Seitz et al. 2008). Because there are several excellent technical reviews on this topic and comparisons of software packages we will only provide a general overview of the SfM process in this book (Turner, Lucieer et al. 2012, Westoby, Brasington et al. 2012, Ingwer, Gassen et al. 2015, Gross and Heumann 2016).

7.2 Basic process

Even though the various software packages that can be used for the SfM workflow differ in their details (Torres, Arroyo et al. 2012, Gini, Pagliari et al. 2013, Sona, Pinto et al. 2014, Gross and Heumann 2016) the basic process is similar (Figure 7.1).

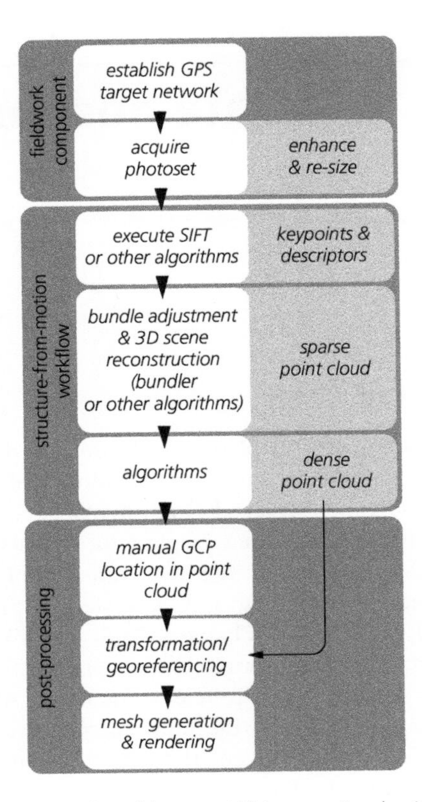

Figure 7.1 Overview of the general SfM process. Based on Westoby, Brasington et al. (2012). © Perry van Duijnhoven.

In Chapter 5 the basics of flying a mapping mission were explained, which included geotagging images. Geotagged images are not necessary for SfM, but it can speed up the process and enhance the quality of the final products. If no geolocation information is available in the images the resulting outputs will have no scale, will not be georeferenced, and will not be oriented. Adding ground control points (GCPs) will solve this issue though. Pose (attitude) data are not necessary for SfM, but if available can improve the quality of the final products. There are various procedures to provide the SfM software with the pose (pitch, roll, yaw) data for each image. For instance Pix4Dmapper allows for importing the datalog files that are saved on the autopilot hardware (e.g. Pixhawk) and will automatically extract the pose data from such a file and link it to individual images. To assure that this works properly it is best if the camera is triggered by the autopilot so that the process will work seamlessly. Triggering the camera through the autopilot can also facilitate that the images have an appropriate interval between them so that sufficient overlap (side and frontal) is achieved. Ground control software such as Mission Planner has straightforward mapping options for such missions (Figure 7.2).

Figure 7.2 Indicating the overlap between the various images during a mapping flight. © Perry van Duijnhoven.

It is important to realize that a fixed wing drone will usually fly with a deviation between the direction of travel and its actual heading (Wolf, Dewitt et al. 2014). This is called the crab angle and is usually the result of side winds (Figure 7.3a). The crab angle has the undesired effect of reducing the actual overlap area that is covered during a flight (Figure 7.3b, c). Manned aircraft systems have gyro-stabilized camera mounts in which the crab angle is being corrected for by rotation of the camera in the mount, but small drones tend not to have such mounts and hence the operator must be aware of this when planning a mission and plan the area covered to be somewhat larger than the area needed for the point cloud and orthomosaic.

Before importing images into a SfM software package it is worth considering whether the full quality images are needed or whether these can be readjusted in size and hence speed up processing. The outputs will be of lower quality when resized images are used but for some projects the full size might be more than required and hence readjusting could be a sensible step.

Once all images are in a SfM software package the SfM process will start with its key process, the extraction of matching features in multiple images that are often obtained at a slightly different angle due to changes in the pose of the drone. A visualization of part of this process for a mapping flight and a flight to reconstruct a single object can be seen in Figure 7.4a and b.

When a gimbal is used the pose changes of the drone are compensated for by the gimbal and hence the variation in angles is much less. A procedure commonly used for the feature extraction is SIFT (Scale Invariant Feature Transform; Lowe 1999, Lowe 2004, Snavely, Seitz et al. 2008), but not all software packages use this algorithm (Dandois and Ellis 2013). SIFT identifies features in images that are invariant to the scale and rotation of the image and somewhat invariant to the orientation of the camera in 3D and the illumination of the image. This process can create large numbers of keypoints (>10,000) in each image, depending on factors such as resolution and image texture. Thus higher image resolution or flying lower to the object of interest will usually lead to more keypoints that can be extracted and hence to a denser point cloud. It is important to realize that flying low over complex landscapes such as tropical rainforests can lead to difficulties for keypoint extraction and the subsequent steps in SfM. Flying higher can then often lead to better results. Homogenous areas such as sand or snow often lead to fewer keypoints. Once the keypoints are identified the descriptors are created which allow for the matching of features between images.

Once the keypoints and descriptors are computed the next step is to make the sparse point cloud through the bundle adjustment system bundler (Snavely, Seitz et al. 2008). Keypoints in multiple images are matched through certain rules which will automatically remove features that are transient such as moving cars, but also features such as the legs of a multirotor. A triangulation procedure is used to estimate (relative) the 3D location of each point. During this process the interior (e.g. focal length, lens distortion) and exterior (position and orientation) camera parameters will also be estimated. In some software packages some of the interior and exterior parameters can be provided before this process starts (e.g. Pix4Dmapper). Once the sparse point cloud has been generated there are several algorithms to compute the dense point cloud and the various software packages differ here (Torres, Arroyo et al. 2012, Westoby, Brasington et al. 2012).

After this step ground control points can be entered to either transform the relative coordinate system to an absolute coordinate system (if no

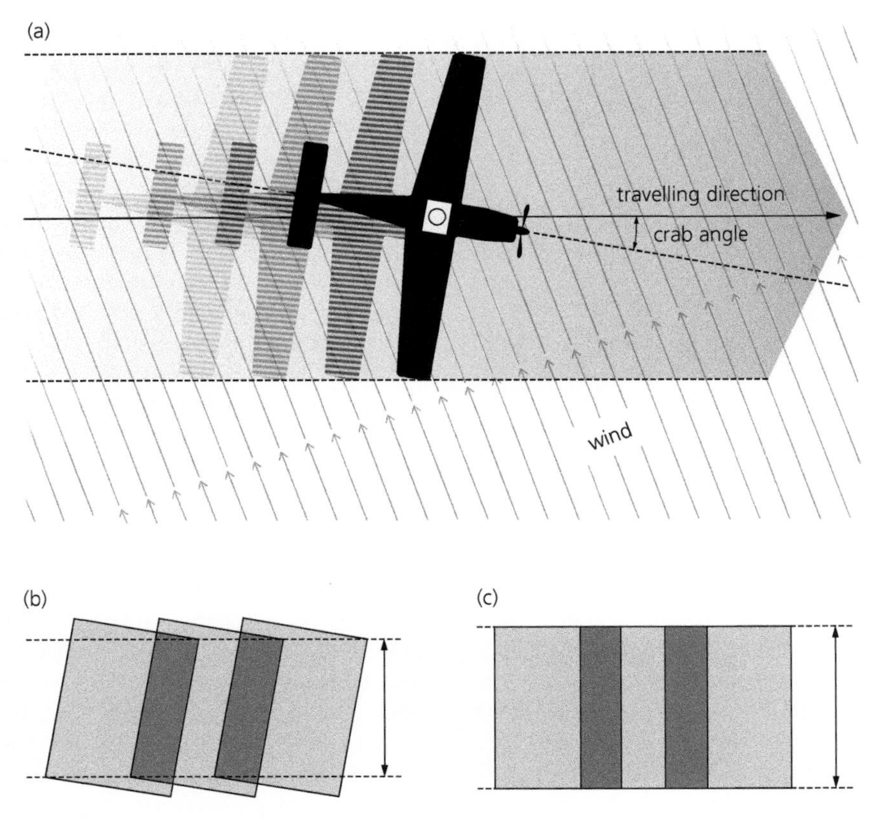

Figure 7.3 a) Showing the influence of the orientation of the plane and the traveling direction. b) Orientation of the images based on a crab angle larger than zero degrees. c) Orientation of images if crab angle were zero. Based on Wolf, Dewitt et al. (2014). © Perry van Duijnhoven.

geolocation data were available in the images) or to improve the accuracy of the point cloud and add an absolute error from the GPS system to the relative location error in the point cloud so that the total error will be available. From the transformed dense point cloud the mesh generation step will generate the full 3D triangle model and the texture between the points.

The digital surface model (DSM), which is a single-band raster, will be created next in which each pixel will have an elevation above the surface value as well as X and Y coordinates. After this the images will be projected onto the DSM in order to create a raster in which each pixel has a value corresponding to the value in the images (RGB, multispectral, thermal).

Once the DSM and orthomosaic have been created the user can generate several other outputs. A DTM (digital terrain model) can be generated in which the vegetation, houses, cars, and other structures on the surface have been removed. The quality of the DTM is dependent on the number of ground hits and given that these can be limited in densely vegetated areas such as forests the applicability of RGB images to create DTMs can be limited compared to airborne LiDAR where the number of ground hits is higher than with a camera mapping mission (White, Wulder et al. 2013). Some researchers have therefore used a DTM based on airborne LiDAR in combination with a DSM generated from the photo mapping mission to derive forest stand characteristics (Lisein, Pierrot-Deseilligny et al. 2013). Generally point clouds based on drone images have less penetration than LiDAR in vegetated areas (Figure 7.5, Lisein, Pierrot-Deseilligny et al. 2013).

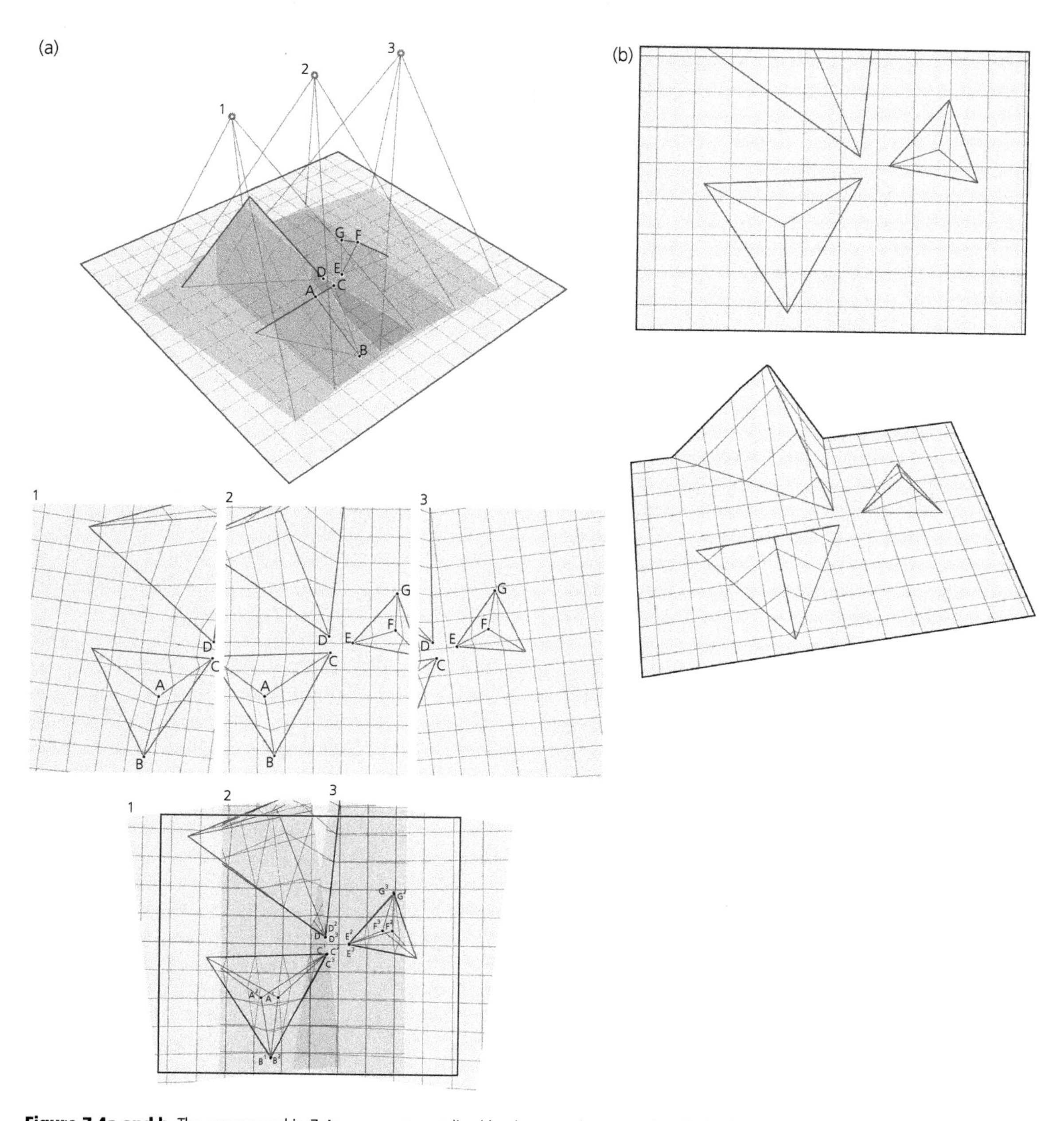

Figure 7.4a and b The upper panel in 7.4a represents a stylized landscape to be mapped, with the footprint of three images acquired from a drone. Marked are some reference points that occur on two or more photographs. The middle panel shows the individual photographs, on the lower panel the overlap without any processing is shown. As shown in Figure 7.4b, the result from SfM can be a corrected flat top view image (e.g. orthomosaic), or a 3D layer (DSM) as in the lower panel. © Perry van Duijnhoven.

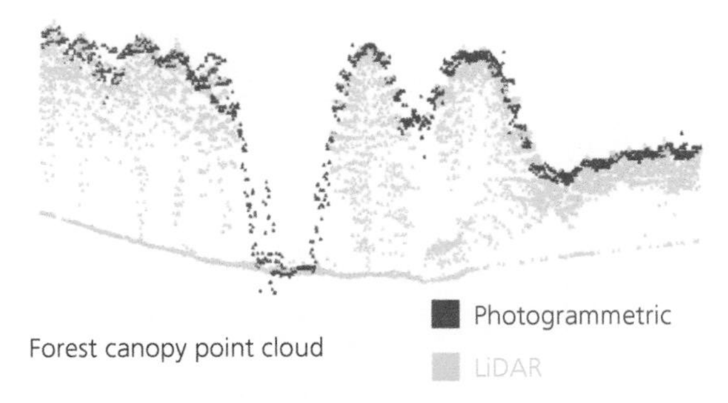

Figure 7.5 The dots in the figure indicate the height for the points in the point cloud from LiDAR and photogrammetry. From Lisein, Pierrot-Deseilligny et al. (2013). Permission through https://creativecommons.org/licenses/by/4.0/.

If converted RGB or true multispectral cameras are used, several software packages also allow for user-friendly workflows to derive metrics that provide information about vegetation such as NDVI, etc. Researchers have also used thermal images as the basis for thermal point clouds. Programs such as Pix4D allow for straightforward workflows to build such thermal landscapes.

7.3 Photogrammetry software packages

There are an increasing number of photogrammetry software packages available that range from being free to costing thousands of US dollars.[1] In addition to their price tag these packages differ in various features that they do or do not offer. Although this chapter does not aim to provide a full overview of features that are important several are mentioned here. Although many programs automatically identify tie points between images some programs (e.g. MapKnitter[2]) offer only manual options to mosaic images. Other programs like Microsoft ICE (image composite editor) do automatically produce a mosaic, but do not produce a 3D point cloud. Another important difference between packages is whether they allow for incorporation of location (geotag) data from images or not. If not (e.g. MapKnitter), then scaling and georeferencing can be achieved with geographic information software (GIS) such as ArcGIS or QGIS, to name but two of the many GIS packages available. The inclusion of ground control points (GCPs) also varies between packages, from some packages having user-friendly interfaces to import the GCPs to there being no options to include these (e.g. Microsoft ICE). The number of cameras for which the information is included in the packages and regularly updated also varies between programs, with the more costly programs having extensive camera databases so camera information can be used in data processing. The overall workflow also differs extensively from easy workflows from image entry to georeferenced DSM and orthomosaics in programmes such as Pix4D and Agisoft Photoscan to more complicated workflows with combinations of open-source software packages. An advantage of open-source packages can be that it can be determined which algorithms are used for certain steps in the overall process. For instance the Bundler open-source package uses the SIFT algorithm for image features (Snavely, Seitz et al. 2008), while in Agisoft Photoscan it is mentioned that it is 'SIFT-like' (Dandois and Ellis 2013). In cases where different results are achieved, and researchers need to understand the factors leading to differences, it is important to understand which algorithms have been used, and in such cases there could be an advantage using open-source packages over commercial packages that do not provide details of the algorithms used. There are several recent papers that provide comparisons between the various programs for the same dataset (Gini, Pagliari et al. 2013, Qin, Grün et al. 2013, Sona, Pinto et al. 2014, Gross 2015, Gross and Heumann 2016) and provide reviews of some of the issues surrounding the various methods and programs used (Colomina and Molina 2014). These comparisons examine differences between programs (Agisoft Photoscan, Pix4Dmapper, ERDAS Leica Photogrammetry Suite, PhotoModeler Scanner, Microsoft ICE, and EyesDEA) for aspects such as the geometric accuracy, visual quality, and the quality of the DSM (see Table 7.1).

[1] https://en.wikipedia.org/wiki/Comparison_of_photogrammetry_software (accessed 8 November 2017)
[2] https://mapknitter.org/ (accessed 8 November 2017)

Table 7.1 Several of the packages used in the studies reviewed in this book or that we have used ourselves

Package	Point cloud	Orthomosaic	Geolocation	Georeferenced	Price category
Pix4Dmapper	x	x	x	x	C
Agisoft Photoscan	x	x	x	x	B
Microsoft ICE		x	x	x	Free
VisualSfM/CMVS	x		x	x	Free
CMPMVS		x	x	x	Free

Prices: Free, A = 1–1000 US$, B = 1001–5000 US$, C = 5001+. Note, some packages can be purchased for monthly usage, which is useful for specific projects that have a limited time span (e.g. Pix4Dmapper). Several packages also offer discounts for educational or non-for-profit usage. Price ranges here are indicative only for the full commercial rate and have been compiled in 2017.

7.4 Analyses

7.4.1 General

The 3D point clouds that are generated by the software packages allow for analyses on a number of variables related to the 3D structure of an area. Before providing some examples it is useful to take note of the fact that various variables can be influenced by the mission itself. In other words, when planning a flight it is important to consider how the planned flight might influence the quality of the point cloud. A recent study examined how flight altitude, weather conditions, and image overlap influenced a range of measures such as point cloud density, canopy penetration, various estimates of canopy structure, and geometric positioning accuracy for a temperate deciduous forest site in the US (Dandois, Olano et al. 2015). Researchers collected ground data on various forest variables (tree height and above-ground biomass density) to compare to the data collected with the drone and with airborne LiDAR. The results indicate that canopy height estimates were not significantly influenced by weather, flying altitude, and side overlap. Canopy penetration was, however, significantly influenced by the weather with flights on cloudy days leading to significantly less canopy penetration. Canopy penetration was not influenced by flying altitude though, but decreased with decreasing forward overlap between images (Figure 7.6). As indicated by other studies (Lisein, Pierrot-Deseilligny et al. 2013), LiDAR penetrated the canopy significantly more than the point cloud from drone images. The authors conclude that for their specific forest characteristics the optimal conditions would be a clear day and flights with at least a sidelap of larger than 60 per cent obtained while flying at 80 m altitude because that yielded the largest field of view. With these settings estimates of canopy height were highly significantly correlated with ground and LiDAR estimates.

Other studies might not necessarily lead to the same results. In contrast to the the study just mentioned, a study in the Arctic found that a reduction in photo overlap or resolution (flying at a higher altitude) led to less accurate height models based on the SfM process (Fraser, Olthof et al. 2016). In addition, the type of vegetation might influence the quality of the orthomosaics and whether these can be processed at all. In our own experience, creating orthomosaics from flights where we flew low (80 m) over dense tropical rainforests generally did not work, even with 70–80 per cent overlap. Increasing flying height to approximately 150 m led to much better results. But for more open forests, such as miombo woodlands in Africa, flying at 80 m does not lead to processing difficulties. It is thus important to sufficiently test what the most suitable flying height is for a specific aim. It is clear from the above that much more research needs to be conducted to examine how basic aspects of data collection such as choice of flying platform, sensor, flying height, weather, software package used for image feature matching, and others influence point clouds. Such research could then lead to a set of best-practice guidelines for data collection.

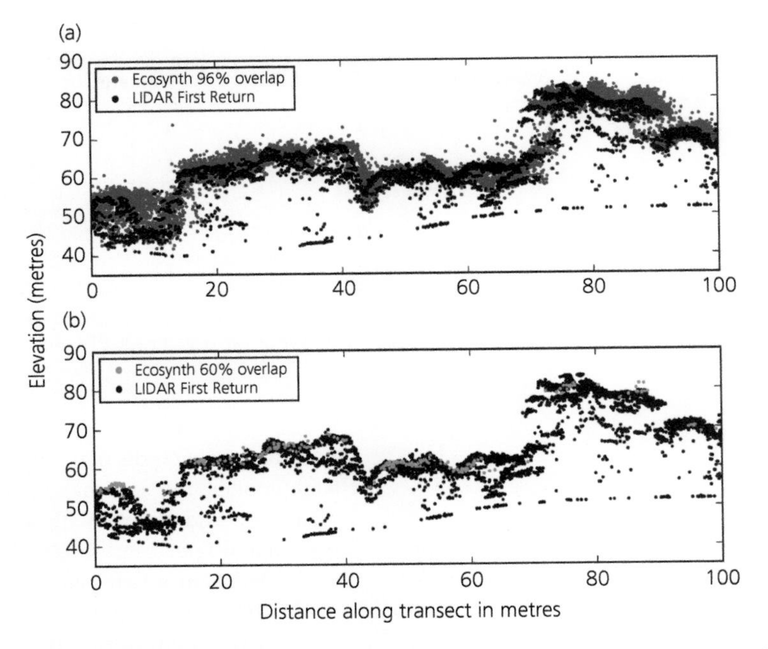

Figure 7.6 Cross-sections of a 100 × 5 m forest area which show the drone-image derived point cloud for different levels of frontal overlap compared to the first return LiDAR point cloud. Figure reproduced with permission of authors from Dandois, Olano et al. (2015).

7.4.2 Three-dimensional quantification of landscape and vegetation using photogrammetry

Traditionally most studies that aim to measure the three-dimensional aspects of a landscape (e.g. vegetation structure, elevation) have used ground surveys, manned aerial surveys, or satellites in combination with a variety of sensors (Horning, Robinson et al. 2010, Wolf, Dewitt et al. 2014). Drones provide researchers, managers, surveyors, and others that need such data for academic research, management, or other purposes with an additional tool to obtain data over relatively large areas with a very high resolution. So far a key aspect of the research that has been conducted using drones has been the validation of the approach by comparing results obtained through photogrammetry from images acquired with a drone to those obtained through other methods (mainly ground measurements or airborne LiDAR). A non-exhaustive summary of such studies is provided in Table 7.2. These studies indicate that the vegetation variables that can be derived from photogrammetry generally correlate well with those measured with other methods that are either ground based or aerial LiDAR (Table 7.2). This indicates that for variables such as vegetation height, leaf-area index, and others drones will allow for frequent monitoring over relatively large areas. The combination of high spatial and high temporal resolution will enable exciting opportunities to monitor vegetation growth for example.

It is important to note, however, that, as indicated earlier, LiDAR penetrates further into the canopy than photogrammetry point clouds and that this poses limitations on how well the digital terrain model can be derived from photogrammetry in densely vegetated areas. Combining low-cost LiDAR systems with RGB sensors might allow for an improvement if such data can be well aligned. Despite this, in some cases the high resolution of the point clouds derived from drone imagery allows for better results than LiDAR data. A study in which the aim was to compare tree detection on a drone image-derived point cloud and

Table 7.2 Examples of studies using photogrammetry to derive 2- and 3-dimensional vegetation characteristics and compare those to ground and aerial surveys

Location	Area (ha)/ GSD (cm)	Variables	Drone data significantly correlated to ground survey?*	Drone data significantly correlated to airborne LiDAR?*	Source
UK	Varying/0.7–1.4	Crop height	Yes (terrestrial laser scanner)		(Holman, Riche et al. 2016)
Germany	175/2–8	Crown height		Yes	(Thiel and Schmullius 2016a)
Peru	516/6.2–7.7	Top-of-canopy height		Yes	(Messinger, Asner et al. 2016)
Peru	516/6.2–7.7	Above-ground carbon density		Yes	(Messinger, Asner et al. 2016)
China	20/4.3	Canopy height, canopy closure			(Zhang, Hu et al. 2016)
Canada	–/0.75–1.5	Vegetation height	Yes (tape measure)		(Fraser, Olthof et al. 2016)
Norway	200	Several vegetation measures	Yes (hypsometer)		(Puliti, Ørka et al. 2015)
Costa Rica	9.75/	Canopy height, canopy openness	Yes (Laser range finder and densiometer)		(Zahawi, Dandois et al. 2015)
Greece	–	Leaf-area index	Yes (manual count of all leaves)		(Kalisperakis, Stentoumis et al. 2015)
Norway	194.7/4.2	Various variables (e.g. dominant height)	Yes (hypsometer)	Yes	(Puliti, Ørka et al. 2015)
UK	–/2.2–8.8	Topography	Yes (total station)		(Tonkin, Midgley et al. 2014)
Spain	148.58/4.5	Tree height	Yes (GPS in RTK mode)		(Zarco-Tejada, Diaz-Varela et al. 2014)
Germany	0.9	3D stem reconstruction	No (terrestrial laser scanner)		(Fritz, Kattenborn et al. 2013)
US	1.9/–	Leaf-area index	Moderate (ceptometer)		(Mathews and Jensen 2013)
US	4.5/	Tree canopy height	Yes (hypsometer)	Yes	(Dandois and Ellis 2010)**

A non-exhaustive overview of several key aspects of studies on vegetation characteristics. GSD = ground sampling distance. – = not provided in the paper. * = not all studies provide p values. ** = kite was used for data acquisition.

airborne LiDAR found that fewer trees (particularly small ones) were missed on the drone point cloud than the airborne LiDAR (Thiel and Schmullius 2016). This difference is due to the much lower number of points per square metre in the LiDAR point cloud (4 points/m^2) compared to the RGB point cloud (310 points/m^2)(Thiel and Schmullius 2016).

7.4.3 Analyses based on orthomosaics: land-cover and feature classification

Mapping areas with drones started in the late 1970s, but did not received much attention in those days or soon thereafter (Przybilla and Wester-Ebbinghaus 1979, Wester-Ebbinghaus 1980). The recent development of SfM software has made drones a very suitable tool for very high resolution land-cover classification studies (Table 3.3). Such studies generally use the orthomosaics from drone flights to classify land-cover types or to classify specific features (e.g. certain plant species) in such mosaics. Even though there have been a relatively large number of studies conducted on land-cover classification with drones, its usage is still limited compared to satellites. This is perhaps not surprising given the much larger areas covered by satellites, the well-developed analytical workflows for land-cover analyses for satellite data, and the ease with which satellite data can be obtained from the internet (Horning, Robinson et al. 2010). Nevertheless the very high spatial resolution of drone-derived orthomosaics allows for much finer-grained analyses than is possible on even the highest resolution satellite images that are commercially available. As a result researchers have been using drones for land-cover studies in a variety of landscapes and with a multitude of sensors (RGB, multispectral, hyperspectral, thermal). These studies have used the orthomosaics from drone images to classify land-cover types or more fine-grained analyses on individual plant species (Table 7.3).

Studies have also used the orthomosaics from drones to classify habitat for salmon (Woll, Prakash et al. 2011), fish nurseries (Ventura, Bruno et al. 2016), or more general land-cover classification in wetland and riparian areas (Dunford, Michel et al. 2009, Chabot and Bird 2013). The use of drones for land-cover classification is occurring in many different climates and vegetation types around the world, from the hot and humid tropical forests to the cold polar regions, which indicates the enormous potential of drones to map areas (Turner, Lucieer et al. 2012, Lucieer, Malenovský et al. 2014, Lucieer, Turner et al. 2014, Turner, Lucieer et al. 2014, Stuchlík, Stachoň et al. 2015, Wich, Koh et al. in press).

The majority of studies on land-cover classification use reflectance values from the pixels or objects to classify land-cover, but studies have also used the height information from the point clouds to classify vegetation types (grasses and shrubs) and differentiate these from bare ground (Cunliffe, Brazier et al. 2016).

An important aspect of classification with very high resolution orthomosaics is that objects, such as trees, are constituted of hundreds or thousands of pixels showing a large variation in reflectance values due to, for instance, small gaps in the canopy exposing bare ground under the canopy. This problem is well-known from classification with very high resolution satellite images as well; in such cases object-based image analyses approaches have been recommended (Riggan and Weih 2009, Whiteside, Boggs et al. 2011). Several comparative studies have been conducted (Table 7.4) in which object-based classification generally improves the accuracy of the classification (Dunford, Michel et al. 2009, Wich, Koh et al. in press). The Wich et al. (in press) study examined whether several forest classes could be distinguished by using pixel- and object-based classifications for an area containing areas of logged, reforested, and oil palm vegetation classes (Figures 7.7 and 7.8). Although the reforested areas were difficult to differentiate for the various years in which these were reforested and had low accuracies, the oil palm areas were classified with high accuracy: maximum producer accuracy of 90.32 per cent and user accuracy of 81.04 per cent. Combining the reforested areas for the various years into one class improved overall accuracy (pixel-based classification: 31.04–57.76 per cent; object-based classification: 34.61–61.94 per cent), but it was still relatively low (Figure 7.8). Although not tested yet, a potential improvement could come from downsampling the very high resolution orthomosaic to a larger pixel size and assess whether better classification results could be obtained.

Land-cover change monitoring has usually been conducted with satellites and allows for important support for conservation such as global-scale analyses on forest loss and gain at medium or high resolution at weekly or monthly intervals (Broich, Hansen et al. 2011, Hansen, Potapov et al. 2013). More recently, researchers have started to use drones for land-cover change monitoring, but surprisingly little has been published yet. An exception is a study examining forest change detection with a LiDAR system in a eucalypt plantation in Tasmania, Australia (Wallace, Lucieer et al. 2012a,b). Studies using low-cost fixed wing platforms with off-the-shelf RGB sensors have started in Indonesia and Tanzania to support conservation. The preliminary results of these studies show that drones can be

Table 7.3 Examples of studies using drone images for land-cover classification studies

Location	Aim	Sensor	Area (ha)/ GSD (cm)	Classifiers	Accuracy*	Source
Indonesia	Forest classification	RGB	522/5.36	Pixel and object-based classification	31.04–61.94%†	(Wich, Koh et al. in press)
Canada	Classifying arctic vegetation	RGB	–/1	Multi-resolution segmentation and See5 decision tree	50–94%	(Fraser, Olthof et al. 2016)
US	Surface cover classification in arid area	RGB	–/0.4	Vegetation canopy height to classify vegetation classes and bare ground	>90%	(Cunliffe, Brazier et al. 2016)
Belgium	Riparian forest classification	RGB and NIR	–/10	Object-based classification	79.5 and 84.1% ***	(Michez, Piégay et al. 2016)
Argentina	Peat bog plant species classification	Modified RGB (NIR)	1.8/–	Object-based classification	86% and 92%	(Lehmann, Münchberger et al. 2016)
Italy	Classifying bush and tree species	RGB and NIR	–/1.33 & 2.63	Unsupervised and supervised classification	50.08 and 79.06% **	(Gini, Passoni et al. 2014)
Germany	Bog vegetation classification	Modified RGB (NIR)	–/~1.5–3	Object-based classification	91%	(Knoth, Klein et al. 2013)
US	Dryland vegetation classification	RGB and hyperspectral	–/4.2 and 110	Unsupervised and supervised classification	88% (unsupervised). Supervised performed poorly and no quantitative assessment provided	(Mitchell, Glenn et al. 2012, Mitchell, Glenn et al. 2016)
Australia	Vegetation classification (invasive species)	RGB	–/4	Supervised machine learning	87 and 89%	(Reid, Ramos et al. 2011)
France	Riparian forest classification	RGB	22.78/6.8–21.8	Pixel and object-based classification	84 and 91%	(Dunford, Michel et al. 2009)
US	Rangeland classification	RGB	–/5	Object-based classification	>90%	(Laliberte and Rango 2009)
US	Classifying arid areas	RGB	–/5	Object-based classification	96%	(Laliberte, Rango et al. 2007)

A non-exhaustive overview of several key aspects of land-cover studies. GSD = ground sampling distance. – = not mentioned in paper. * = often, but not always, the user's and producer's accuracy is provided by papers. ** = depending on classification type. *** = depending on area. † = accuracies are given for a variety of classification methods.

an effective tool to monitor land-cover change due to the fact that no expert knowledge is needed for the interpretation of the images. This contrasts with satellite imagery where experts are often needed. Figure 7.9 shows three images from the eastern part of the Leuser ecosystem (Sumatra, Indonesia). The first image shows an intact forest area. During the second aerial survey this area clearly had been logged, and had started to regrow when the third mission was conducted. Because the images are in RGB, non-experts can easily interpret these images, which is very useful for protected area managers and other decision makers who might not always be trained in GIS/remote sensing or have the technical

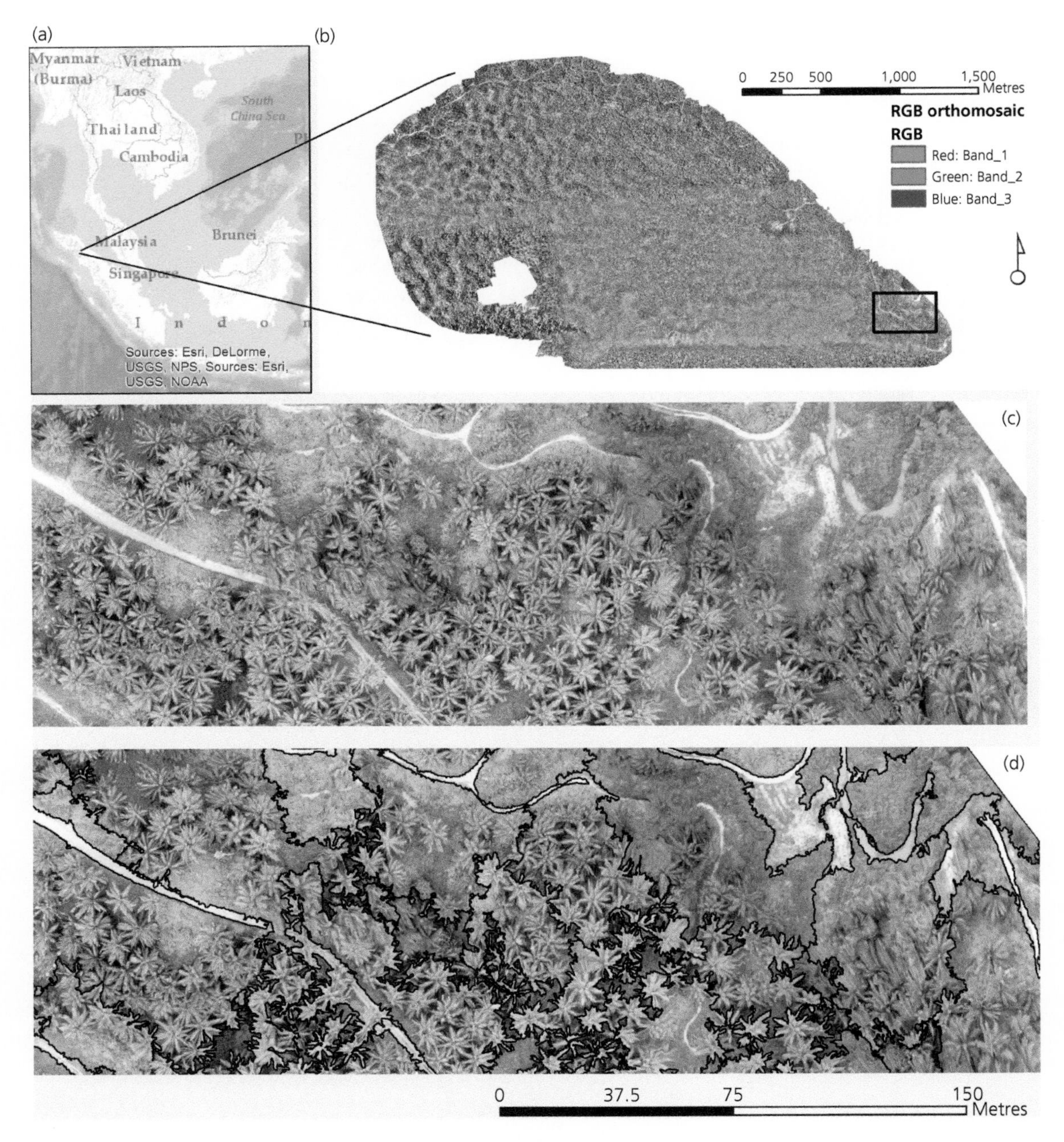

Figure 7.7 a) The location of the study area referred to in the text; b) detail of orthomosaic; c), d) example of segments on the detailed area. From Wich, Koh et al. (in press). © Serge Wich.

teams and budget available to use high resolution satellite imagery.

A second pilot study has started in Tanzania where researchers have started to monitor community-managed forest reserves surrounding the Gombe National Park. Fixed wing flights allow for complete coverage of a single forest reserve in one flight and allow for monitoring of land-cover

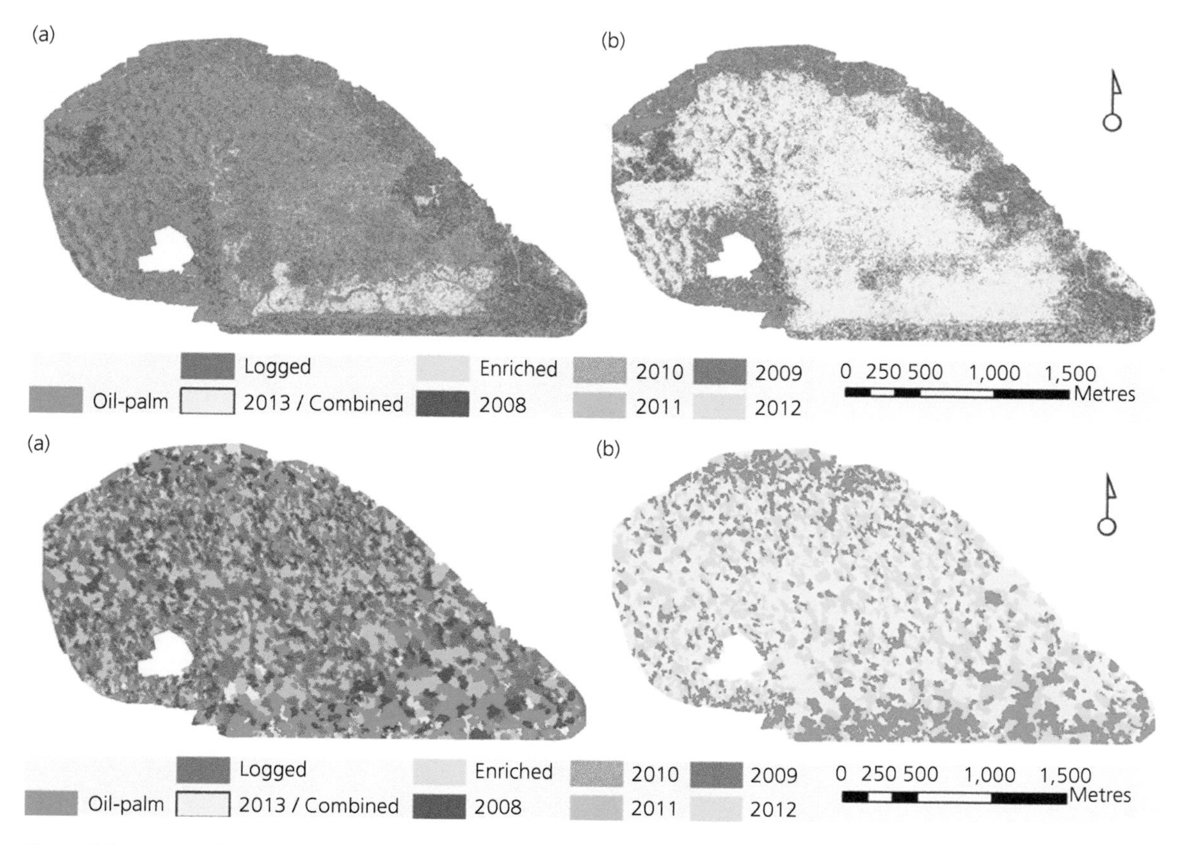

Figure 7.8 Upper panel shows results for the pixel-based analyses (a = full set of classes, b = reforested classes for various years combined into one class). Lower panel shows the results for the object-based analyses (a = full set of classes, b = reforested classes for various years combined into one class). From Wich, Koh et al. (in press). © Serge Wich.

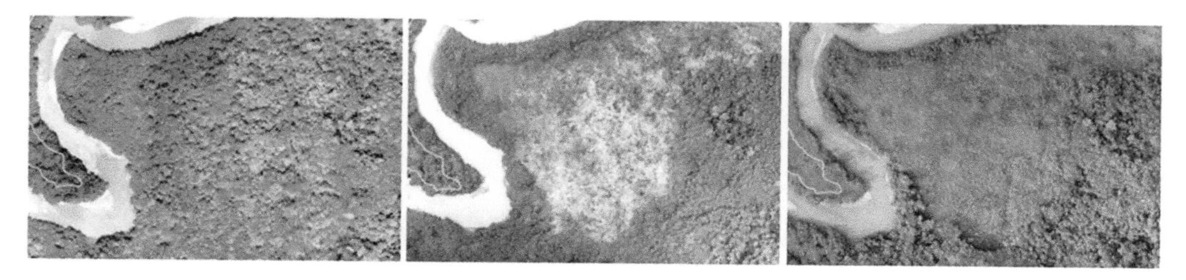

Figure 7.9 Leuser ecosystem in Sumatra (Indonesia). Upper panel: November 2013, middle panel: May 2014, lower panel: January 2015. All images were obtained with fixed wing systems. © conservationdrones.

change between years (Figure 7.10). Because forest reserves are monitored by a small number of staff, yearly land-cover change monitoring by drones can be of enormous support to such staff to monitor their reserves effectively.

7.4.4 Combining drone and satellite data

Even though the flight duration of drones is increasing, the area that most drones can cover during a mapping mission remains much smaller than that

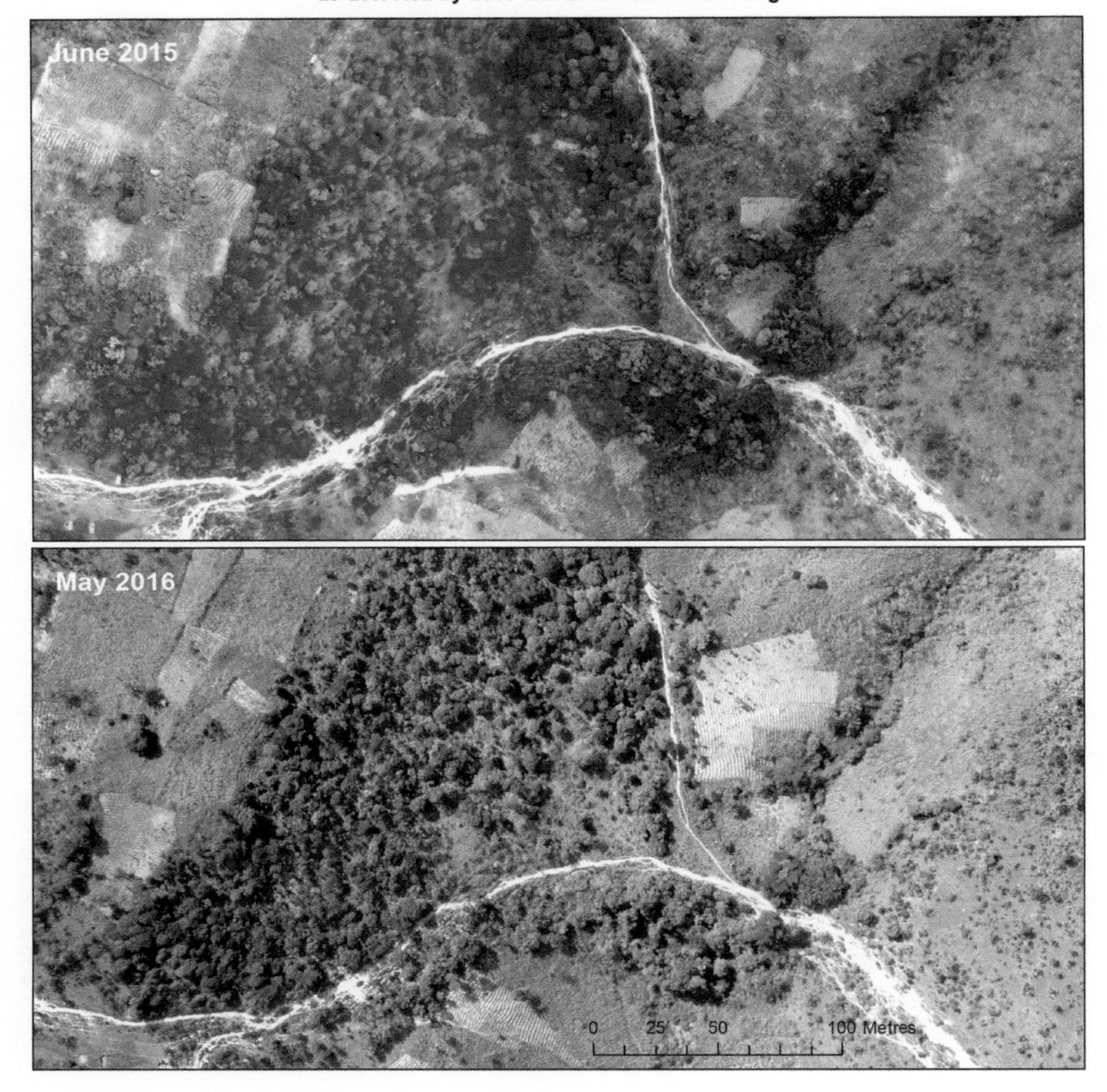

Figure 7.10 Images from pilot study monitoring forest reserves in Tanzania (see text). Flights were conducted by Jeff Kirby and processed by ESRI. The project is a collaboration between the Jane Goodall Institute, ESRI, Liverpool John Moores University, and conservationdrones. The project is funded by US Fish and Wildlife Services, USAID, and supported by the Tanzanian Government. © Lilian Pintea.

of satellites. As a result satellites remain the remote sensing platform of choice for analyses at a large spatial scale. Land-cover classification studies with satellites rely on ground truth data to train and validate the algorithms and resulting classifications (Horning, Robinson et al. 2010). Obtaining ground truth data over large areas can be a time-consuming and costly effort, particularly when areas are difficult to access. The very high resolution images that drones provide can potentially be used as ground-truthing data because land-cover classes can be easily determined on those images. Because drones can cover large areas much faster than humans can and generally have geotagged images this method would allow obtaining training and validation data for large areas much faster and at much reduced costs than obtaining the same data on foot. A recent study that used this 'drone-truthing' approach used seven land-cover classes (Figure 7.11) that were discernable from drone images to train classification on Landsat satellite images (Szantoi, Smith et al. 2017).

Using the drone training data yielded an overall classification accuracy of 75.82 per cent with several classes having high producer accuracies: oil palm (89 per cent), reforestation (76 per cent) and logged areas (76 per cent). The classification accuracies reported by this study were higher than other studies that classified some of the same land-cover classes, indicating that this is a promising approach to improve classification based on freely available medium resolution satellite images. In another study using satellite images and drones, researchers using Landsat satellite images to classify land-cover before and after fires in the Indonesian province of Riau used a fixed wing drone to obtain data to assess the accuracy of the Landsat-based classification (Gaveau, Salim et al. 2014). There is scope for many more studies where drones and satellite data are combined for training and validation of classifiers. Particularly with the increasing amount of drone data being collected worldwide there is an option to potentially use drone data at a global scale to assist land-cover analyses if the necessary platforms could be set up or existing platforms could be integrated.

7.4.5 Automated analyses to detect and identify plant species

To increase the efficiency of vegetation surveys with drones increased automation is imperative. Although the number of studies focusing on this aspect of vegetation analyses is still limited the results are promising. A recent study used the point cloud from SfM to automate detection of oil palm trees. Flying height influenced the accuracy of the detection with lower (70 m) leading to less accurate detection than higher flying height (100 m). Out of 615 palm trees, 72 per cent were automatically detected at 70 m while 91 per cent were detected at 100 m. Trees that were omitted in the detection were those for which the tree crown was not reconstructed during the SfM process. These tended to be trees exposed to wind due to their location in the plantation (Kattenborn, Sperlich et al. 2014). Other studies have focused on segregating tree crowns to facilitate detection and identification of different species (Hung, Bryson et al. 2011, Hung, Bryson et al. 2012). These studies have shown that several species can be differentiated using a statistical model that uses colour and texture features of the target trees for detection and then uses image segmentation to classify the species using the tree crowns with an overall detection accuracy of 80 per cent (Hung, Bryson et al. 2011). Although this study was conducted in a relatively open area with a limited number of species (five) to be differentiated, the method is promising and can be adapted to more complex settings. A study in a much more diverse cloud rainforest in the Western Ecuadorian Andes used an RGB camera fitted to a remote controlled helicopter to study automated species detection, with promising results (Peck, Mariscal et al. 2012). The researchers used several decision-tree based predictive models to correctly identify tree species. The best model predicted the correct species with 47 per cent accuracy for 41 species. Results were similar for the genus and family levels. Even though 47 per cent might seem relatively low it is important to appreciate that these results are from a diverse and hilly tropical cloud forest, so it is still promising.

7.4.6 Automated animal detection, identification, and tracking

As mentioned in Table 3.2 and the introduction to this chapter, the increase in efficiency that drones offer on the data acquisition side leads to large volumes of data (stills/video) that need to be examined for the object of interest (e.g. plants, animals). At present most researchers visually scan images one by one for the object of interest. With single missions often leading to hundreds or thousands of

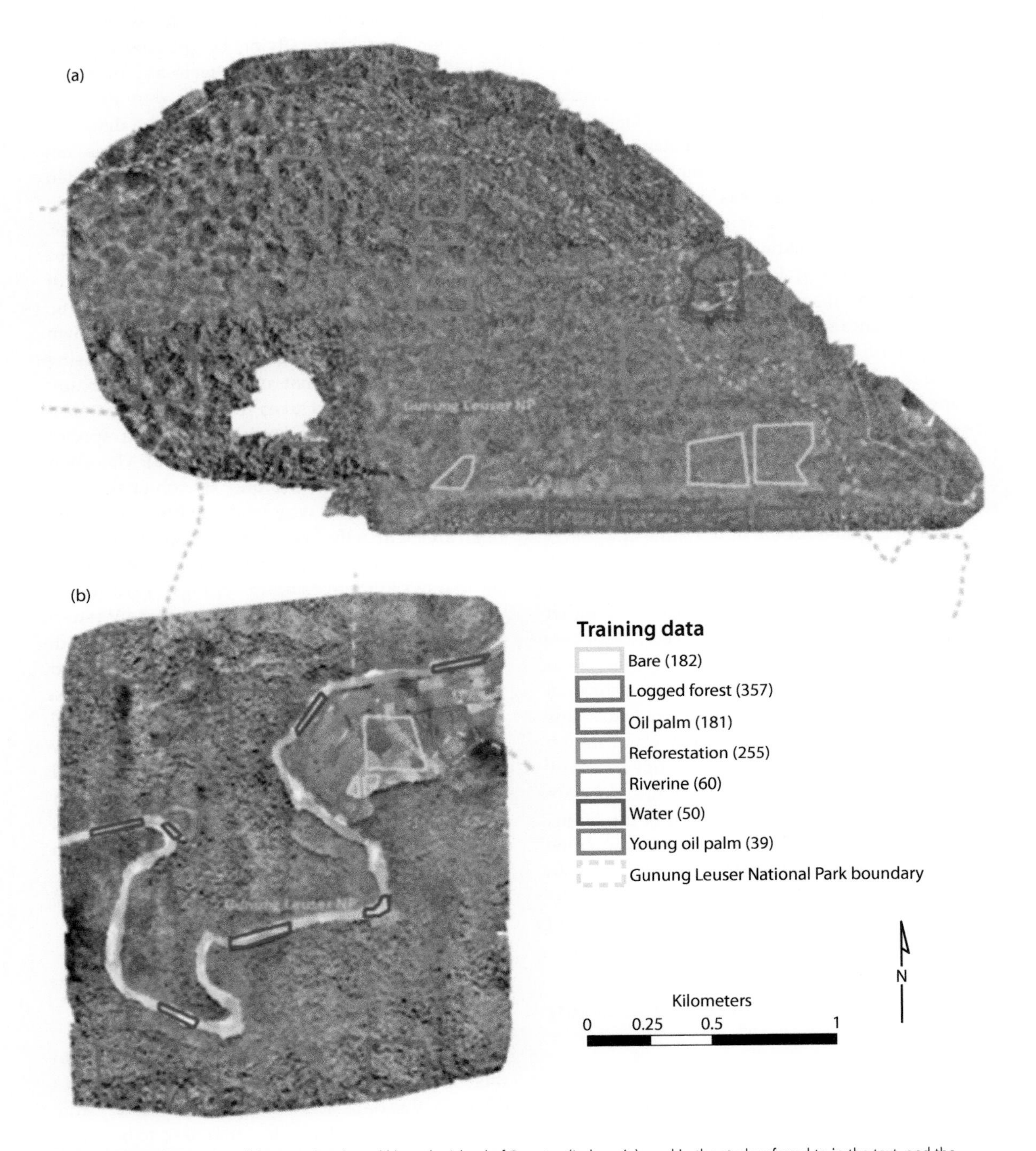

Figure 7.11 Orthomosaics of the two sites (a and b) on the island of Sumatra (Indonesia) used in the study referred to in the text, and the corresponding training data polygons. Total area of each class is expressed in number of Landsat pixels (Szantoi, Smith et al. 2017). Reprinted with permission from Taylor & Francis Ltd (www.tandfonline.com).

images, manually sifting through images is a huge time investment that reduces the efficiency of using drones. There is thus a strong need to automate as much of the process as possible. The development of computer algorithms allowing for object detection and tracking in combination with the increase and lower costs of processing power on computers is favoring automated detection, identification, and tracking of wildlife from drone imagery. Although there have been decades of research on object detection, identification, and tracking, the majority of this work has been conducted in situations in which there are stationary cameras with objects moving over a stable background. These algorithms generally do not work well with drones due to the fact that the camera on the drones moves as well as the animals. In addition animals often lack a stable contour due to the movements of various body parts in relation to each other and often are camouflaged, which makes detection more challenging (Fang, Du et al. 2016). Nevertheless there are a modest, but promising, number of studies that have focused on automating detection, identification, or tracking of animals in drone images from various sensors (mainly RGB and thermal). There are at least three main aspects that would benefit from automation. First, the detection of animals in images would increase effectiveness of surveys. Second, in cases where multiple species could be detected there is in most cases a need to be able to identify the various species. Third, with the increased flight times of drones tracking of animals becomes of interest so after the detection and potentially the identification steps there is a need to develop algorithms that can track individual animals based on the real-time data from a sensor. Tracking of animals also needs the integration of the tracking algorithm with the flight controller so that the drone can track the animal of interest.

The majority of studies compared manual counts either from the ground or from the images to the automatically detected count (Table 7.4) and thus were mainly concerned with the first aspect as outlined above. A variety of different algorithms have been used to detect animals (Table 7.4). Quite basic methods to detect animals are to use spectral thresholding in which individual pixels of a certain reflectance are separated from the other pixels or to use an interval between two values which should be isolated. Such approaches are most applicable when the reflectance difference between the animal and background is high or when the spectral interval of the animal is not part of the background. Such methods can be very accurate as indicated by research on snow geese (Chabot and Bird 2012). These and other pixel-based methods (Table 7.4) do not work well if the background and the animals or their signs (e.g. nests) are not clearly separated spectrally. Object-based image analyses (OBIA) in which an image is segmented in objects (see Figure 7.7 for a land-cover example) and the analyses incorporate both spectral and spatial information might then improve detection. This approach has been used to detect white-tailed deer in comparison to (un)supervised pixel-based classification methods (Chrétien, Théau et al. 2016). The results indicated that the (un)supervised pixel-based method had a high number of false positives, even with RGB and thermal combined, but that these approaches did detect all detectable deer (i.e. excluding those under the canopy) in the case of the unsupervised approach and 9–10 deer out of 15 with the supervised approach. The high number of false positives in the pixel-based approach was due to areas of the forest floor that had similar spectral signatures to the deer and made the pixel-based approach not useful. The OBIA (RGB and thermal combined) performed best and all detectable deer were detected with no false positives or false negatives. Nevertheless this method did not lead to the detection of all animals in the study because animals under the canopy were not detected. Thus overall detection rate was 0.52 (16 out of 31 deer). This study also examined the importance of spatial resolution on the detection rate when different bands and combinations thereof were used. Generally the pixel-based approached performed better at a lower pixel resolution and the OBIA approach at a higher pixel resolution. Although the OBIA approach had a higher accuracy the approach itself is time intensive due to the time needed to find the most suitable threshold settings, which can take several days (Chrétien, Théau et al. 2016). A promising approach to combine a segmentation step with recent developments in classification is to use Deep Convolutional Neural Networks (DCNNs) to classify the outputs from the segmentation step

(Maire, Alvarez et al. 2015). Maire, Alvarez et al.'s (2015) study used this approach to improve the recall and precision obtained with other methods (Maire, Mejias et al. 2013).

As indicated above, pixel-based approaches can lead to false positives if pixels in the background have the same reflectance as the reflectance of the animal or sign of interest. Using a different or larger part of the electromagnetic spectrum could improve detection and therefore a number of studies have used thermal sensors in isolation or in combination with RGB because the animals of interest were homeothermic endotherms which radiate heat due to the thermoregulation of their body temperature (Chrétien, Théau et al. 2016, Gonzalez, Montes et al. 2016, Ward, Hensler et al. 2016, Longmore, Collins et al. 2017; Figure 7.12).

Although the thermal part of the spectrum can facilitate detection, it is not without challenges. Other objects in the area such as rocks, tree trunks, or other vegetation can have a similar radiative temperature and thus at certain times of a day at least can lead to false positives (Franke, Goll et al. 2012, Lhoest, Linchant et al. 2015). In some cases researchers have clipped the parts of the image containing the animals to facilitate detection (Lhoest, Linchant et al. 2015), but this of course increases the time input from humans and ideally this part would be automated as well. Other aspects that can influence the emission of heat are physical aspects of the animal such as their activity and thickness of the coat (winter vs summer coat) and meteorological conditions such as temperature, cloud cover, and exposure to sunlight (Croon, McCullough et al. 1968). In some cases also the orientation of parts of the body such as the head in relation to the body can lower detection success (Lhoest, Linchant et al. 2015).

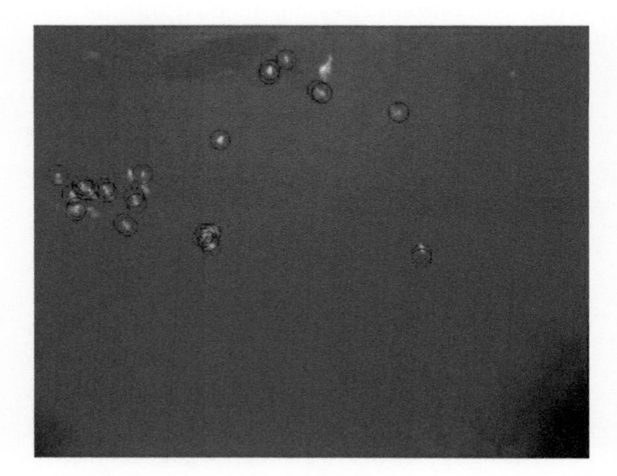

Figure 7.12 Example of automated detection of baboons at Knowsley Safari, UK using the algorithms described in Longmore, Collins et al. (2017). © Steve Longmore and Maisie Rashman.

However, as with manual counts of animals from the air, the most challenging aspect of using thermal and/or RGB sensors remains that animals under dense vegetation or under water cannot be detected and that even relatively open vegetation can reduce detection because the animal is not complete on the image, which poses difficulties for 'reconstruction' of the animal from the parts that are detected. Even in open areas two animals might be detected as one or one animal might be divided into two (Fang, Du et al. 2016). The fact that animals or their signs are missed under thick vegetation or under water has led researchers to develop correction factors to compensate for the missing fraction (Buckland, Anderson et al. 2001, Buckland, Anderson et al. 2004) which can also be applied to automated counts from drones (Lhoest, Linchant et al. 2015).

Almost all studies using automated detection of animals have focused on single species. Although this is a logical starting point, the next step is to explore identification after the initial detection step. Many species occur in areas where multiple species occur and in most cases researchers and conservationists will be interested in detecting and identifying the various species. This will certainly be challenging as identifying the varying species will require much more fine-grained analyses where more variables that characterize animals are incorporated. It is likely that a combination of sensors will be most successful to achieve this goal due to more bands in the electromagnetic spectrum being available for analyses.

Another important aspect of animal monitoring is tracking of animals for behavioral studies. This is now mostly done on foot or car, but with the increasing flight duration of drones it is possible that in the future drones will follow animals for prolonged periods of time. To achieve this drones

Table 7.4 Examples of studies using automated detection for animals

Type of object(s)	Sensor	Analyses/software	Automated count errors*	Source
Cows and humans	Thermal	HoG (histogram of oriented gradients) and SVM (Support Vector Machine)/Python	80 m: 30% 120 m: 90%****	(Longmore, Collins et al. 2017)
Grey seal	Thermal	Spectral thresholds, pixel cluster size sorting for individuals and integrated object recognition and high pass filtering (i.e. edge detection) to discriminate individuals within closely packed aggregations/ArcGIS	2–5%	(Seymour, Dale et al. 2017)
Koalas	Thermal	Pixel Intensity Threshold and Template Matching Binary Mask algorithms	20 m: 0% 30 m: 0% 60 m: 0–16.7%	(Gonzalez, Montes et al. 2016)
White-tailed deer	Visible and thermal	(Un)supervised pixel-based and object-based classification/eCognition	All flights combined: 48% for (RGB + thermal and OBIA)	(Chrétien, Théau et al. 2016)
Zebras and antelopes	RGB	Optical flow algoritm/NA	Zebras: 18% Antelope: 11%	(Fang, Du et al. 2016)
Dog	Thermal	Blob detection/Python	0%***	(Ward, Hensler et al. 2016)
Hippopotamus	Thermal	Local maxima and isolines/QGIS	2.3% (range −9.8 to +13.7) depending on flight altitude	(Lhoest, Linchant et al. 2015)
Spoonbills	RGB	Unsupervised spectral classification/ENVI	50 m: 0% 200 m: 0.7% 300 m 0.7%	(Liu, Chen et al. 2015)
Dugongs	RGB	Deep Convolutional Neural Networks (DCNNs)/Python	Recall: 80% Precision: 27%	(Maire, Alvarez et al. 2015)
Cows	RGB	Several algorithms/Matlab	40.0–97.6% recall, 26.0–66.0% precision	(van Gemert, Verschoor et al. 2014)
Orangutan nests	RGB	Active detection framework/NA	~72–96% recall, 13–33% precision	(Chen, Shioi et al. 2014)
Chickens	Thermal	Template matching/NA	3–10 m: 0.1–6.0% 10–20 m: 25.1–29.0%	(Christiansen, Steen et al. 2014)
Deer, cattle, and humans	RGB	Algorithm to detect moving animals: feature point extraction, corresponding point identification, and detection of wild moving animals using image comparison/ENVI	Cows: 16.7% Deer: 50%	(Oishi and Matsunaga 2014)
Cattle	RGB	Convolutional Neural Network/NA	2.5%	(Chamoso, Raveane et al. 2014)
Common gull	RGB	Supervised pixel classification/ArcGIS	Year 1: 4.8% Year 2: 2.4%	(Grenzdörffer 2013)

(continued)

Table 7.4 Continued

Type of object(s)	Sensor	Analyses/software	Automated count errors*	Source
Dugongs	RGB	Two step approach 1) colour and morphological determination; 2) comparison with shape templates/Matlab	69.4–75.4% recall, 30–87.5% precision depending on sea conditions	(Maire, Mejias et al. 2013)
Humans	RGB	Algorithm to detect moving animals: artificial and real optical flow differentiating algorithm	Not provided	(Rodríguez-Canosa, Thomas et al. 2012)
Snow geese	RGB	Spectral thresholding/Photoshop	Flock 1: 0.9% Flock 2: 0.8% Sample total: 28.5%	(Chabot and Bird 2012)
Whales	RGB	Object recognition based on hue and saturation/NA	98.9% recall**	(Selby, Corke et al. 2011)
Wading birds	BW	Template matching/Intel Open Source Computer Vision Library and C++	Sample mean: 10.5% Sample total: 6.3%	(Abd-Elrahman 2005)

* = where feasible automated count errors followed the procedure from Chabot and Francis in which the percentage difference of automated and manual counts is calculated (Chabot and Francis 2016). We also followed their sample mean as being the mean count error in multiple images and their sample total as being the overall error of the total count summed from all samples. ** = if standardizing was not feasible the numbers presented in the papers were used. These are then indicated by **. NA = not mentioned. Two flights were conducted in which one dog was in a field. **** = the accuracies are for single frames. For the complete sequence of frames all cows were detected.

need to be able to detect the individual animal (or in some cases herds of animals) and then follow the animal as it moves around its habitat. Several studies have focused on the detection and tracking aspect (Olivares-Mendez, Fu et al. 2015, Ward, Hensler et al. 2016) and even off-the-shelf multirotor systems now allow for a pilot to lock in on an object it would like to follow and the drone will follow that object. It is therefore expected that active tracking of animals (or other moving objects) with drones will become available on more systems in the near future, allowing researchers and conservationists to track moving animals in open areas.

CHAPTER 8

Future casting

8.1 Power systems

As discussed in Chapter 2, a key component of a drone system is its power source, which is usually a rechargeable Li-ion or LiPo battery. The energy density of current batteries and thus the drone's flight endurance is still a major constraint on the ability of drones to realize their potential, such as accessing remote locations, patrolling large tracts of protected areas, and spotting sparsely distributed wildlife species. Although lithium battery technology has steadily improved over the past few decades, there is growing consensus that there might be a limit to the improvement in energy storage density of lithium batteries (Schlachter 2013).

There are significant developments in the use of solar cells to supplement batteries for powering drones. The most famous of such drones arguably is the Solara 50 that has been developed by Titan Aerospace. This drone was designed to fly at an altitude of 65,000 feet to collect scientific data, functioning as an 'atmospheric satellite'. Another notable example is a small 6.9 kg hand-launchable drone developed by a Swiss group (Figure 8.1). The AtlantikSolar completed an 81-hour continuous flight, covering a total distance of 2338 km (Oettershagen, Melzer et al. 2017). The ability of such drones for sustained flight also makes them amenable to act as communications platforms to bring the internet to remote locations in various parts of the world. Although solar powered drones are currently prohibitively expensive for most conservation applications, there is hope that their development for industrial applications will have trickle-down benefits for the conservation sector in the near future.

The drawback of solar powered drones is that their endurance would suffer when operated at low altitudes under less than ideal weather conditions. Another power technology being looked at to fill this niche is hydrogen fuel cells (Gong and Verstraete 2017). Several research laboratories around the world are actively developing this technology, including the American University of Sharjah (AUS 2016). The hydrogen fuel cell system typically comprises four components: the hydrogen tank, control system, fuel reactor, and a LiPo battery (Gong and Verstraete 2017). A battery is still needed to provide power during periods of high energy requirement such as take-off and high speed flights. Hydrogen fuel cell technology enables multirotor drones to extend their flight times to up to four hours.

8.2 Autonomous systems

Conservation drones are most effective if they can perform missions with some degree of autonomy. At the moment, that often involves mainly programming a set of predetermined geographical waypoints (coordinates) that the drone would fly to as an 'autonomous' mission. In some cases, the height and speed of the flight can also be programmed into the mission. Even then these are not truly autonomous missions as all decision making with regards to the mission still rests with the ground operator.

Conservation Drones: Mapping and Monitoring Biodiversity. Serge A. Wich & Lian Pin Koh. Oxford University Press (2018).
 DOI:10.1093/oso/9780198787617.001.0001

Figure 8.1 The AtlantikSolar fixed wing drone developed by the Swiss Federal Institute of Technology, ETH Zurich. © AtlantikSolar.ethz.ch.

For conservation drones to realize their true potential, they need to be capable of navigating difficult terrain and other challenging environmental conditions without human input. Recent developments in the field of Computer Science for Simultaneous Localization and Mapping (SLAM) are pushing the technical boundaries for achieving autonomous flight (López, García et al. 2017). These developments are particularly important in terms of enabling reliable navigation in GPS-denied environments, which would open up a new world of drone applications under the forest canopy.

The technology for the simultaneous and autonomous operation of multiple units of drones is also rapidly progressing. Such swarming systems are already being deployed in the entertainment industry (Kim, Jeong et al. 2017), and actively being developed for humanitarian applications (Yanmaz, Quaritsch et al. 2017). In the context of conservation, a swarm of drones could increase the efficiency of habitat mapping, wildlife spotting, and even cooperative tracking of radio-collared animals (Koohifar, Kumbhar et al. 2017).

8.3 Platform integration

At the end of the day, conservation drones are just one of the many tools available for the conservation biologist to achieve her objectives. Traditional sampling methods, such as field inventory plots and ground-based transect surveys, will continue to be important for biologists collecting local-scale data. At the same time, aside from conservation drones, there is an increasing number of emerging technologies that are actively being used for conservation research and applications, including smart parks (Toor 2017), satellite and airborne remote sensing (Asner, Knapp et al. 2014, Skidmore, Pettorelli et al. 2015), radio telemetry (Cagnacci, Boitani et al. 2010, Kays, Crofoot et al. 2015), camera trapping (Lynam, Jenks et al. 2013, Burton, Neilson et al. 2015), wireless sensor networks (Collins,

Bettencourt et al. 2006, Benson, Bond et al. 2010), and mobile device-based citizen science (Teacher, Griffiths et al. 2013, Olson, Bissonette et al. 2014).

Each platform has its limitations, especially in terms of the spatial and temporal scales at which they can be deployed most cost effectively. For example, in the context of monitoring landscapes to keep track of ecological dynamics, satellite imagery could be used for generating coarse resolution land-cover maps and data, while airborne imaging systems can produce detailed baseline vegetation maps to identify specific areas of interest to deploy conservation drones to acquire high resolution imagery containing valuable ecology information (Marvin, Koh et al. 2016). In this way, the sum of these technologies is likely to be greater than its parts, ultimately resulting in more cost-effective conservation strategies.

References

Aasen, H., A. Burkart, A. Bolten, and G. Bareth (2015). 'Generating 3D hyperspectral information with lightweight UAV snapshot cameras for vegetation monitoring: From camera calibration to quality assurance.' *ISPRS Journal of Photogrammetry and Remote Sensing* **108**: 245–59.

Abd-Elrahman, A., L. Pearlstine, and F. Percival (2005). 'Development of pattern recognition algorithm for automatic bird detection from unmanned aerial vehicle imagery.' *Surveying and Land Information Science* **65**: 37–45.

Acevedo-Whitehouse, K., A. Rocha-Gosselin, and D. Gendron (2010). 'A novel non-invasive tool for disease surveillance of free-ranging whales and its relevance to conservation programs.' *Animal Conservation* **13**(2): 217–25.

Al-Bakri, J., J. Taylor, and T. Brewer (2001). 'Monitoring land use change in the Badia transition zone in Jordan using aerial photography and satellite imagery.' *The Geographical Journal* **167**(3): 248–62.

Ambrosia, V., S. Wegener, T. Zajkowski, D. Sullivan, S. Buechel, F. Enomoto, B. Lobitz, S. Johan, J. Brass, and E. Hinkley (2011). 'The Ikhana unmanned airborne system (UAS) western states fire imaging missions: from concept to reality (2006–2010).' *Geocarto International* **26**(2): 85–101.

Andersen, D. E., O. J. Rongstad, and W. R. Mytton (1989). 'Response of nesting red-tailed hawks to helicopter overflights.' *Condor* **91**: 296–9.

Anderson, K. (2016). 'Integrating multiple scales of remote sensing measurement – from satellites to kites.' *Progress in Physical Geography* **40**(2): 187–95.

Anderson, K., D. Griffiths, L. DeBell, S. Hancock, J. P. Duffy, J. D. Shutler, W. J. Reinhardt, and A. Griffiths (2016). 'A grassroots remote sensing toolkit using live coding, smartphones, kites and lightweight drones.' *PLoS ONE* **11**(5): e0151564.

Arias, A. and R. L. Pressey (2016). 'Combatting illegal, unreported, and unregulated fishing with information: A case of probable illegal fishing in the tropical Eastern Pacific.' *Frontiers in Marine Science* **3**(13). doi.org/10.3389/fmars.2016.00013

Aryal, A., K. P. Acharya, U. B. Shrestha, M. Dhakal, D. Raubenhiemer, and W. Wright (2017). 'Global lessons from successful rhinoceros conservation in Nepal.' *Conservation Biology* **31**(6): 1494–7.

Asner, G. P., D. E. Knapp, R. E. Martin, R. Tupayachi, C. B. Anderson, J. Mascaro, F. Sinca, K. D. Chadwick, M. Higgins, W. Farfan, W. Llactayo, and M. R. Silman (2014). 'Targeted carbon conservation at national scales with high-resolution monitoring.' *Proceedings of the National Academy of Sciences USA* **111**(47): E5016–22.

AUS (2016). 'AUS conducts hydrogen fuel cell multi-rotor drone flight.' from https://www.aus.edu/media/news/aus-conducts-hydrogen-fuel-cell-multi-rotor-drone-flight.

Baer, D., M. Gupta, J. B. Leen, and E. Berman (2012). 'Environmental and atmospheric monitoring using off-axis integrated cavity output spectroscopy (OA-ICOS).' *Am Lab* **44**(10).

Baluja, J., M. P. Diago, P. Balda, R. Zorer, F. Meggio, F. Morales, and J. Tardaguila (2012). 'Assessment of vineyard water status variability by thermal and multispectral imagery using an unmanned aerial vehicle (UAV).' *Irrigation Science* **30**(6): 511–22.

Barasona, J. A., M. Mulero-Pázmány, P. Acevedo, J. J. Negro, M. J. Torres, C. Gortázar, and J. Vicente (2014). 'Unmanned aircraft systems for studying spatial abundance of ungulates: relevance to spatial epidemiology.' *PLoS ONE* **9**(12): e115608.

Barazzetti, L., F. Remondino, M. Scaioni, and R. Brumana (2010). *Fully automatic UAV image-based sensor orientation*. Proceedings of the 2010 Canadian Geomatics Conference and Symposium of Commission I, Calgary, AB, Canada.

Batut, A. (1890). *La photographie aérienne par cerf-volant*. Paris, Gauthier-Villars.

Bendig, J., A. Bolten, S. Bennertz, J. Broscheit, S. Eichfuss, and G. Bareth (2014). 'Estimating biomass of barley using crop surface models (CSMs) derived from UAV-based RGB imaging.' *Remote Sensing* **6**(11): 10395.

Bennett, A., D. Barrett, V. Preston, J. Woo, S. Chandra, D. Diggins, R. Chapman, A. Wee, Z. Wang, and M. Rush (2015). *Autonomous vehicles for remote sample collection: Enabling marine research.* OCEANS 2015-Genova, IEEE.

Benson, B. J., B. J. Bond, M. P. Hamilton, R. K. Monson, and R. Han (2010). 'Perspectives on next-generation technology for environmental sensor networks.' *Frontiers in Ecology and the Environment* **8**(4): 193–200.

Berni, J. A. J., P. J. Zarco-Tejada, G. Sepulcre-Cantó, E. Fereres, and F. Villalobos (2009a). 'Mapping canopy conductance and CWSI in olive orchards using high resolution thermal remote sensing imagery.' *Remote Sensing of Environment* **113**(11): 2380–8.

Berni, J. A. J., P. J. Zarco-Tejada, L. Suarez, and E. Fereres (2009b). 'Thermal and narrowband multispectral remote sensing for vegetation monitoring from an unmanned aerial vehicle.' *IEEE Transactions on Geoscience and Remote Sensing* **47**(3): 722–38.

Berra, E., S. Gibson-Poole, A. MacArthur, R. Gaulton, and A. Hamilton (2015). 'Estimation of the spectral sensitivity functions of un-modified and modified commercial off-the-shelf digital cameras to enable their use as a multispectral imaging system for UAVs.' *International Archives of Photogrammetry, Remote Sensing and Spatial Information Sciences* **40**(1): 207.

Bevan, E., T. Wibbels, B. M. Najera, M. A. Martinez, L. A. Martinez, F. I. Martinez, J. M. Cuevas, T. Anderson, A. Bonka, and M. H. Hernandez (2015). 'Unmanned aerial vehicles (UAVs) for monitoring sea turtles in near-shore waters.' *Marine Turtle Newsletter* **145**: 19–22.

Biggs, D., F. Courchamp, R. Martin, and H. P. Possingham (2013). 'Legal trade of Africa's rhino horns.' *Science* **339**(6123): 1038–9.

Bleich, V. C., R. T. Bowyer, A. M. Pauli, M. C. Nicholson, and R. W. Anthes (1994). 'Mountain sheep *Ovis canadensis* and helicopter surveys: ramifications for the conservation of large mammals.' *Biological Conservation* **70**(1): 1–7.

Bogacki, M., M. Giersz, P. Przadka-Giersz, W. Malkowski, and K. Misiewicz (2010). *GPS RTK mapping, kite aerial photogrammetry, geophysical survey and GIS based analysis of surface artifact distribution at the pre-Hispanic site of the Castillo de Huarmey, north coast of Peru.* Proceedings of 30th EARSeL Symposium: Remote Sensing for Science, Education and Culture.

Boike, J. and K. Yoshikawa (2003). 'Mapping of periglacial geomorphology using kite/balloon aerial photography.' *Permafrost and Periglacial Processes* **14**(1): 81–5.

Born, E. W., F. F. Riget, R. Dietz, and D. Andriashek (1999). 'Escape responses of hauled out ringed seals (*Phoca hispida*) to aircraft disturbance.' *Polar Biology* **21**(3): 171–8.

Borrelle, S. B. and A. T. Fletcher (2017). 'Will drones reduce investigator disturbance to surface-nesting seabirds?' *Marine Ornithology* **45**: 89–94.

Bosak, K. (2013). *Secrets of photomapping*. Technical report, Pteryx UAV by Trigger Composites.

Breckenridge, R. P. and M. E. Dakins (2011). 'Evaluation of bare ground on rangelands using unmanned aerial vehicles: a case study.' *GIScience & Remote Sensing* **48**(1): 74–85.

Bréda, N. J. J. (2003). 'Ground-based measurements of leaf area index: a review of methods, instruments and current controversies.' *Journal of Experimental Botany* **54**(392): 2403–17.

Broich, M., M. Hansen, F. Stolle, P. Potapov, B. A. Margono, and B. Adusei (2011). 'Remotely sensed forest cover loss shows high spatial and temporal variation across Sumatera and Kalimantan, Indonesia 2000–2008.' *Environmental Research Letters* **6**(1): 014010.

Brooke, S., D. Graham, T. Jacobs, C. Littnan, M. Manuel, and R. O'Conner (2015). 'Testing marine conservation applications of unmanned aerial systems (UAS) in a remote marine protected area.' *Journal of Unmanned Vehicle Systems* **3**(4): 237–51.

Brush, J. and A. Watts (2008). *An assessment of autonomous unmanned aircraft systems (UAS) for avian surveys. Florida Fish and Wildlife Conservation Commission.* Fish and Wildlife Research Institute, Wildlife Research Section, Gainesville, USA.

Bryson, M., M. Johnson-Roberson, R. J. Murphy, and D. Bongiorno (2013). 'Kite aerial photography for low-cost, ultra-high spatial resolution multi-spectral mapping of intertidal landscapes.' *PLoS ONE* **8**(9): e73550.

Bryson, M., A. Reid, F. Ramos, and S. Sukkarieh (2010). 'Airborne vision-based mapping and classification of large farmland environments.' *Journal of Field Robotics* **27**(5): 632–55.

Buckland, S. T., D. R. Anderson, K. P. Burnham, J. L. Laake, D. L. Borchers, and L. Thomas (2001). *Introduction to Distance Sampling*. Oxford, Oxford University Press.

Buckland, S. T., D. R. Anderson, K. P. Burnham, J. L. Laake, D. L. Borchers, and L. Thomas (2004). *Advanced Distance Sampling: Estimating Abundance of Biological Populations*. Oxford, Oxford University Press.

Burton, A. C., E. Neilson, D. Moreira, A. Ladle, R. Steenweg, J. T. Fisher, E. Bayne, and S. Boutin (2015). 'Wildlife camera trapping: a review and recommendations for linking surveys to ecological processes.' *Journal of Applied Ecology* **52**(3): 675–85.

Cagnacci, F., L. Boitani, R. A. Powell, and M. S. Boyce (2010). 'Animal ecology meets GPS-based radiotelemetry: a perfect storm of opportunities and challenges.' *Philosophical Transactions of the Royal Society B: Biological Sciences* **365**(1550): 2157–62.

Calderón, R., J. A. Navas-Cortés, C. Lucena, and P. J. Zarco-Tejada (2013). 'High-resolution airborne hyperspectral and thermal imagery for early detection of

Verticillium wilt of olive using fluorescence, temperature and narrow-band spectral indices.' *Remote Sensing of Environment* **139**: 231–45.

Campos Cerqueira, M. and M. T. Aide (2016). 'Improving distribution data of threatened species by combining acoustic monitoring and occupancy modeling.' *Methods in Ecology and Evolution* **7**: 1340–8. doi: 10.1111/2041-210X.12599

Canal, D., M. Mulero-Pázmány, J. J. Negro, and F. Sergio (2016). 'Decoration increases the conspicuousness of raptor nests.' *PLoS ONE* **11**(7): e0157440.

Caughley, G. (1977). 'Sampling in aerial survey.' *The Journal of Wildlife Management* **41**: 605–15.

Cawthorn, M. (1993). *Census and Population Estimation of Hooker's Sea Lions at the Auckland Islands, December 1992 – February 1993.* Wellington, New Zealand, Department of Conservation, Technical series No. 2.

Chabot, D. (2009). *Systematic evaluation of a stock unmanned aerial vehicle (UAV) for small-scale wildlife survey applications.* PhD thesis, McGill University, Montreal.

Chabot, D. and D. M. Bird (2012). 'Evaluation of an off-the-shelf unmanned aircraft system for surveying flocks of geese.' *Waterbirds* **35**(1): 170–4.

Chabot, D. and D. M. Bird (2013). 'Small unmanned aircraft: precise and convenient new tools for surveying wetlands.' *Journal of Unmanned Vehicle Systems* **1**(1): 15–24.

Chabot, D. and D. M. Bird (2015). 'Wildlife research and management methods in the 21st century: Where do unmanned aircraft fit in?' *Journal of Unmanned Vehicle Systems* **3**(4): 137–55.

Chabot, D., V. Carignan, and D. M. Bird (2014). 'Measuring habitat quality for least bitterns in a created wetland with use of a small unmanned aircraft.' *Wetlands* **34**(3): 527–33.

Chabot, D., S. R. Craik, and D. M. Bird (2015). 'Population census of a large common tern colony with a small unmanned aircraft.' *PLoS ONE* **10**(4): e0122588.

Chabot, D. and C. M. Francis (2016). 'Computer-automated bird detection and counts in high-resolution aerial images: a review.' *Journal of Field Ornithology* **87**(4): 343–59. doi: 10.1111/jofo.12171

Chamoso, P., W. Raveane, V. Parra, and A. González (2014). UAVs applied to the counting and monitoring of animals. *Ambient Intelligence: Software and Applications.* Cham, Springer: 71–80.

Chen, J., S. Yi, Y. Qin, and X. Wang (2016). 'Improving estimates of fractional vegetation cover based on UAV in alpine grassland on the Qinghai–Tibetan Plateau.' *International Journal of Remote Sensing* **37**(8): 1922–36.

Chen, J. M., G. Pavlic, L. Brown, J. Cihlar, S. Leblanc, H. White, R. Hall, D. Peddle, D. King, and J. Trofymow (2002). 'Derivation and validation of Canada-wide coarse-resolution leaf area index maps using high-resolution satellite imagery and ground measurements.' *Remote Sensing of Environment* **80**(1): 165–84.

Chen, Y., H. Shioi, C. F. Montesinos, L. P. Koh, S. Wich, and A. Krause (2014). *Active detection via adaptive submodularity.* Proceedings of The 31st International Conference on Machine Learning.

Chisholm, R. A., J. Cui, S. K. Lum, and B. M. Chen (2013). 'UAV LiDAR for below-canopy forest surveys.' *Journal of Unmanned Vehicle Systems* **1**(1): 61–8.

Chrétien, L., J. Théau, and P. Ménard (2015). 'Wildlife multispecies remote sensing using visible and thermal infrared imagery acquired from an unmanned aerial vehicle (UAV).' *International Archives of Photogrammetry, Remote Sensing and Spatial Information Sciences* **40**(1): 241.

Chrétien, L. P., J. Théau, and P. Ménard (2016). 'Visible and thermal infrared remote sensing for the detection of white-tailed deer using an unmanned aerial system.' *Wildlife Society Bulletin* **40**: 181–91.

Christiansen, P., K. A. Steen, R. N. Jørgensen, and H. Karstoft (2014). 'Automated detection and recognition of wildlife using thermal cameras.' *Sensors* **14**(8): 13778–93.

Christie, K. S., S. L. Gilbert, C. L. Brown, M. Hatfield, and L. Hanson (2016). 'Unmanned aircraft systems in wildlife research: current and future applications of a transformative technology.' *Frontiers in Ecology and the Environment* **14**(5): 241–51.

Cliff, O. M., R. Fitch, S. Sukkarieh, D. L. Saunders, and R. Heinsohn (2015). 'Online localization of radio-tagged wildlife with an autonomous aerial robot system.' *Proceedings of Robotics Science and Systems XI*: 13–17.

Coetzee, B. W. T. and S. L. Chown (2015). 'A meta-analysis of human disturbance impacts on Antarctic wildlife.' *Biological Reviews* **91**: 578–96.

Collins, S. L., L. M. A. Bettencourt, A. Hagberg, R. F. Brown, D. I. Moore, G. Bonito, K. A. Delin, S. P. Jackson, D. W. Johnson, S. C. Burleigh, R. R. Woodrow, and J. M. McAuley (2006). 'New opportunities in ecological sensing using wireless sensor networks.' *Frontiers in Ecology and the Environment* **4**: 402–7.

Colomina, I. and P. Molina (2014). 'Unmanned aerial systems for photogrammetry and remote sensing: A review.' *ISPRS Journal of Photogrammetry and Remote Sensing* **92**: 79–97.

Cooke, S. J., V. M. Nguyen, S. T. Kessel, N. E. Hussey, N. Young, and A. T. Ford (2017). 'Troubling issues at the frontier of animal tracking for conservation and management.' *Conservation Biology* **31**(5):1205–7. doi: 10.1111/cobi.12895

Córcoles, J. I., J. F. Ortega, D. Hernández, and M. A. Moreno (2013). 'Estimation of leaf area index in onion (*Allium cepa* L.) using an unmanned aerial vehicle.' *Biosystems Engineering* **115**(1): 31–42.

Croon, G. W., D. R. McCullough, C. E. Olson Jr, and L. M. Queal (1968). 'Infrared scanning techniques for big game censusing.' *The Journal of Wildlife Management* **32**: 751–9.

Crutsinger, G. M., J. Short, and R. Sollenberger (2016). *The future of UAVs in ecology: an insider perspective from the Silicon Valley drone industry.* Ottawa, NRC Research Press.

Cunliffe, A. M., R. E. Brazier, and K. Anderson (2016). 'Ultra-fine grain landscape-scale quantification of dryland vegetation structure with drone-acquired structure-from-motion photogrammetry.' *Remote Sensing of Environment* **183**: 129–43.

Dandois, J. P. and E. C. Ellis (2010). 'Remote sensing of vegetation structure using computer vision.' *Remote Sensing* **2**(4): 1157–76.

Dandois, J. P. and E. C. Ellis (2013). 'High spatial resolution three-dimensional mapping of vegetation spectral dynamics using computer vision.' *Remote Sensing of Environment* **136**: 259–76.

Dandois, J. P., M. Olano, and E. C. Ellis (2015). 'Optimal altitude, overlap, and weather conditions for computer vision UAV estimates of forest structure.' *Remote Sensing* **7**(10): 13895–920.

Davies, A. B. and G. P. Asner (2014). 'Advances in animal ecology from 3D-LiDAR ecosystem mapping.' *Trends in Ecology & Evolution* **29**(12): 681–91.

De Biasio, M., T. Arnold, R. Leitner, G. McGunnigle, and R. Meester (2010). *UAV-based environmental monitoring using multi-spectral imaging.* SPIE Defense, Security, and Sensing, International Society for Optics and Photonics.

Delaney, D. K., T. G. Grubb, P. Beier, L. L. Pater, and M. H. Reiser (1999). 'Effects of helicopter noise on Mexican spotted owls.' *The Journal of Wildlife Management* **63**(1): 60–76.

Díaz-Varela, R. A., R. de la Rosa, L. León, and P. J. Zarco-Tejada (2015). 'High-resolution airborne UAV imagery to assess olive tree crown parameters using 3D photo reconstruction: Application in breeding trials.' *Remote Sensing* **7**(4): 4213–32.

Ditmer, M. A., J. B. Vincent, L. K. Werden, J. C. Tanner, T. G. Laske, P. A. Iaizzo, D. L. Garshelis, and J. R. Fieberg (2015). 'Bears show a physiological but limited behavioral response to unmanned aerial vehicles.' *Current Biology* **25**(17): 2278–83.

Duan, S.-B., Z.-L. Li, B.-H. Tang, H. Wu, L. Ma, E. Zhao, and C. Li (2013). 'Land surface reflectance retrieval from hyperspectral data collected by an unmanned aerial vehicle over the Baotou test site.' *PLoS ONE* **8**(6): e66972.

Duan, S.-B., Z.-L. Li, H. Wu, B.-H. Tang, L. Ma, E. Zhao, and C. Li (2014). 'Inversion of the PROSAIL model to estimate leaf area index of maize, potato, and sunflower fields from unmanned aerial vehicle hyperspectral data.' *International Journal of Applied Earth Observation and Geoinformation* **26**: 12–20.

Duffy, J. P. and K. Anderson (2016). 'A 21st-century renaissance of kites as platforms for proximal sensing.' *Progress in Physical Geography* **40**(2): 352–61.

Duffy, J. P., A. M. Cunliffe, L. DeBell, C. Sandbrook, S. A. Wich, J. D. Shutler, I. H. Myers-Smith, M. R. Varela, and K. Anderson (2017). 'Location, location, location: Considerations when using lightweight drones in challenging environments.' *Remote Sensing in Ecology and Conservation*. doi: 10.1002/rse2.58

Duffy, R. (2016a). '6 The illegal wildlife trade in global perspective.' *Handbook of Transnational Environmental Crime*. L. Elliott and W. Schaedla. Cheltenham, UK, Edward Elgar: 109–28.

Duffy, R. (2016b). 'War, by conservation.' *Geoforum* **69**: 238–48.

Dulava, S., W. T. Bean, and O. M. Richmond (2015). 'Applications of unmanned aircraft systems (UAS) for waterbird surveys.' *Environmental Practice* **17**(03): 201–10.

Dunford, R., K. Michel, M. Gagnage, H. Piégay, and M.-L. Trémelo (2009). 'Potential and constraints of Unmanned Aerial Vehicle technology for the characterization of Mediterranean riparian forest.' *International Journal of Remote Sensing* **30**(19): 4915–35.

Durban, J., H. Fearnbach, L. Barrett-Lennard, W. Perryman, and D. Leroi (2015). 'Photogrammetry of killer whales using a small hexacopter launched at sea 1.' *Journal of Unmanned Vehicle Systems* **3**(3): 131–5.

Essen, H., W. Johannes, S. Stanko, R. Sommer, A. Wahlen, and J. Wilcke (2012). *High resolution W-band UAV SAR.* 2012 IEEE International Geoscience and Remote Sensing Symposium, IEEE.

Evans, L. J., T. H. Jones, K. Pang, M. N. Evans, S. Saimin, and B. Goossens (2015). 'Use of drone technology as a tool for behavioral research: A case study of crocodilian nesting.' *Herpetological Conservation and Biology* **10**(1): 90–8.

Everaerts, J. (2008). 'The use of unmanned aerial vehicles (UAVs) for remote sensing and mapping.' *International Archives of the Photogrammetry, Remote Sensing and Spatial Information Sciences* **37**: 1187–92.

Fa, J. E. and D. Brown (2009). 'Impacts of hunting on mammals in African tropical moist forests: a review and synthesis.' *Mammal Review* **39**(4): 231–64.

Fang, F., T. H. Nguyen, R. Pickles, W. Y. Lam, G. R. Clements, B. An, A. Singh, M. Tambe, and A. Lemieux (2016). *Deploying PAWS: Field Optimization of the Protection Assistant for Wildlife Security.* AAAI.

Fang, Y., S. Du, R. Abdoola, K. Djouani, and C. Richards (2016). 'Motion based animal detection in aerial videos.' *Procedia Computer Science* **92**: 13–17.

Finn, R. L. and D. Wright (2016). 'Privacy, data protection and ethics for civil drone practice: A survey of industry, regulators and civil society organisations.' *Computer Law & Security Review* **32**(4): 577–86.

Fleming, P. J. and J. P. Tracey (2008). 'Some human, aircraft and animal factors affecting aerial surveys: how to enumerate animals from the air.' *Wildlife Research* **35**(4): 258–67.

Floreano, D. and R. J. Wood (2015). 'Science, technology and the future of small autonomous drones.' *Nature* **521**(7553): 460–6.

Flynn, K. F. and S. C. Chapra (2014). 'Remote sensing of submerged aquatic vegetation in a shallow non-turbid river using an unmanned aerial vehicle.' *Remote Sensing* **6**(12): 12815–36.

Fornace, K. M., C. J. Drakeley, T. William, F. Espino, and J. Cox (2014). 'Mapping infectious disease landscapes: unmanned aerial vehicles and epidemiology.' *Trends in Parasitology* **30**(11): 514–19.

Franke, U., B. Goll, U. Hohmann, and M. Heurich (2012). 'Aerial ungulate surveys with a combination of infrared and high-resolution natural colour images.' *Animal Biodiversity and Conservation* **35**(2): 285–93.

Fraser, R., I. Olthof, T. C. Lantz, and C. Schmitt (2016). 'UAV photogrammetry for mapping vegetation in the low-Arctic.' *Arctic Science* **2**(3): 79–102.

Fraser, W. R., J. C. Carlson, P. A. Duley, E. J. Holm, and D. L. Patterson (1999). 'Using kite-based aerial photography for conducting Adelie penguin censuses in Antarctica.' *Waterbirds* **22**: 435–40.

Fretwell, P. T., M. A. LaRue, P. Morin, G. L. Kooyman, B. Wienecke, N. Ratcliffe, A. J. Fox, A. H. Fleming, C. Porter, and P. N. Trathan (2012). 'An emperor penguin population estimate: the first global, synoptic survey of a species from space.' *PLoS ONE* **7**(4): e33751.

Fretwell, P. T., I. J. Staniland, and J. Forcada (2014). 'Whales from space: counting southern right whales by satellite.' *PLoS ONE* **9**(2): e88655.

Frid, A. (2003). 'Dall's sheep responses to overflights by helicopter and fixed-wing aircraft.' *Biological Conservation* **110**(3): 387–99.

Fritz, A., T. Kattenborn, and B. Koch (2013). 'UAV-based photogrammetric point clouds—Tree stem mapping in open stands in comparison to terrestrial laser scanner point clouds.' *International Archives of the Photogrammetry, Remote Sensing and Spatial Information Sciences* **40**: 141–6.

Gago, J., C. Douthe, I. Florez-Sarasa, J. M. Escalona, J. Galmes, A. R. Fernie, J. Flexas, and H. Medrano (2014). 'Opportunities for improving leaf water use efficiency under climate change conditions.' *Plant Science* **226**: 108–19.

Gatziolis, D., J. F. Lienard, A. Vogs, and N. S. Strigul (2015). '3D tree dimensionality assessment using photogrammetry and small unmanned aerial vehicles.' *PLoS ONE* **10**(9): e0137765.

Gaveau, D. L., M. A. Salim, K. Hergoualc'h, B. Locatelli, S. Sloan, M. Wooster, M. E. Marlier, E. Molidena, H. Yaen, and R. DeFries (2014). 'Major atmospheric emissions from peat fires in Southeast Asia during non-drought years: evidence from the 2013 Sumatran fires.' *Scientific Reports* **19**(4): 6112.

Gay, A. P., T. P. Stewart, R. Angel, M. Easey, A. J. Eves, N. J. Thomas, D. A. Pearce, and A. I. Kemp (2009). *Developing unmanned aerial vehicles for local and flexible environmental and agricultural monitoring*. Proceedings of RSPSoc 2009 Annual Conference. RSPSoc.

GBRMPA. (2017). 'Flying home: using drones to care for country.' from http://www.gbrmpa.gov.au/media-room/latest-news/sea-country-partnerships/2017/flying-home-using-drones-to-care-for-country.

Gerard, F., S. Petit, G. Smith, A. Thomson, N. Brown, S. Manchester, R. Wadsworth, G. Bugar, L. Halada, and P. Bezak (2010). 'Land-cover change in Europe between 1950 and 2000 determined employing aerial photography.' *Progress in Physical Geography* **34**(2): 183–205.

Getzin, S., R. S. Nuske, and K. Wiegand (2014). 'Using unmanned aerial vehicles (UAV) to quantify spatial gap patterns in forests.' *Remote Sensing* **6**(8): 6988–7004.

Getzin, S., K. Wiegand, and I. Schöning (2012). 'Assessing biodiversity in forests using very high-resolution images and unmanned aerial vehicles.' *Methods in Ecology and Evolution* **3**(2): 397–404.

Giese, M. and M. Riddle (1999). 'Disturbance of emperor penguin *Aptenodytes forsteri* chicks by helicopters.' *Polar Biology* **22**(6): 366–71.

Gini, R., D. Pagliari, D. Passoni, L. Pinto, G. Sona, and P. Dosso (2013). 'UAV photogrammetry: Block triangulation comparisons.' *International Archives of the Photogrammetry, Remote Sensing and Spatial Information Sciences* Volume XL-1/W2: 157–62.

Gini, R., D. Passoni, L. Pinto, and G. Sona (2012). 'Aerial images from an UAV system: 3d modeling and tree species classification in a park area.' *International Archives of the Photogrammetry, Remote Sensing and Spatial Information Sciences* **39**(B1): 361–6.

Gini, R., D. Passoni, L. Pinto, and G. Sona (2014). 'Use of Unmanned Aerial Systems for multispectral survey and tree classification: a test in a park area of northern Italy.' *European Journal of Remote Sensing* **47**: 251–69.

Goebel, M. E., W. L. Perryman, J. T. Hinke, D. J. Krause, N. A. Hann, S. Gardner, and D. J. LeRoi (2015). 'A small unmanned aerial system for estimating abundance and size of Antarctic predators.' *Polar Biology* **38**(5): 619–30.

Gong, A. and D. Verstraete (2017). 'Fuel cell propulsion in small fixed-wing unmanned aerial vehicles: Current status and research needs.' *International Journal of Hydrogen Energy* **42**(33): 21311–33.

Gonzalez-Dugo, V., P. Zarco-Tejada, E. Nicolás, P. A. Nortes, J. J. Alarcón, D. S. Intrigliolo, and E. Fereres (2013). 'Using high resolution UAV thermal imagery to assess the variability in the water status of five fruit tree species within a commercial orchard.' *Precision Agriculture* **14**(6): 660–78.

Gonzalez, L. F., G. A. Montes, E. Puig, S. Johnson, K. Mengersen, and K. J. Gaston (2016). 'Unmanned Aerial Vehicles (UAVs) and artificial intelligence revolutionizing wildlife monitoring and conservation.' *Sensors* **16**(1): 97.

Götmark, F. (1992). The effects of investigator disturbance on nesting birds. *Current ornithology*. D. M. Power. Boston, MA, Springer: 63–104.

Grenzdörffer, G. (2013). 'UAS-based automatic bird count of a common gull colony.' *International Archives of the Photogrammetry, Remote Sensing and Spatial Information Sciences* **1**(W2). doi: 10.5194/isprsarchives-XL-1-W2-169-2013

Griffiths, M. and C. P. Van Schaik (1993). 'Camera trapping: a new tool for the study of elusive rain forest mammals.' *Tropical Biodiversity* **1**: 131–5.

Gross, J. W. (2015). *'A comparison of orthomosaic software for use with ultra high resolution imagery of a wetland environment.'* Center for Geographic Information Science and Geography Department, Central Michigan University, Mt. Pleasant, MI, USA. Available from: http://www.imagin.org/awards/sppc/2015/papers/john_gross_paper.pdf.

Gross, J. W. and B. W. Heumann (2016). 'A statistical examination of image stitching software packages for use with unmanned aerial systems.' *Photogrammetric Engineering & Remote Sensing* **82**(6): 419–25.

Groves, P. A., B. Alcorn, M. M. Wiest, J. M. Maselko, and W. P. Connor (2016). 'Testing unmanned aircraft systems for salmon spawning surveys.' *FACETS* **1**: 187.

Guy, P. (1932). 'Balloon photography and archaeological excavation.' *Antiquity* **6**(22): 148–55.

Haas, T. C. and S. M. Ferreira (2017). 'Optimal patrol routes: interdicting and pursuing rhino poachers.' *Police Practice and Research*. doi.org/10.1080/15614263.2017.1295243

Hansen, M. C., P. V. Potapov, R. Moore, M. Hancher, S. A. Turubanova, A. Tyukavina, D. Thau, S. V. Stehman, S. J. Goetz, T. R. Loveland, A. Kommareddy, A. Egorov, L. Chini, C. O. Justice, and J. R. G. Townshend (2013). 'High-resolution global maps of 21st-century forest cover change.' *Science* **342**(6160): 850–3.

Hansen, M. C., D. P. Roy, E. Lindquist, B. Adusei, C. O. Justice, and A. Altstatt (2008). 'A method for integrating MODIS and Landsat data for systematic monitoring of forest cover and change in the Congo Basin.' *Remote Sensing of Environment* **112**(5): 2495–513.

Hanson, L., C. L. Holmquist-Johnson, and M. L. Cowardin (2014). *Evaluation of the Raven sUAS to detect and monitor greater sage-grouse leks within the Middle Park population*. US Geological Survey.

Hedge, P., F. Molloy, H. Sweatman, K. R. Hayes, J. M. Dambacher, J. Chandler, N. Bax, M. Gooch, K. Anthony, and B. Elliot (2017). 'An integrated monitoring framework for the Great Barrier Reef World Heritage Area.' *Marine Policy* **77**(Supplement C): 90–6.

Heffner, R. S. (2004). 'Primate hearing from a mammalian perspective.' *The Anatomical Record Part A: Discoveries in Molecular, Cellular, and Evolutionary Biology* **281**(1): 1111–22.

Hennekam, M. E. (2015). *Are drones the answer? Effective orangutan population monitoring in Southeast Asia*. Honours Environmental Biology, University of Adelaide.

Hinkley, E. A. and T. Zajkowski (2011). 'USDA forest service–NASA: unmanned aerial systems demonstrations–pushing the leading edge in fire mapping.' *Geocarto International* **26**(2): 103–11.

Hird, J., A. Montaghi, G. McDermid, J. Kariyeva, B. Moorman, S. Nielsen, and A. McIntosh (2017). 'Use of unmanned aerial vehicles for monitoring recovery of forest vegetation on petroleum well sites.' *Remote Sensing* **9**(5): 413.

Hodgson, A., N. Kelly, and D. Peel (2013). 'Unmanned aerial vehicles (UAVs) for surveying marine fauna: a dugong case study.' *PLoS ONE* **8**(11): e79556.

Hodgson, J. C., S. M. Baylis, R. Mott, A. Herrod, and R. H. Clarke (2016). 'Precision wildlife monitoring using unmanned aerial vehicles.' *Scientific Reports* **6**: 22574.

Hodgson, J. C. and L. P. Koh (2016). 'Best practice for minimising unmanned aerial vehicle disturbance to wildlife in biological field research.' *Current Biology* **26**(10): R404-R405.

Holman, F. H., A. B. Riche, A. Michalski, M. Castle, M. J. Wooster, and M. J. Hawkesford (2016). 'High throughput field phenotyping of wheat plant height and growth rate in field plot trials using UAV based remote sensing.' *Remote Sensing* **8**(12): 1031.

Honkavaara, E., R. Arbiol, L. Markelin, L. Martinez, M. Cramer, S. Bovet, L. Chandelier, R. Ilves, S. Klonus, and P. Marshal (2009). 'Digital airborne photogrammetry—A new tool for quantitative remote sensing?—A state-of-the-art review on radiometric aspects of digital photogrammetric images.' *Remote Sensing* **1**(3): 577–605.

Honkavaara, E., H. Saari, J. Kaivosoja, I. Pölönen, T. Hakala, P. Litkey, J. Mäkynen, and L. Pesonen (2013). 'Processing and assessment of spectrometric, stereoscopic imagery collected using a lightweight uav spectral camera for precision agriculture.' *Remote Sensing* **5**(10): 5006.

Horning, N., J. Robinson, E. Sterling, W. Turner, and S. Spector (2010). *Remote sensing for ecology and conservation.* Oxford, Oxford University Press.

Hruska, R., J. Mitchell, M. Anderson, and N. F. Glenn (2012). 'Radiometric and geometric analysis of hyperspectral imagery acquired from an unmanned aerial vehicle.' *Remote Sensing* **4**(9): 2736–52.

Humber, F., E. T. Andriamahaino, T. Beriziny, R. Botosoamananto, B. J. Godley, C. Gough, S. Pedron, V. Ramahery, and A. C. Broderick (2017). 'Assessing the small-scale shark fishery of Madagascar through community-based monitoring and knowledge.' *Fisheries Research* **186**(Part 1): 131–43.

Humle, T., R. Duffy, D. L. Roberts, C. Sandbrook, F. A. St John, and R. J. Smith (2014). 'Biology's drones: Undermined by fear.' *Science* **344**(6190): 1351.

Hung, C., M. Bryson, and S. Sukkarieh (2011). *Vision-based shadow-aided tree crown detection and classification algorithm using imagery from an unmanned airborne vehicle.* Proceedings of 34th International Symposium on Remote Sensing of Environment (ISRSE), Sydney, Australia.

Hung, C., M. Bryson, and S. Sukkarieh (2012). 'Multi-class predictive template for tree crown detection.' *ISPRS Journal of Photogrammetry and Remote Sensing* **68**: 170–83.

Hunt, E. R., W. D. Hively, S. J. Fujikawa, D. S. Linden, C. S. Daughtry, and G. W. McCarty (2010). 'Acquisition of NIR-green-blue digital photographs from unmanned aircraft for crop monitoring.' *Remote Sensing* **2**(1): 290–305.

Hutt, M. (2011). 'USGS takes to the sky.' *E Journal* September/October **2011**: 54–6. Available online at: rmgsc.cr.usgs.gov/UAS/pdf/sandhillcranes/ejournal_SeptOct_2011_birdsandUAVs.pdf;

Ingwer, P., F. Gassen, S. Püst, M. Duhn, M. Schälicke, K. Müller, H. Ruhm, J. Rettig, E. Hasche, and A. Fischer (2015). *Practical usefulness of structure from motion (SfM) point clouds obtained from different consumer cameras.* SPIE/IS&T Electronic Imaging, International Society for Optics and Photonics.

Inoue, T., S. Nagai, S. Yamashita, H. Fadaei, R. Ishii, K. Okabe, H. Taki, Y. Honda, K. Kajiwara, and R. Suzuki (2014). 'Unmanned aerial survey of fallen trees in a deciduous broadleaved forest in eastern Japan.' *PLoS ONE* **9**(10): e109881.

Inoue, Y., S. Morinaga, and A. Tomita (2000). 'A blimp-based remote sensing system for low-altitude monitoring of plant variables: a preliminary experiment for agricultural and ecological applications.' *International Journal of Remote Sensing* **21**(2): 379–85.

Israel, M. (2011). *A UAV-based roe deer fawn detection system.* Proceedings of the International Conference on Unmanned Aerial Vehicle in Geomatics (UAV-g), H. Eisenbeiss, M. Kunz, and H. Ingensand, Eds.

Jaakkola, A., J. Hyyppä, A. Kukko, X. Yu, H. Kaartinen, M. Lehtomäki, and Y. Lin (2010). 'A low-cost multi-sensoral mobile mapping system and its feasibility for tree measurements.' *ISPRS journal of Photogrammetry and Remote Sensing* **65**(6): 514–22.

Jachmann, H. (2002). 'Comparison of aerial counts with ground counts for large African herbivores.' *Journal of Applied Ecology* **39**(5): 841–52.

Jannoura, R., K. Brinkmann, D. Uteau, C. Bruns, and R. G. Joergensen (2015). 'Monitoring of crop biomass using true colour aerial photographs taken from a remote controlled hexacopter.' *Biosystems Engineering* **129**: 341–51.

Jensen, A. and Y. Chen (2013). *Tracking tagged fish with swarming unmanned aerial vehicles using fractional order potential fields and Kalman filtering.* Unmanned Aircraft Systems (ICUAS), 2013 International Conference on, IEEE.

Jensen, A. M., D. K. Geller, and Y. Chen (2014). 'Monte Carlo simulation analysis of tagged fish radio tracking performance by swarming unmanned aerial vehicles in fractional order potential fields.' *Journal of Intelligent & Robotic Systems* **74**(1–2): 287–307.

Jensen, T., A. Apan, F. Young, and L. Zeller (2007). 'Detecting the attributes of a wheat crop using digital imagery acquired from a low-altitude platform.' *Computers and Electronics in Agriculture* **59**(1–2): 66–77.

Jhan, J.-P., J.-Y. Rau, and C.-Y. Huang (2016). 'Band-to-band registration and ortho-rectification of multi-lens/multispectral imagery: A case study of MiniMCA-12 acquired by a fixed-wing UAS.' *ISPRS Journal of Photogrammetry and Remote Sensing* **114**: 66–77.

Job, J. R., S. L. Kohler, and S. A. Gill (2016). 'Song adjustments by an open habitat bird to anthropogenic noise, urban structure, and vegetation.' *Behavioral Ecology* **27**(6), 1734–44.

Johnson, L., S. Herwitz, S. Dunagan, B. Lobitz, D. Sullivan, and R. Slye (2003). *Collection of ultra high spatial and spectral resolution image data over California vineyards with a small UAV.* Proceedings of the 30th International Symposium on Remote Sensing of Environment.

Jones, G. P. I. V., L. G. Pearlstine, and H. F. Percival (2006). 'An assessment of small unmanned aerial vehicles for wildlife research.' *Wildlife Society Bulletin* **34**(3): 750–8.

Jones, R. (2014). 'Trends in plant virus epidemiology: opportunities from new or improved technologies.' *Virus research* **186**: 3–19.

Junda, J., E. Greene, and D. M. Bird (2015). 'Proper flight technique for using a small rotary-winged drone aircraft to safely, quickly, and accurately survey raptor nests.' *Journal of Unmanned Vehicle Systems* **3**(4): 222–36.

Junda, J. H., E. Greene, D. Zazelenchuk, and D. M. Bird (2016). 'Nest defense behaviour of four raptor species (osprey, bald eagle, ferruginous hawk, and red-tailed hawk) to a novel aerial intruder–a small rotary-winged drone.' *Journal of Unmanned Vehicle Systems* **4**(4): 217–27.

Kalan, A. K., A. K. Piel, R. Mundry, R. M. Wittig, C. Boesch, and H. S. Kühl (2016). 'Passive acoustic monitoring reveals group ranging and territory use: a case study of wild chimpanzees (*Pan troglodytes*).' *Frontiers in Zoology* **13**(1): 34.

Kalisperakis, I., C. Stentoumis, L. Grammatikopoulos, and K. Karantzalos (2015). 'Leaf Area Index estimation in vineyards from UAV hyperspectral data, 2D image mosaics and 3D canopy surface models.' *International Archives of Photogrammetry, Remote Sensing and Spatial Information Sciences* **40**(1): 299.

Kar, D., B. Ford, S. Gholami, F. Fang, A. Plumptre, M. Tambe, M. Driciru, F. Wanyama, and A. Rwetsiba (2017). 'Cloudy with a chance of poaching: adversary behavior modeling and forecasting with real-world poaching data.' *Proceedings of the 16th International Conference on Autonomous Agents and Multiagent Systems (AAMAS 2017)*, S. Das, E. Durfee, K. Larson, M. Winikoff (eds.), May 8–12, 2017, São Paulo, Brazil. International Foundation for Autonomous Agents and Multiagent Systems (www.ifaamas.org).

Karanth, K. U. (1995). 'Estimating tiger *Panthera tigris* populations from camera-trap data using capture-recapture models.' *Biological Conservation* **71**: 333–8.

Kattenborn, T., M. Sperlich, K. Bataua, and B. Koch (2014). 'Automatic single palm tree detection in plantations using UAV-based photogrammetric point clouds.' *International Archives of the Photogrammetry, Remote Sensing & Spatial Information Sciences*. Volume XL-3, 2014, ISPRS Technical Commission III Symposium, 5 – 7 September 2014, Zurich, Switzerland.

Kays, R., M. C. Crofoot, W. Jetz, and M. Wikelski (2015). 'Terrestrial animal tracking as an eye on life and planet.' *Science* **348**(6240): aaa2478.

Kenward, R. E. (2001). *A manual for wildlife radio tagging*. London, Academic Press.

Kersnovski, T., F. Gonzalez, and K. Morton (2017). 'A UAV system for autonomous target detection and gas sensing.' 2017 IEEE Aerospace Conference, 4–11 March 2017, Yellowstone Conference Center, Big Sky, Montana.

Kim, S. J., Y. Jeong, S. Park, K. Ryu, and G. Oh (2017). A survey of drone use for entertainment and AVR (Augmented and Virtual Reality). *Augmented Reality and Virtual Reality: Empowering Human, Place and Business*. T. Jung and M. C. tom Dieck. Cham, Springer International Publishing: 339–52.

Kirkman, S. P., D. Yemane, W. Oosthuizen, M. Meÿer, P. Kotze, H. Skrypzeck, F. Vaz Velho, and L. Underhill (2013). 'Spatio-temporal shifts of the dynamic Cape fur seal population in southern Africa, based on aerial censuses (1972–2009).' *Marine Mammal Science* **29**(3): 497–524.

Kiszka, J. J., J. Mourier, K. Gastrich, and M. R. Heithaus (2016). 'Using unmanned aerial vehicles (UAVs) to investigate shark and ray densities in a shallow coral lagoon.' *Marine Ecology Progress Series* **560**: 237–42.

Knoth, C., B. Klein, T. Prinz, and T. Kleinebecker (2013). 'Unmanned aerial vehicles as innovative remote sensing platforms for high-resolution infrared imagery to support restoration monitoring in cut-over bogs.' *Applied Vegetation Science* **16**(3): 509–17.

Koh, L. P. and S. A. Wich (2012). 'Dawn of drone ecology: low-cost autonomous aerial vehicles for conservation.' *Tropical Conservation Science* **5**(2): 121–32.

Koo, V. C., Y. K. Chan, G. Vetharatnam, M. Y. Chua, C. H. Lim, C.-S. Lim, C. Thum, T. S. Lim, Z. bin Ahmad, and K. A. Mahmood (2012). 'A new unmanned aerial vehicle synthetic aperture radar for environmental monitoring.' *Progress in Electromagnetics Research* **122**: 245–68.

Koohifar, F., A. Kumbhar, and I. Guvenc (2017). 'Receding horizon multi-UAV cooperative tracking of moving RF source.' *IEEE Communications Letters* **21**: 1433–6.

Korner, F., R. Speck, A. Goktogan, and S. Sukkarieh (2010). *Autonomous airborne wildlife tracking using radio signal strength*. Intelligent Robots and Systems (IROS), 2010 IEEE/RSJ International Conference on, IEEE.

Koski, W., P. Abgrall, and S. Yazvenko (2010). 'An inventory and evaluation of unmanned aerial systems for offshore surveys of marine mammals.' *J. Cetacean Res. Manage* **11**(3): 239–47.

Koski, W. R., T. Allen, D. Ireland, G. Buck, P. R. Smith, A. M. Macrander, M. A. Halick, C. Rushing, D. J. Sliwa, and T. L. McDonald (2009). 'Evaluation of an unmanned airborne system for monitoring marine mammals.' *Aquatic Mammals* **35**(3): 347–57.

Koski, W. R., G. Gamage, A. R. Davis, T. Mathews, B. LeBlanc, and S. H. Ferguson (2015). 'Evaluation of UAS for photographic re-identification of bowhead whales, *Balaena mysticetus*.' *Journal of Unmanned Vehicle Systems* **3**(1): 22–9.

Koski, W. R., T. A. Thomas, D. W. Funk, and A. M. Macrander (2013). 'Marine mammal sightings by analysts of digital imagery versus aerial surveyors: a preliminary comparison.' *Journal of Unmanned Vehicle Systems* **1**(01): 25–40.

Kudo, H., Y. Koshino, A. Eto, M. Ichimura, and M. Kaeriyama (2012). 'Cost-effective accurate estimates of adult chum salmon, *Oncorhynchus keta*, abundance in a Japanese river using a radio-controlled helicopter.' *Fisheries Research* **119**: 94–8.

Kühl, H., F. Maisels, M. Ancrenaz, and E. A. Williamson (2008). *Best practice guidelines for surveys and monitoring of great ape populations*. Gland, Switzerland, IUCN/SSC Primate Specialist Group.

Laliberte, A. S., M. A. Goforth, C. M. Steele, and A. Rango (2011). 'Multispectral remote sensing from unmanned aircraft: Image processing workflows and applications for rangeland environments.' *Remote Sensing* **3**(11): 2529–51.

Laliberte, A. S. and A. Rango (2009). 'Texture and scale in object-based analysis of subdecimeter resolution unmanned aerial vehicle (UAV) imagery.' *Geoscience and Remote Sensing, IEEE Transactions on* **47**(3): 761–70.

Laliberte, A. S., A. Rango, and J. Herrick (2007). *Unmanned aerial vehicles for rangeland mapping and monitoring: a comparison of two systems*. ASPRS Annual Conference Proceedings.

Laliberte, A. S. and W. J. Ripple (2003). 'Automated wildlife counts from remotely sensed imagery.' *Wildlife Society Bulletin* **31**(2): 362–71.

Le Maho, Y., J. D. Whittington, N. Hanuise, L. Pereira, M. Boureau, M. Brucker, N. Chatelain, J. Courtecuisse, F. Crenner, and B. Friess (2014). 'Rovers minimize human disturbance in research on wild animals.' *Nature Methods* **11**(12): 1242–4.

Lebourgeois, V., A. Bégué, S. Labbé, B. Mallavan, L. Prévot, and B. Roux (2008). 'Can commercial digital cameras be used as multispectral sensors? A crop monitoring test.' *Sensors* **8**(11): 7300–22.

Lee, J. S. H., Z. Jaafar, A. K. J. Tan, L. R. Carrasco, J. J. Ewing, D. P. Bickford, E. L. Webb, and L. P. Koh (2016). 'Toward clearer skies: Challenges in regulating transboundary haze in Southeast Asia.' *Environmental Science & Policy* **55, Part 1**: 87–95.

Lega, M. and R. M. Napoli (2010). 'Aerial infrared thermography in the surface waters contamination monitoring.' *Desalination and Water Treatment* **23**(1–3): 141–51.

Lehmann, J. R., W. Münchberger, C. Knoth, C. Blodau, F. Nieberding, T. Prinz, V. A. Pancotto, and T. Kleinebecker (2016). 'High-resolution classification of south Patagonian peat bog microforms reveals potential gaps in up-scaled CH_4 fluxes by use of unmanned aerial system (UAS) and CIR imagery.' *Remote Sensing* **8**(3): 173.

Leonardo, M., A. M. Jensen, C. Coopmans, M. McKee, and Y. Chen (2013). *A miniature wildlife tracking UAV payload system using acoustic biotelemetry*. ASME 2013 International Design Engineering Technical Conferences and Computers and Information in Engineering Conference, American Society of Mechanical Engineers.

Lhoest, S., J. Linchant, S. Quevauvillers, C. Vermeulen, and P. Lejeune (2015). 'HOW MANY HIPPOS (HOMHIP): Algorithm for automatic counts of animals with infrared thermal imagery from UAV.' *International Archives of Photogrammetry, Remote Sensing and Spatial Information Sciences* **40**(3): 355.

Li, C. J. and H. Ling (2015). Synthetic aperture radar imaging using a small consumer drone. *IEEE International Symposium on Antennas and Propagation*. Vancouver, Canada: 685–6.

Li, C. J. and H. Ling (2016). High-resolution, downward-looking radar imaging using a small consumer drone. *IEEE International Symposium on Antennas and Propagation*. Fajardo, Puerto Rico.

Linchant, J., J. Lisein, J. Semeki, P. Lejeune, and C. Vermeulen (2015). 'Are unmanned aircraft systems (UASs) the future of wildlife monitoring? A review of accomplishments and challenges.' *Mammal Review* **45**(4): 239–52.

Lisbona, D. and T. Snee (2011). 'A review of hazards associated with primary lithium and lithium-ion batteries.' *Process Safety and Environmental Protection* **89**: 434–42.

Lisein, J., A. Michez, H. Claessens, and P. Lejeune (2015). 'Discrimination of deciduous tree species from time series of unmanned aerial system imagery.' *PLoS ONE* **10**(11): e0141006.

Lisein, J., M. Pierrot-Deseilligny, S. Bonnet, and P. Lejeune (2013). 'A photogrammetric workflow for the creation of a forest canopy height model from small unmanned aerial system imagery.' *Forests* **4**(4): 922–44.

Liu, C.-C., Y.-H. Chen, and H.-L. Wen (2015). 'Supporting the annual international black-faced spoonbill census with a low-cost unmanned aerial vehicle.' *Ecological Informatics* **30**: 170–8.

Longmore, S., R. Collins, S. Pfeifer, S. Fox, M. Mulero-Pazmany, F. Bezombes, A. Goodwind, M. Ovelar, J. Knapen, and S. Wich (2017). 'Adapting astronomical source detection software to help detect animals in thermal images obtained by unmanned aerial systems.' *International Journal of Remote Sensing* **38**: 2623–38.

López, E., S. García, R. Barea, L. Bergasa, E. Molinos, R. Arroyo, E. Romera, and S. Pardo (2017). 'A multi-sensorial simultaneous localization and mapping (SLAM) system for low-cost micro aerial vehicles in GPS-denied environments.' *Sensors* **17**(4): 802.

Lowe, D. G. (1999). *Object recognition from local scale-invariant features*. Computer vision, 1999, The proceedings of the seventh IEEE international conference on, IEEE.

Lowe, D. G. (2004). 'Distinctive image features from scale-invariant keypoints.' *International Journal of Computer Vision* **60**(2): 91–110.

Lucieer, A., Z. Malenovský, T. Veness, and L. Wallace (2014). 'HyperUAS: Imaging spectroscopy from a multi-rotor unmanned aircraft system.' *Journal of Field Robotics* **31**(4): 571–90.

Lucieer, A., S. Robinson, D. Turner, S. Harwin, and J. Kelcey (2012). *Using a micro-UAV for ultra-high resolution multi-sensor observations of Antarctic moss beds*. XXII ISPRS Congress, Melbourne, Australia, Netherlands: ISPRS.

Lucieer, A., S. A. Robinson, and D. Turner (2010). 'Using an unmanned aerial vehicle (UAV) for ultra-high resolution mapping of Antarctic moss beds.' 15th Australasian Remote Sensing & Photogrammetry Conference. Alice Springs. 14-16th Sept 2010. University of Wollongong Library: research-pubs@uow.edu.au

Lucieer, A., D. Turner, D. H. King, and S. A. Robinson (2014). 'Using an Unmanned Aerial Vehicle (UAV) to capture micro-topography of Antarctic moss beds.' *International Journal of Applied Earth Observation and Geoinformation* **27**: 53–62.

Lynam, A. J., K. E. Jenks, and N. Tantipisanuh (2013). 'Terrestrial activity patterns of wild cats from camera-trapping.' *Raffles Bulletin of Zoology* **61**: 407–15.

MacArthur, R. A., V. Geist, and R. H. Johnston (1982). 'Cardiac and behavioral responses of mountain sheep to human disturbance.' *The Journal of Wildlife Management* 46: 351–8.

MacKenzie, D. I. (2006). *Occupancy estimation and modeling: inferring patterns and dynamics of species occurrence*. London, Academic Press.

Maire, F., L. M. Alvarez, and A. Hodgson (2015). *Automating marine mammal detection in aerial images captured during wildlife surveys: A deep learning approach*. Australasian Joint Conference on Artificial Intelligence, Springer.

Maire, F., L. Mejias, A. Hodgson, and G. Duclos (2013). *Detection of dugongs from unmanned aerial vehicles*. Intelligent Robots and Systems (IROS), 2013 IEEE/RSJ International Conference on, IEEE.

Marques, P. and A. D. Ronch (2017). *Advanced UAV aerodynamics, flight stability and control: novel concepts, theory and applications*. Chichester, Wiley.

Martin, J., H. H. Edwards, M. A. Burgess, H. F. Percival, D. E. Fagan, B. E. Gardner, J. G. Ortega-Ortiz, P. G. Ifju, B. S. Evers, and T. J. Rambo (2012). 'Estimating distribution of hidden objects with drones: from tennis balls to manatees.' *PLoS ONE* **7**(6): e38882.

Martínez-de-Dios, J., L. Merino, A. Ollero, L. M. Ribeiro, and X. Viegas (2007). Multi-UAV experiments: application to forest fires. *Multiple Heterogeneous Unmanned Aerial Vehicles*, A. Ollero and I. Maza (Eds.), Heidelberg, Springer-Verlag: 207–28.

Marvin, D. C., L. P. Koh, A. J. Lynam, S. Wich, A. B. Davies, R. Krishnamurthy, E. Stokes, R. Starkey, and G. P. Asner (2016). 'Integrating technologies for scalable ecology and conservation.' *Global Ecology and Conservation* **7**: 262–75.

Mathews, A. J. and J. L. Jensen (2013). 'Visualizing and quantifying vineyard canopy LAI using an unmanned aerial vehicle (UAV) collected high density structure from motion point cloud.' *Remote Sensing* **5**(5): 2164–83.

McClelland, G. T. W., A. L. Bond, A. Sardana, and T. Glass (2016). 'Rapid population estimate of a surface-nesting seabird on a remote island using a low-cost unmanned aerial vehicle.' *Marine Ornithology* **44**: 215–20.

McEvoy, J. F., G. P. Hall, and P. G. McDonald (2016). 'Evaluation of unmanned aerial vehicle shape, flight path and camera type for waterfowl surveys: disturbance effects and species recognition.' *PeerJ* **4**: e1831.

McGwire, K. C., M. A. Weltz, J. A. Finzel, C. E. Morris, L. F. Fenstermaker, and D. S. McGraw (2013). 'Multiscale assessment of green leaf cover in a semi-arid rangeland with a small unmanned aerial vehicle.' *International Journal of Remote Sensing* **34**(5): 1615–32.

McMahon, C. R., H. Howe, J. van den Hoff, R. Alderman, H. Brolsma, and M. A. Hindell (2014). 'Satellites, the all-seeing eyes in the sky: counting elephant seals from space.' *PLoS ONE* **9**(3): e92613.

Meek, P., G. Ballard, P. Fleming, and G. Falzon (2016). 'Are we getting the full picture? Animal responses to camera traps and implications for predator studies.' *Ecology and Evolution* **6**(10): 3216–25.

Meek, P. D., G.-A. Ballard, and P. J. Fleming (2015). 'The pitfalls of wildlife camera trapping as a survey tool in Australia.' *Australian Mammalogy* **37**(1): 13–22.

Meek, P. D., G.-A. Ballard, P. J. S. Fleming, M. Schaefer, W. Williams, and G. Falzon (2014). 'Camera traps can be heard and seen by animals.' *PLoS ONE* **9**(10): e110832.

Merino, L., F. Caballero, J. R. Martínez-de-Dios, I. Maza, and A. Ollero (2012). 'An unmanned aircraft system for automatic forest fire monitoring and measurement.' *Journal of Intelligent & Robotic Systems* **65**(1–4): 533–48.

Messinger, M., G. P. Asner, and M. Silman (2016). 'Rapid assessments of Amazon forest structure and biomass using small unmanned aerial systems.' *Remote Sensing* **8**(8): 615.

Michez, A., H. Piégay, L. Jonathan, H. Claessens, and P. Lejeune (2016a). 'Mapping of riparian invasive species with supervised classification of Unmanned Aerial System (UAS) imagery.' *International Journal of Applied Earth Observation and Geoinformation* **44**: 88–94.

Michez, A., H. Piégay, J. Lisein, H. Claessens, and P. Lejeune (2016b). 'Classification of riparian forest species and health condition using multi-temporal and hyperspatial imagery from unmanned aerial system.' *Environmental monitoring and assessment* **188**(3): 1–19.

Mitchell, J., N. Glenn, M. Anderson, and R. Hruska (2016). 'Flight considerations and hyperspectral image classifications for dryland vegetation management from a fixed-wing UAS.' *Environmental Management and Sustainable Development* **5**(2): 17.

Mitchell, J. J., N. F. Glenn, M. O. Anderson, R. C. Hruska, A. Halford, C. Baun, and N. Nydegger (2012). *Unmanned aerial vehicle (UAV) hyperspectral remote sensing for dryland vegetation monitoring*. 2012 4th Workshop on Hyperspectral Image and Signal Processing: Evolution in Remote Sensing (WHISPERS), IEEE.

Miyamoto, M., K. Yoshino, K. Kushida, and T. Nagano (2005). 'Use of balloon aerial photography and airborne color near infrared (CNIR) video image for mapping vegetation in Kushiro wetland, northeast Japan.' *International Journal of Geoinformatics* **1**(2).

Miyamoto, M., K. Yoshino, T. Nagano, T. Ishida, and Y. Sato (2004). 'Use of balloon aerial photography for classification of Kushiro wetland vegetation, northeastern Japan.' *Wetlands* **24**(3): 701–10.

Moreland, E. E., M. F. Cameron, R. P. Angliss, and P. L. Boveng (2015). 'Evaluation of a ship-based unoccupied aircraft system (UAS) for surveys of spotted and ribbon seals in the Bering Sea pack ice 1.' *Journal of Unmanned Vehicle Systems* **3**(3): 114–22.

Morgan, J. L., S. E. Gergel, and N. C. Coops (2010). 'Aerial photography: a rapidly evolving tool for ecological management.' *BioScience* **60**(1): 47–59.

Morton, S. and N. Papanikolopoulos (2016). *Two meter solar UAV: Design approach and performance prediction for autonomous sensing applications*. 2016 IEEE/RSJ International Conference on Intelligent Robots and Systems (IROS).

Mulaca, B., R. Storvoldb, and E. Weatherhead (2011). *Remote sensing in the Arctic with unmanned aircraft: Helping scientists to achieve their goals*. Proceedings of 34th International Symposium on Remote Sensing of Environment (ISRSE), Sydney, Australia.

Mulero-Pázmány, M., S. Jenni-Eiermann, N. Strebel, T. Sattler, J. J. Negro, and Z. Tablado (2017). 'Unmanned aircraft systems as a new source of disturbance for wildlife: A systematic review.' *PLoS ONE* **12**(6): e0178448.

Mulero-Pázmány, M., J. J. Negro, and M. J. o. U. V. S. Ferrer (2014). 'A low cost way for assessing bird risk hazards in power lines: Fixed-wing small unmanned aircraft systems.' *Journal of Unmanned Vehicle Systems* **2**: 5–15.

Mulero-Pázmány, M., R. Stolper, L. Van Essen, J. J. Negro, and T. Sassen (2014). 'Remotely piloted aircraft systems as a rhinoceros anti-poaching tool in Africa.' *PLoS ONE* **9**(1): e83873.

Murray, J. C., M. J. Neal, and F. Labrosse (2013). 'Development and deployment of an intelligent kite aerial photography platform (iKAPP) for site surveying and image acquisition.' *Journal of Field Robotics* **30**(2): 288–307.

Naidoo, R., B. Fisher, A. Manica, and A. Balmford (2016). 'Estimating economic losses to tourism in Africa from the illegal killing of elephants.' *Nature Communications* **7**: 13379.

Newhall, B. (1982). *The History of Photography*. New York, Museum of Modern Art..

Nex, F. and F. Remondino (2014). 'UAV for 3D mapping applications: a review.' *Applied Geomatics* **6**: 1–15.

Nguy-Robertson, A. L. and A. A. Gitelson (2015). 'Algorithms for estimating green leaf area index in C3 and C4 crops for MODIS, Landsat TM/ETM+, MERIS, Sentinel MSI/OLCI, and Venμs sensors.' *Remote Sensing Letters* **6**(5): 360–9.

Nilssen, K. T., R. Storvold, D. Stødle, S. A. Solbø, K.-S. Johansen, M. Poltermann, and T. Haug (2014). 'Testing UAVs to perform aerial photo-graphic survey of harp and hooded seals in the West Ice area.' Survey report – KV Svalbard 16–26 March 2014, Havforskningsinstituttet.

Noyes, J. H., B. K. Johnson, R. A. Riggs, M. W. Schlegel, and V. L. Coggins (2000). 'Assessing aerial survey methods to estimate elk populations: a case study.' *Wildlife Society Bulletin* 28: 636–42.

O'Donoghue, P. and C. Rutz (2016). 'Real-time anti-poaching tags could help prevent imminent species extinctions.' *Journal of Applied Ecology* **53**(1): 5–10.

Oettershagen, P., A. Melzer, T. Mantel, K. Rudin, T. Stastny, B. Wawrzacz, T. Hinzmann, S. Leutenegger, K. Alexis, and R. Siegwart (2017). 'Design of small hand-launched solar-powered UAVs: From concept study to a multi-day world endurance record flight.' *Journal of Field Robotics* **34**(7): 1352–77.

Ogden, L. E. (2013). 'Drone ecology.' *BioScience* **63**: 776.

Oishi, Y. and T. Matsunaga (2014). 'Support system for surveying moving wild animals in the snow using aerial remote-sensing images.' *International Journal of Remote Sensing* **35**(4): 1374–94.

Olivares-Mendez, M. A., T. F. Bissyandé, K. Somasundar, J. Klein, H. Voos, and Y. Le Traon (2013). *The NOAH project: Giving a chance to threatened species in Africa with UAVs*. International Conference on e-Infrastructure and e-Services for Developing Countries, Cham, Springer.

Olivares-Mendez, M. A., C. Fu, P. Ludivig, T. F. Bissyandé, S. Kannan, M. Zurad, A. Annaiyan, H. Voos, and P. Campoy (2015). 'Towards an autonomous vision-based unmanned aerial system against wildlife poachers.' *Sensors* **15**(12): 31362–91.

Olson, D. D., J. A. Bissonette, P. C. Cramer, A. D. Green, S. T. Davis, P. J. Jackson, and D. C. Coster (2014). 'Monitoring wildlife–vehicle collisions in the information age: how smartphones can improve data collection.' *PLoS ONE* **9**: e98613.

Ore, J. P., S. Elbaum, A. Burgin, and C. Detweiler (2015). 'Autonomous aerial water sampling.' *Journal of Field Robotics* **32**(8): 1095–113.

Page, S. E. and A. Hooijer (2016). 'In the line of fire: the peatlands of Southeast Asia.' *Philosophical Transactions of the Royal Society B: Biological Sciences* **371**(1696).

Paine, D. P. and J. D. Kiser (2012). *Aerial photography and image interpretation*. New Jersey, USA, John Wiley & Sons.

Pajares, G. (2015). 'Overview and current status of remote sensing applications based on unmanned aerial vehicles (UAVs).' *Photogrammetric Engineering & Remote Sensing* **81**(4): 281–329.

Park, N., E. Serra, T. Snitch, and V. S. Subrahmanian (2015a). 'APE: A data-driven, behavioral model based anti-poaching engine.' *IEEE Transactions on Computational Social Systems* **2**: 15–37.

Park, N., E. Serra, and V. Subrahmanian (2015b). 'Saving rhinos with predictive analytics.' *IEEE Intelligent Systems* **30**(4): 86–8.

Patenaude, N. J., W. J. Richardson, M. A. Smultea, W. R. Koski, G. W. Miller, B. Würsig, and C. R. Greene (2002). 'Aircraft sound and disturbance to bowhead and beluga whales during spring migration in the Alaskan Beaufort Sea.' *Marine Mammal Science* **18**(2): 309–35.

Pater, L. L., T. G. Grubb, and D. K. Delaney (2009). 'Recommendations for improved assessment of noise impacts on wildlife.' *The Journal of Wildlife Management* **73**(5): 788–95.

Patterson, C., W. Koski, P. Pace, B. McLuckie, and D. M. Bird (2016). 'Evaluation of an unmanned aircraft system for detecting surrogate caribou targets in Labrador.' *Journal of Unmanned Vehicle Systems* **4**(1): 53–69.

Peck, M., A. Mariscal, M. Padbury, T. Cane, D. Kniveton, and M. A. Chinchero (2012). 'Identifying tropical Ecuadorian Andean trees from inter-crown pixel distributions in hyperspatial aerial imagery.' *Applied Vegetation Science* **15**(4): 548–59.

Pimm, S. L., S. Alibhai, R. Bergl, A. Dehgan, C. Giri, Z. Jewell, L. Joppa, R. Kays, and S. Loarie (2015). 'Emerging technologies to conserve biodiversity.' *Trends in Ecology & Evolution* **30**(11): 685–96.

Pitt, D., G. Glover, and R. Jones (1996). 'Two-phase sampling of woody and herbaceous plant communities using large-scale aerial photographs.' *Canadian Journal of Forest Research* **26**(4): 509–24.

Plieninger, T. (2006). 'Habitat loss, fragmentation, and alteration–quantifying the impact of land-use changes on a Spanish dehesa landscape by use of aerial photography and GIS.' *Landscape Ecology* **21**(1): 91–105.

Pomeroy, P., L. O'Connor, and P. Davies (2015). 'Assessing use of and reaction to unmanned aerial systems in gray and harbor seals during breeding and molt in the UK.' *Journal of Unmanned Vehicle Systems* **3**(3): 102–13.

Posch, A. and S. Sukkarieh (2009). *UAV based search for a radio tagged animal using particle filters*. Australasian Conference on Robotics and Automation (ACRA), Sydney, Australia, Dec, Citeseer.

Post, E., U. S. Bhatt, C. M. Bitz, J. F. Brodie, T. L. Fulton, M. Hebblewhite, J. Kerby, S. J. Kutz, I. Stirling, and D. A. Walker (2013). 'Ecological consequences of sea-ice decline.' *Science* **341**(6145): 519–24.

Post, E., M. C. Forchhammer, M. S. Bret-Harte, T. V. Callaghan, T. R. Christensen, B. Elberling, A. D. Fox, O. Gilg, D. S. Hik, and T. T. Høye (2009). 'Ecological dynamics across the Arctic associated with recent climate change.' *Science* **325**(5946): 1355–8.

Potapov, E., I. Utekhina, M. McGrady, and D. Rimlinger (2013). 'Usage of UAV for surveying Steller's sea eagle nests.' *Raptors Conservation* **27**: 253–260.

Poulsen, J. R., S. E. Koerner, S. Moore, V. P. Medjibe, S. Blake, C. J. Clark, M. E. Akou, M. Fay, A. Meier, and J. Okouyi (2017). 'Poaching empties critical Central African wilderness of forest elephants.' *Current Biology* **27**(4): R134–5.

Primicerio, J., S. F. Di Gennaro, E. Fiorillo, L. Genesio, E. Lugato, A. Matese, and F. P. Vaccari (2012). 'A flexible unmanned aerial vehicle for precision agriculture.' *Precision Agriculture* **13**(4): 517–23.

Professional Aerial Photographers Association. (2016). 'History of aerial photography.' Retrieved 05/02/2016, from http://professionalaerialphotographers.com/.

Przybilla, H. and W. Wester-Ebbinghaus (1979). 'Bildflug mit ferngelenktem Kleinflugzeug. Bildmessung und Luftbildwessen.' *Bildmessung und Luftbildwessen* **47**: 137–42.

Puliti, S., H. O. Ørka, T. Gobakken, and E. Næsset (2015). 'Inventory of small forest areas using an unmanned aerial system.' *Remote Sensing* **7**(8): 9632–54.

Puttock, A., A. Cunliffe, K. Anderson, and R. E. Brazier (2015). 'Aerial photography collected with a multirotor drone reveals impact of Eurasian beaver reintroduction on ecosystem structure.' *Journal of Unmanned Vehicle Systems* **3**(3): 123–30.

Qin, R., A. Grün, and X. Huang (2013). 'UAV project—Building a reality-based 3D model.' *Coordinates* **9**: 18–26.

Rabatel, G., N. Gorretta, and S. Labbe (2014). 'Getting simultaneous red and near-infrared band data from a single digital camera for plant monitoring applications: Theoretical and practical study.' *Biosystems Engineering* **117**: 2–14.

Rango, A., A. Laliberte, J. E. Herrick, C. Winters, K. Havstad, C. Steele, and D. Browning (2009). 'Unmanned aerial vehicle-based remote sensing for rangeland assessment, monitoring, and management.' *Journal of Applied Remote Sensing* **3**(1): 033542. doi:10.1117/1.3216822

Ratcliffe, N., D. Guihen, J. Robst, S. Crofts, A. Stanworth, and P. Enderlein (2015). 'A protocol for the aerial survey of penguin colonies using UAVs.' *Journal of Unmanned Vehicle Systems* **3**: 95–101.

Reid, A., F. Ramos, and S. Sukkarieh (2011). *Multi-class classification of vegetation in natural environments using an unmanned aerial system*. Robotics and Automation (ICRA), 2011 IEEE International Conference on, IEEE.

Remy, M. A., K. A. de Macedo, and J. R. Moreira (2012). *The first UAV-based P-and X-band interferometric SAR system*. 2012 IEEE International Geoscience and Remote Sensing Symposium, IEEE.

Riggan, N. and R. Weih (2009). 'A comparison of pixel-based versus object-based land use/land-cover classification methodologies.' *Journal of the Arkansas Academy of Science* **63**: 145–52.

Rodríguez-Canosa, G. R., S. Thomas, J. del Cerro, A. Barrientos, and B. MacDonald (2012). 'A real-time method to detect and track moving objects (DATMO) from unmanned aerial vehicles (UAVs) using a single camera.' *Remote Sensing* **4**(4): 1090–111.

Rodríguez-Veiga, P., J. Wheeler, V. Louis, K. Tansey, and H. Balzter (2017). 'Quantifying forest biomass carbon stocks from space.' *Current Forestry Reports* **3**(1): 1–18.

Rodríguez, A., J. J. Negro, M. Mulero, C. Rodríguez, J. Hernández-Pliego, and J. Bustamante (2012). 'The eye in the sky: combined use of unmanned aerial systems and GPS data loggers for ecological research and conservation of small birds.' *PLoS ONE* **7**(12): e50336.

Rufino, G. and A. Moccia (2005). 'Integrated VIS-NIR hyperspectral/thermal-IR electro-optical payload system for a mini-UAV.' Infotech@ Aerospace, 26–29 September 2005, Arlington, Virginia.

Rümmler, M.-C., O. Mustafa, J. Maercker, H.-U. Peter, and J. Esefeld (2016). 'Measuring the influence of unmanned aerial vehicles on Adélie penguins.' *Polar Biology* **39**(7): 1329–34.

Rump, M., A. Zinke, and R. Klein (2011). *Practical spectral characterization of trichromatic cameras*. ACM Transactions on Graphics (TOG), ACM.

Salamí, E., C. Barrado, and E. Pastor (2014). 'UAV flight experiments applied to the remote sensing of vegetated areas.' *Remote Sensing* **6**(11): 11051–81.

Sandbrook, C. (2015). 'The social implications of using drones for biodiversity conservation.' *Ambio* **44**(4): 636–47.

Santos, G. A. M. d., Z. Barnes, E. Lo, B. Ritoper, L. Nishizaki, X. Tejeda, A. Ke, H. Lin, C. Schurgers, and A. Lin (2014). *Small unmanned aerial vehicle system for wildlife radio collar tracking*. Mobile Ad Hoc and Sensor Systems (MASS), 2014 IEEE 11th International Conference on, IEEE.

Sardà-Palomera, F., G. Bota, C. Vinolo, O. Pallares, V. Sazatornil, L. Brotons, S. Gomariz, and F. Sarda (2012). 'Fine-scale bird monitoring from light unmanned aircraft systems.' *Ibis* **154**(1): 177–83.

Sasse, D. B. (2003). 'Job-related mortality of wildlife workers in the United States, 1937–2000.' *Wildlife Society Bulletin* **31**(4): 1015–20.

Schlachter, F. (2013). 'No Moore's Law for batteries.' *Proceedings of the National Academy of Sciences USA* **110**: 5273.

Schofield, G., K. A. Katselidis, M. K. Lilley, R. D. Reina, and G. C. Hays (in press). 'Detecting elusive aspects of wildlife ecology using drones: new insights on the mating dynamics and operational sex ratios of sea turtles.' *Functional Ecology*.

Schofield, G., K. Papafitsoros, R. Haughey, and K. Katselidis (2017). 'Aerial and underwater surveys reveal temporal variation in cleaning-station use by sea turtles at a temperate breeding area.' *Marine Ecology Progress Series* **575**: 153–64.

Schoonmaker, J., T. Wells, G. Gilbert, Y. Podobna, I. Petrosyuk, and J. Dirbas (2008). *Spectral detection and monitoring of marine mammals*. SPIE Defense and Security Symposium, International Society for Optics and Photonics..

Scobie, C. A. and C. H. Hugenholtz (2016). 'Wildlife monitoring with unmanned aerial vehicles: Quantifying distance to auditory detection.' *Wildlife Society Bulletin* **40**: 781–5.

Scoffin, T. (1982). 'Reef aerial photography from a kite.' *Coral Reefs* **1**(1): 67–9.

Selby, W., P. Corke, and D. Rus (2011). *Autonomous aerial navigation and tracking of marine mammals*. Proceedings of the 2011 Australasian Conference on Robotics and Automation, Melbourne, Australia.

Serreze, M., A. Barrett, J. Stroeve, D. Kindig, and M. Holland (2009). 'The emergence of surface-based Arctic amplification.' *The Cryosphere* **3**(1): 11–19.

Seymour, A. C., J. Dale, M. Hammill, P. N. Halpin, and D. W. Johnston (2017). 'Automated detection and enumeration of marine wildlife using unmanned aircraft systems (UAS) and thermal imagery.' *Scientific Reports* **7**: 45127.

Shaffer, M. J. and J. A. Bishop (2016). 'Predicting and preventing elephant poaching incidents through statistical analysis, GIS-based risk analysis, and aerial surveillance flight path modeling.' *Tropical Conservation Science* **9**(1): 525–48.

Shahbazi, M., J. Théau, and P. Ménard (2014). 'Recent applications of unmanned aerial imagery in natural resource management.' *GIScience & Remote Sensing* **51**(4): 339–65.

Shannon, G., M. F. McKenna, L. M. Angeloni, K. R. Crooks, K. M. Fristrup, E. Brown, K. A. Warner, M. D. Nelson, C. White, and J. Briggs (2016). 'A synthesis of two decades of research documenting the effects of noise on wildlife.' *Biological Reviews* **91**(4): 982–1005.

Silveira, L., A. T. A. Jácomo, and J. A. F. Diniz-Filho (2003). 'Camera trap, line transect census and track surveys: a comparative evaluation.' *Biol. Cons.* **114**: 351–5.

Skidmore, A. K., N. Pettorelli, N. C. Coops, G. N. Geller, M. Hansen, R. Lucas, C. A. Mücher, B. O'Connor, M. Paganini, H. M. Pereira, M. E. Schaepman, W. Turner, T. Wang, and M. Wegmann (2015). 'Agree on biodiversity metrics to track from space.' *Nature* **523**: 403–5.

Slabbekoorn, H. and A. den Boer-Visser (2006). 'Cities change the songs of birds.' *Current Biology* **16**(23): 2326–31.

Slabbekoorn, H. and M. Peet (2003). 'Ecology: Birds sing at a higher pitch in urban noise.' *Nature* **424**(6946): 267.

Sleno, G. A. and A. W. Mansfield (1978). *Aerial photography of marine mammals using a radio-controlled model aircraft.* Fish. Mar. Serv. MS Rep. 1457: 7 p.

Smith, C. E., S. T. Sykora-Bodie, B. Bloodworth, S. M. Pack, T. R. Spradlin, and N. R. LeBoeuf (2016). 'Assessment of known impacts of unmanned aerial systems (UAS) on marine mammals: data gaps and recommendations for researchers in the United States.' *Journal of Unmanned Vehicle Systems* **4**(1): 31–44.

Smith, M. J., J. Chandler, and J. Rose (2009). 'High spatial resolution data acquisition for the geosciences: kite aerial photography.' *Earth Surface Processes and Landforms* **34**(1): 155–61.

Snavely, N., S. M. Seitz, and R. Szeliski (2008). 'Modeling the world from internet photo collections.' *International Journal of Computer Vision* **80**(2): 189–210.

Snitch, T. (2013). Leveling the playing field: Employing high technology to combat poachers. *Protecting Threatened Wildlife in Africa with Technology and Training*. Washington, D. C., Richardson Center for Global Engagement, World Wildlife Fund and African Parks.

Sona, G., L. Pinto, D. Pagliari, D. Passoni, and R. Gini (2014). 'Experimental analysis of different software packages for orientation and digital surface modelling from UAV images.' *Earth Science Informatics* **7**(2): 97–107.

Soriano, P., F. Caballero, A. Ollero, and C. A. de Tecnologıas Aeroespaciales (2009). *RF-based particle filter localization for wildlife tracking by using an UAV*. 40th International Symposium of Robotics, Barcelona, España.

Soubry, I., P. Patias, and V. Tsioukas (2017). 'Monitoring vineyards with UAV and multi-sensors for the assessment of water stress and grape maturity.' *Journal of Unmanned Vehicle Systems* **5**(2): 37–50.

Spillmann, B., M. A. van Noordwijk, E. P. Willems, T. Mitra Setia, U. Wipfli, and C. P. van Schaik (2015). 'Validation of an acoustic location system to monitor Bornean orangutan (*Pongo pygmaeus wurmbii*) long calls.' *American Journal of Primatology* **77**(7): 767–76.

Stagakis, S., V. González-Dugo, P. Cid, M. Guillén-Climent, and P. J. Zarco-Tejada (2012). 'Monitoring water stress and fruit quality in an orange orchard under regulated deficit irrigation using narrow-band structural and physiological remote sensing indices.' *ISPRS Journal of Photogrammetry and Remote Sensing* **71**: 47–61.

Stankowich, T. (2008). 'Ungulate flight responses to human disturbance: A review and meta-analysis.' *Biological Conservation* **141**(9): 2159–73.

Stuchlík, R., Z. Stachoň, K. Láska, and P. Kubíček (2015). 'Unmanned Aerial Vehicle–Efficient mapping tool available for recent research in polar regions.' *Czech Polar Reports* **5**: 210–21. doi: 10.5817/CPR2015-2-18.

Sugiura, R., N. Noguchi, and K. Ishii (2005). 'Remote-sensing technology for vegetation monitoring using an unmanned helicopter.' *Biosystems Engineering* **90**(4): 369–79.

Sullivan, D. G., J. P. Fulton, J. N. Shaw, and G. Bland (2007). 'Evaluating the sensitivity of an unmanned thermal infrared aerial system to detect water strees in a cotton canopy.' *Transactions of the American Society of Agricultural Engineers* **50**: 1995–62.

Suomalainen, J., J. Franke, N. Anders, S. Iqbal, P. Wenting, R. Becker, and L. Kooistra (2014). *Lightweight hyperspectral mapping system and a novel photogrammetric processing chain for UAV-based sensing*. EGU General Assembly Conference Abstracts, Vol. 16, EGU2014-14473.

Szantoi, Z., S. E. Smith, G. Strona, L. P. Koh, and S. A. Wich (2017). 'Mapping orangutan habitat and agricultural areas using Landsat OLI imagery augmented with unmanned aircraft system aerial photography.' *International Journal of Remote Sensing* **38**(8–10): 2231–45.

Talukdar, B. K., R. Emslie, S. S. Bist, A. Choudhury, S. Ellis, B. S. Bonal, M. C. Malakar, B. N. Talukdar, and M. Barua. (2008). '*Rhinoceros unicornis*.' *The IUCN Red List of Threatened Species 2008* Retrieved Downloaded on 07 April, 2017.

Teacher, A. G. F., D. J. Griffiths, D. J. Hodgson, and R. Inger (2013). 'Smartphones in ecology and evolution: a guide for the app-rehensive.' *Ecology and Evolution* **3**(16): 5268–78.

Thiel, C. and C. Schmullius (2016a). 'Comparison of UAV photograph-based and airborne lidar-based point clouds over forest from a forestry application perspective.' *International Journal of Remote Sensing*. doi.org/10.1080/01431161.2016.1225181

Thiel, C. and C. Schmullius (2016b). *Derivation of forest parameters from stereographic UAV data - A comparison with airborne LiDAR data*. Proceedings CD of ESA Living Planet Symposium, Prague, Czech Republic.

Thomas, B., J. D. Holland, and E. O. Minot (2012). 'Wildlife tracking technology options and cost considerations.' *Wildlife Research* **38**(8): 653–63.

Thomas, L., S. T. Buckland, E. A. Rexstad, J. L. Laake, S. Strindberg, S. L. Hedley, J. R. B. Bishop, T. A. Marques, and K. P. Burnham (2009). 'Distance software: design and analysis of distance sampling surveys for estimating population size.' *Journal of Applied Ecology*. doi: 10.1111/j.1365-2664.2009.01737.x

Thornton, P. K., R. H. Fawcett, J. B. Dent, and T. J. Perkins (1990). 'Spatial weed distribution and economic thresholds for weed control.' *Crop Protection* **9**(5): 337–42.

Tonkin, T. N., N. G. Midgley, D. J. Graham, and J. C. Labadz (2014). 'The potential of small unmanned aircraft systems and structure-from-motion for topographic surveys: A test of emerging integrated approaches at Cwm Idwal, North Wales.' *Geomorphology* **226**: 35–43.

Toor, A. (2017). 'This African park has a high-tech plan to combat poachers.' from https://www.theverge.com/2017/7/20/16002752/smart-park-rwanda-akagera-poaching-lorawan.

Torres, J., G. Arroyo, C. Romo, and J. De Haro (2012). *3D Digitization using Structure from Motion*. CEIG-Spanish Computer Graphics Conference.

Toth, C. and G. Jóźków (2016). 'Remote sensing platforms and sensors: A survey.' *ISPRS Journal of Photogrammetry and Remote Sensing* **115**: 22–36.

Tremblay, J. A., A. Desrochers, Y. Aubry, P. Pace, and D. M. Bird (2017). 'A low-cost technique for radio-tracking wildlife using a small standard unmanned aerial vehicle.' *Journal of Unmanned Vehicle Systems* **5**(3): 102–8.

Turner, D., A. Lucieer, Z. Malenovský, D. H. King, and S. A. Robinson (2014). 'Spatial co-registration of ultra-high resolution visible, multispectral and thermal images acquired with a micro-UAV over Antarctic Moss Beds.' *Remote Sensing* **6**(5): 4003–24.

Turner, D., A. Lucieer, and C. Watson (2012). 'An automated technique for generating georectified mosaics from ultra-high resolution unmanned aerial vehicle (UAV) imagery, based on structure from motion (SfM) point clouds.' *Remote Sensing* **4**(5): 1392–410.

Twidwell, D., C. R. Allen, C. Detweiler, J. Higgins, C. Laney, and S. Elbaum (2016). 'Smokey comes of age: unmanned aerial systems for fire management.' *Frontiers in Ecology and the Environment* **14**(6): 333–9.

van Andel, A., S. A. Wich, C. Boesch, L. P. Koh, M. M. Robbins, J. Kelly, and H. S. Kühl (2015). 'Locating chimpanzee nests and identifying fruiting trees with an Unmanned Aerial Vehicle.' *Americal Journal of Primatology* **77**(1122–34).

van Blyenburgh, P. (2013). 2013–2014 RPAS Yearbook: Remotely Piloted Aircraft Systems: The Global Perspective 2013/2014. Paris, France, UVS International.

van der Werf, G. R., J. Dempewolf, S. N. Trigg, J. T. Randerson, P. S. Kasibhatla, L. Giglio, D. Murdiyarso, W. Peters, D. Morton, and G. Collatz (2008). 'Climate regulation of fire emissions and deforestation in equatorial Asia.' *Proceedings of the National Academy of Sciences USA* **105**(51): 20350–5.

van Gemert, J. C., C. R. Verschoor, P. Mettes, K. Epema, L. P. Koh, and S. Wich (2014). 'Nature conservation drones for automatic localization and counting of animals.' *ECCV Workshops* 1: 255–270.

van Schaik, C. P., A. Priatna, and D. Priatna (1995). Population estimates and habitat preferences of orang-utans based on line transects of nests. *The neglected ape*. R. D. Nadler, B. M. F. Galdikas, L. K. Sheeran and N. Rosen. New York, Plenum Press: 129–47.

Vas, E., A. Lescroël, O. Duriez, G. Boguszewski, and D. Grémillet (2015). 'Approaching birds with drones: first experiments and ethical guidelines.' *Biology Letters* **11**(2): 20140754.

Ventura, D., M. Bruno, G. J. Lasinio, A. Belluscio, and G. Ardizzone (2016). 'A low-cost drone based application for identifying and mapping of coastal fish nursery grounds.' *Estuarine, Coastal and Shelf Science* **171**: 85–98.

Verhoeven, G. J. (2009). 'Providing an archaeological bird's-eye view–an overall picture of ground-based means to execute low-altitude aerial photography (LAAP) in Archaeology.' *Archaeological Prospection* **16**(4): 233–49.

Vericat, D., J. Brasington, J. Wheaton, and M. Cowie (2009). 'Accuracy assessment of aerial photographs acquired using lighter-than-air blimps: low-cost tools for mapping river corridors.' *River Research and Applications* **25**(8): 985–1000.

Vermeulen, C., P. Lejeune, J. Lisein, P. Sawadogo, and P. Bouché (2013). 'Unmanned aerial survey of elephants.' *PLoS ONE* **8**(2): e54700.

Villa, T. F., F. Gonzalez, B. Miljievic, Z. D. Ristovski, and L. Morawska (2016). 'An overview of small unmanned aerial vehicles for air quality measurements: present applications and future prospectives.' *Sensors* **16**(7): 1072.

Wackrow, R., J. H. Chandler, and P. Bryan (2007). 'Geometric consistency and stability of consumer-grade digital cameras for accurate spatial measurement.' *The Photogrammetric Record* **22**(118): 121–34.

Wallace, L., A. Lucieer, and C. Watson (2012a). 'Assessing the feasibility of UAV-based LiDAR for high resolution forest change detection.' *Proc. ISPRS, Int. Archives Photogramm., Remote Sens. Spatial Inf. Sci* **38**: B7.

Wallace, L., A. Lucieer, C. Watson, and D. Turner (2012b). 'Development of a UAV-LiDAR system with application to forest inventory.' *Remote Sensing* **4**(6): 1519–43.

Ward, S., J. Hensler, B. Alsalam, and L. F. Gonzalez (2016). *Autonomous UAVs wildlife detection using thermal imaging, predictive navigation and computer vision*. Proceedings of the 2016 IEEE Aerospace Conference, IEEE, Yellowstone Conference Center, Big Sky, Montana: 1–8.

Watts, A. C., W. S. Bowman, A. H. Abd-Elrahman, A. Mohamed, B. E. Wilkinson, J. Perry, Y. O. Kaddoura, and K. Lee (2008). 'Unmanned Aircraft Systems (UASs) for ecological research and natural-resource monitoring (Florida).' *Ecological Restoration* **26**(1): 13–14.

Watts, A. C., J. H. Perry, S. E. Smith, M. A. Burgess, B. E. Wilkinson, Z. Szantoi, P. G. Ifju, and H. F. Percival (2010). 'Small unmanned aircraft systems for low-altitude aerial surveys.' *The Journal of Wildlife Management* **74**(7): 1614–19.

Wegge, P., C. P. Pokheral, and S. R. Jnawali (2004). 'Effects of trapping effort and trap shyness on estimates of tiger abundance from camera trap studies.' *Animal Conservation* **7**(3): 251–6.

Weissensteiner, M. H., J. W. Poelstra, and J. B. Wolf (2015). 'Low-budget ready-to-fly unmanned aerial vehicles: An effective tool for evaluating the nesting status of canopy-breeding bird species.' *Journal of Avian Biology* **46**(4): 425–30.

Wester-Ebbinghaus, W. (1980). 'Aerial photography by radio controlled model helicopter.' *Photogrammetric Record* **10**(55): 85–92.

Westoby, M., J. Brasington, N. Glasser, M. Hambrey, and J. Reynolds (2012). ''Structure-from-Motion' photogrammetry: A low-cost, effective tool for geoscience applications.' *Geomorphology* **179**: 300–14.

White, J. C., M. A. Wulder, M. Vastaranta, N. C. Coops, D. Pitt, and M. Woods (2013). 'The utility of image-based point clouds for forest inventory: A comparison with airborne laser scanning.' *Forests* **4**(3): 518–36.

Whitehead, K. and C. H. Hugenholtz (2014). 'Remote sensing of the environment with small unmanned aircraft systems (UASs), part 1: a review of progress and challenges 1.' *Journal of Unmanned Vehicle Systems* **2**(3): 69–85.

Whitehead, K., C. H. Hugenholtz, S. Myshak, O. Brown, A. LeClair, A. Tamminga, T. E. Barchyn, B. Moorman, and B. Eaton (2014). 'Remote sensing of the environment with small unmanned aircraft systems (UASs), part 2: scientific and commercial applications.' *Journal of Unmanned Vehicle Systems* **02**(03): 86–102.

Whiteside, T. G., G. S. Boggs, and S. W. Maier (2011). 'Comparing object-based and pixel-based classifications for mapping savannas.' *International Journal of Applied Earth Observation and Geoinformation* **13**(6): 884–93.

Wich, S., D. Dellatore, M. Houghton, R. Ardi, and L. P. Koh (2016). 'A preliminary assessment of using conservation drones for Sumatran orang-utan (*Pongo abelii*) distribution and density.' *Journal of Unmanned Vehicle Systems* **4**: 45–52.

Wich, S., L. P. Koh, and Z. Szantoi (in press). Classifying very-high resolution land cover mosaics acquired from drones. *New Geospatial Approaches in Anthropology*. R. Anemone and G. C. Conroy. Santa Fe, New Mexico, School of Advanced Research Press.

Wich, S., L. Scott, and L. P. Koh (2016). Wings for wildlife: the use of conservation drones, challenges and opportunities. *The Good Drone*. K. B. Sandvik and M. G. Jumbert. Oxon, UK, Routledge: 153–67.

Wich, S. A. (2015). Drones and conservation. *Drones and aerial observation: New technologies for property rights, human rights, and global development. A primer*, K. Kakaes. Washington, D.C., New America: 63–71.

Wich, S. A., I. Singleton, M. G. Nowak, S. S. Utami Atmoko, G. Nisam, S. M. Arif, R. H. Putra, R. Ardi, G. Fredriksson, G. Usher, D. L. A. Gaveau, and H. S. Kühl (2016). 'Land-cover changes predict steep declines for the Sumatran orangutan (*Pongo abelii*).' *Science Advances* **2**(3).

Wiegmann, D. A. and N. Taneja (2003). 'Analysis of injuries among pilots involved in fatal general aviation airplane accidents.' *Accident Analysis & Prevention* **35**(4): 571–7.

Wilkinson, B. E. (2007). *The design of georeferencing techniques for unmanned autonomous aerial vehicle video for use with wildlife inventory surveys: A case study of the National Bison Range, Montana.* MSc thesis, University of Florida.

Wilkinson, B. E., B. A. Dewitt, A. C. Watts, A. H. Mohamed, and M. A. Burgess (2009). 'A new approach for pass-point generation from aerial video imagery.' *Photogrammetric Engineering & Remote Sensing* **75**(12): 1415–23.

Wilson, A. M., J. Barr, and M. Zagorski (2017). 'The feasibility of counting songbirds using unmanned aerial vehicles.' *The Auk* **134**(2): 350–62.

Wilson, R. P., B. Culik, R. Danfeld, and D. Adelung (1991). 'People in Antarctica—how much do Adélie Penguins *Pygoscelis adeliae* care?' *Polar Biology* **11**(6): 363–70.

Winter, I. I. d. (2012). *The behaviour, activity pattern and substrate use of green turtles (*Chelonia mydas*) in a heavily grazed seagrass meadow, East Kalimantan, Indonesia.* MSc thesis, Wageningen University and Research Centre and Radboud University Nijmegen.

Winter, S. R., S. Rice, G. Tamilselvan, and R. Tokarski (2016). 'Mission-based citizen views on UAV usage and privacy: an affective perspective.' *Journal of Unmanned Vehicle Systems* **4**: 125–35.

Wittemyer, G., J. M. Northrup, J. Blanc, I. Douglas-Hamilton, P. Omondi, and K. P. Burnham (2014). 'Illegal killing for ivory drives global decline in African elephants.' *Proceedings of the National Academy of Sciences USA* **111**(36): 13117–21.

Wolf, P. R., B. A. Dewitt, and B. E. Wilkinson (2014). *Elements of Photogrammetry: with applications in GIS*. New York, McGraw-Hill.

Woll, C. L., A. Prakash, and T. Sutton (2011). 'A case-study of in-stream juvenile salmon habitat classification using decision-based fusion of multispectral aerial images.' *Applied Remote Sensing Journal* **2**(1): 37–46.

Wright, J. P. and C. G. Jones (2006). 'The concept of organisms as ecosystem engineers ten years on: progress, limitations, and challenges.' *BioScience* **56**(3): 203–9.

Yang, Z., T. Wang, A. K. Skidmore, J. de Leeuw, M. Y. Said, and J. Freer (2014). 'Spotting east African mammals in open savannah from space.' *PLoS ONE* **9**(12): e115989.

Yanmaz, E., M. Quaritsch, S. Yahyanejad, B. Rinner, H. Hellwagner, and C. Bettstetter (2017). 'Communication and coordination for drone networks.' *Ad Hoc Networks: 8th International Conference, ADHOCNETS 2016, Ottawa, Canada, September 26–27, 2016, Revised Selected Papers*. Y. Zhou and T. Kunz. Cham, Springer International Publishing: 79–91.

Yuan, C., Y. Zhang, and Z. Liu (2015). 'A survey on technologies for automatic forest fire monitoring, detection, and fighting using unmanned aerial vehicles and remote sensing techniques.' *Canadian Journal of Forest Research* **45**(7): 783–92.

Zahawi, R. A., J. P. Dandois, K. D. Holl, D. Nadwodny, J. L. Reid, and E. C. Ellis (2015). 'Using lightweight unmanned aerial vehicles to monitor tropical forest recovery.' *Biological Conservation* **186**: 287–95.

Zaman, B., A. M. Jensen, and M. McKee (2011). *Use of high-resolution multispectral imagery acquired with an autonomous unmanned aerial vehicle to quantify the spread of an invasive wetlands species*. Geoscience and Remote Sensing Symposium (IGARSS), 2011 IEEE International, IEEE.

Zarco-Tejada, P. J., R. Diaz-Varela, V. Angileri, and P. Loudjani (2014). 'Tree height quantification using very high resolution imagery acquired from an unmanned aerial vehicle (UAV) and automatic 3D photo-reconstruction methods.' *European Journal of Agronomy* **55**: 89–99.

Zarco-Tejada, P. J., M. Guillén-Climent, R. Hernández-Clemente, A. Catalina, M. González, and P. Martín (2013). 'Estimating leaf carotenoid content in vineyards using high resolution hyperspectral imagery acquired from an unmanned aerial vehicle (UAV).' *Agricultural and Forest Meteorology* **171**: 281–94.

Zhang, J., J. Hu, J. Lian, Z. Fan, X. Ouyang, and W. Ye (2016). 'Seeing the forest from drones: Testing the potential of lightweight drones as a tool for long-term forest monitoring.' *Biological Conservation* **198**: 60–9.

Zhou, G., C. Li, and P. Cheng (2005). *Unmanned aerial vehicle (UAV) real-time video registration for forest fire monitoring*. Proceedings, 2005 IEEE International Geoscience and Remote Sensing Symposium, 2005, IGARSS'05, IEEE.

Index

Tables and figures are indicated by an italic *t* or *f* following the page number.

The Complete IDIOT'S Guide to CorelDRAW!

by Jenna & Michael Howard

alpha books

A Division of Macmillan Computer Publishing
201 West 103rd Street, Indianapolis, Indiana 46290 USA

We dedicate this book to our wonderful nephews and nieces: Nicholas, Emile, Andrew, Meahgan, and Michael.

International Standard Book Number: 1-56761-429-9
Library of Congress Catalog Card Number: 94-70519

96 95 94 8 7 6 5 4 3 2 1

Interpretation of the printing code: the rightmost number of the second series of numbers is the number of the book's printing. For example, a printing code of 94-1 shows that the first printing of the book occurred in 1994.

Printed in the United States of America

Screen reproductions in this book were created by means of the program Collage Plus from Inner Media, Inc., Hollis, NH.

Publisher
Marie Butler-Knight

Managing Editor
Elizabeth Keaffaber

Product Development Manager
Faithe Wempen

Acquisitions Manager
Barry Pruett

Senior Development Editor
Seta Frantz

Manuscript Editor
Audra Gable

Book Designer
Barbara Webster

Production Team
Gary Adair, Dan Caparo, Brad Chinn, Kim Cofer, Lisa Daugherty, Jennifer Eberhardt, Greg Eldred, Mark Enochs, Jenny Kucera, Beth Rago, Bobbi Satterfield, Kris Simmons, Greg Simsic, Carol Stamile, Robert Wolf

Special thanks to Christopher Denny for ensuring the technical accuracy of this book.

Contents at a Glance

1 The Top 10 Things You Need to Know About CorelDRAW! 3
Knowing this stuff will make your time with CorelDRAW! so much more enjoyable.

2 CorelDRAW!: More for Your Money 7
An overview of the mammoth CorelDRAW! program.

3 Taking Your First Step 13
Starting, exiting, undoing, getting help—you know, the mundane essentials.

4 Juggling Drawings 25
You can't get anywhere without knowing this stuff.

5 Getting into Shape 35
Draw better shapes than you did in elementary school.

6 To Select or Not to Select 43
Try to do something to an object that's not selected and see how far you get.

7 Getting a Good Look 51
Making your screen smile and do other things.

8 Right Down to the Line 63
The ABC's of drawing lines—and the XYZ's too.

9 If You Don't Like It, Rearrange It 77
Nobody ever likes their stuff the first time around. Now you can rearrange your objects.

10 Nodes, Nerds, and Segments: Reshaping Your Objects 87
Everything you want to know, and don't want to know, about reshaping your objects.

11 Color Me Beautiful 97
This is not about Avon products; it's about coloring your objects.

12 What a Transformation! 105
Doing this and that, yes and even that, to your objects.

13 The Grab Bag: Pages, Master Layers, and Symbols 113
What we couldn't logically fit into any of the other chapters.

14 Rolling the Presses 123
So you want to know how to print, huh?

15 Painting Pictures with Words 131
You still can't get away from word processing—even in a graphics program.

16 Revamping Text 143
The second round with your words.

17 **Twisted Text** 155
Doing all sorts of unspeakable things to the text you added to your drawing.

18 **The Envelope, Please** 165
Who won for best supporting actress. (Not. More like, molding your objects into some kinky shapes.)

19 **Blend for 30 Seconds** 179
Sticking two of your objects into a blender and blending them to death.

20 **Getting Deep: Extruding Objects** 187
Making your objects look like they have depth, even though you and everyone else know they don't.

21 **A Smuggler's Dream: Importing and Exporting Your Goods** 201
How to snag stuff from other programs and sneak it into CorelDRAW!, and vice-versa.

22 **Linking and Embedding, OLE!** 207
Sharing your masterpiece with other programs and vice versa.

23 **Charting Your Way to the Top** 221
A more acceptable way to climb the corporate ladder . . . with snazzy looking charts.

24 **The Coolest Feature of All: CorelMOVE!** 237
Making the objects on your screen move in funny ways.

25 **Watch Out, Picasso** 251
Adding some really cool effects to images.

26 **SHOW and Tell** 279
Taking all your cool stuff from the other CorelDRAW! programs and showing them off to your colleagues via a screen show.

27 **Blast Off with the Corel 5.0 Engine** 297
Skip this if you have no interest in CorelDRAW! 5.

28 **A Crash Course on Windows** 313
Provided for our less-experienced Windows users (yes, there are still some of you out there).

29 **Put It to Work** 323
All sorts of cool things you can do with the CorelDRAW! programs.

Installing CorelDRAW! 339
Take a guess.

Speak Like a Geek: The Complete Archive 349
Otherwise known as a glossary.

Index 359

Contents

Introduction

Don't be insulted by the title of this book. Most people feel like an idiot the first time they use CorelDRAW!. It's not a simple program. But don't let that scare you off. This book makes it simple for you to learn. So we'd like to congratulate you on your wise investment—it proves you're no idiot!

For Whom the Book Tolls

If you've never even looked at CorelDRAW!, this book is for you. If you've looked at CorelDRAW! and shied away, this book is for you. Basically, this book is for all beginning users of CorelDRAW!. It will help greatly if you have had some experience with Microsoft Windows. But for those of you who are new to Windows as well as CorelDRAW!, you'll need to start near the end of this book, with Chapter 28, "A Crash Course on Windows." This chapter will fill you in on some of the Windows lingo and Windows' way of doing things.

What's Inside This Book

This book is not meant to be comprehensive documentation on all you can possibly know about CorelDRAW!. You're lucky: we've picked what are generally the most used features CorelDRAW! has to offer, so that by the time you're done with this book, you'll be 100% familiar with what you'll use 90% of the time in CorelDRAW!.

This book is organized in a logical manner (after all, it's not written for idiots). You start with the basics in Part I, like how to open CorelDRAW!, retrieve and create new files, save your drawings, select objects—you know, the kind of stuff you need to know before you can do anything meaty. Part II gets you into the heart of drawing, and includes instruction on how to draw lines and shapes, reshape, resize, rotate, and flip objects, work with colors, and print your drawings.

CorelDRAW! isn't exclusively a graphics program. You can create complete documents with CorelDRAW!, including brochures, flyers, and manuals. But to do so, you need to know something about CorelDRAW!'s text features. Part III is entirely devoted to working with text in CorelDRAW!.

Part IV is the fun part. You learn how to blend your objects, mold them into weird shapes, and stretch them into a 3D look. Part IV also includes help on how to use files created in other software programs in CorelDRAW!, and how to use CorelDRAW! drawings in other programs.

You'll soon find out that CorelDRAW! is not a single program. It's a schizophrenic program, if you will, and actually has seven accessory programs. Part V is devoted to giving you the basics on some of the more useful accessory programs, including CorelCHART!, CorelPHOTO-PAINT!, CorelMOVE!, and CorelSHOW!.

The last part, Part VI, includes a chapter exclusively for users of CorelDRAW!'s latest release, version 5 (though would-be version 5 users should feel free to check out some of the new features here). Part VI also contains chapters on Windows basics and a chapter called "Put It to Work," which includes ideas on how to use CorelDRAW! in the real world.

Also provided, towards the end of the book (not really stuck in a part), is a section on installing CorelDRAW! and a section called "Speak Like a Geek: The Complete Archive," otherwise known as a glossary of terms.

How to Use This Book

Even though you were always taught to read a book from start to finish, don't do that with this one. Not only should you not feel guilty for skipping around, we encourage it! Probably the best way to find what you need is to look through the Table of Contents and Index to find the chapter or page that discusses the topic you need help on.

As far as conventions go, who needs them anyway? Well, even though we live in a liberated age, this book still holds to some conventions (go ahead, call us old-fashioned). For example, when you're supposed to press a key on your keyboard, the name of the key will appear in bold, such as:

Press **Enter** to continue.

Sometimes you'll need to press two keys at a time, which some people call *shortcut keys* and others call a *key combination*. Whatever your fancy, you'll know when to use two keys because a plus sign appears between the two keys. Always hold down the first key as you press the second. For example, if you see:

Press **Alt+F** to open the **F**ile menu.

it means to hold down the **Alt** key while you press the **F** key.

You can open any menu by holding down the **Alt** key and pressing the selection letter in the menu name. You'll know what the *selection letter* is in CorelDRAW! because it is always underlined. In this book, we help you out by bolding the selection letter in a menu name, and in other function names, such as buttons and check boxes.

What's with All the Boxes?

There are some boxed notes in this book that point out special points:

Speak Like a Geek Simple definitions to help you wade through the technical lingo.

This will attract your attention to special hints from the know-it-alls who wrote this book.

Check out an easier way to do something here.

This information will help when things don't go the way you planned.

This is stuff that all users of version 5 should know.

This is stuff that all users of version 3 should know.

TECHNO NERD TEACHES...

This information is for techno nerds: only read these if you plan on being a CorelDRAW! know-it-all.

Acknowledgments

Special thanks are due to Seta Frantz for her steadfastness and patience, to Barry Pruett for helping the project take off and stay going, and to publisher Marie Butler-Knight.

We'd also like to thank each other for the incredible restraint we exercised when things got hectic. Like good citizens, we refrained from tearing each other's hair out and pouncing on each other's stomachs (as much as we wanted to at times).

Trademarks

Part I
Even a Child Can Draw This

Would you believe it if we told you it was harder for you to learn how to draw your first rectangles and squares back in Mrs. Dillweed's second grade class than it will be for you to learn how to draw shapes in CorelDRAW!? If you don't believe us, plop one of your kids (or a neighbor's kid) in front of your computer, and see what happens. After a couple of minutes, you'll probably see rectangles, circles, and lines drawn all over the page. But in case you don't have any kids around to show you how to do these things, we wrote this Part for you, so you can learn how to get up and running with CorelDRAW! quickly, and draw your first shapes and lines in a matter of moments.

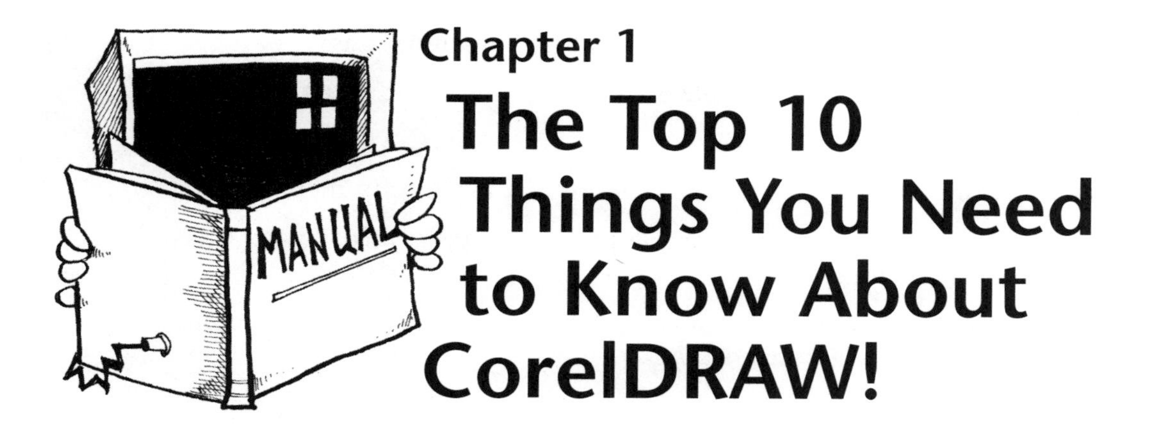

Chapter 1

The Top 10 Things You Need to Know About CorelDRAW!

How many times have you said to yourself, "If I had only known this when I was younger. . . ." Unfortunately, too few of us get a list of "the top ten things you really need to know about life" when we need it most—in our youth (though would we have paid attention to that list then?).

But luckily you do get a list of the ten things you need to know most about CorelDRAW! Okay, maybe it's not as significant as life itself, but it's certainly useful. And presumably, you are an adult now—so we expect you to use this list!

1. CorelDRAW! is not one, but actually eight separate software programs.

The creators of CorelDRAW! really weren't trying to confuse you when they decided to call their entire graphics package, and as well as a single program in that package, by the same name: CorelDRAW!. The following list will help to keep things straight:

CorelDRAW! 1) The name of the whole software package, and 2) a program for creating beautiful drawings.

CorelCHART! A program for creating awesome data charts.

CorelPHOTO-PAINT! A program for touching up scanned photos or graphics.

CorelMOVE! A program for creating fun animations.

CorelSHOW! A program for creating captivating presentations.

CorelTRACE! A program for scanning images.

CorelMOSAIC! A file manager for holding everything together.

CCapture! A program that captures the appearance of your screen, creating a file.

The first five programs are major programs, while the last three are minor (sometimes called *utilities*).

2. CorelDRAW! runs under Windows.

Learning CorelDRAW! is a big enough challenge without trying to learn Windows at the same time. If you aren't yet familiar with how the Windows environment works, review Chapter 28, "A Crash Course on Windows," before you start using CorelDRAW!. Be of good cheer though; after you get used to Windows, you'll agree with us that Windows is easy. Perhaps you'll chime in to the old tune: "Working in DOS programs is like driving in a car without windows."

3. You will use your mouse for almost all of the work you do.

Some of you are probably cringing, while others are rejoicing. For keyboard devotees, hopping onto a chair won't help you get away from this mouse. But you'll probably learn to like the mouse more when you see what it can do in CorelDRAW!. Mouse-lovers, sigh with satisfaction: here's a program that lets you do everything using the mouse (you'll only use the keyboard to type in text).

4. Press F1 for help.

You'll fall in love with on-line help. Let's say you're stuck trying to perform a function, but you don't feel like searching though a huge table of contents or glossary for help. Well, you don't need to; just press **F1**, and the help system displays information on the very function you're using. It's smart—most of the time you don't have to tell it what you need help on, it already knows.

5. Save often!

There's nothing more frustrating than spending a couple of hours on a project only to lose everything you've done because you didn't save to disk before some tragedy occurred.

6. A blank drawing page is ready and waiting for you whenever you start CorelDRAW!.

If you grew up on some of the older word processing programs, after you start CorelDRAW! you may think, "How do I start a new file to begin?" You don't have to start anything: you're ready to go as soon as you start CorelDRAW!. A new file is created automatically for you when you start the program. Just start drawing and save your file!

7. CorelDRAW! is a fat cat; it takes up loads of room and can create humongous files.

Depending on your computer system, you may have enough time to eat breakfast before CorelDRAW! completely opens up. And you may be able to take a good solid nap while waiting for some of your more hefty files to open up. Solution? Buy more RAM, or a faster computer!

8. The CorelDRAW! toolbox has two types of tool icons.

What you see is not what you get when it comes to CorelDRAW!'s toolbox. You get more. Some of the tools have additional subtools that will be displayed when you select the tool. But here's the trick: for some of the tools, the subtools appear when you click on the tool; for others you have to click and hold the mouse button over the tool until the subtools appear. So if you know a tool has subtools, but nothing happens when you click on the tool, try holding down the mouse button on the tool instead.

9. Learn to use those roll-ups.

Roll-ups can do wonders for you. These windows stay open on your screen until you decide to close them. They're full of all sorts of commands that you'd otherwise have to use repeated menu selections in order to access. Roll-ups give you immediate access to some of the commands you'll use most. So experiment with them until you become comfortable enough to use roll-ups on a regular basis.

10. Always remember to select an object before you try to do anything with it.

You can't get anywhere in editing an object if you don't select that object first. Whether you want to copy it, move it, stretch it, color it, outline it, or delete it, you must select the object first with the Selection tool at the top of the toolbox.

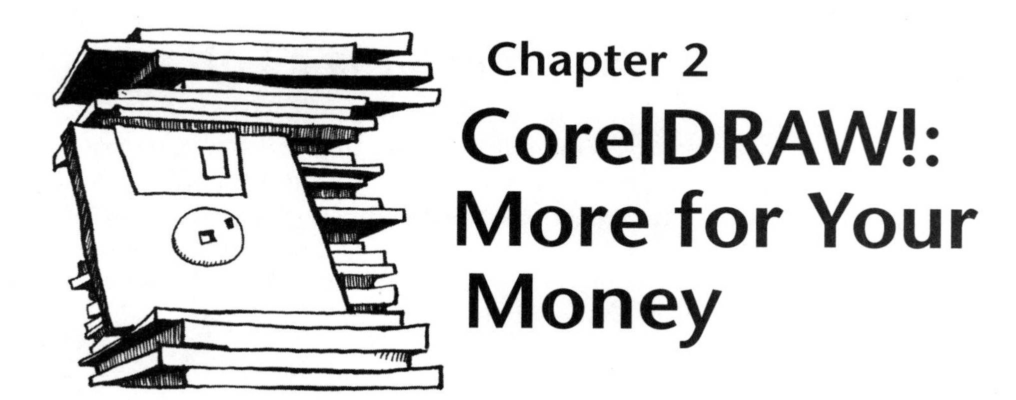

Chapter 2
CorelDRAW!: More for Your Money

In This Chapter

- The amazing CorelDRAW! package
- The best of the best: CorelDRAW! and CHART!
- PHOTO-PAINT! and TRACE!: For true techno nerds
- MOSAIC! and CCapture: Take 'em or leave 'em

You saw them if you installed all CorelDRAW! programs, didn't you: eight little icons tidily arranged in a Corel group window. But you had expected only one icon. You thought you were buying just one software program, albeit with lots of features; after all, there was only *one* box, *one* set of disks, and *one* name. Instead, after you installed your nifty new CorelDRAW! disks, you found eight new program icons sitting on your computer desktop. Seven miscreants you hadn't heard of, never mind invited. CorelTRACE!? CorelCHART!? CCapture? "What are all these icons for?" you may have asked yourself rather nervously . . . or impatiently (depending on your personality). Well, you can relax. This chapter will tell you all about these accessories.

You get even more with CorelDRAW! 5, including a complete desktop publisher (Corel Ventura), a font manager, and a database utility that allows you to extract information for use in Ventura or Chart.

What You Got Is a Lot

You must have guessed when CorelDRAW! needed 37 megs of your hard disk that this was one massive program. You were right that it's massive, but you were wrong that it's one program. CorelDRAW! actually consists of eight separate programs, ranging from powerful applications, to small, take-em-or-leave-em applications.

The Work Horse: CorelDRAW!

This is the head honcho. El Capitan. Mr. Big. You'll probably work with CorelDRAW! more than any of the other programs. Why? Because you probably bought a graphics program to create and edit graphics. And if that's what you want to do, CorelDRAW! is what you'll use to do it. You can draw buttons and bowknots, cake and ice cream, planes, trains, and automobiles, or a sleepless single (in Seattle)—to name a few. Basically, you can draw anything with CorelDRAW!.

What's more, Corel supplies you with tons of clip art and hundreds of fonts for use in your drawings. Rather than struggle with designing your own plane, check out the different planes Corel has already drawn and provides with the program. And for those individualists out there, you can even edit Corel's pre-drawn clip art to suit your own tastes.

If you purchased CorelDRAW! 4 or 5, the tons of clip art Corel supplies comes on a CD-ROM, along with hundreds of fonts, animations, sounds, and more. There's a wealth of material, but you must have a CD drive to access it. If you don't have one, you can do one of a number of things: buy a CD drive, ask a friend or co-worker who has a CD drive to copy the files you want to a disk, see if a local computer store will download the files onto a floppy for you, or throw away the CD-ROM disc (however, we don't recommend this, as you may win the lotto soon and could then get that CD drive). If you do have a CD drive, consider running CorelDRAW! from the CD to save some valuable hard drive space. Read Chapter 28, "Installing CorelDRAW!," for details.

It goes even further! You can use basically any clip art created in virtually any program, thanks to Corel's ability to import all sorts of different files (see Chapter 21, "Importing and Exporting").

Clip art Graphics you can import in CorelDRAW!. CorelDRAW! comes with its own clip art, but you can use clip art created in almost any graphics program.

For those of you who'll be using CorelDRAW! to create brochures and flyers, you'll be amazed by how professional your documents will look. And be sure to check out Corel's templates (files already set up for you by CorelDRAW!); they include some creative looking flyers, brochures, and newsletters that you can type your own information into.

The Brilliance of CorelCHART!

If you've ever worked with a presentation program such as Freelance Graphics, Microsoft Powerpoint, Aldus Persuasion, or Harvard Graphics, you'll be right at home with CorelCHART!. But don't expect CHART! to do everything these other programs do, because it doesn't run screen shows or let you create slides. CHART! does exactly what its name implies: it lets you create awesome charts. We'd even venture to say that CHART! lets you create more types of graphs, more complex graphs, and more funky-looking graphics than any of the presentation programs we just mentioned. (If you wanted to create slide shows, never fear—CorelDRAW! has not let you down. But if you want to include these graphs in a slide show presentation, you'll have to turn to CorelSHOW!.)

CHART! creates just about any kind of graph you can imagine, and then some. Floating bars, 3-D, California style—you name it. Chapter 22 in this book introduces you to CHART!.

The Picasso of Graphic Programs: CorelPHOTO-PAINT!

PHOTO-PAINT! is used primarily for what its name hints at: photos. Using PHOTO-PAINT!, you can touch up photos so that they have a different background, unusual colors, captivating textures, and more.

"But how do you get a photograph into your computer," one asks with bewilderment. Quite easily, actually. All you need to do is get your hands on a scanned image of the photograph. You don't even need a scanner to do this, though a scanner would make your life a whole lot easier if you intend to include photographs in your documents or presentations on a regular basis. Your local print shop or a service bureau may be able to scan your photographs for you. Or, if you want to use generic photos, you can purchase CD-ROMs that contain dozens of photographic images.

Scanned photos generate large disk files. That's why it's helpful to have a CD-ROM drive if you plan to work with PHOTO-PAINT! on a regular basis. Also, a Photo CD drive opens the door to better quality and more easily accessed photo images; you can transfer undeveloped rolls of film onto a CD-ROM.

So let's say you went to the trouble of having your photograph scanned at a local service bureau, only to find that it doesn't look like a photo at all on-screen. Instead, it looks like a drawing that some five year-old created in CorelDRAW!. What's the problem? Probably an inadequate video card. You may want to invest in a new graphic card (such as ATI's Ultra) to make your photos look on-screen just as they do in your hand.

Don't let it be said that we implied PHOTO-PAINT! can only be used for editing photographic images. You can create pictures of your own from scratch. It's a very simple process; so simple in fact that you should be able to figure it out on your own just by playing around with the different painting tools. Nevertheless, we do help you out in Chapter 24, "PHOTO-PAINT!ing Pretty Pictures."

CorelTRACE!

TRACE! serves one primary purpose: it converts scanned images for use in CorelDRAW!. You cannot edit scanned images in CorelDRAW! without first converting the images in TRACE!. For example, if you wanted to bring a blueprint into CorelDRAW! to edit and print, you would first scan the blueprint (or have it scanned at a service bureau), then bring it into TRACE! to clean it up and convert it to a drawing, and finally open the file in CorelDRAW!.

Move It with CorelMOVE!

One of the latest crazes is animation: making computer objects come alive on-screen captivates an audience and keeps them interested in what you have to say (unless what you have to say is really boring). It can also keep the creator interested as he or she works on an otherwise dull project. MOVE! makes creating full-motion objects easy. When you read Chapter 23, "On the MOVE!," you'll be surprised by just how simple it is to make your objects come alive.

No, you weren't gypped. CorelMOVE! does not come with CorelDRAW! 3.0. However, if you want to pick up a copy of CorelMOVE!, Corel sells it as a separate program. Check with your local computer software store.

Show and Tell with CorelSHOW!

It's almost as if Corel split up a full presentation program into several different pieces: CHART!, CorelDRAW!, PHOTO-PAINT!, MOVE!, and SHOW!. SHOW! is the one that pulls it all together and runs the show. You can bring CHART! graphs, CorelDRAW! drawings, MOVE! animations, and PHOTO-PAINT! images into a SHOW! presentation to run on your screen or someone else's computer. You don't have to press the Enter key or click the mouse to move from slide to slide: a SHOW! presentation can run on its own. Read Chapter 25, "SHOW! and Tell," for all the dirt on CorelSHOW!.

The Beauty of CorelMOSAIC!

This program can be fun—in a mindless sort of way. If you like looking at itsy bitsy pictures of the contents of your files, you'll like MOSAIC!. It can actually be quite useful to see the contents of your files (despite the fact that they appear in a size that demands reading glasses, even for people with 20-20 vision). Think of how much time you can save by not having to start up an application and open a file just to see what's inside that file. Primarily, MOSAIC! is a visual file manager that lets you manage your files easily. Its drag and drop feature lets you move or copy files between directories very quickly. Consider playing with this program when you get a free moment.

Catch It with CCapture

CCapture captures your screen configuration and saves it to a file. Usually only people who produce manuals or books on software (such as this one) will need screen shots of this sort. So, unless you're producing a manual on a software product, you won't need to use this program. And if you're nerdy enough to be writing a manual or book on an application, you certainly don't need us telling you how to use CCapture—indeed, you may be even more nerdy than we are! (Hey, we're writing a book on an application! Does this make us nerds?)

The Least You Need to Know

You're no idiot when it comes to CorelDRAW!. You now know what it's all about.

- CorelDRAW! is actually a bundled package of eight software programs: CorelDRAW!, CHART!, PHOTO-PAINT!, TRACE!, MOVE!, SHOW!, MOSAIC!, and CCapture.
- You'll probably use CorelDRAW! and CHART! the most.
- Unless your system is loaded with high-tech equipment, you may not use TRACE! very often, perhaps not at all.
- You'll have fun bringing your screen to life with MOVE! and SHOW!.
- You may find MOSAIC! and CCapture helpful.

Chapter 3
Taking Your First Step

In This Chapter

- Start it up!
- Dissecting the CorelDRAW! screen
- Windows that roll up and down
- Calling in the Red Cross
- Where's help when you need it?
- Closing up for the night

Have you ever watched a baby take its first step? Wobbly, uncertain—maybe the little tyke even falls. But then the second try comes and, lo and behold, this time, although baby is still wobbly and a bit uncertain, baby makes it. By the next day, baby gets to wherever baby wants to go (which might make Mommy and Daddy wish baby still crawled).

Assumptions are nasty things to make, but the editors twisted our arms and forced us to make the following assumptions in this book:

- You followed the typical, full installation.
- You have a printer available for your use.
- (Here's the trouble-maker): You use the mouse more than the keyboard for menu commands and operations. (For info on using the mouse, read Chapter 28, "A Crash Course on Windows.")
- This book is based on version 4.0, which most of our readers probably use, we're guessing. However, we'll provide help for 3.0 and 5.0 users along the way too.

While you're certainly no baby (unless you have an amazingly high IQ and can read this at age one), you may feel like one when you work with CorelDRAW! for the first time. But here's the clincher: the first time. CorelDRAW! is so logically designed and so user-friendly that you'll already feel you've made a lot of progress when you use CorelDRAW! the second time. And like baby, in no time at all, you'll get to wherever you want to go with CorelDRAW!—and fast.

All Aboard! Starting CorelDRAW!

Starting out is never very much fun, but neither is procrastinating (at least in the end). Believe us, starting CorelDRAW! is easy. Before you do anything else, make sure Windows is running (type **WIN** at your C:\> prompt or read Chapter 28, "A Crash Course on Windows," if you're desperate). Now, from the Program Manager, open the Corel program group. How? Just double-click on the **Corel** program group icon. Then double-click on the **CorelDRAW!** program icon.

You can start any of the Corel programs in the same way. Just double-click on the appropriate program icon. For now, start CorelDRAW! by double-clicking on the **CorelDRAW!** icon.

We suggest you leave the Corel 4.0 program group window open so that whenever you want to start CorelDRAW! in the future, all you need to do is double-click on the CorelDRAW! program icon (this lets you skip a step).

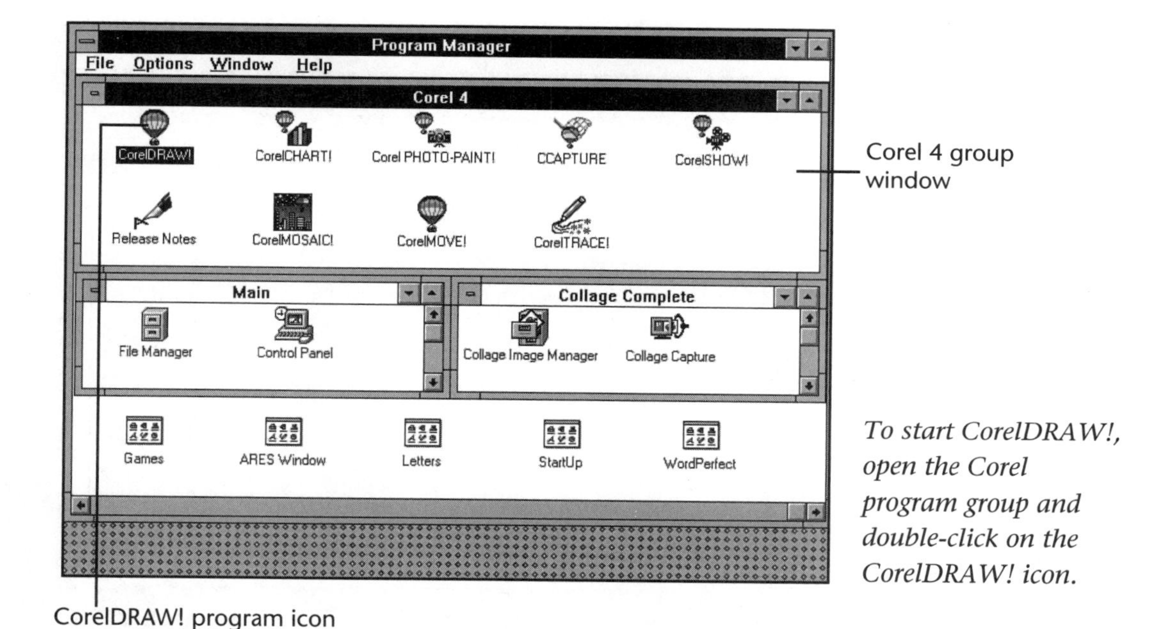

To start CorelDRAW!, open the Corel program group and double-click on the CorelDRAW! icon.

A Grand Tour of the CorelDRAW! Screen

Now to the basics. Take a quick look at the CorelDRAW! screen in the figure before moving on to the next paragraph.

Program group A window that holds icons for all seven of the Corel 4.0 programs.

Program icon An icon (located inside the program group window) that starts a program.

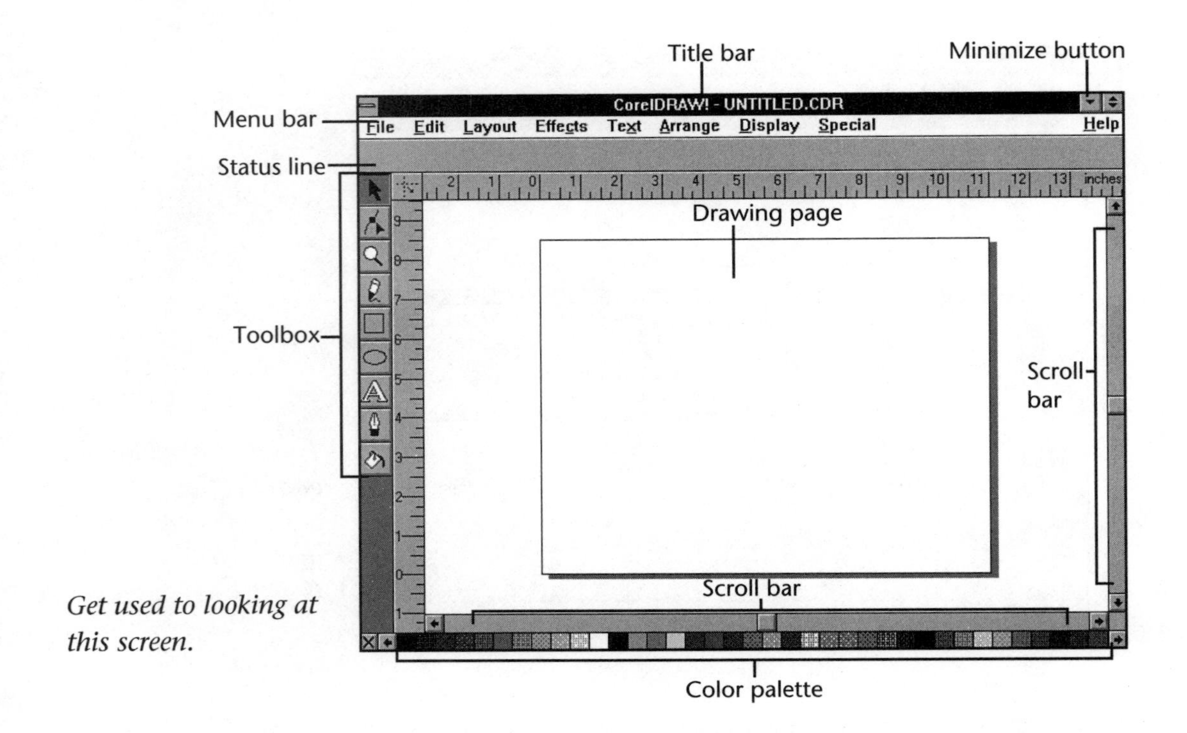

Get used to looking at this screen.

Whether you happen to like it or not, whenever you start CorelDRAW!, this screen appears. See the menus, icons, toolbox, and scroll bars on this screen? Get used to looking at them because they're your ticket to the performance of your life (that is, performing CorelDRAW! functions).

- The **title bar** at the top of the screen tells you the name of the drawing. "Untitled.CDR" appears here until you save the drawing.
- The **menu bar**, just below the title bar, holds the names of all those very important menus you'll be using.
- The **drawing page** in the middle of the window is where you'll draw objects and insert text. Everything that appears on the drawing page will be printed if you choose to print.
- We'll get into the **toolbox** in a minute.
- The **status line** appears towards the top of the screen. This line displays messages (dull ones), the coordinates of the insertion point, and other (I suppose, useful) stuff.

- The **scroll bars** at the right edge and bottom of the screen let you maneuver your way around the document. To scroll one line at a time, click the scroll arrow at the top or bottom of the vertical scroll bar. To scroll continuously, click the scroll arrow and hold down the mouse button. To scroll more quickly, drag the scroll box (the box that appears on the scroll bar somewhere between the scroll arrows) up or down the scroll bar.
- The **color palette** at the bottom of the screen has more colors than you'll know what to do with. You'll have fun filling your objects with these lovely, brilliant colors (but remember, no color printer—no pretty colors in print!).

If you've fooled around with the menus, you may have accidentally removed the status line from the screen without knowing it. Make sure there's a check mark next to **Show** Status Line on the **Display** menu if you want to see the status line.

The CorelDRAW! 5 screen contains a Ribbon bar full of icons. You can use these icons to perform many of your CorelDRAW! functions, such as opening and saving a drawing or adding a symbol to your drawing.

Shuffling Through Dad's Toolbox

You'll have so much more fun using CorelDRAW! tools than you did trying to use Dad's oversized tools when you were a kid. CorelDRAW! tools don't weigh a lot, don't slice you when you're not looking, and certainly don't pound on your thumb to make it black and blue. CorelDRAW!'s toolbox is much more tame, giving you access to the most common operations in CorelDRAW!. To use a tool, simply click on it.

Tool	Name	Function
	Selection	Selects and manipulates objects.
	Shape	Reshapes objects.

continues

continued

Tool	Name	Function
	Zoom	Changes how you view a drawing (how much of it you see).
	Pencil	Draws lines and curves.
	Rectangle	Draws rectangles and squares.
	Ellipse	Draws ellipses and circles.
	Text	Adds text and symbols.
	Outline	Changes outline settings.
	Fill	Changes fill settings.

The Symbols tool no longer appears as a subtool on the Text fool flyout menu. Instead, use the Symbols icon on the Ribbon bar.

BY THE WAY

Some CorelDRAW! tools have subtools attached. To display the subtools, sometimes all you need to do is click on the tool. However, for other tools (like that dern Pencil tool), you have to hold down the mouse button on the tool until the subtools appear.

Up They Go and Down They Go

For those of you who are experienced at doing Windows (you'd never admit to it), you may see something in CorelDRAW! you've never seen before: a *roll-up*. A roll-up does just what its name says: it rolls up and it rolls down. To display a roll-up, you have to select the menu command that opens it (for example, the Layers Roll-Up command on the Layout menu displays the Layers roll-up).

So what's so special about roll-ups? In a nutshell, they remain on-screen so that you can use them whenever you want. They contain loads of functions so that you don't have to keep using menu commands, and keep using menu commands, and . . . to do something like shape objects. What's also great is that you can roll up the window when you're not using it to clear more space on the screen, and roll it down when you're ready to use it again.

You can display a roll-up when the mood suits you, or you can make it so that a roll-up opens every time you start CorelDRAW!. Don't worry, you can always change your mind later and not have the window open.

Here's how you display a roll-up at startup (that is, until you change your mind about it).

5.0

CorelDRAW! 5 users can select which roll-ups to display using the **R**oll-Ups command on the **V**iew menu.

1. Display all the roll-ups you want to use for each session (select the roll-ups from the Effects, Layout, and Text menus).

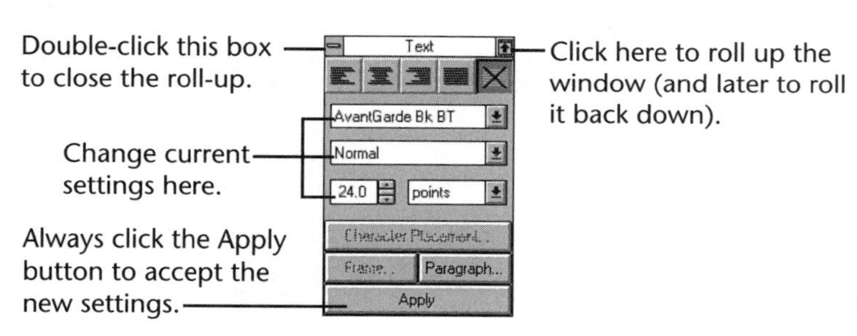
Double-click this box to close the roll-up.
Click here to roll up the window (and later to roll it back down).
Change current settings here.
Always click the Apply button to accept the new settings.
Text
AvantGarde Bk BT
Normal
24.0
points
Character Placement...
Frame...
Paragraph...
Apply

The Text roll-up.

2. Select the **S**pecial **P**references command. The Preferences dialog box appears on-screen.
3. Click on the **Roll-Ups** button. The Preferences Roll-Ups dialog box appears.
4. Mark the **C**urrent Appearance of Roll-Ups option. This forces CorelDRAW! (don't you like being a bully?) to open the roll-ups every time you start CorelDRAW! from now on.
5. Click on **OK**. Now, even if you close one of the roll-ups before you exit, it will still open the next time you start CorelDRAW!.

BY THE WAY

There's a certain pattern you need to follow when you use a roll-up. First select the object you want to change, then set the options in the roll-up, and finally, click on **Apply** (an important step) to make the changes take effect.

Unlike the Stoics, roll-ups can be easily moved. To move a roll-up to a new location, click on the roll-up's title bar and drag it to where you want it.

Undo to the Rescue

You just deleted the wrong object that you spent 20 hours working on. Don't panic! CorelDRAW!'s Undo feature comes to the rescue. And believe it or not, you can undo up to 99 actions! Don't you wish you had that same ability for undoing what comes out of your mouth?

And for those of you who are finicky, you can even redo what you've undone if you change your mind and want to do again the thing you just undid without having to do it all over again. Huh? In other words, you can redo what you've undone.

It's really simple to undo something: just display the **E**dit menu and click on **U**ndo. CorelDRAW! helps you remember what it is you're undoing: right after the word Undo, a description of the action appears. For example, if you just deleted an object, the Undo command will read **U**ndo Delete Object. And the same goes for R**e**do (which happens to appear oh-so conveniently beneath the **U**ndo command on the **E**dit menu).

Use the **S**pecial Preferences command to specify the number of Undo levels you want (up to 99). Beware CorelDRAW! 3 users! You're stuck with only one level of undo, so exercise caution!

Now pay special attention to this tidbit: not all actions can be undone. For example, you cannot undo zooming, scrolling, or opening, saving, or importing files. (CorelDRAW! isn't God, after all.) "Can't Undo" appears on the **E**dit menu for operations you can't undo.

Can Anyone Help Me?

Probably not. But if you need help when you're working in CorelDRAW!, you can get your hands on it quickly and easily. CorelDRAW! has what's called *context-sensitive help* or *on-line help*: helpful information that specifically applies to what you're working on at the moment or to anything you choose on-screen (such as menu commands, tools, or any object). To access this on-line help, press **Shift+F1**. The cursor becomes an arrow with a question mark. At this point, you can click on what you need information about, such as an ellipse you just drew. Or, you can click on a menu command, a tool, the ruler, the status line, and so on. CorelDRAW! then opens a window with help on the topic you specified.

Press Shift+F1 to display the Help cursor, and then click on what you need help on.

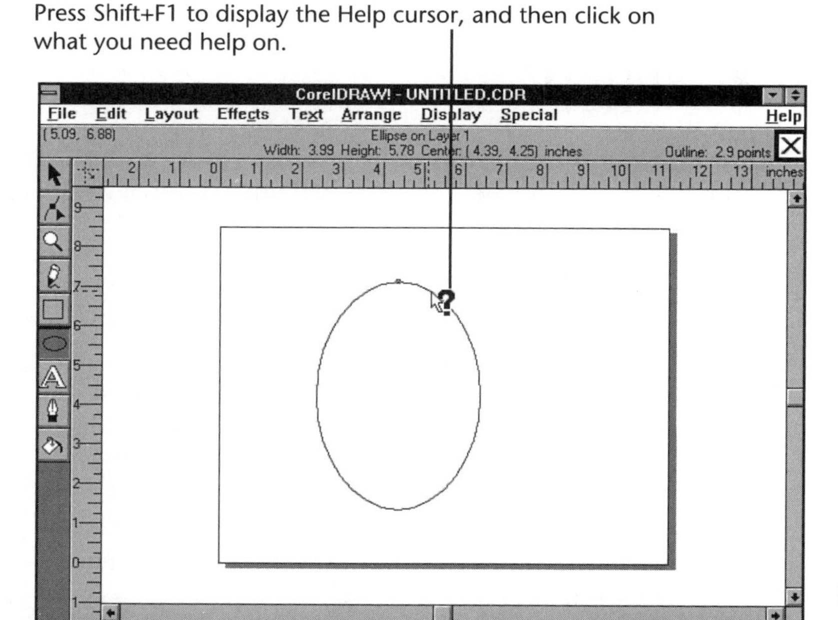

Using CorelDRAW!'s on-line help.

You can also use CorelDRAW!'s Help menu to access DRAW's entire help system. Press **F1**, or choose Contents from the Help menu to bring up the Help window. This window contains eight icons, which give you help for the following eight categories:

Using help

Screen

Commands

Tools

How to . . .

Keyboard

Glossary

Reference

Just click on one of the eight icons to bring up a window for that category. For example, if you click on the **Glossary** icon, a list of terms and definitions appears.

Making a Graceful Exit

Parting is such sweet sorrow—especially with CorelDRAW!. And if you want to start it up again later, it can take what seems like forever on some computers to start up. So don't exit CorelDRAW! until you are finished working on it for the day.

If you want to use one of the other Corel programs (CorelPHOTO-PAINT!, CorelCHART!, and so on) or another Windows application, you don't have to shut down CorelDRAW!. Instead, you can run CorelDRAW! in the background while you work on the other program. Just click on the **Minimize** button (the down arrow) in the upper right corner of the screen to run CorelDRAW! in the background. Then later when you want to work in CorelDRAW! again, double-click on the minimized **CorelDRAW!** icon at the bottom of your screen. Or, you can click on the **Control-menu box** of the program you're currently working in, select the Switch To command, and then double-click on **CorelDRAW!**.

Don't run CorelDRAW! in the background if you don't have a whole lot of computer memory. In this case, you should exit CorelDRAW! before you start another program.

To exit CorelDRAW!, select Exit from the File menu, press **Alt+F4**, or double-click on the **Control-menu box** in the upper left corner of the screen (how's that for options?). If you've saved recent changes made to your drawing, or if you didn't make any changes, CorelDRAW! closes immediately.

On the other hand, if you didn't save changes to your drawing before exiting, CorelDRAW! asks if you want to save the changes now. Click on **Y**es to save the changes or **N**o to dump the changes.

The Least You Need to Know

Well, how does it feel? You made it through the first few chapters, and you're still smiling . . . and breathing. Imagine that.

- To start CorelDRAW! or any of the CorelDRAW! programs, open the Corel program group, and double-click on the program icon.
- To display roll-ups at startup: open the roll-ups, select the **S**pecial **P**references command, click on the **Roll-ups** button, mark the **C**urrent Appearance of Roll-Ups option, and click on **OK**.
- To undo an action, select the **E**dit **U**ndo command.
- To redo an action you've undone, select the **E**dit R**e**do command.
- To get help, press **Shift+F1** or use the Help menu.
- To close CorelDRAW! or any other of the CorelDRAW! programs, select the **F**ile E**x**it command or press **Alt+F4**.

Recycling tip:
tear this page out and photocopy it.

Chapter 4
Juggling Drawings

In This Chapter

- Saving your work so you don't have to kick and scream later
- Opening the files you so wisely saved earlier
- Creating a brand new drawing
- Pirating CorelDRAW!'s stuff
- Cheating with CorelDRAW! templates

I warned him. I told him to save the new drawing. "Yeah, yeah," he said impatiently, and in such a tone that "blah, blah," would have been more to the point. Lo and behold about 45 minutes later, his computer froze, flashing that wonderful little Unrecoverable Application Error (UAE) message that Windows likes to throw at your face all too often. "Unrecoverable Application Error?" he cringed. Yes. His only options now were to close his application and lose all his work, or reboot his system and lose all his work. What a choice. Suffice it to say, he saved regularly from that point on.

This faithful chapter covers what all the other chapters insisted on shunning: those boring tidbits you really don't want to hear about, let alone learn about. Yet, mundane daily functions like saving, opening, and creating new files are crucial to your work in CorelDRAW!. So let's bite the bullet and get it over with.

Save It for Later

So often we like to procrastinate. "I'll do it later." "Tomorrow will be a better day for me." "I can't do it now, I have a headache." Great excuses for not doing the laundry, but not for saving your files in CorelDRAW!.

No procrastination allowed here: putting it off for a few minutes all too often turns into a few hours, and the entire time you're working on your drawing while it is unsaved, you're at risk of losing everything you've done. At least once you've saved your file, you have something to turn to later, even if your system crashes. At worst, you'll lose only the changes you made since you last saved; but you won't be at risk of losing everything. Your new motto: "Save during the first five minutes; save every ten minutes thereafter."

Saving Your Drawing for the First Time

Saving a new drawing is simple. Just open the **F**ile menu and choose the **S**ave command, or press **Ctrl+S**. The Save dialog box appears on your screen. Type a name for the file. You can enter up to eight characters, and then DRAW automatically adds a .CDR extension to the end of the file name. How's that for easy?

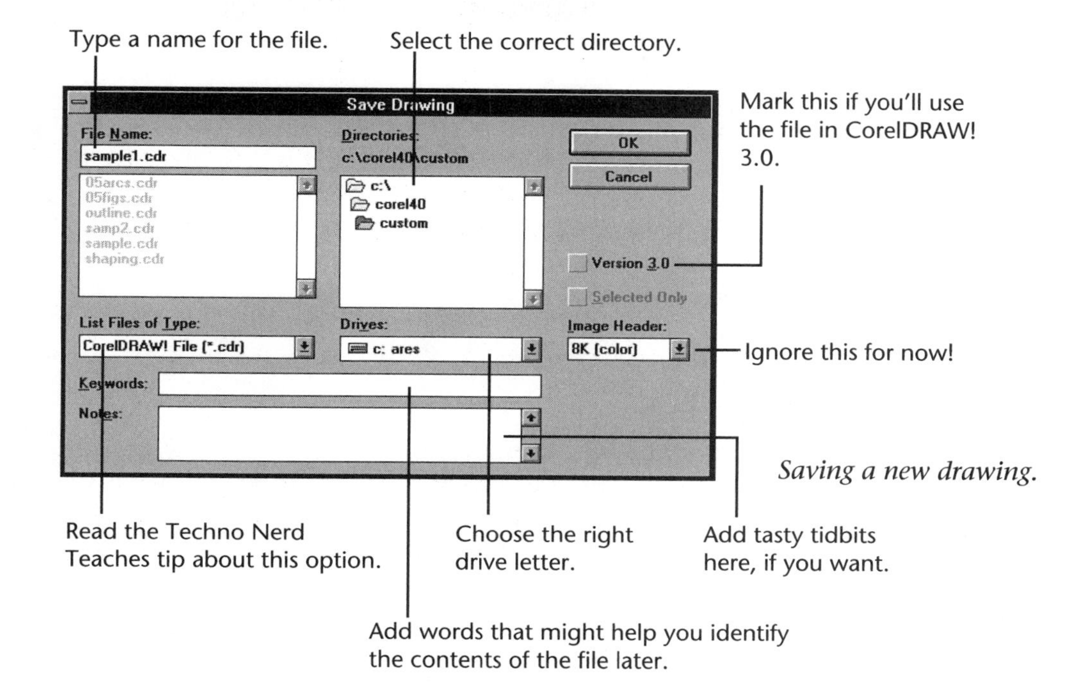

Saving a new drawing.

If you plan to use a certain document, such as a flyer or brochure, on a weekly or monthly basis where only some of the information changes, you should consider saving the document as a *template*. By doing so, you can create a new document each week or month using the template as the foundation. Rather than starting from scratch, you only have to make the necessary changes to the template, and then save the modified template as a new drawing, leaving the template intact for the next time you need it.

TECHNO NERD TEACHES...

To save a drawing as a template, display the drop-down list for the List Files of Type option. Then choose **CorelDRAW! Template (*.cdt)**. To use the template for a new document later, open the File menu and select the New From Template command.

Saving Your Drawing Over and Over Again

Every time you happen to think about it, press **Ctrl+S** to save your most recent changes. This sounds like an order from your Mom, but after a while it becomes second nature (like driving a stick-shift or like arriving at home after work and not remembering the familiar commute at all). It's the same with saving files. After a few days, you'll be pressing Ctrl+S without realizing it. Just make sure you get into the habit early on of saving frequently and regularly.

Open It, Will You?

So you've earned your brownie points and saved your file. The next time you want to work on it, you'll need to open it first, right? But what was its name again? CorelDRAW! really helps you out here. It displays a miniature picture of each file to help you locate the one you want. To open a drawing, press **Ctrl+O**. Or, for masochists who want to take the longer approach, select the **O**pen command from the **F**ile menu. Then select the file you want in the Open Drawing dialog box.

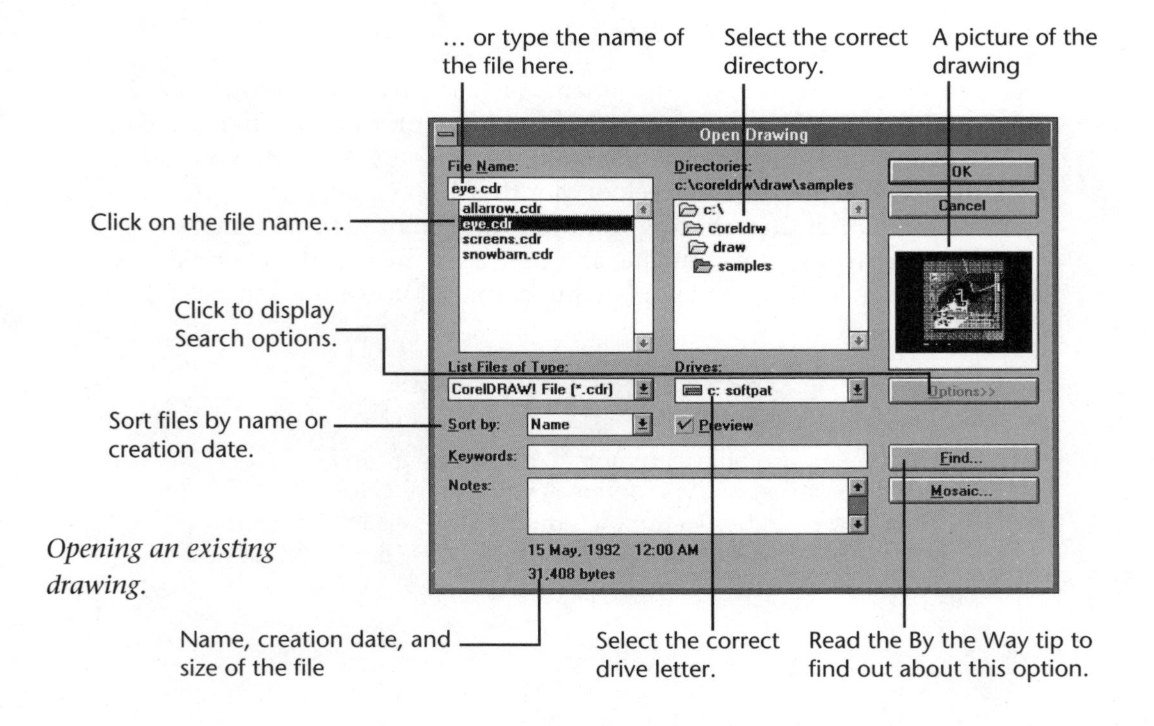

Opening an existing drawing.

If you worked on a file recently, you may be able to open it quickly by selecting it at the bottom of the **F**ile menu, where CorelDRAW! lists the names of the four most recently opened files.

Snagging CorelDRAW!'s Clip Art

CorelDRAW! comes loaded with drawings saved in separate files that you can add to your own drawings, or use as is. These CorelDRAW! drawings are called clip art, and are arranged by topic in different subdirectories of the CLIPART directory. For example, you'll find drawings of all sorts of bugs and animals in the CREATURE subdirectory.

You can open clip art the same way you open any existing file. Select the **F**ile **O**pen command to display the Open Drawing dialog box. Double-click on the **CLIPART** subdirectory of the Corel directory. This displays a list of subdirectories under the CLIPART directory. The name of a subdirectory gives you an indication of what kind of art is located in that subdirectory. To access a subdirectory's file, double-click on the subdirectory.

BY THE WAY

If you assigned keywords to a file when you saved it, now's the time to reap the benefits of your thoroughness. You can search for your file by specifying one of the keywords you entered when you saved the file. Click on the **Find** button in the Open Drawing dialog box. Type the keyword and then select the **Search** button.

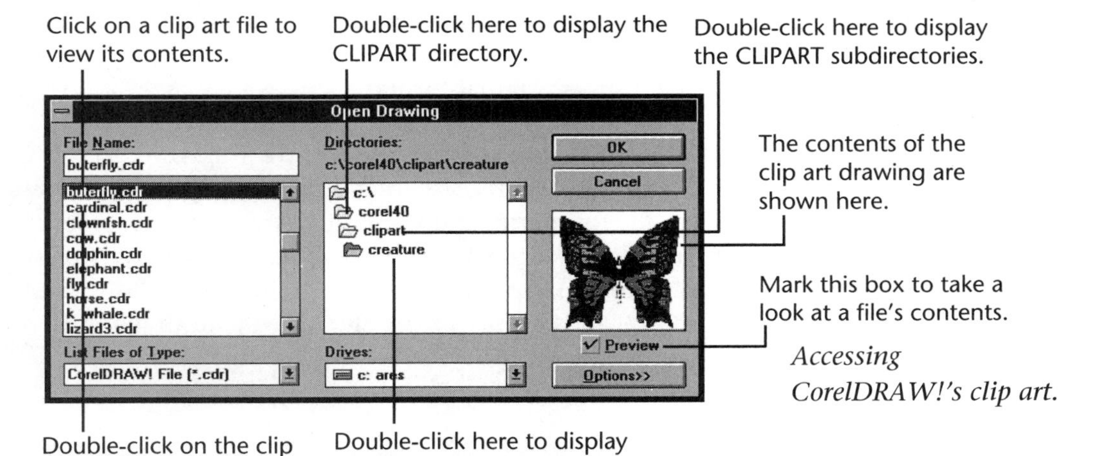

Accessing CorelDRAW!'s clip art.

Click once on a clip art file to see what the drawing looks like (make sure the **Preview** box is marked). A small picture of the drawing is shown in the lower right corner of the Open Drawing dialog box. If you want to open the clip art file, double-click on its file name.

Opening the BUTERFLY.CDR clipart file.

After you open a clip art file, you can print it or edit it to suit your own tastes. However, if you choose to alter the drawing, save it under a different file name first, using the **File Save As** command (see the first section in this chapter for help). This preserves the original clip art drawing in case you want to use it later in its original format.

There may be times when you want to add a clip art drawing to one of your own existing drawings. In this case, first open the clip art file. Then select the **Edit Select All** command and the **Edit Copy** command. (For help on selecting and copying, refer to Chapters 6 and 9, respectively.) Open the drawing you want to add the clip art to, and then use the **Edit Paste** command. CorelDRAW! adds a copy of the clip art to your drawing.

I Want a New One

Too often in life, getting something new, whether it's 1) a car, 2) an outfit, 3) a job, or 4) a date, involves 1) penny-pinching, 2) losing weight, 3) fibbing a bit, or 4) fibbing a bit. What a relief then that starting a new drawing in CorelDRAW! is as simple as asking for one—er, almost.

To start a new drawing, press **Ctrl+N**; you masochists can select **New** from the **File** menu. And voilà! A new drawing file is opened on the screen just like that . . .unless you haven't saved your changes to the previous drawing. In this case, DRAW pauses and asks you whether you want to save changes to Untitled.CDR. Say "No" if you have a SoundBlaster Card, voice recognition software, and a microphone, and you don't want to save the changes. If you don't have all that, just select **No**. On the other hand, if you're a diehard pack-rat and want to save the changes to your first drawing (even though it stinks), select **Yes**, and then enter a name for the drawing following the instructions in "Saving Your Drawing for the First Time," earlier in this chapter.

Worshipping at the Template

Why draw a new drawing if the drawing you want to draw has already been drawn? Lucky you: those nerdy men and women at Corel have already labored over designing business cards, flyers, brochures, price lists, newsletters, Christmas cards, and menus, so why should you? Unless you enjoy reinventing the wheel, you should check out CorelDRAW!'s templates before you labor to create your own stuff. The table below lists the name of each CorelDRAW! template and offers suggestions on how the templates can be used.

Template name	What you can do with it	Idiotic comments
3_fold2.cdt	Great for brochures; designed to be folded into 3 parts	Looks like it came out of a print shop!
b_day.cdt	We're really not sure . . .a calendar? monthly planner?	Corel may have goofed up on this one.

continues

continued

Template name	What you can do with it	Idiotic comments
bank_bc.cdt	Business cards	Very helpful if you're into making your own business cards.
catalog1.cdt	Price lists, catalogs	Only 1 page.
coreldrw.cdt	The default template	What you see whenever you start CorelDRAW!.
cover.cdt	Cover page for reports, booklets, manuals	Cool looking.
job.cdt	1/6 page newspaper ad soliciting applicants for a job opening	How convenient for you managers out there!
lunch.cdt	Invitation to seminars, company parties; announcements	Where did they get the name "lunch?" They must have been hungry.
meeting1.cdt	Bulletins, flyers	Why "1?" Where's "meeting2?"
menuleaf.cdt	Restaurant menus	Er . . .is it dinner yet?
news_box.cdt	Newsletters	You may have time to eat a full breakfast while you wait for this hog to load.
party.cdt	Business advertisements, flyers	Let's party.
peace.cdt	Christmas cards	Could be modified for other holidays.
spec.cdt	Spec sheets, announcements, press releases	What is a spec sheet, anyway?
sports.cdt	Business advertisements, flyers	Love those rackets.

Choosing the Template You Want

Roll your sleeves up, grab the reigns, and take a deep breath because we're ready to get down to the nitty gritty of templates. From the File menu, choose the New From Template command. The New From Template dialog box appears. Are you beginning to get familiar with how CorelDRAW! works? Menu command . . . dialog box, menu command . . . dialog box, and so on. But don't get too cocky. Just when you're feeling comfortable, CorelDRAW! will spring a new result on you.

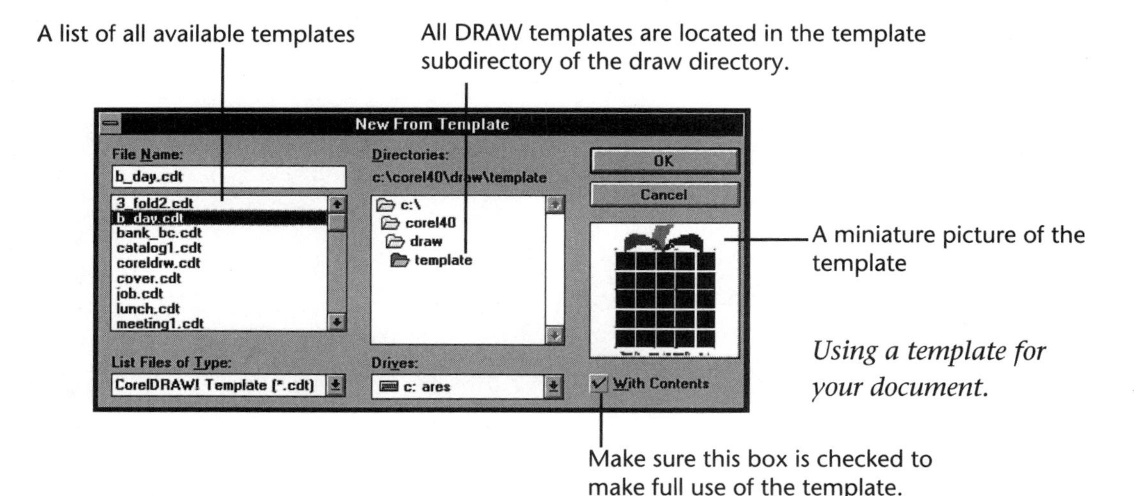

Using a template for your document.

If you're a visual learner, take a scroll through the files in the New From Template dialog box to see a tiny picture of each template. Let us emphasize *tiny picture;* sometimes you may need to open the template completely to get a better idea of what the template really looks like.

We can tell you're starting to panic; we're at the end of the chapter and nobody mentioned how to close a drawing! Relax. Unlike most Windows programs, there isn't a Close command on the File menu in CorelDRAW!. And you can only have one drawing open at a time. So to close a drawing, simply start a new drawing with File New or File New From Template, open an existing drawing using the File Open command, or exit CorelDRAW! altogether following the instructions in the section, "All Good Things Must Come to an End," in Chapter 3.

The Least You Need to Know

You may not have been dazzled by what you learned in this chapter, but hey, you just got through some of the essentials.

- Press **Ctrl+S** to save your drawing.
- Save your drawing as often as you think about it. And if you don't tend to think about it much, force yourself to save every 10 minutes.
- Press **Ctrl+O** to open a drawing you already worked on.
- To use CorelDRAW!'s clip art, select the **F**ile **O**pen command and display the Corel\CLIPART directory. Double-click on the clip art subdirectory for the category of art you want, and then double-click on the clip art file you need.
- Press **Ctrl+N** to create a new drawing.
- Select New **F**rom Template from the **F**ile menu to use one of CorelDRAW!'s templates for your document.

Chapter 5
Getting into Shape

In This Chapter

- Learn how to draw a rectangle as an adult
- Computerized squares are more fun
- And you thought you couldn't draw an ellipse
- Have fun drawing circles
- Build your own arc
- A bit about nodes and segments

When the word "shape" is mentioned, most people cringe. *Breathe in, lift, hold, release . . . I can't get into a size seven anymore . . . one more mile to go . . . do you really need a third donut, hon? . . . give us a week, and we'll get rid of your weight (along with your health) . . . you have high bad cholesterol and low good cholesterol . . . stretch after not before you run . . . caffeine's bad . . . no, caffeine's good.* It's amazing what nightmarish images a simple, innocent word like "shape" can invoke. But before your blood pressure gets too high, let's clarify the contents of this chapter. It's not another discussion on the fit vs. fat issue. Rather, you'll learn all about drawing CorelDRAW! shapes, like rectangles, squares, and circles. Whew!

BY THE WAY

CorelDRAW! provides rulers to help you as you draw. So before you begin drawing objects, display the rulers for extra guidance. Choose **S**how Rulers from the **D**isplay menu.

It Takes Two to Rectangle

Once you got out of elementary school, you probably thought that you'd never have to be taught again how to draw a rectangle. Well, you're wrong. This will be a little more exciting though. This time you get to draw rectangles on a computer screen.

The ease of drawing a rectangle will blow you away (if it doesn't, you should consider shedding a few pounds). Click on the **Rectangle** tool (the square) on the toolbox. Your cursor becomes a cross. Wake up and pay attention to this, it's important. Whenever you see a cross cursor, it means you're about to create something, not select something.

Decide where on your screen you want the upper left corner of the rectangle to be. Press the mouse button there, and don't release it till we tell you to. Drag the cursor to where you want the bottom right corner of the rectangle. All right, now you can release the mouse button. Voilà! Your first drawn object.

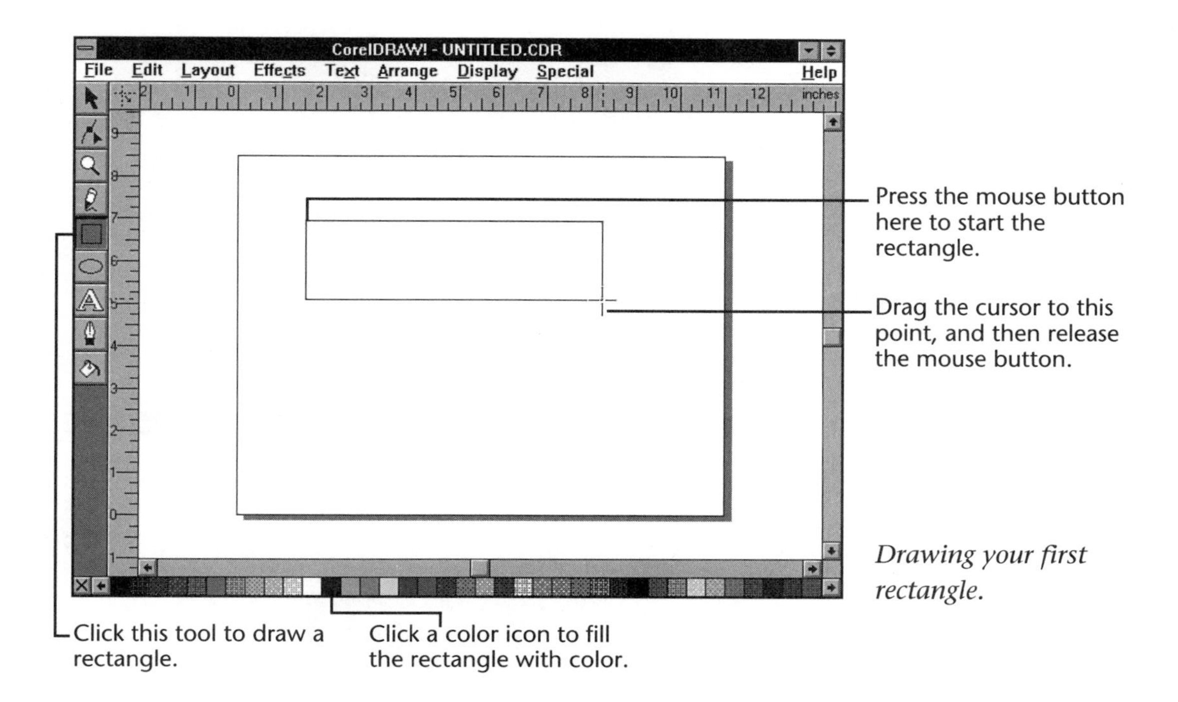
CorelDRAW! - UNTITLED.CDR
File Edit Layout Effects Text Arrange Display Special Help
Press the mouse button here to start the rectangle.
Drag the cursor to this point, and then release the mouse button.
Click this tool to draw a rectangle.
Click a color icon to fill the rectangle with color.

Drawing your first rectangle.

What a Square

There's a square deal, a square meal, and then just a plain square. We're interested in the latter, and do we have good news. Drawing a square is virtually as simple as drawing a rectangle; you even use the same tool. One exception though: you have to draw a square one-handed. Then again, this shouldn't be too difficult since most people do manipulate the mouse with only one hand, don't they? Your other hand will be pressing down the Ctrl key.

To instantly fill the rectangle and all other objects you create with color, click on one of the color icons on the Color Palette at the bottom of the screen immediately after you finish drawing the object. To fill the rectangle at a later time, read Chapter 11, "Color Me Beautiful."

Be sure to release the Ctrl key *after* you release the mouse button when drawing squares and circles. Otherwise, you may end up with a rectangle or an ellipse after all!

For all of you who need it spelled out, click on the **Rectangle** tool (the square) on the toolbox, hold down the **Ctrl** key, move the cursor to the upper left corner of the square to be drawn, and drag the cursor diagonally downward to the right. When the square reaches the right size, release the mouse button and the Ctrl key. Presto! Your second drawn object.

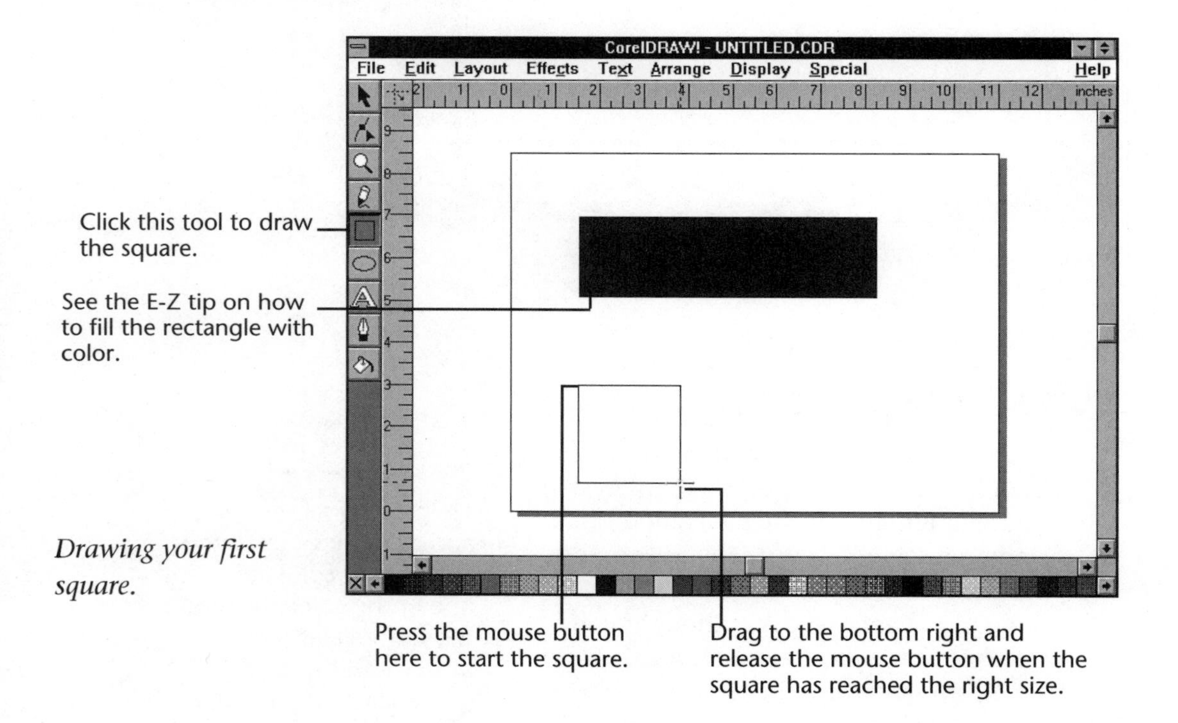

Drawing your first square.

A Solar Ellipse

What is an ellipse again? No, not a car. And it doesn't have anything to do with the sun. An ellipse is an oval by any other name. An oblong circle. An ellipse can never have perfect dimensions, because if it does, it's no longer an ellipse, but a circle. Kind of like a person who leads a perfect life is no longer a human, but a saint.

You have to use your imagination when you draw an ellipse. In order to position the ellipse just where you want it, and to get it just the right size, you sort of have to conjure up an imaginary rectangle that would hold it, and then draw as if you were drawing the rectangle itself. (I promise: imaginary rectangles do not cause lasting psychological damage.) Either that, or just experiment until you get the ellipse right.

Click on the **Ellipse** tool (the oblong circle) on the toolbox. The cursor changes to a cross. Get your imagination going and move the cursor to the upper left corner of your imaginary rectangle. Drag the cursor down and to the right until the ellipse is the size you want. Release the mouse button.

SPEAK LIKE A GEEK

Ellipse An oblong circle.

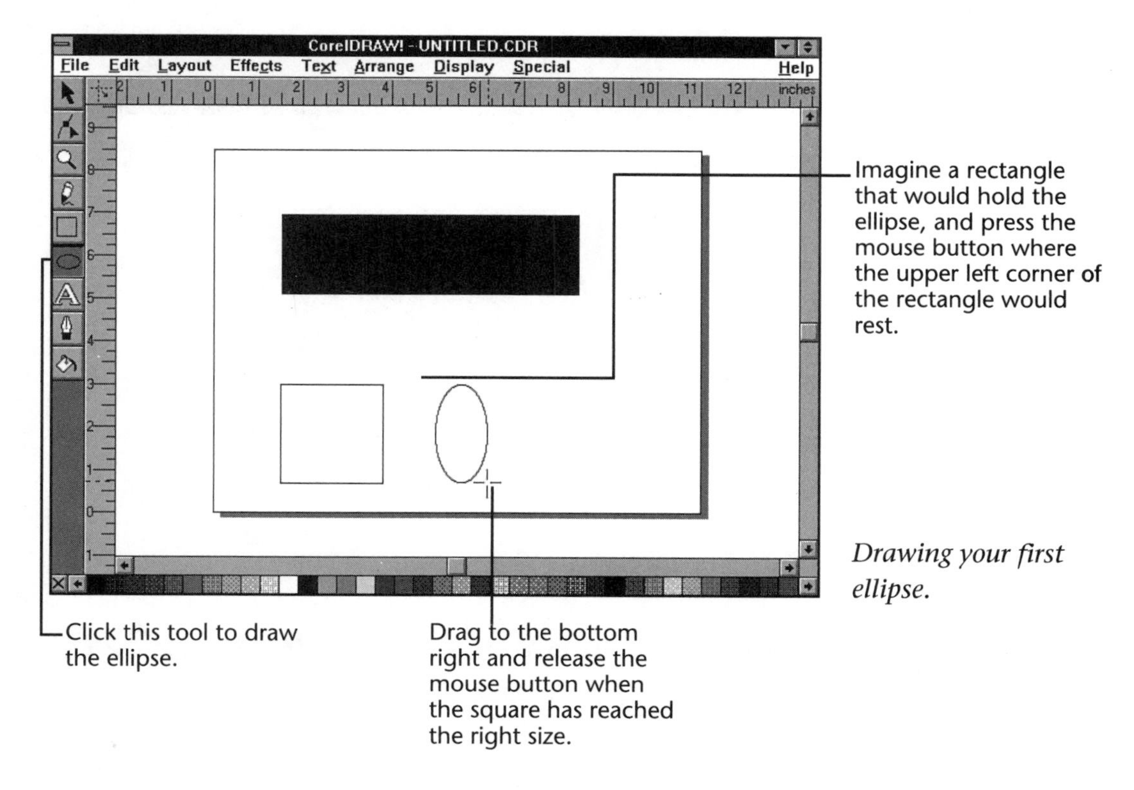
CorelDRAW! - UNTITLED.CDR
File Edit Layout Effects Text Arrange Display Special Help
Imagine a rectangle that would hold the ellipse, and press the mouse button where the upper left corner of the rectangle would rest.
Click this tool to draw the ellipse.
Drag to the bottom right and release the mouse button when the square has reached the right size.

Drawing your first ellipse.

Going in Circles

By now you've probably realized how much simpler it is to draw objects than you had thought. And you're surprised by how similar the process is for each type of object. Well, circles are no different. If you were hoping for some new, radically different way to conjure up a circle on your drawing page, you'll have to keep looking. As with the old rectangles and squares, drawing a circle is just like drawing an ellipse, except that you hold down the Ctrl key as you draw. And instead of imagining a rectangle as you draw (as you did for an ellipse), imagine a square.

We won't bore you with any more details. If you have a problem getting the circle right, read the instructions on drawing an ellipse and chant to yourself or aloud, "Hold down the Control key, imagine a square. . . ." We will be generous, though, and provide you with at least a picture of a circle.

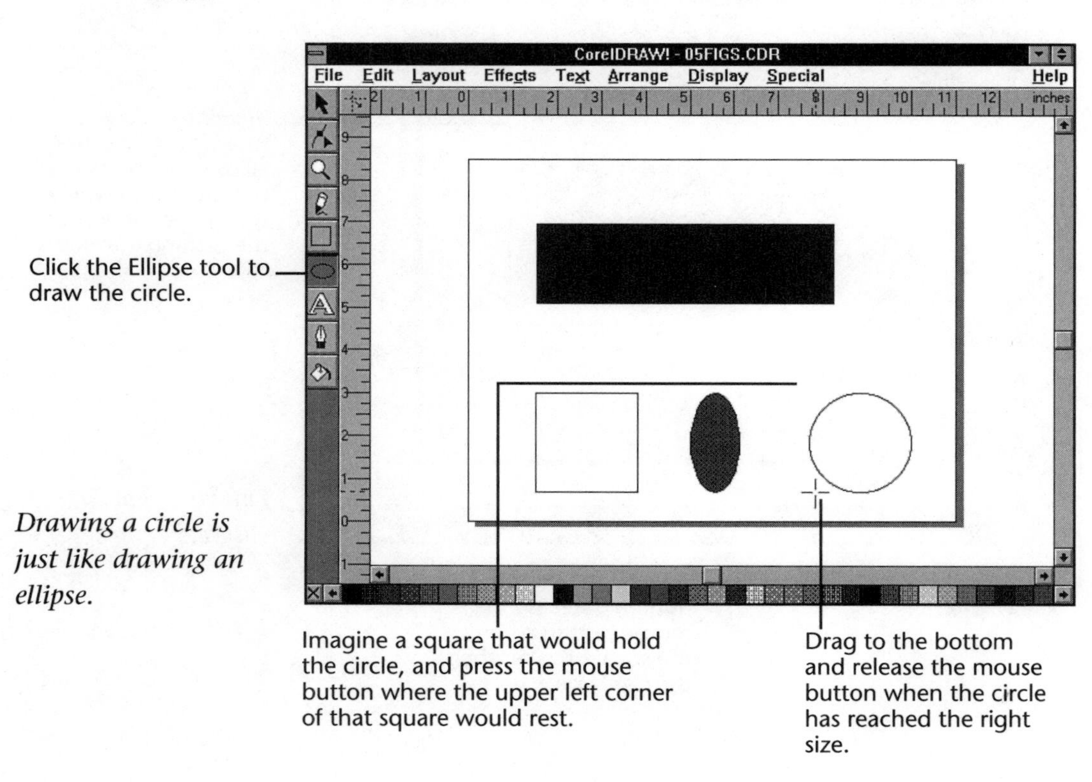
CorelDRAW! - 05FIGS.CDR
File Edit Layout Effects Text Arrange Display Special Help
inches
Click the Ellipse tool to draw the circle.
Imagine a square that would hold the circle, and press the mouse button where the upper left corner of that square would rest.
Drag to the bottom and release the mouse button when the circle has reached the right size.

Drawing a circle is just like drawing an ellipse.

Building an Arc

Okay, now we're getting to the tricky stuff. If you get through this section all right, you're on your way to becoming a pro. Drawing an arc is a two step process. First you draw a circle or an ellipse, and then you use the Shape tool to convert the circle or ellipse into an arc.

When you click on the **Shape** tool (the second tool from the top on the toolbox), the cursor turns into a dark, thick arrow. Each circle has a node at the top or bottom. A *node* is an endpoint of an object, with which you control the shape of the object. Drag this node around the *outside* edge of the circle. As you drag, the circle opens up into an arc. Later, if you want to change the size of the arc, drag either node at the end of the arc.

You can also create a pie or pie wedge using the Shape tool. Instead of dragging outside of the circle, drag around the *inside* of the circle to create a pie. And keep dragging to create a pie wedge.

Node An endpoint of an object. The number and position of nodes determine the shape of an object.

Segment A line or curve between two nodes.

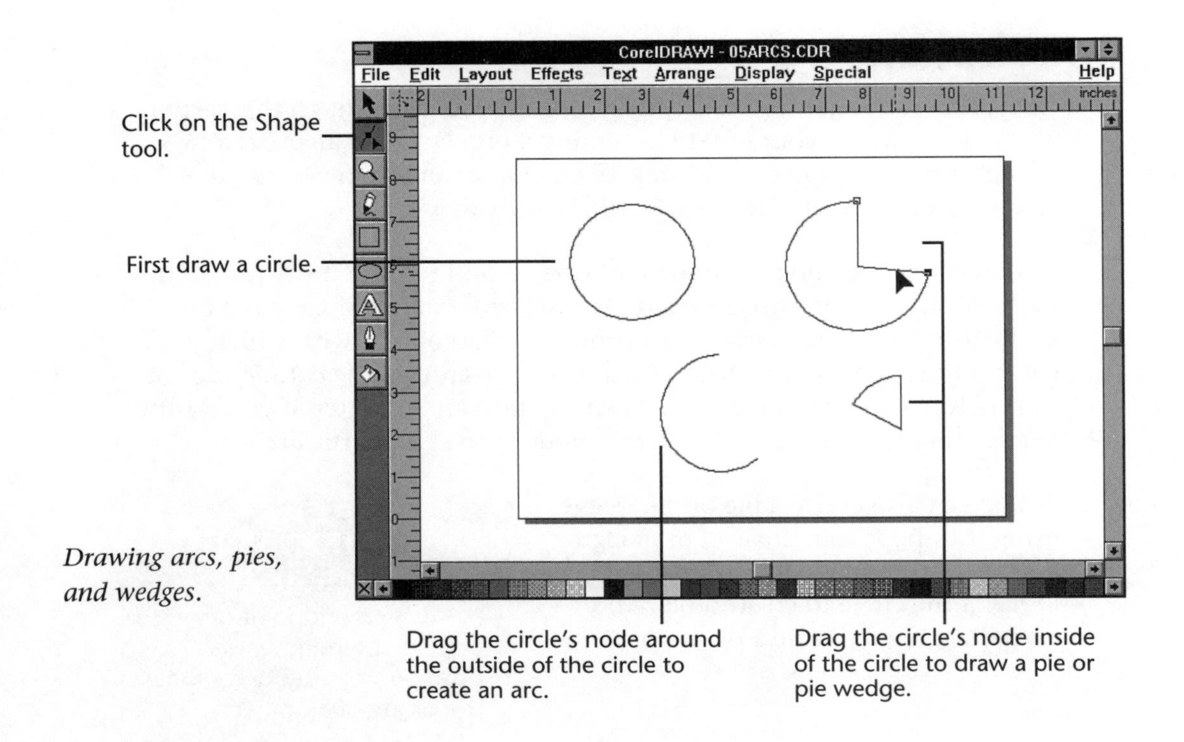
CorelDRAW! - 05ARCS.CDR
File Edit Layout Effects Text Arrange Display Special Help
Click on the Shape tool.
First draw a circle.
Drag the circle's node around the outside of the circle to create an arc.
Drag the circle's node inside of the circle to draw a pie or pie wedge.

Drawing arcs, pies, and wedges.

The Least You Need to Know

You should be in pretty good shape after reading this chapter—or at least your objects should.

- Click on the **Rectangle** tool, and then drag the cursor on the drawing page to create a rectangle.
- Do the same to draw a square, except keep the Ctrl key pressed down as you drag. Release the Ctrl key after you release the mouse button.
- Click on the **Ellipse** tool, and then drag the cursor on the drawing page to create an ellipse.
- Draw a circle the same way you draw an ellipse, except keep the Ctrl key pressed down as you drag. Release the Ctrl key after you release the mouse button.

Chapter 6
To Select or Not to Select

In This Chapter

- Only one for me
- One's never enough
- It's more fun in a group
- Choosing one out of the bunch
- I don't want any, thanks

Some of your most important decisions in life deal with selections. Like selecting the right college to attend. Selecting your first car. Selecting whom to date, and eventually, selecting someone to spend the rest of your life with. Even selecting what to eat for breakfast can be tricky for some of us. Thank goodness selecting objects in CorelDRAW! has nothing to do with life-long effects, good or bad decisions, or desires. What it has to do with is selecting an object so you can work with it—to edit, enhance, or move it. It's a logistical matter, not a subjective matter. But it is an important thing to do in CorelDRAW! because you can't do anything to an object unless you select it first.

For Singles Only

It's almost possible to say: "Click on an object to select it." But then life's never that simple. We do need to put a few stipulations on that statement. First of all, make sure you've clicked on the **Selection** tool (the arrow at the top of the toolbox) before you click on an object. Then we also have to say that if an object is filled with a color or pattern, you can click on it anywhere, but if it isn't filled, you have to click on the outline of the object to select it.

A few more things. You don't even need to select an object if you just finished drawing it. The Corel program nerds were smart enough to realize that we'd probably want to edit objects right after we finished creating them. So consider an object that you just drew already selected. One last item: You can tell when an object's selected because big black boxes border it. You can't miss them. And they're actually named: alone, a box is called a *selection handle*; together the boxes comprise the *highlighting box*.

Selection handles Boxes that border a selected object. You'll use these boxes to size, transform, and skew objects.

Highlighting box An invisible rectangle formed by the selection handles of a selected object.

That's about it. But take a look at the figure for visual reinforcement. Wait. We almost forgot: the Status Line. When you select an object, the Status Line tells you the type of object that's selected. For example, the Status Line will display "rectangle, ellipse, circle, square, curve, bitmap, grouped, or child object" when one of those object types is selected. (You'll find out about grouped and child objects in just a moment.) The Status Line will also tell you about the outline and fill of an object; for example, it specifies colors, patterns, dashed lines, and so on.

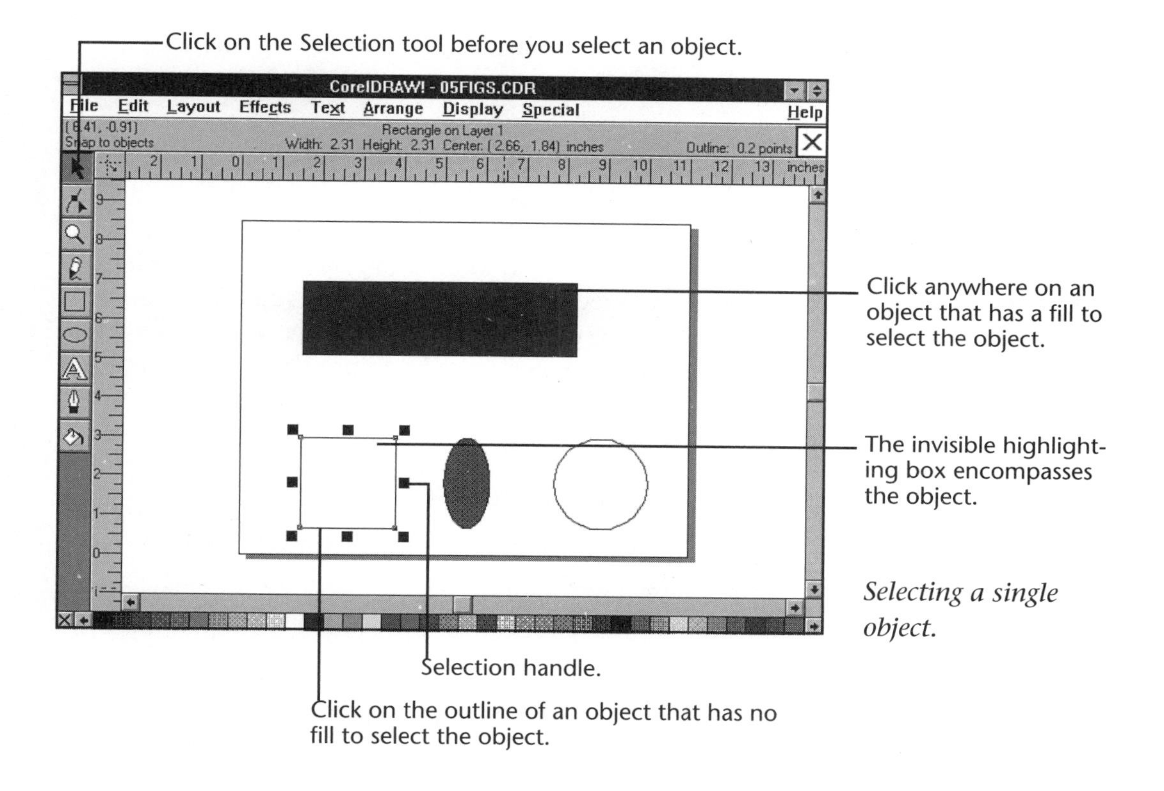

Selecting a single object.

Multiple Selection

So you want more than one object. Just remember, greed never pays off—but the movie *Wall Street* taught you that. You have your choice of three ways to select multiple objects.

- You can drag the Selection tool over the objects you want.
- You can hold down the Shift key and pick 'em one by one.
- If you want to select all objects, use the **S**elect All command on the **E**dit menu.

BY THE WAY

If you're working with a complex drawing, it may get *very* tricky trying to select just the one object you want. In this case, press the **Tab** key to select the next object in the drawing (make sure the Selection tool is active). If you want to move back to previous objects, press **Shift+Tab**.

The first method is faster than the second, but the second method is necessary when you want to select only certain objects scattered all over the drawing page.

Marquee box A dotted rectangle that appears as you drag the Selection tool over objects to select them.

The Drag Queen

For some reason known only to whomever, the dotted rectangle that appears as you drag the Selection tool over the objects you want to select is called a *marquee box*. If you use the marquee approach to select multiple objects, keep a few things in mind:

- The starting point needs to be on white space.
- Make sure each object you want to select is entirely enclosed by the marquee box; otherwise, an object won't be selected.
- When you release the mouse button, the enclosed objects become selected.
- Remember: marquee boxes only appear when you're selecting more than one object. If you only need to select one object, click on it.

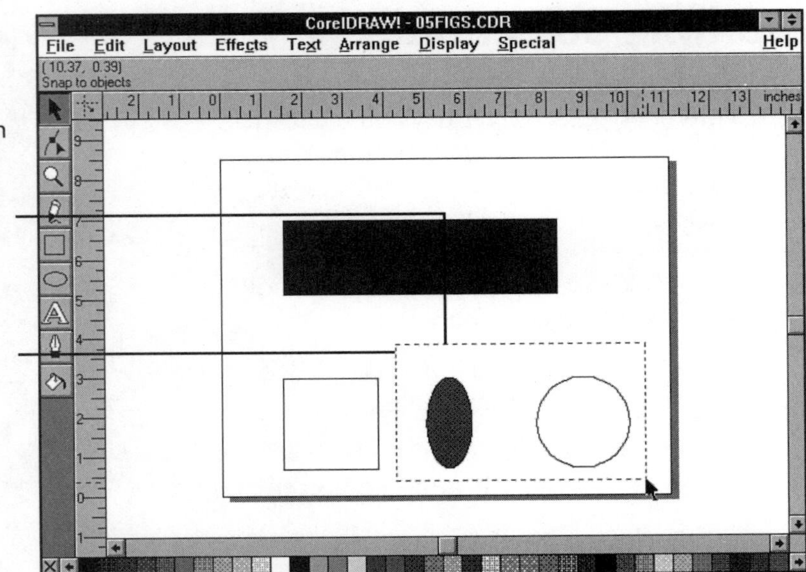

Always start the drag on white space and completely enclose all objects to be selected.

The dotted marquee box appears as you drag.

Selecting multiple objects with a marquee box.

Shift and Click and Click

If you want to select a bunch of objects that are spread all over the drawing page, the marquee approach won't work very well. You'll end up selecting objects you don't want to select. In this case, use the Shift+Click approach. It's very simple. Just hold down the **Shift** key and click on each object you want to select. As you click on the objects, you'll notice that the highlighting box enlarges to encompass all of the objects. Though the highlighting box may cover objects you didn't select, don't panic. Those objects haven't become part of the selection; rather, the highlighting box simply rests on top of them.

If, in your attempt to get all of the objects you wanted to select completely inside the marquee box, you got some other objects in there that you didn't want, there is a solution. We suggest, in this case, that you use the other method for selecting: the Shift+Click method, discussed in the next section.

True Groupies

Imagine that you're drawing a graphic that contains several small objects, curves, and lines. Now imagine trying to move the graphic as a whole to a different location on the page. You'd be out of luck if it weren't for CorelDRAW!'s grouping feature. What you would do to move that graphic is select all of the objects that comprise the graphic, group them into one object, and then easily move that object to wherever you wanted.

I selected the wrong object. Do I have to start selecting all over again? No. Just click again on the object you don't want (making sure to continue holding down the Shift key). This will deselect that object only.

Beyond that, grouping objects makes it easy to format them or edit them all at once. For example, if you wanted to enlarge ten objects, you could do so in one operation (maintaining the scale) instead of ten operations.

To group objects together so that they work as one entity, select all of the objects. Then choose **Group** from the **Arrange** menu, or press **Ctrl+G**. From now on you can select the group by clicking anywhere on it.

If you want to ungroup grouped objects, select the group, and then select Arrange Ungroup or press **Ctrl+U**. The group breaks apart, with all of the objects that were in the group currently selected.

Pick One out of the Group

Okay, so you've grouped a whole mess of objects together, but you notice that one tiny part needs to be changed. Do you have to ungroup all the objects and then regroup them later just to get at that one problem child? No. And that one tyke you want to get at is called a *child object*, because it's a single object that is a part of a group.

To select a child object, hold down the **Ctrl** key as you click on the object. To help you know whether you selected the right object, CorelDRAW! puts a highlighting box around the object. To distinguish this as a child object, the highlighting box has round handles rather than the usual boxes.

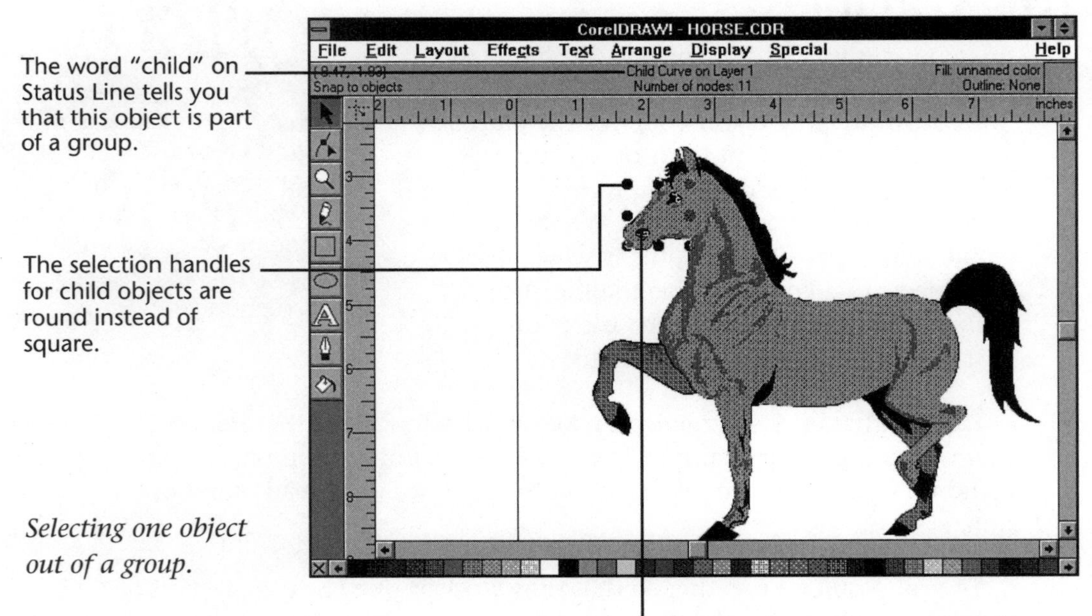
The word "child" on Status Line tells you that this object is part of a group.
The selection handles for child objects are round instead of square.
Hold down the Ctrl key as you click on the object (the horse's nostril).
CorelDRAW! - HORSE.CDR
File Edit Layout Effects Text Arrange Display Special Help
Child Curve on Layer 1
Number of nodes: 11
Snap to objects
Fill: unnamed color
Outline: None
inches

Selecting one object out of a group.

Changing Your Mind

It happens to all of us at some point: the ultimate rejection. The big drop. The break-up, split, I'm outta here thing. Well, it's your turn to do the rejecting. You get the pleasure of deselecting those same objects you just got through selecting.

Child object A single object that is part of a grouped object.

Grouped object Two or more objects that have been grouped together using the **A**rrange **G**roup command. With a few exceptions, CorelDRAW! treats a grouped object as if it were a single object, meaning you can move, transform, rotate, or size the group of objects as one object.

Follow these guidelines when deselecting objects:

- Deselecting a single object is kinda the opposite of selecting a single object: you must *not* click on the outline to deselect an object. Click anywhere else (except on one of the selection handles), and you're okay. So basically we're saying, click on any white space.
- To deselect multiple objects, just click anywhere on the screen that's blank.
- To deselect a child object (how cruel of you!), hold down the **Shift** key and click on the child (but not on its outline!).

The Least You Need to Know

Don't forget these rules of thumb when it comes to selecting and deselecting:

- Click on an object's outline to select it.
- Don't click on an object's outline to deselect it. Click anywhere else on the object.
- To select multiple objects, drag the Selection tool over the objects, or hold down the **Shift** key and click on each object.

continues

continued

- To select all objects, use the **E**dit **S**elect All command.
- To group objects, select all the objects to be included in the group, and then press **Ctrl+G**.
- Select a grouped object as you would any single object.
- To select a single object in a grouped object, hold down the **Ctrl** key and then click on the object.
- To deselect multiple objects, click on any blank portion of the screen.

Chapter 7
Getting a Good Look

In This Chapter

- Up close and far out
- Down to the wire
- What wonders: rulers, status lines, palettes, and toolboxes
- Grid and bear it
- Let CorelDRAW! be your guide

Retailers spend lots of time and money getting their displays just right because they know how important alluring displays are in attracting the amount and type of customers they want. Well, you might spend a good deal of time initially setting up your CorelDRAW! screen display, not to attract customers of course, but to make your time working with CorelDRAW! easier, more functional, and more enjoyable. This chapter tells you about the different display options available in CorelDRAW!.

Closer . . . Closer . . . Stop! Now Back Up!

Life would not be easy if you had to try to draw complicated objects d'art in CorelDRAW!'s regular view. You'd have to put a magnifying glass to your computer screen. That would look a bit odd. Fortunately those smart Corel devils designed a Zoom tool so you can work with your drawings up close.

Hold down the mouse button on the **Zoom** tool (the magnifying glass on the toolbox). This displays the five subtools of the Zoom tool. You can use these tools to adjust your view of the drawing page.

Zoom out
View all objects
Zoom tool
Return to entire drawing page
Zoom in
View actual size

The Zoom subtools.

Up Close and Personal

If you work with complex objects or lengthy paragraph text, you will find the Zoom tool indispensable. This tool lets you adjust the viewing area of the editing window, so that you can zoom in on any part of the drawing page you want. This is essential when you are drawing a complex object of small dimensions.

When you click on the Zoom subtool (the first subtool), the cursor changes to a magnifying glass with a plus sign. You can zoom in on your drawing in two ways. First, you can simply click on the area you want to magnify. Or, to view a more precise portion of your drawing up close, you can place a marquee around the area. To do this, drag the magnifying glass over the area you want to view in detail. As you drag, a dotted box appears. When you release the mouse button, CorelDRAW! magnifies the portion of the drawing that fell inside the dotted marquee box.

Your Zoom flyout menu has six subtools instead of five. The extra tool (which looks like a group of selected objects) lets you view all currently selected objects.

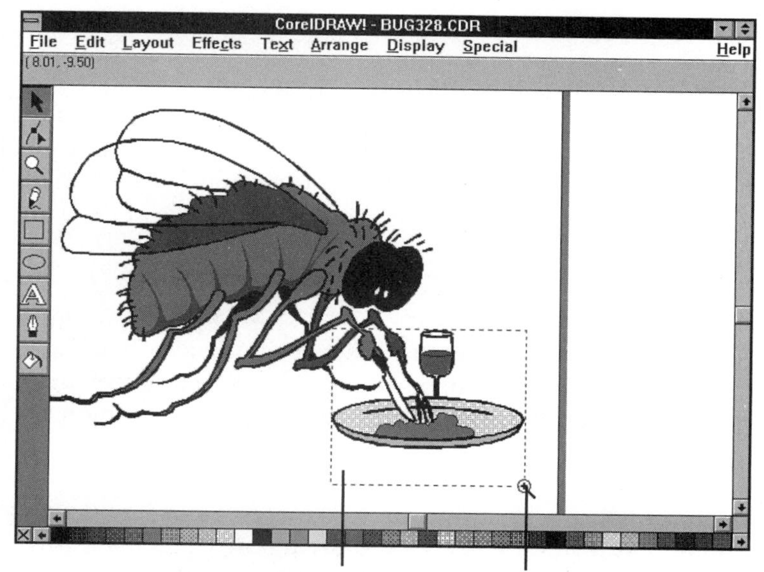

Selecting an area to magnify.

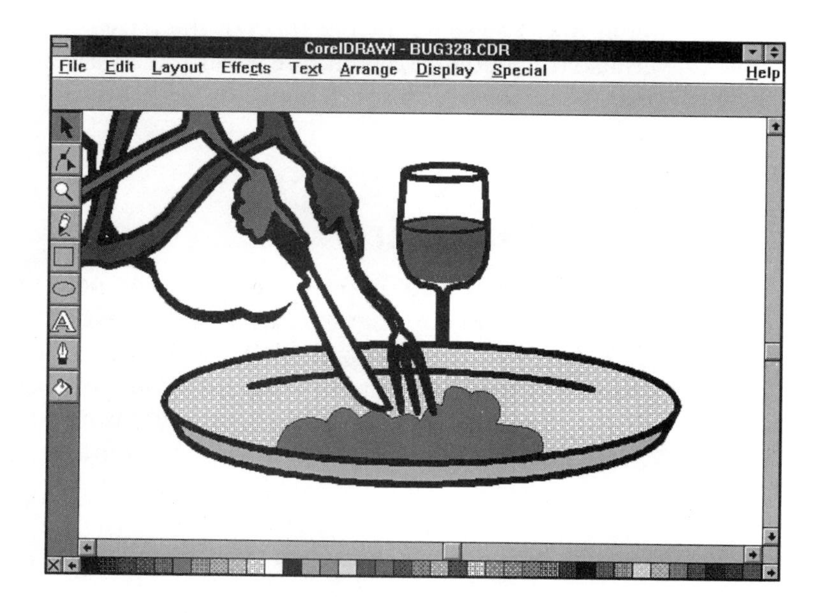

A portion of the drawing is magnified.

Getting the Big Picture

After you use the Zoom tool to magnify objects, you can zoom out again to see the objects at their original size or you can return to the previous zoom view. To see the entire drawing page, click on the subtool at the far right (the full-size blank sheet of paper) of the Zoom subtools flyout menu. CorelDRAW! instantly displays the entire drawing page at its normal size.

To return to the previous view before your last zoom-in, click on the second subtool on the flyout menu (the magnifying glass with a minus sign). The cursor changes to a magnifying glass with a minus sign inside. Click on the drawing page, and you'll zoom back to the previous view before your last zoom-in.

Initially, your drawing isn't shown at its actual size—as it will appear when printed. If you want to see your drawing on-screen as it will appear on the printed page, inch for inch, click on the **Actual Size** tool on the Zoom tool flyout menu. CorelDRAW! then displays your drawing at its actual size. You see only a portion of the page because the whole page will not fit on-screen. How much of the page you see depends on your monitor.

You don't have a Display menu! So to switch to wireframe view, choose Wireframe from the **View** menu.

Thin As a Wire

Have you ever stayed up to watch the stupid pet tricks on David Letterman and been sorry the next day because you were dragging? If CorelDRAW! runs as sluggishly on your computer as you feel, you may find it helpful to edit your drawings in Wireframe view. This view shows only the outlines of your objects. No fills, colors, or other attributes are shown. Since this view takes up less memory than the default view, wireframe view can be helpful when you are working on a long, multi-page document that would otherwise tax your computer's memory. Also, you may find that editing wireframes is easier than editing full-color objects.

Editable view CorelDRAW!'s normal view, which displays graphics as they will appear printed, with fill, colors, and attributes.

Wireframe view Shows only the outlines of objects, making it somewhat easier and faster to edit your objects.

By default, your drawings are displayed in Editable view. To switch to wireframe view, open the **D**isplay menu and click on **E**dit Wireframe. A check mark now appears to the left of the menu command. This command works as a toggle. When you want to return to Editable view, simply choose the **D**isplay **E**dit Wireframe command again.

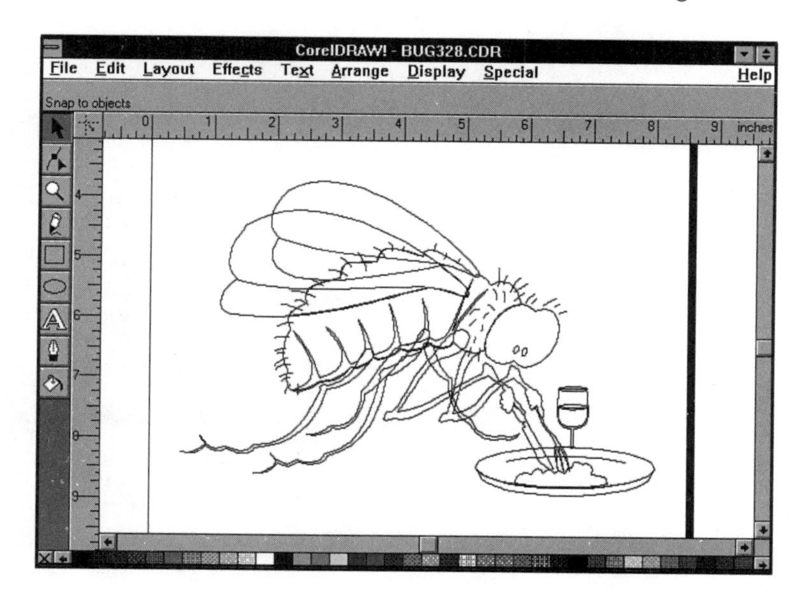

A drawing in Wireframe view.

Admiring the View

CorelDRAW! gives you a lot of say as to how you want your screen display set up. Here's what you can choose to display or not to display, and how to do it:

- **Rulers** Choose the Show **R**ulers command on the **D**isplay menu. A check mark appears to show you that the option is currently selected. This displays rulers at the top and left sides of the screen. You can use these rulers to help size and position objects.
- **Status line** Choose the Sh**o**w Status Line command on the **D**isplay menu. Again, a check mark indicates that the option is currently selected. This command shows the status line at the top of the screen. The status line displays the exact coordinates of the mouse cursor and gives you information about selected objects. If you do not need such detailed information, you should consider hiding the status line since it takes up valuable screen space.

Instead of the Display menu (which isn't there), use the **V**iew menu for rulers, color palette, and floating toolbox options. To show or hide the status line, use the **S**pecial Preferences command and click on the **View** tab.

- **Floating Toolbox** Choose the Floating Toolbox command on the **D**isplay menu. You've got it: a check mark indicates that the option is activated. This command frees the toolbox, but does not move it from its fixed position at the left side of the screen. You can now move the toolbox to any position on-screen that your little heart desires. To do so, drag the upper right corner of the toolbox.
- **Color Palette** If you want a new set of colors, choose the **D**isplay **C**olor Palette command and then select a different color palette from the submenu. If you want to remove the Color palette altogether, choose **N**o Palette.

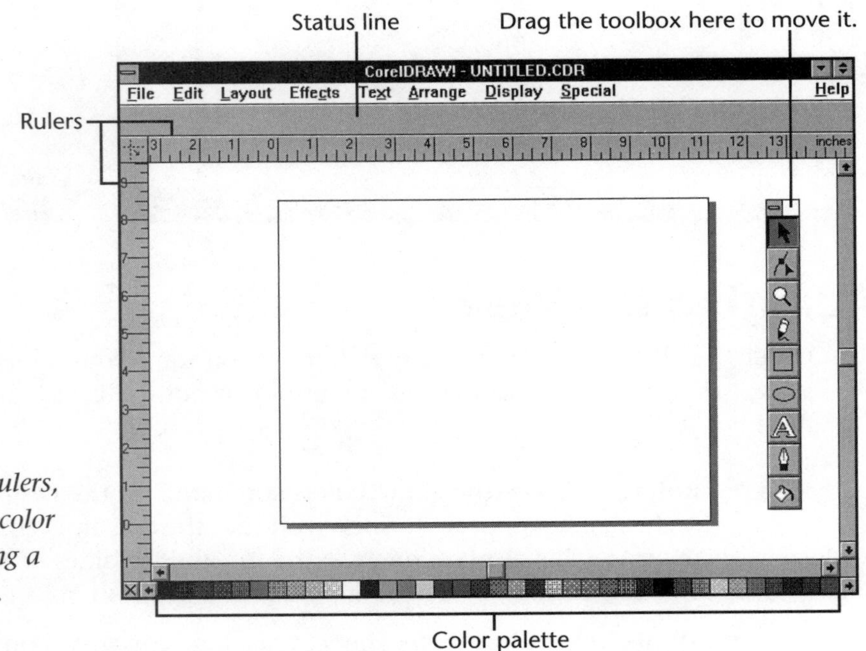

Displaying the rulers, status line, and color palette, and using a floating toolbox.

The Nitty Griddy

To bring up the Grid Setup dialog box, choose the Grid Setup command from the Display menu.

Connect the dots was always a fun game, so you might want to try setting up a grid on your screen for old time's sake. In fact, we strongly urge you to when you need to draw objects with exact measurements at precise locations on the page. Good luck doing it without a grid. And if you're drawing rectangles, squares, or polygons, or reshaping any object, CorelDRAW!'s Snap To feature should come in very handy. You'll draw straight lines like you've never drawn them before.

First you need to set up the grid. Choose the Grid Setup command on the Layout menu. This displays the Grid Setup dialog box.

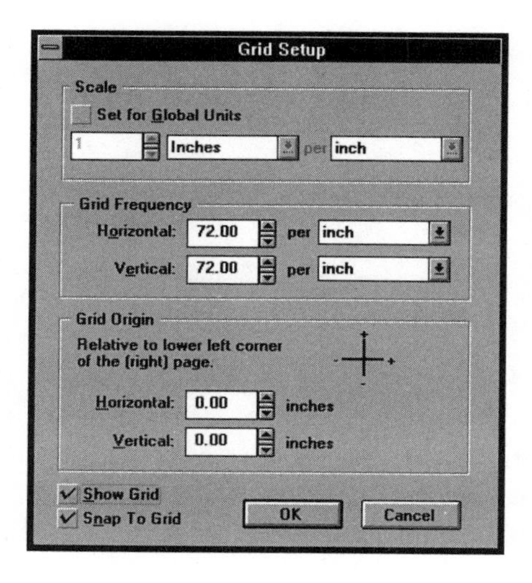

Mark the Show Grid option to display the grid on screen.

Don't let this dialog box scare you. It's really not as nasty as it looks. In fact, if you're satisfied with CorelDRAW!'s preset grid, all you need to do is mark the Show Grid option in the lower left corner of the dialog box, and then click on **OK**. A lovely set of evenly aligned dots (otherwise called a grid) now embellish your screen. You can just leave it at that.

However, you go-getters might want to know a bit more about the grid options. As mentioned earlier, if you want to have your objects snapped to the nearest grid point, mark the **Snap** To Grid option in the Grid Setup dialog box as well. The status line now reminds you that the option is activated (look at the left corner). With this feature turned on, selection handles, nodes, and control points will "snap" to the nearest grid point.

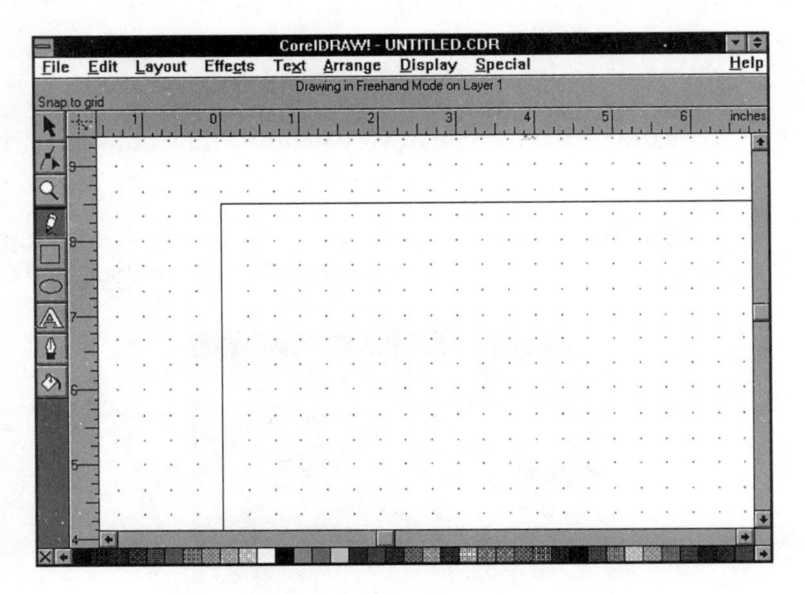

The grid, with the Snap To feature turned on.

There's a quicker way to access the Grid Setup dialog box. If you have rulers displayed on screen, double-click on a ruler to open the Grid Setup dialog box. Or, if you have the Layers roll-up opened, double-click on **Grid** in the roll-up and click the **Setup** button. And there she is . . . the Grid Setup dialog box.

If that's still not enough, you can change the grid frequency settings. The numbers in the Horizontal and Vertical boxes in the Grid Frequency area determine the grid's scale. In other words, the settings determine the number of grid lines per inch. When you want to space the lines more than a whole unit apart, use a fractional value. For example, a setting of .5 in both boxes spaces the grid lines two inches apart. The smaller the number, the farther apart the grid points are from each other. CorelDRAW!'s default grid may be too small for you. If so, enter smaller numbers in the Grid Frequency boxes.

Lost Without Guidelines

Here's yet another way to dress up the screen display and give you more control when drawing: guidelines. These are dashed lines that stretch across the screen, but are not part of your drawing. They serve only one purpose: to guide you as you draw. You can add as many horizontal and vertical guidelines as you need. As with the grid, you can have your objects snap to the guidelines.

To display guidelines, choose the **L**ayout G**u**idelines Setup command. The Guidelines dialog box appears. Choose the type of guideline you want, either **H**orizontal or **V**ertical, in the Guideline type section. Then enter the point on the ruler where you want the guideline to be drawn, for example, 2.3 inches. Enter a positive number if you want the guideline to appear above or to the right of the zero point on the rulers; enter a negative number to place the guidelines below or to the left of the ruler's zero points. Mark the **S**nap To Guides box if you want your object to snap to the guideline. Finally, click on the **A**dd button. The dialog box closes and the guideline appears across the screen. Follow the same steps to set up additional guidelines. If you want to remove a guideline later, open the Guidelines dialog box and press the **D**elete button.

You can also drag guidelines right off from the CorelDRAW! rulers. To display rulers, select **D**isplay Show **R**ulers. Then click on either ruler and drag onto the drawing page. Dragging from the top ruler creates a horizontal guideline, while dragging from the side ruler creates a vertical guideline.

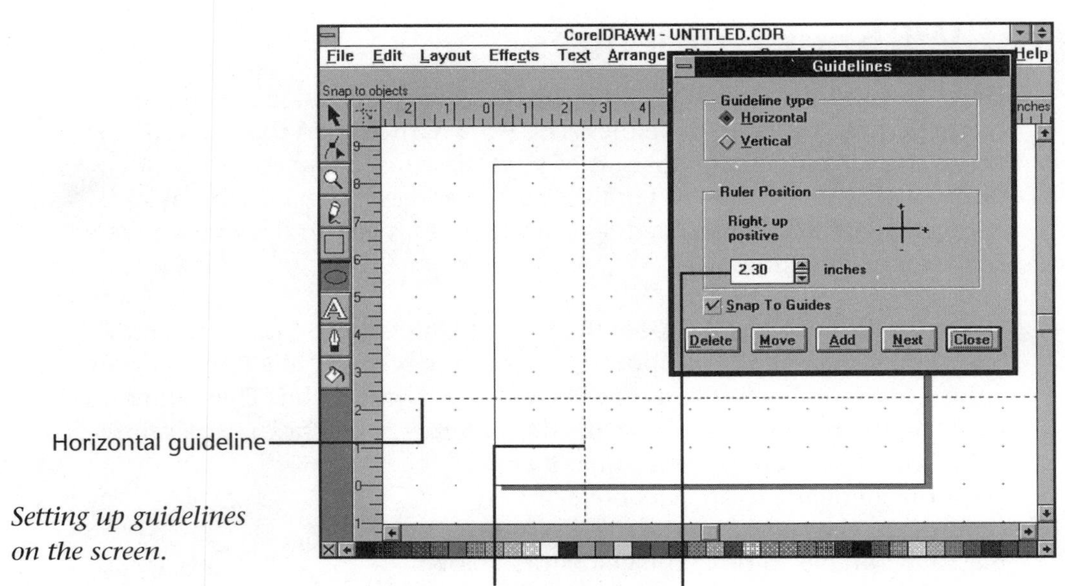
CorelDRAW! - UNTITLED.CDR
File Edit Layout Effects Text Arrange
Help
Guidelines
Guideline type
Horizontal
Vertical
Ruler Position
Right, up positive
2.30
inches
Snap To Guides
Delete Move Add Next Close
Snap to objects
Horizontal guideline
Vertical guideline
Type the position of the guideline here. Add a positive number to display a guideline above or to the right of zeros on ruler. Add a negative number to display a guideline below or to the left of zeros on ruler.

Setting up guidelines on the screen.

The Least You Need to Know

You have a lot of freedom in deciding how to set up your screen display. Various options give you added control as you draw and edit objects; but keep in mind that the display can become rather cluttered when everything is turned on!

- Use the Zoom In tool to see your drawing in better detail.
- Use the **D**isplay menu to display rulers, the status line, and/or the color palette if you so desire.
- If you don't like where the toolbox is located, make it floating using the **D**isplay **F**loating Toolbox command, and then drag the toolbox to wherever you want it.

- Set up a grid in the Grid Setup dialog box, which can be accessed by double-clicking on a ruler or by using the **L**ayout G**r**id Setup command. Turn on the Snap To feature for the greatest control when drawing and editing.
- Display horizontal and/or vertical guidelines if you need additional guidance as you draw and edit.

Special bonus: virtual text page.
(There's virtually no text on it.)

Chapter 8
Right Down to the Line

In This Chapter

- Drawing the line of lines
- Freeing your hand with freehand objects
- Appreciating those curved lines
- A dashing line
- Forging arrowheads
- Lines as thick as pea soup: outlining

Grocery lines. Lines on your face. Incredulous lines slurred to you at the local bar. Incredulous lines you slur at the local bar. Lines you have to wait in just to see a movie. . . . You probably like to steer away from lines as a rule, but lines are harmless in CorelDRAW!. As a matter of fact, lines are essential to your work in CorelDRAW!. And here's some good news: drawing lines in CorelDRAW can be fun. These lines are definitely not distressing or irritating like those other lines in life.

Drawing Straight Lines

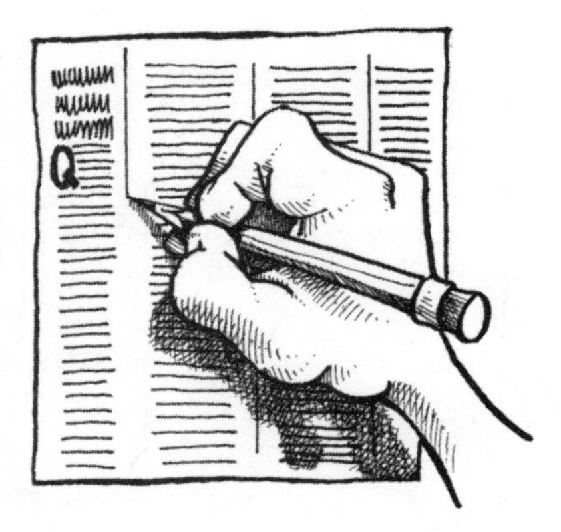

Before you draw a line, you have to make a decision. Do you want to draw straight lines or freehand shapes? To do the former, click on the **Pencil** on the toolbox, and the cursor becomes a cross. Move the cursor to where you want the line to begin and click once. To finish your line, move the cursor (do *not* drag it—meaning don't press the mouse button) to where you want the line to end and click.

You may notice some jaggedness in your line. To ensure a true straight line, it is necessary to press and hold the Ctrl key while moving the cursor. This will constrain the line to a horizontal or vertical, or to an angle of fifteen degree increments. It is important to click at the end of your line before releasing the Ctrl key; otherwise your jagged friends will return.

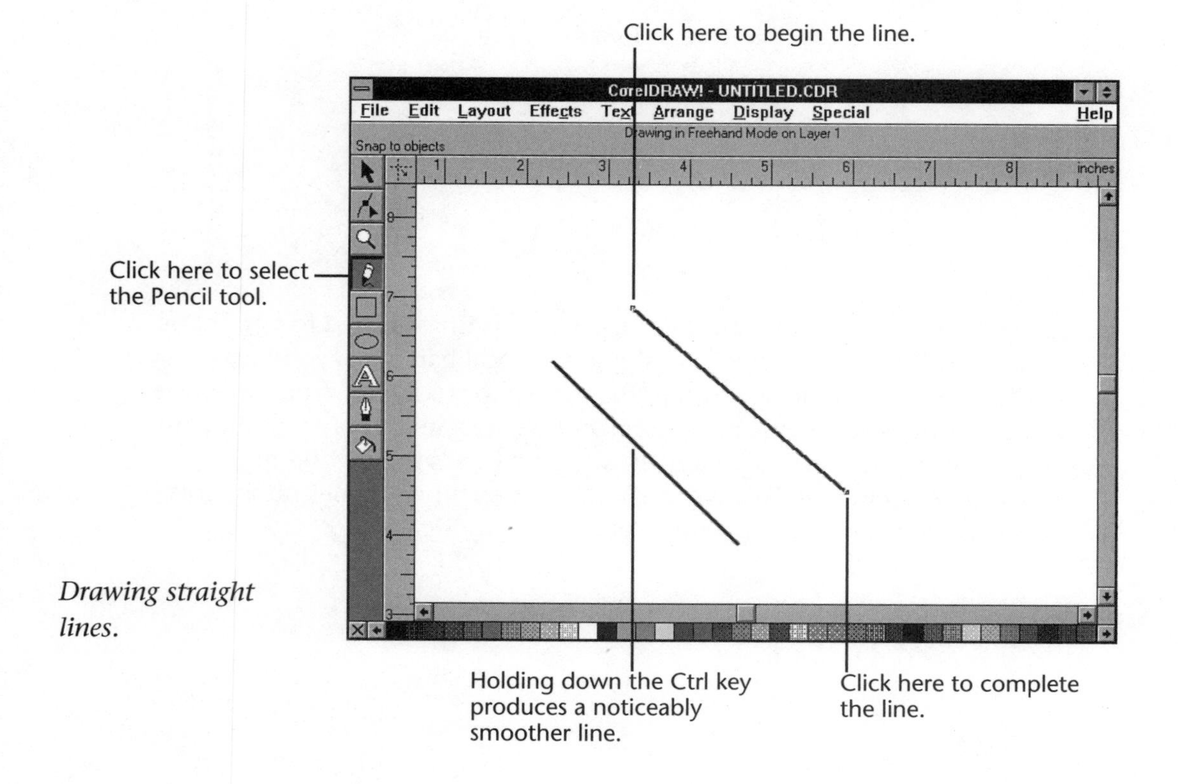

Drawing straight lines.

To draw objects with connecting straight lines, such as triangles and polygons, follow these steps:

1. Click to start the first segment.
2. Double-click to end the first segment and begin the second.
3. Double-click to end the second segment and begin the third.
4. Continue double-clicking to draw all the segments you need. However, to complete the last segment and finish the drawing, click only once.

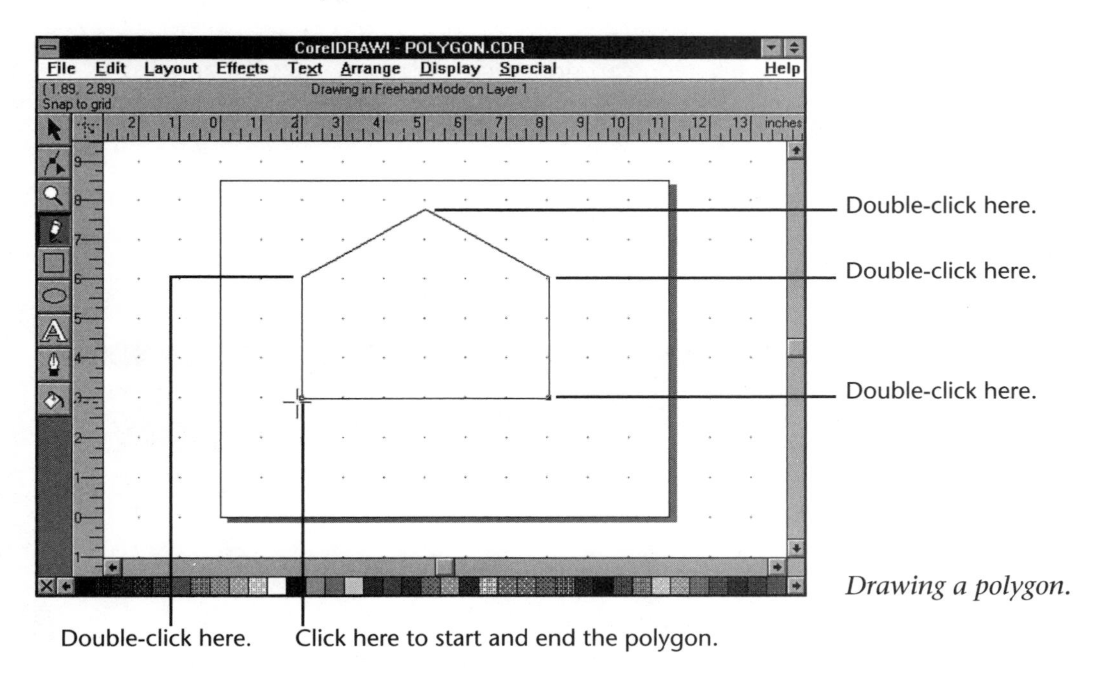

Drawing a polygon.

I messed up on the last segment of a complex drawing. Can I do undo the last segment without erasing the entire drawing? Yes. The Undo feature undoes one step at a time. So go ahead and choose Edit Undo to get rid of the last segment. Later, if you want to redo any portion of the drawing, click on the segment you want to change with the **Selection** tool (the arrow). Press **Delete**, and only that segment vanishes. Now you can redraw it.

Drawing with a Free Hand

Although drawing with a free hand sounds liberating, it requires a delicate and precise touch. To begin you must once again select the **Pencil** from the toolbox. After deciding where to begin your drawing, press and hold down the mouse button at the starting point. Now drag the mouse to create your shape (making sure to keep the mouse button depressed). Notice that any movement of the mouse draws part of the object. Releasing the mouse button completes that segment of your drawing. As you can probably tell by your results, drawing a smooth freehand shape is not an easy task, but like all good things, it takes time and practice.

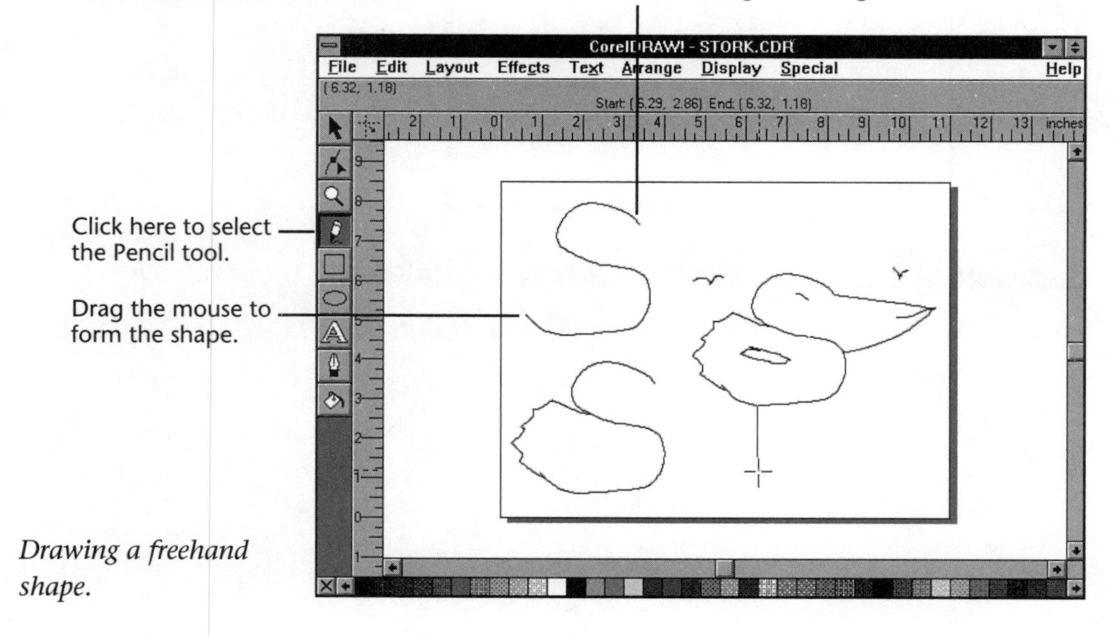

Drawing a freehand shape.

Would You Look at Those Curves?

If only it were as easy to develop curves on our bodies as it is to draw curves in CorelDRAW!. Oh well, c'est la vie. Anyway, it would take the fun out of sweating, weight lifting, and running. (Not.)

You can draw curves in Freehand mode, but those curves might be a little too squiggly and rough for your taste. There is another, more accurate way to draw curves, and that's with a different pencil tool. Unfortunately, using this tool is going to force us to slam a bunch of terms down on the table—so pay close attention! Click on the **Pencil** tool, but this time, continue to hold down the mouse button until a *flyout menu* appears. A flyout menu contains additional tools relating to the tool that displayed the flyout menu. For example, the Pencil tool flyout menu has subtools similar to the Pencil tool. You should see a tool with a smaller pencil and a dotted line or a flyout menu. Click on the pencil and dotted line tool. Okay, here's the first term: this new mode you're drawing in is called *Bézier mode*.

Bézier mode A connect-the-dots approach to drawing curves. Bézier mode is a more accurate way to draw curves than Freehand mode.

Freehand mode Using the Pencil tool, you just drag the mouse across the screen to draw an object the same way you would drag a pencil across a piece of paper.

Control points The points at the end of a dotted line that appear when you drag over a node. Control points determine the direction and shape of a segment.

Bézier mode is a connect-the-dots type of drawing style. You put down the nodes, and CorelDRAW! connects the nodes with a curve. Well, did you catch the second term? *Nodes* are the endpoints of an object, but we already told you about nodes when you learned how to draw arcs and pies.

If the curve isn't the shape you want, you can change it by dragging the node's control points in a new direction. *Control points*? Yes, those are the points at the end of the dotted line that appear when you drag over a node. Control points determine the direction and shape of the curved segments. Take a look at the figure if you're not quite catching this.

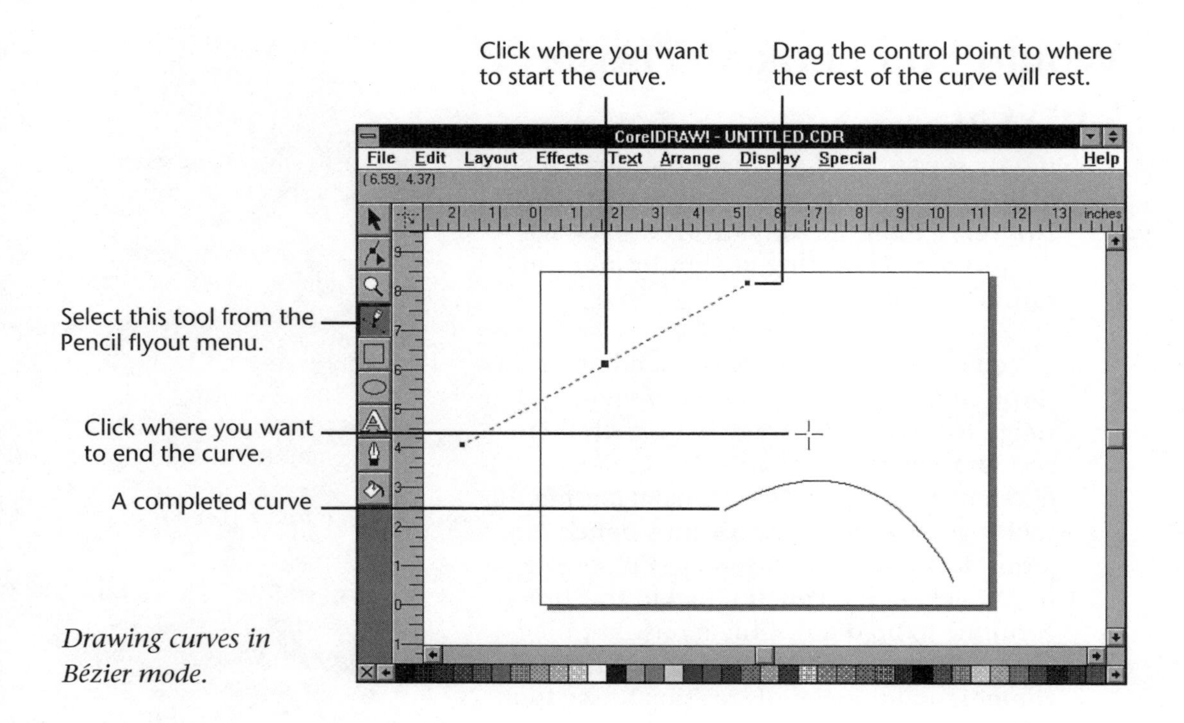

Drawing curves in Bézier mode.

So, to draw a curve in Bézier mode, follow these steps:

1. Click on the Pencil tool and hold the mouse button.
2. From the flyout menu that appears, click on the pencil and dotted line tool.
3. Click where you want to start the curve.
4. Drag towards where you want the crest of the curve to lay. As you begin to drag, a dotted line with two control points appears.
5. Once you've positioned the control point where you want the top of the curve to rest, release the mouse button.
6. Click where you want the end of the curve. CorelDRAW! automatically draws a curve between the two nodes.
7. If the curve isn't round enough or in the direction you wanted, simply drag one of the node's control points until the curve is the shape you want.

Jazzing Things Up with Lines

So much for the shapes of lines and curves, let's move onto the type of line you want. We're talking about all kinds of lines: straight lines, freehand shapes, curves, outlines of objects, squares, circles, etc. Basically, anything drawn. How's that for a simple definition. You don't have to stick with the boring, thin, plain line that CorelDRAW! uses as a default. You can make your lines thick, dotted, or dashed, and you can add arrowheads or other funky endpoints like hands, balls, or rounded corners—and the list goes on. To modify the line style, you'll use a new tool called the Outline Pen.

A Dash of This

Let's start by changing the outline of a simple drawing from plain to dashed. Always remember that you must select an object before you can do anything with it (or did you skip the last chapter?). With that said, let's open the Pen roll-up: click on the **Outline Pen** tool on the toolbox (second tool from the bottom). A flyout menu containing several more tools appears. Click on the tool that looks like a miniature window with an arrow. If you chose the right tool, the Pen roll-up should now be sitting pretty on your screen. If not, try, try again.

The Outline Pen flyout menu contains preset line widths, patterns, and shades you can use to instantly change an object's outline. Select the object, and then click on one of these tools.

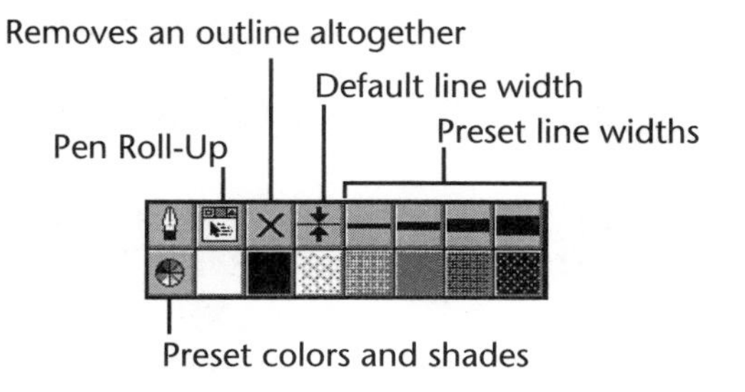

The Outline Pen flyout menu.

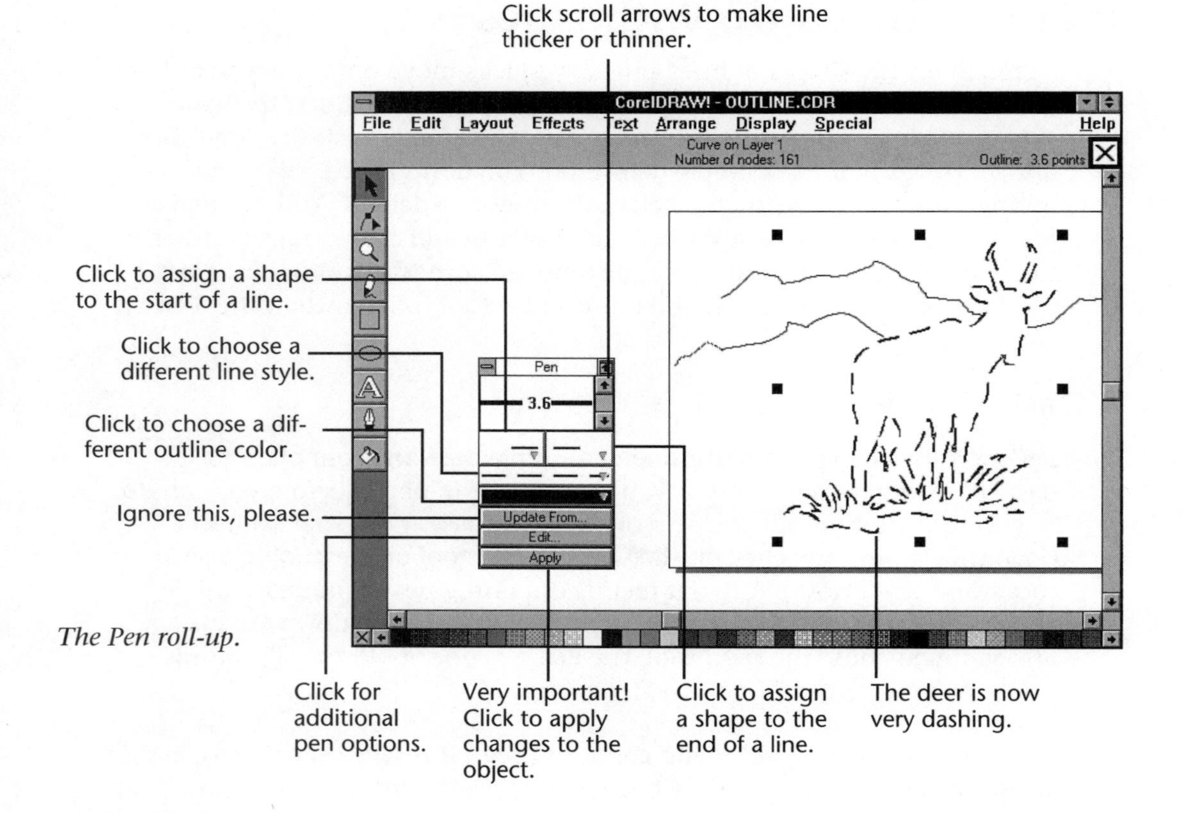

The Pen roll-up.

So with the drawing selected, click on the line style option in the Pen roll-up. Next you'll see a drop-down list of funky lines styles, some mixtures of dashes and dots, others equally-spaced dots or dashes. Pick a dashed line, and presto! The object's line style changes.

What? Yours didn't change? That's because you didn't choose Apply yet! And we purposefully didn't tell you to choose it so that you might remember the next time you don't get an instant result that it's because you didn't click on the Apply button. Okay, now click on **Apply**. This time the outline does change to the new style.

TECHNO NERD TEACHES...

Only nerds would be concerned with such minute details as line corners and caps. But if you're drawing shapes with especially sharp outlines, you may want to play around with corners and end points (also called *caps*).

Click on the **Edit** button in the Pen roll-up to display the Outline Pen dialog box. In the Corners box, try switching to the rounded or beveled corner. In the Line Caps box, try the rounded line cap for a smoother looking shape. Keep in mind line caps includes all segments of dashed lines.

The End of the Line

Now we're getting to some fun stuff. You knew it had to come eventually. You get such choices as knick knacks, tidbits, and wacky willies to stick on the ends of your lines. You might feel like a kid again playing around with the selections.

With the object selected and the Pen roll-up open, click on the left box in the Pen roll-up that has a short line and small arrow. Now scroll through the list of endpoints that appears. You should see arrowheads, arrow tails, a devil's pitchfork, hands, an airplane, circles, squares, and all sorts of other weird shapes. Select the one you want. We're selecting the airplane. Don't forget to click on the **Apply** button. Now it looks like there's a plane soaring above the mountains in the background. How majestic.

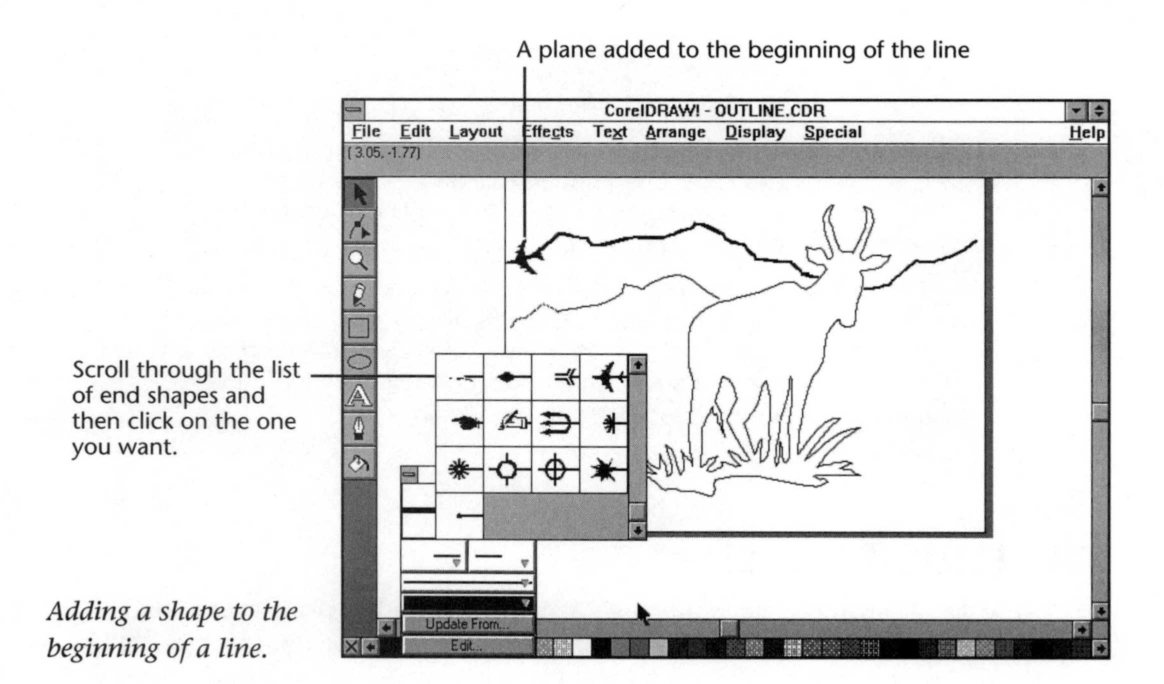

Adding a shape to the beginning of a line.

The Thick and Thin of It

For some strange reason known only to the brainy programmers at Corel way over in the Canadian realm, the default line thickness is just too thin for anyone to take seriously. However, you can do something about it. Change the thickness of the line so you can actually see it without having to buy a pair of cheap reading classes.

One more time . . .1) object is selected and 2) Pen roll-up is open. Now click on the scroll button in the upper right corner of the roll-up. Keep clicking until you reach the thickness you want. As you click, the number increases in the box to the left. This number measures the width of the line in inches. Usually .03 to .06 inches is a good choice, though for some objects, like the mountain backdrop in the figure, a thicker line is appropriate. We used .07 inches for the mountains. Unless you're totally brain-dead, you should be able to tell whether a line is either too thick or too thin. . . . Oh yeah, click on that **Apply** button if you want to see anything happen.

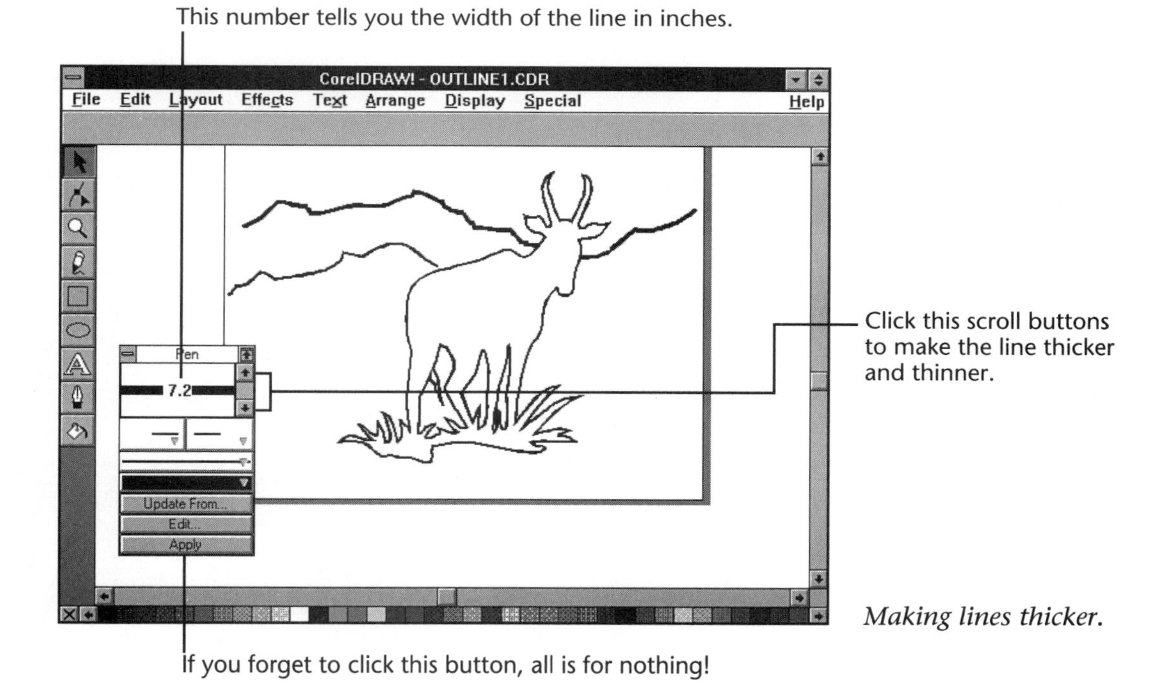

Making lines thicker.

The Least You Need to Know

Pat yourself on the back, if you can reach it. You now know how to draw straight lines, curves, and freehand shapes in various widths and styles and with end shapes.

- To draw a smoother line, hold down the Ctrl key before you start drawing the line, and release the key as the final step.
- Drawing freehand objects is just like sketching on a piece of paper with a pencil, except that you're using a mouse and computer instead.
- Try drawing curves in Bézier mode (using the Bézier pencil tool) to draw cleaner curves.
- When modifying lines and outlines: 1) select the object, 2) display the Pen roll-up, 3) select the new settings, and 4) click on the **Apply** button in the Pen roll-up!

This blank isn't blank—it just looks that way.

Part II

Is There More to Drawing Than This?

I guess I was about 20 when I asked myself that grandiose, mind-boggling question everyone usually comes around to asking himself at some point in his life: Is there more to life than this? Ten years later, I still haven't found the answer to that question.

After learning about the basics in Part I, you may have asked yourself a similar question: Isn't there more to CorelDRAW! than this? Well, I know the answer to that one . . . yes! And this Part will tell you all about some of CorelDRAW!'s more advanced (but not too advanced) features.

Chapter 9
If You Don't Like It, Rearrange It

In This Chapter

- Just the right size
- Movers and shakers
- Ditto that: copying, duplicating, and cloning
- I like the look of that object
- Getting rid of the bums

Now comes the fun part. You get to play around with your objects—and you'll find they are quite submissive little things. They won't bite you when you try to move them, sue you when you copy them, or fight back when you attempt to delete them. You can stretch, shrink, duplicate, and clone them without any argument. This chapter shows you just how to get away with all of this.

Getting into That Size 7

Imagine how much money you could make if you discovered a formula for resizing bodies that's as simple as resizing objects in CorelDRAW!. Perhaps some day . . . but for now, you can get pleasure from your results with your resized objects in CorelDRAW!. CorelDRAW! lets you stretch or shrink an object to virtually any size you want just by dragging the selection handles on a selected object. As you drag, a dotted outline shows you what the new dimensions will be if you release the mouse button at that point.

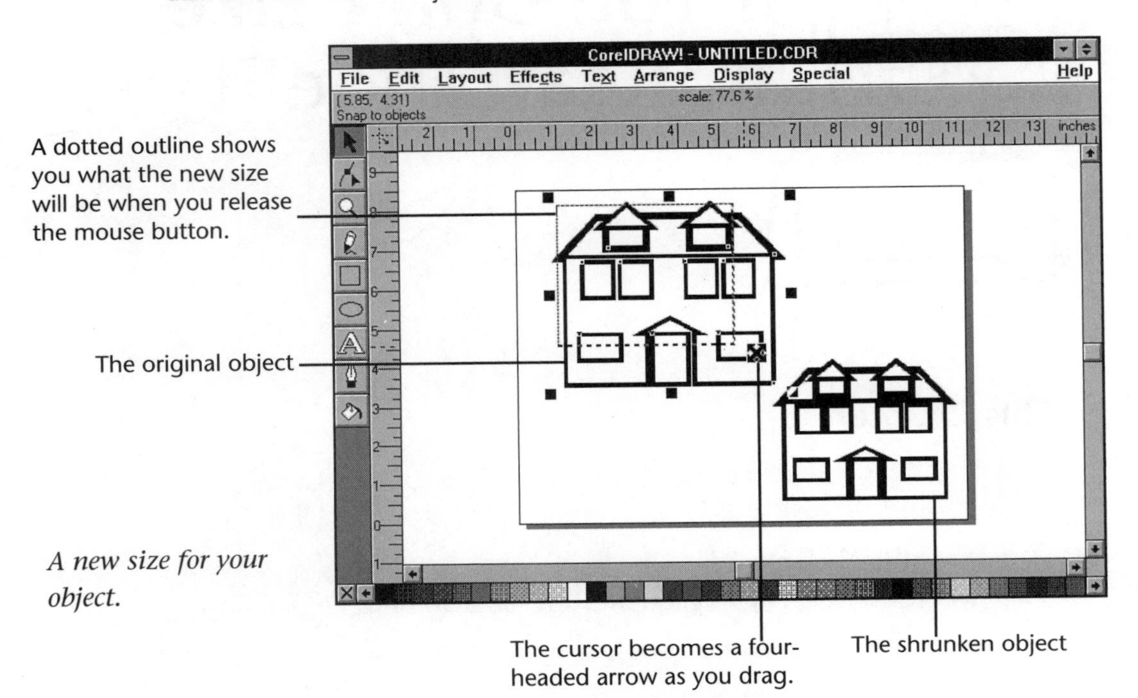
A dotted outline shows you what the new size will be when you release the mouse button.
The original object
CorelDRAW! - UNTITLED.CDR
File Edit Layout Effects Text Arrange Display Special Help
(5.85, 4.31)
Snap to objects
scale: 77.6 %
inches
The cursor becomes a four-headed arrow as you drag.
The shrunken object

A new size for your object.

BY THE WAY

To resize or reshape individual lines and curves, use the Shape tool instead of resizing the whole object. See Chapter 5, "Getting Into Shape," for details. Also, for exact size measurements, use the Effects Stretch & Mirror command instead of the mouse. For more on this command, refer to Chapter 13, "What a Transformation!"

The selection handle you choose to move greatly determines the outcome of the object's new size. To widen an object, drag the handles on the left or right sides. To lengthen an object, drag the handles on the top or bottom. If you want to shrink or enlarge an object while maintaining the current proportions, drag one of the corner handles. Experiment with different handles until you come up with the size you want.

Movers and Shakers

There's nothing to it. Just drag an object to wherever you want it on the page (or off the page, for that matter)—without even hurting your back. In order to move an object that has no fill, you *must* drag its outline.

When you drag, the cursor changes to a four-headed arrow. This lets you know that you're moving the object, and not resizing it (remember that the cursor changes to a cross when you resize an object). As you drag, CorelDRAW! displays a dotted box representing the object. When the dotted box is where you want the object to rest, release the mouse button.

A few other tidbits to keep in mind:

- To constrain the movement of an object you are moving to horizontal or vertical, hold down **Ctrl** as you drag the object.
- You can also move an object in 1/10 inch increments by pressing one of the arrow keys after you've selected the object.
- Make sure you don't drag one of the selection handles; doing so will change the size of the object instead of moving it.
- The object is automatically selected after you move it.

Now, this is all fine and dandy when the new position of your object doesn't have to be precise. But what if you need to place your object exactly 2.056 inches from the top, and .862 inches from the left. You won't be a happy camper if you try to get the object there by trial and error. But you don't have to. CorelDRAW! gives you a Move dialog box that lets you enter in exact coordinates for your object's position on the page. Use the Arrange Move command to bring up the Move dialog box.

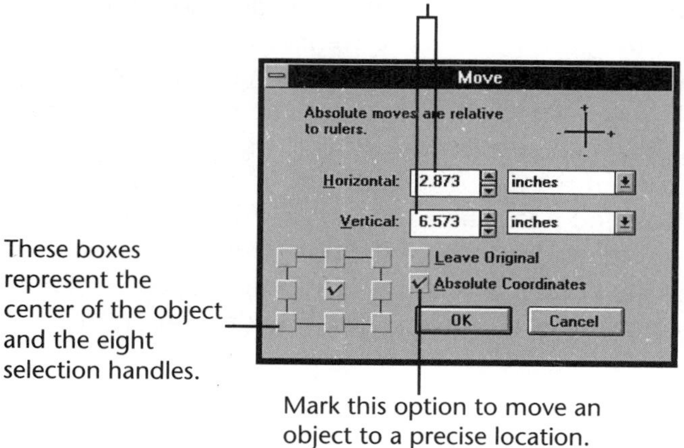

For an exact position, use the Move dialog box.

There is no **A**rrange **M**ove command, you say? And so it is. Use the **M**ove command from the **T**ransform menu instead.

There's no **A**rrange **M**ove command for you either? Geez, can't Corel make up its mind? You get a totally different way of moving an object. Choose **T**ransform Roll-Up from the Effe**c**ts menu. Your moving options are displayed in the Transform roll-up.

If you want to move an object a specific distance from its current position, you can go ahead and type the amount of inches in the **H**orizontal and **V**ertical boxes. But (and that's a big but), if you want to position an object at exact coordinates that have nothing to do with the object's current position, you have to mark the Absolute Coordinates check box. When you do, nine small boxes appear at the left side of the dialog box. These boxes represent the selection handles and the center point of the currently selected object.

Now you need to decide which part of the object will be the reference point for the coordinates you enter: one of the eight selection handles or the center of the object. After you decide, click on the corresponding box in the dialog box so that a check mark appears. You can only choose one box at a time. Then type in the coordinates you want in the **H**orizontal and **V**ertical boxes. The amount of inches are measured from the ruler's zero point. After you click on **OK**, CorelDRAW! moves the object to the position you specified.

Conformist Objects

We're sure you techno nerds can handle the Align dialog box, so we'll introduce it to you. The day may come when you want to align a bunch of objects horizontally or vertically. Well, when that day comes, you'll be prepared! Select the objects you want to align, making sure that you select *last* the object with which you want to align the others. Then choose the Arrange Align command to open the dialog box. CorelDRAW! gives you three choices each for vertical and horizontal alignment. You can align the objects vertically so that the tops or bottoms of the objects line up, or so that their centers are all on the same line. Horizontally, you can set up the objects so that their left or right sides align, or so that their center points fall on the same path.

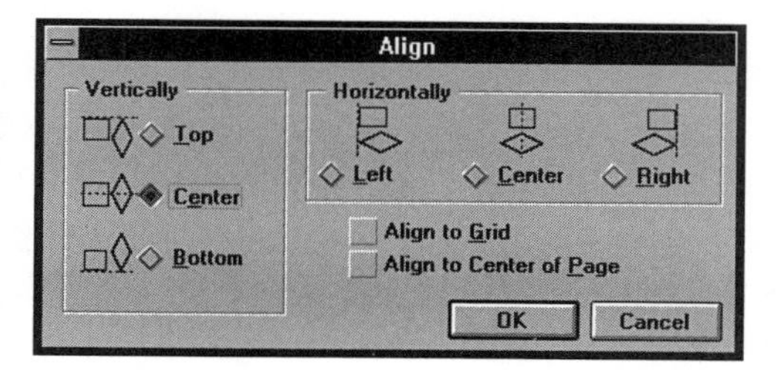

Move a bunch of objects to line up with each other.

You Copycat

Your teachers always told you not to copy, but try to block that rule out of your mind for a moment here. It's perfectly ethical to copy DRAW objects. No one will sue you. And you won't get expelled for plagiarism.

Copying an object in CorelDRAW! can be more than just copying an object. You get three choices: you can either 1) copy the object, 2) duplicate the object, or 3) clone the object. All three choices may sound the same to you, but they're not. If you want a duplicate of an object on the same page, you'll make a *duplicate*. If you want to use a copy of an object on another page, in another drawing, or in another Windows application, you'll *copy* it. Or, if you want a true clone (which mimics all changes you make to the original object), you'll *clone* it.

Plain Old Copies

This is copying the way you know it. Almost every Windows application has an Edit Copy command. You know: you select something, choose the **E**dit **C**opy command, put the cursor where you want to place the copy, and then choose the **E**dit **P**aste command. Well, you can do that in CorelDRAW!, with one exception. It doesn't matter where you put the cursor, because CorelDRAW! always places the copy on top of the original (that is, if you're pasting onto the same page). In fact, you'd never know you made a copy because the copy sits so perfectly on top of the original, you can't tell there are two objects there instead of one. You can then drag the copied object to wherever you want, trusting that the original will remain in place—and it will.

SPEAK LIKE A GEEK

Clipboard A Windows utility that allows you to transfer information to a new location. When you cut or copy an item, the item is placed on the Clipboard so you can paste it to another location on the current page, to a new page, to a new file, or to a file in an entirely different Windows application.

In addition to plain old copying, you can make a clone of an object. Cloning not only creates a copy of the first object, but automatically incorporates any changes you make to the original object into the clone object as well. For details, read the section "Cloneheads: More Than Duplicates."

CorelDRAW!'s **E**dit **C**opy command allows you to do more than just copy onto the current page. (In fact, if that's all you're going to do, you should use the **E**dit **D**uplicate command, which is discussed in the next section.) Sometimes you'll want to copy an object to another page, another file, or even to a document in another Windows application. Well, you can do so very easily with the **E**dit **C**opy command. Just copy the object, and switch to the different page, open the other file, or start the other application. Then use the **E**dit **P**aste command, or press **Shift+Insert**.

A Perfect Duplicate

The **D**uplicate command is perfect when you want to place a copy of an object onto the same page. You actually skip a step: there's no need to use the **E**dit **P**aste command, because CorelDRAW! automatically places the duplicate onto the page immediately after you select **E**dit **D**uplicate. The duplicate object sits on top of the original, but not exactly, so you can see it (unlike the hidden copy you get when using the **E**dit **C**opy command). Just drag the duplicate to where you want it on the page.

The duplicate object is completely separate from the original, which means that, if you make changes to the original object, the duplicate object does *not* change. Use the **E**dit **C**lone command when you want to create a copy of an object that changes anytime you change the original object.

Attention all Version 3 users: no cloning at all for you. CorelDRAW! 3 just doesn't do it.

Cloneheads: More Than Duplicates

I bet you thought cloning was something only scientists dabbled in. Not so. You can become an experienced cloner in CorelDRAW!, and you don't even have to deal with the ethical ramifications of cloning—after all, creating a cloned object isn't anything close to creating a living organism.

When you clone an object, the original object is called the *master*, and the duplicate object is called the *clone*. Most changes you make to the master are automatically applied to the clone. However, certain special effects will not be applied, such as Blend, Extrude, and Contour. If you make changes directly to the clone, this does not change the master. In fact, it breaks the link for the attribute you changed. For example, if you stretch a clone, the next time you stretch the master, the clone will not change. Similarly, if you change the clone's outline width, the width will no longer change when you later change the master's outline width. So try to make changes to the master only, unless you want to purposefully sever the ties of some attributes.

Clone A duplicate object linked to a master object. Most changes you make to the master object will automatically be applied to the clone.

Master An object that has been cloned. Most changes you make to the master are applied to the clone automatically.

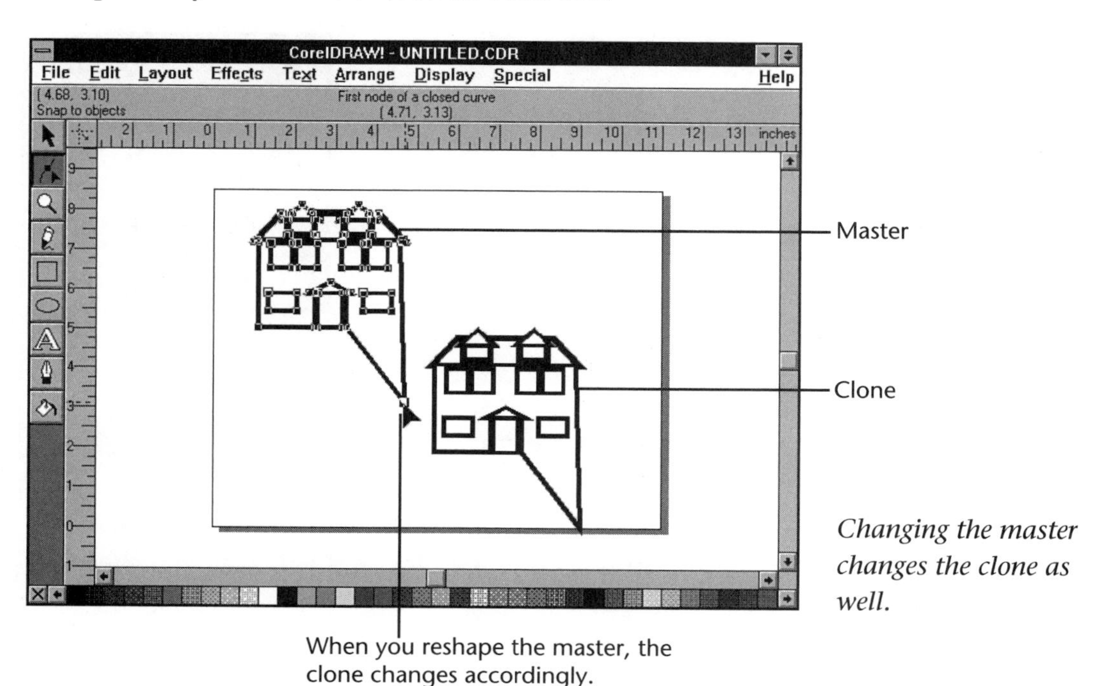

Changing the master changes the clone as well.

You cannot clone a clone, but you can make a duplicate of a clone, and when you do, the duplicate clone works the same as a regular clone. If you cut and paste clones, you break all ties to the master; the clone becomes a regular object. Unfortunately, this means clones are confined to the same page as the master.

Copying the Look of an Object

Maybe you've spent a long time formatting an object with just the right fill, and outline width, style, and color. You want the same setting for a bunch of other objects too. Good news: you don't have to format each of those objects separately. Instead, select the objects to receive the new attributes and choose Copy **A**ttributes From from the **E**dit menu. The Copy Attributes dialog box appears.

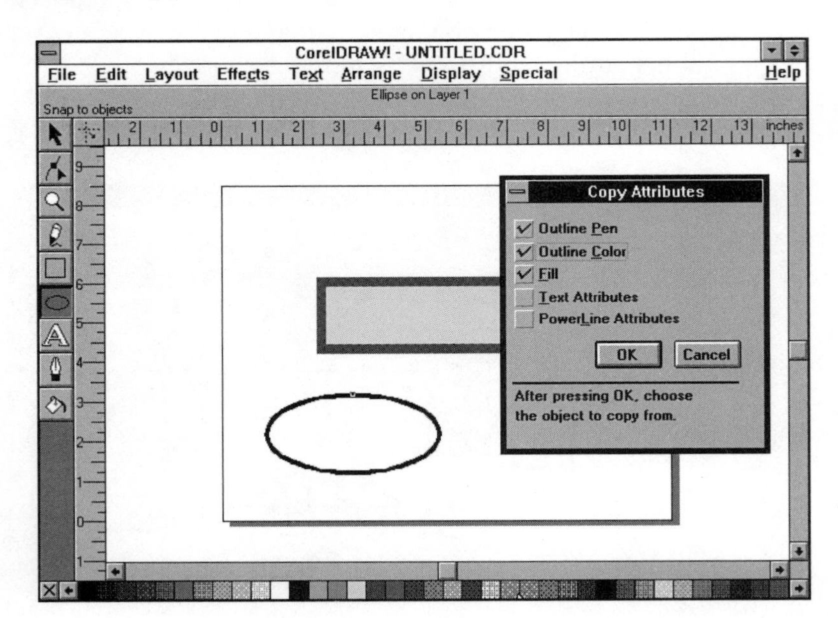

Copying the look of one object to another object.

Mark the attributes you want to copy, in this case, Outline **P**en (for the width and style of the outline), Outline **C**olor, and **F**ill. After you click **OK**, the dialog box closes, and the cursor changes to a funky little "From?" tucked inside an arrow. Click on the object whose attributes you want to copy. Voilà! All of the objects now look alike!

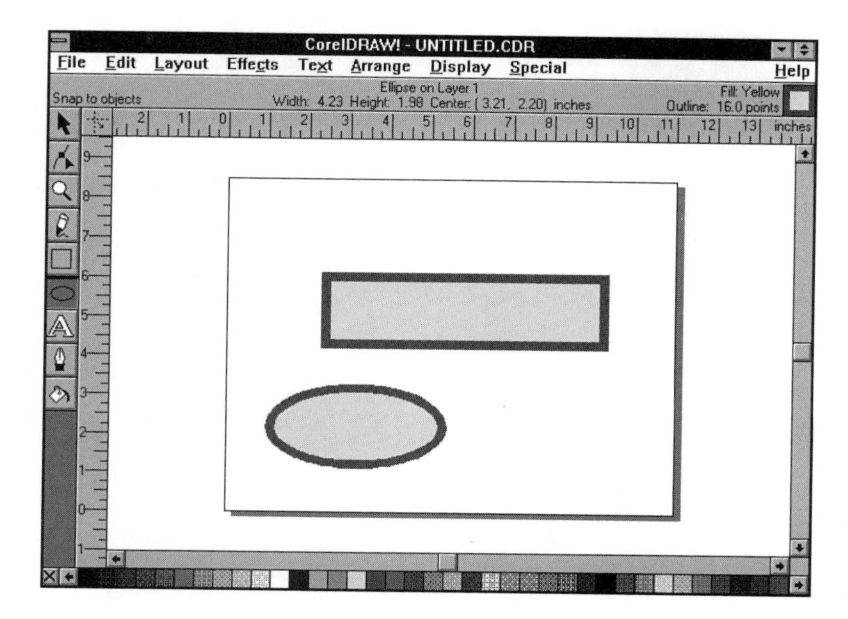

Both objects now have the same attributes.

Nuking an Object

We live in a disposable society. If you don't like something, you dispose of it, right? Well, it's no different in CorelDRAW!, sorry to say. If at any time you feel the desire to terminate the life of one of your poor little objects, just delete it. Select it, and press the **Delete** key or, for you menu-hungry people, select the object and choose Edit Delete. The object is all gone now.

Feeling remorseful? Like you really shouldn't have exterminated that innocent, helpless object? You have the power to bring it back to life! To restore the deleted object, just click on Edit **R**edo before you do anything else.

The Least You Need to Know

You can stretch, shrink, move, copy, duplicate, clone, and delete objects in a jiffy.

- Drag an object's selection handles to stretch or shrink the object.
- Drag the outline of a hollow object to move the object.
- To move an object to a precise location on the page, use the **A**rrange **M**ove command.
- To copy an object to another page, file, or into another application, use the **E**dit **C**opy command.
- To copy an object onto the same page, use the **E**dit **D**uplicate command.
- To make a copy of an object that changes when you change the original object, use the **E**dit **C**lone command.
- To copy the look of one object to another, use the **E**dit Copy **A**ttributes From command.
- To delete an object, select it and press **Delete**.

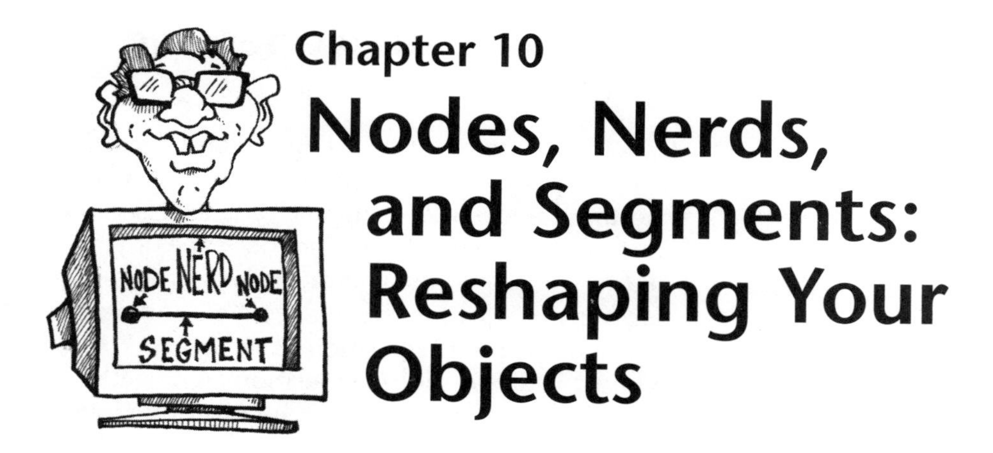

Chapter 10

Nodes, Nerds, and Segments: Reshaping Your Objects

In This Chapter

- Nodes and segments shape things up
- Insisting on curves
- Pushing segments and nodes around
- What the Node Edit roll-up can do for you
- Investing in more nodes
- Get them nodes outta here

Between every two nodes is a segment, and between every two segments is a nerd. Well, the first statement's true at least. Actually, nerds shouldn't be included in this section, but we couldn't resist since what we're going to talk about takes a nerd-mentality to truly appreciate. All others may question, "Why do we have to know about this node malarkey?" Trust us. It will make working in CorelDRAW! a little easier and a little more logical down the road. In fact, you'll probably find yourself reshaping objects on a regular basis—and you need to know about nodes to do that.

Node An end point on an object. The number and position of nodes determine the shape of an object.

Segment A line or curve between two nodes.

The Scoop on Nodes and Segments

A *node* is an end point on an object, and a *segment* is any line or curve between two nodes. "Why do I care about nodes and segments," you ask? Because nodes and segments determine the shape of your objects. Consequently, you can reshape your objects by readjusting the arrangement and number of nodes, and the position of segments. You can use the Shape tool to reposition an object's nodes and segments, and you can use the Node Edit roll-up to adjust the number of nodes an object contains.

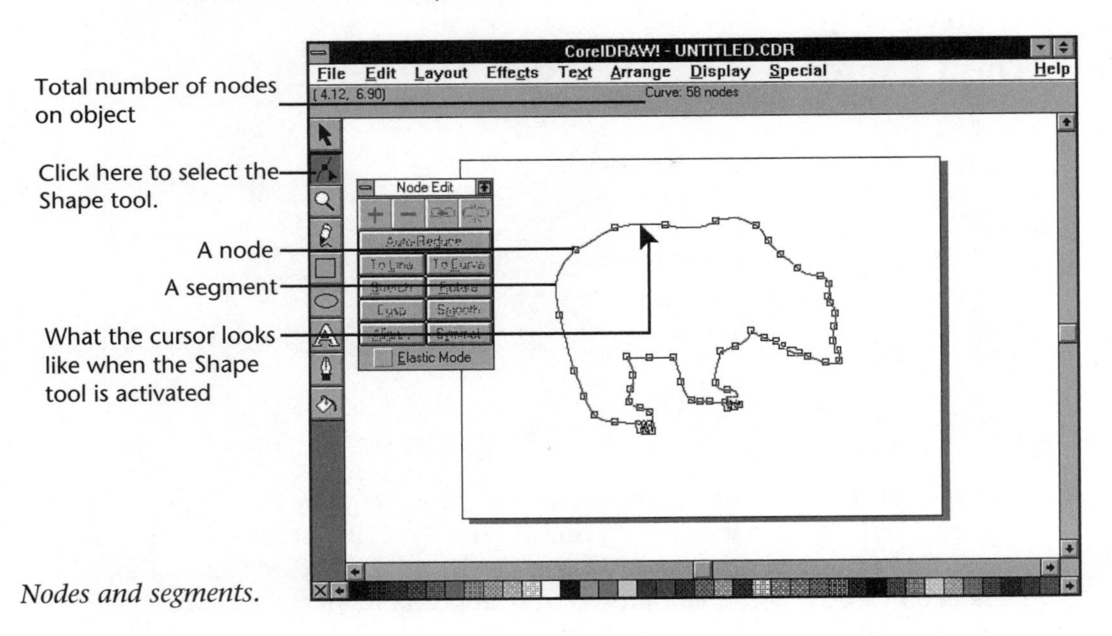

Nodes and segments.

Abracadabra! Converting an Object to Curves

Ladies, wouldn't it be nice if we could push a button to instantly transform those flatter and straighter portions of our bodies to curves? We're not so lucky, but CorelDRAW! is. In order to shape certain types of objects, you have to change the object to curves. This is the only way you can reshape the object's nodes with the Shape tool.

All objects drawn with the Pencil tool are automatically drawn as curves, so you don't have to worry about CorelDRAW! symbols or your own spiffy designs. Additionally, you also don't have to convert arcs and pies to curves. And if you want to round the corners of rectangles, go right ahead. But all other objects, including artistic text, circles, and squares, must be converted to curves before you can reshape them with the Shape tool.

If you're unsure whether an object needs to be converted to curves, click on the Shape tool and then click on the outline of the object. If nodes appear on the object, you don't have to convert the object to curves. It's ready to be reshaped.

If all else fails and you're not sure whether the object needs to be converted to curves, just select the object and then check to see whether the Convert to Curves command is available to you. If so, the object needs to be converted.

To convert an object to curves, first select the object. Then display the Arrange menu and choose the Convert To Curves command. You'll see several small squares pop up along the outline of the object. These are the nodes we discussed in the previous section. The object is now ready to be reshaped.

Your Segments Are All Bent Out of Shape

You'll find reshaping an object is much simpler and certainly less painful than trying to reshape your own body. Once an object has been converted to curves, click on the Shape tool (the second tool on the toolbox). The cursor changes to a thick, black arrow. Click anywhere on the object's outline, and suddenly you'll see a bunch of nodes all over the object's outline. Decide what area of the object you want to reshape. Next, click on the segment that will change that area, and then drag the segment to a new location.

If you do need to restore the object to its original shape, select the **E**dit **U**ndo command immediately after you reshape the object.

Generally, you'll drag segments for coarser adjustments, and drag nodes for finer adjustments. As you drag, the object's shape changes. You might have to experiment quite a bit before you get it right. Use the Undo feature to make your attempts less regrettable.

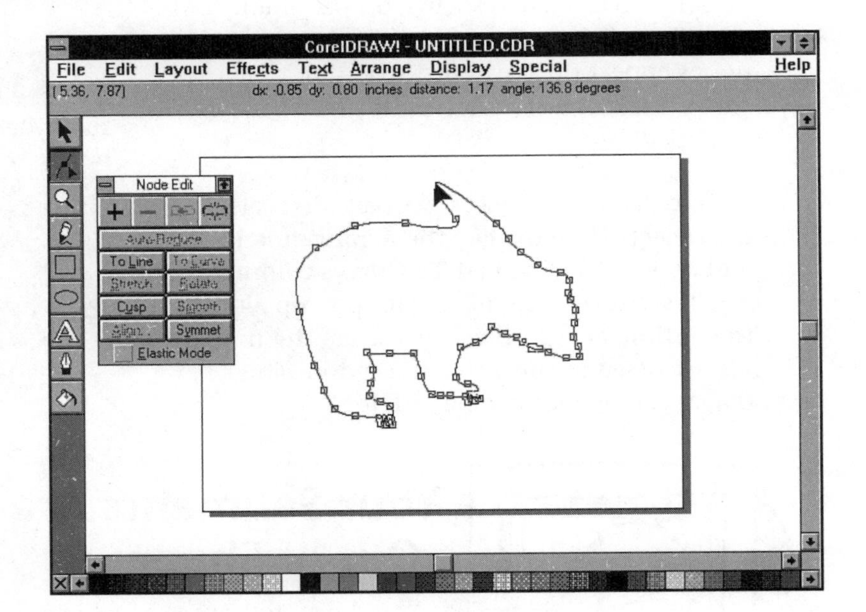

Dragging a segment into a new shape.

The Nodes Know

Okay, so moving some segments around doesn't cut it. That means you need to go after the nodes. So try moving some nodes around instead. Still doesn't cut it? In that case, you'll have to go after the big guys: the control points. *Control points* control the curvature of a segment.

When you click on a node of a curved segment, control points appear for the node you clicked on and other nodes as well. (A control point is a small box at the end of a dotted line that intersects a node.) These control points give you added control in shaping the object because they do not rest on the object. So instead of dragging a node, drag one of its control points and see what happens. Wow, you can really distort an object this way!

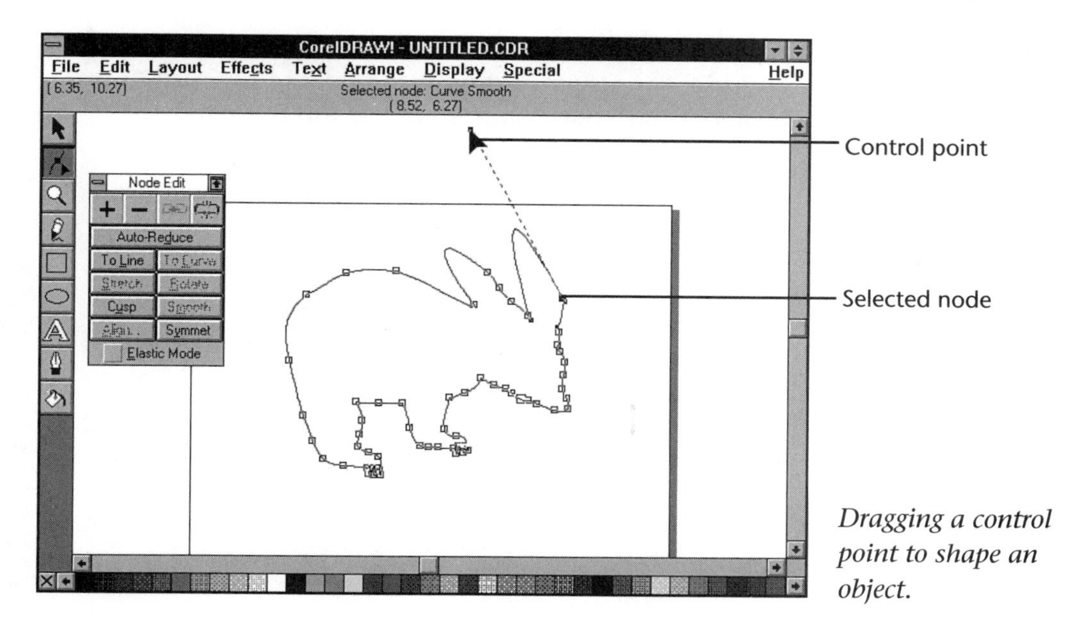

Dragging a control point to shape an object.

No One Knows Nodes Like the Node Edit Roll-Up

They often say it's quality not quantity that matters, but not so with CorelDRAW! nodes. The number of nodes your object has can determine its shape almost as much as how the nodes are positioned. You can really change the look of an object just by adding or deleting nodes. But you can't do any of this without using one of those fine CorelDRAW! roll-ups: this time, the Node Edit roll-up.

You can bring up the Node Edit roll-up in a couple of ways: Double-click on any node, or press **Ctrl+F10**. Easy, huh? There's something missing from the Node Edit roll-up. Have all of you observant types found it yet? You got it: no Apply button to click on! That's right, whatever option you choose in the Node Edit roll-up is immediately applied to the object.

Instead of a Node Edit roll-up, you get a Node Edit dialog box. Most of the options are the same, but you can't leave the dialog box on the screen like you can the roll-up. So what can you do in CorelDRAW! 4 that you can't do in CorelDRAW! 3? Auto-reduce, stretch, and rotate nodes, as well as use Elastic Mode for segments. So no big deal, really.

One less step for you to trouble yourself with. Speaking of options, the Node Edit roll-up has tons of them. But we won't go into them all. If you want to get techy about the Node Edit roll-up, just flip through your voluminous CorelDRAW! documentation for all the details.

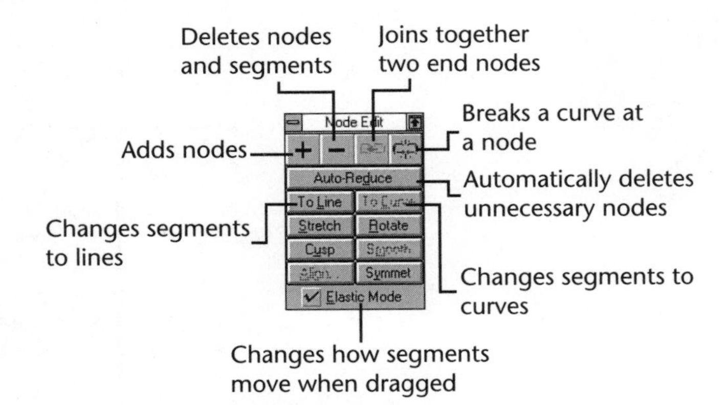

The Node Edit roll-up.

TECHNO NERD TEACHES...

Okay, here's something *you* might want to know about. There are two types of segments: a line segment and a curve segment. Curve segments have control points, line segments don't. A curve object can have a mixture of line and curve segments. Sometimes line segments make an object look rough and jaggedy. You can do something about that fast. Simply drag the Shape tool over the entire object, and then click on the To **C**urve button in the Node Edit roll-up. CorelDRAW! transforms the jagged edges of your object into curves, giving an overall smooth, rounded look.

More Please

Sometimes you want to get the curve of an object to go just right there . . . no THERE . . . but you can't seem to get it there. When this happens, try adding a node or two. It's possible you'll never be able to get the shape you want without those additional nodes. Remember, the more nodes you have, the more definition you can get. However, this does not mean the more nodes the merrier. Not at all. A simple object with too many nodes can look, well, like something a five-year old drew. In this case, you would want to delete some nodes. But a complex object might need more nodes for better definition.

Thanks to the Node Edit roll-up, adding nodes is simple. With the Shape tool, click on the node nearest to where you need more control over the shape. Then click on the plus sign in the Node Edit roll-up. The genius CorelDRAW! adds a node on the segment just where it thinks you need it (and CorelDRAW!'s usually right). Now try to get the shape you need by manipulating the new node or its control point. If you still can't quite get what you want, add another node.

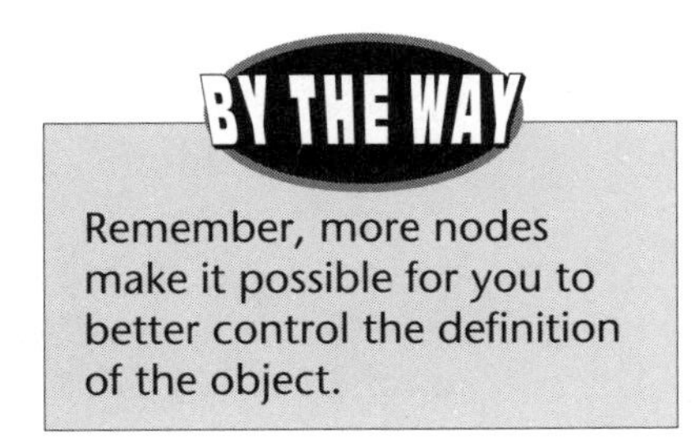

BY THE WAY
Remember, more nodes make it possible for you to better control the definition of the object.

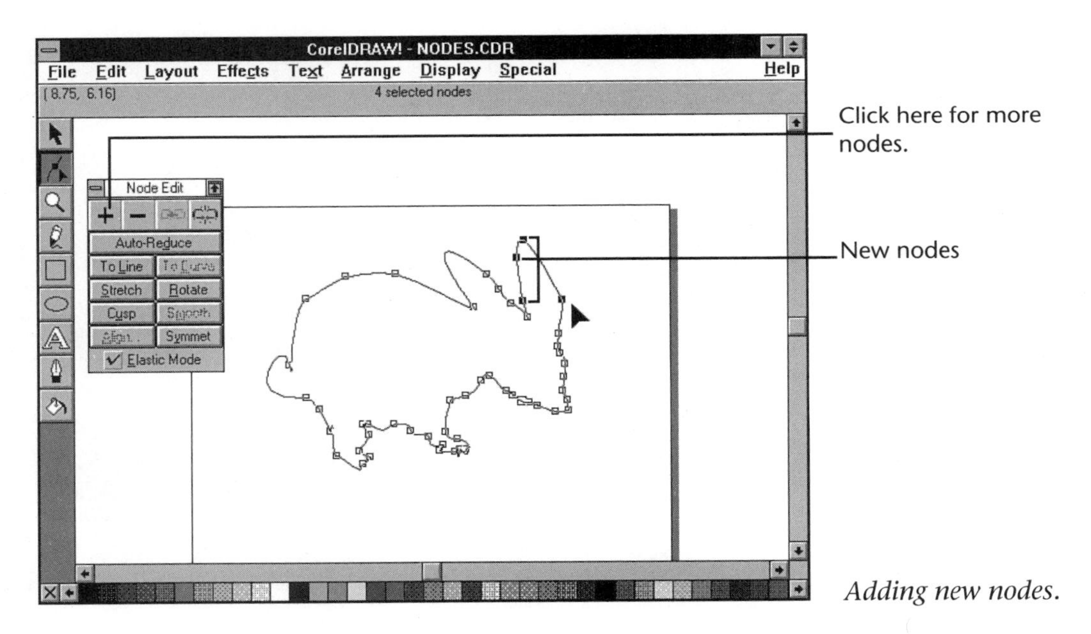

Adding new nodes.

Vote No on Nodes

When you draw in Freehand mode, CorelDRAW! adds nodes to the path as you go along. The speed at which you draw affects the number and placement of nodes. This means that if you take your sweet time getting across the drawing page, you'll probably end up with more nodes than you really need. And too many nodes on a simple object can give the object a sickly, squiggly appearance.

You can smooth out an object by deleting some of its nodes. And, depending on the node you remove, you can also give an object a brand new look by deleting a node. To delete a node, select it and then click on the minus sign in the Node Edit roll-up. To smooth an object without changing its shape, delete only nodes that fall in the middle of curves. If you delete a node at the end of a curve, you may drastically alter the object's shape. For example, deleting the node at the end of the animal's ear in the figure below really changes the shape of the ear (compared with the previous figures).

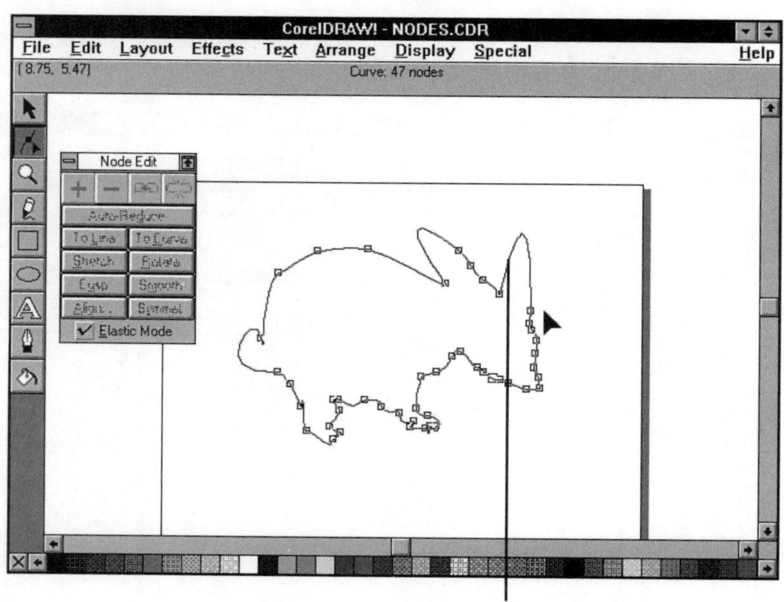

Deleting nodes.

If you're nervous about deleting nodes but want a smoother object, let CorelDRAW! do the work. When you click on the Auto-Reduce button in the Node Edit roll-up, CorelDRAW!

removes extra nodes without dramatically altering the object's shape. First drag the Shape tool over the part of the object you want to smooth out, and then click on the Auto-Reduce button. The curve is redrawn without the extraneous nodes.

You can change the Auto-Reduce setting so that you get a more drastic or less drastic effect on your object when you use the Auto-Reduce button. Choose Preferences from the Special menu. In the Curves section of the Preferences dialog box, increase the Auto Reduce setting (up to 10) to intensify the effect on the curve, or reduce the setting number to lessen the effect.

The Least You Need to Know

Now do you see why nerds appreciate nodes? Nodes can get rather tedious—yet they are essential to your editing capabilities in CorelDRAW!.

- Click on an object's outline with the Shape tool to display an object's nodes.
- If necessary, first change an object to curves using the Arrange Convert to Curves command.
- Drag a segment with the Shape tool to reshape an object.
- Drag nodes and/or control points to redefine the shape even more.
- Add nodes if you can't achieve the look you want.
- Delete nodes to smooth or reshape an object.

What a waste it is to lose one's mind...

Chapter 11
Color Me Beautiful

In This Chapter

- Stuff that object with color
- Feel the texture
- Following the pattern
- What do uniforms and fountains have in common?

You ordered a shrimp Caesar's salad with extra tomatoes ten minutes ago. Your stomach grumbles in anticipation as the waiter finally approaches. He places the salad before you with great fanfare and a distinguished smile. You smile back and give a quick thank you nod, but your eyes suddenly pop open and you gawk as you stare incredulously at your meal. Everything is gray: the shrimp, the lettuce, the croutons, the dressing, even the tomatoes. What would life be like without color?

In Living Color

You'd better set aside a good chunk of time to cover this section. Not because it's complicated or slow-moving, but because it's *fun* working with color. If you're like me, you'll want to experiment with each of the different fill colors, textures, and patterns, testing out all sorts of different color combinations. Since there are hundreds of fills, and eons of color possibilities, plan to be stuck in front of your computer for a while.

Colors, textures, and patterns—what is the difference? Well, all of them are fills in that they fill the inside of objects. Technically speaking, a fill is a color or pattern inside a closed object. But there are different types of fills, such as texture fills, pattern fills, uniform fills, and fountain fills. Each possesses its own distinct qualities.

Fill A color or pattern inside a closed object.

Texture fill Predesigned fills CorelDRAW! provides that are mixtures of colors and gradients. Texture fills resemble designs such as water colors, gravel, clouds, satellite photography, recycled paper, minerals, and more.

Fountain fill A fill that blends two colors or tints of colors.

Uniform fill A solid color fill, including white, black, and shades of gray.

This Will Fill You Up

Regardless of the type of fill you choose, you'll use either the Fill tool, the color palette, or the Fill roll-up to fill your objects. You'll follow this basic procedure: select the object you want to fill, and then choose the fill. This makes using the color palette really simple: just select an object and then click on one of the color icons on the palette. Bingo: your object is now bursting with color. Unfortunately, the other ways to fill objects involve more steps.

The Fill tool has a flyout menu that gives you access to these things: the Fill roll-up; dialog boxes for fountain and uniform fills, patterns, and texture fills; and icons for white, black, transparent, and four shades of gray. If you want to fill an object with white, black, or one of the four shades of gray on the flyout, just select the object and click on the appropriate color tool. Do the same to make an object transparent. A transparent object has no fill, so you can see what lies beneath it. If you want an object to look transparent but you don't want to show what's behind the object, fill the object with white (if you're working on a white background).

The Fill roll-up in CorelDRAW! 3 doesn't have the advanced fill options that the CorelDRAW! 4 Fill roll-up has (such as textures and full color patterns). So sorry!

To display the Fill tool flyout menu, click on the **Fill** tool, which is the bottom tool on the toolbox.

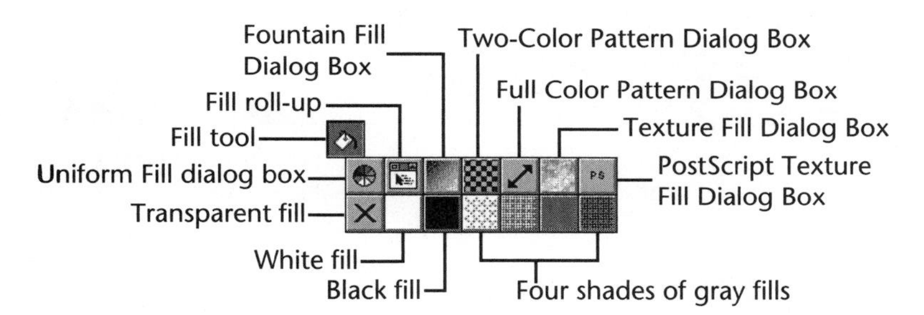

The Fill tool flyout menu.

To display the Fill roll-up, click on the **Fill roll-up** tool on the Fill flyout menu. The Fill roll-up gives you essentially the same stuff the Fill flyout menu does, except for the instant color tools and the PostScript Texture and Uniform Fill dialog boxes. Instead of bringing up dialog boxes for pattern, fountain, and bitmap texture fills, the Fill roll-up just gives you preset choices. This means you can't fool around with the settings, as you can in the dialog boxes. But if you're satisfied with the patterns and other fills just the way they are, go ahead and use the Fill roll-up. Otherwise, if you want more control over the colors and designs of the fills, use the dialog boxes accessed from the Fill flyout menu.

The Fill roll-up in CorelDRAW! 5 looks different from the one in CorelDRAW! 4, but all of the options are the same—they're just arranged differently. One tip: to display fill selections, click on the appropriate fill icon at the top of the window.

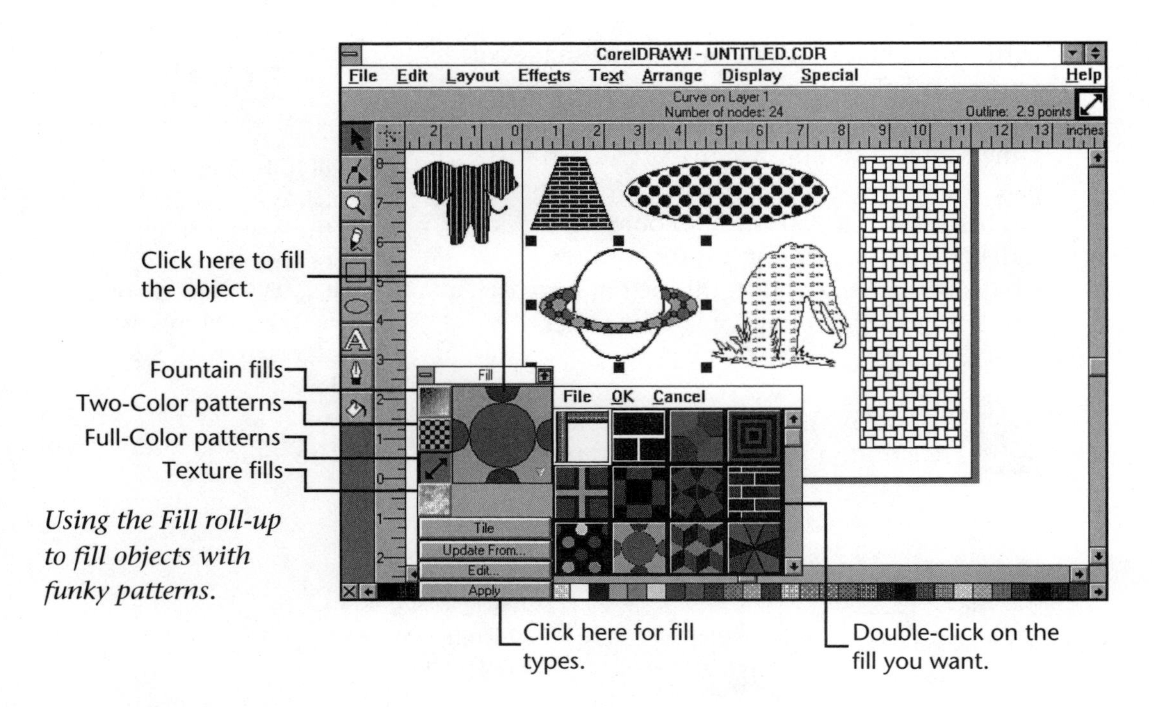

Using the Fill roll-up to fill objects with funky patterns.

Fill It Up with Texture, Please

To make life even more confusing, there are two types of texture fills: bitmap and PostScript. You must have a PostScript printer to use the PostScript textures, so we'll concentrate on bitmap textures since everyone can use them. CorelDRAW! gives you more than one hundred textures to choose from. *Texture fills* are basically mixtures of colors and gradients which resemble nifty designs such as clouds, nebulas, recycled paper, gravel, minerals, satellite photography, water colors, and more. So if you want to add a cloudy background to one of your drawings, try one of the cloudy textures—don't waste half a day designing your own background (unless, of course, you are one of those I-have-to-do-it-myself kind of people).

You can give a plain, transparent object a whole new look just by filling the object with a texture. For example, look at the animal in the figure below; filling it with Mineral Cloud texture gives it an almost 3-D look. Textures can do wonders for your objects, but you have to play around with the textures to see just how they affect the look of your object. Some textures will do wonders, while others will hurt your eyes. It all depends on the object. What looks terrible inside of one form may look lovely inside of another.

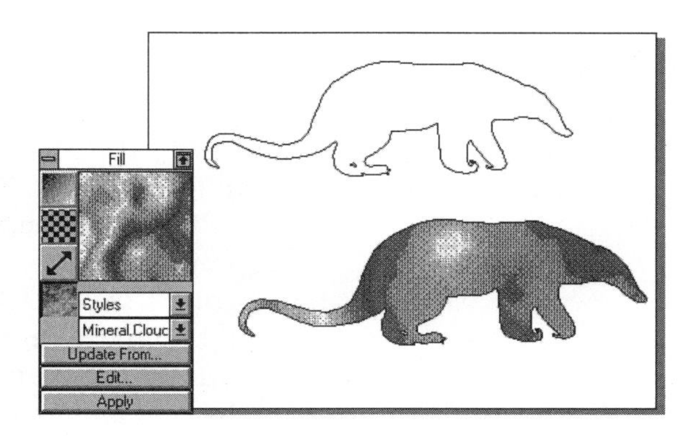

One texture fill can make an animal come alive.

If you want to change the colors of a texture, you have to use the Texture Fill dialog box. To open this dialog box, click on the **Texture Fill** tool on the Fill flyout menu. The Texture Fill dialog box has 1st Color and 2nd Color drop-down boxes, which allow you to set up a different color combination. To see what the texture will look like with the new colors, click the **P**review button. To save the new color combination, overwriting the original, choose **S**ave. To save the new color combination but keep the original intact, choose Save **A**s and assign a new name to the texture.

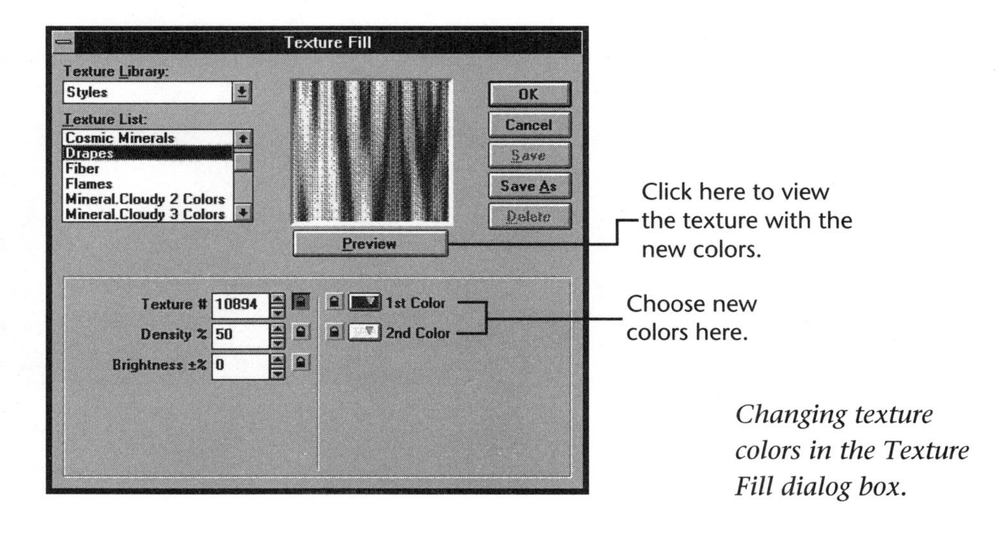

Changing texture colors in the Texture Fill dialog box.

If you want to view some sample textures, select Samples from the Texture Library drop-down list. Then choose a texture in the Texture List box.

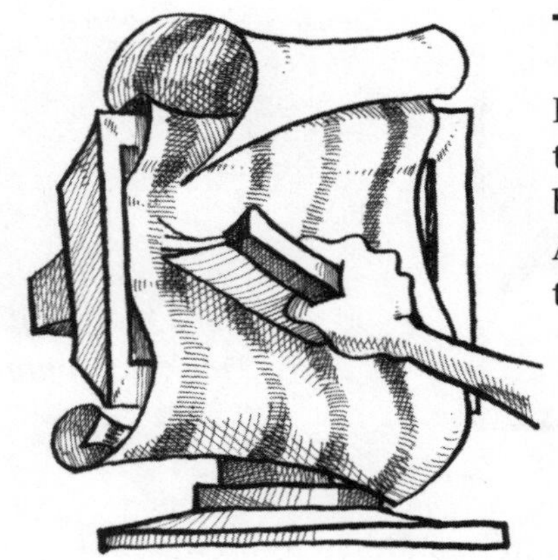

This Is Becoming a Pattern

Have you ever gone to the wallpaper store and tried to pick a pattern from the hundreds of books lined up on the wall? It can be confusing. And to keep things confusing, there are two types of pattern fills: two-color patterns and full-color patterns. Two-color pattern fills are simple black on white patterns (you can add color to them), while full-color pattern fills burst with color.

You can use the Two-Color or Full-Color icon in the Fill roll-up to select the pattern you want. Or you can open the Two-Color or Full-Color dialog box from the Fill tool flyout menu and then make your selection in the dialog box. Both dialog boxes work similarly, so we'll just take a look at the Two-Color Pattern dialog box. To display it, select an object and click on the **Two-Color Pattern Dialog Box** tool on the Fill tool flyout menu.

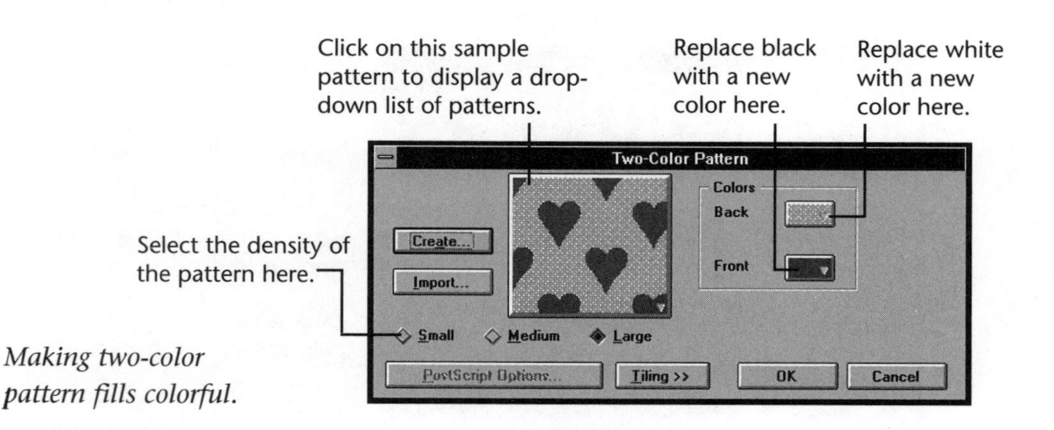

Making two-color pattern fills colorful.

Select the pattern you want by clicking on the sample pattern in the dialog box. A drop-down list of patterns appears. Double-click on the one you want. In the Colors area, assign a new background color by clicking on the Back icon, and then double-clicking on the color you want. Incidentally, this color replaces the white in the original pattern. To select a new foreground color (which will replace black), click on the Front icon and then double-click on a new color. If you want to increase or decrease the density of the pattern, click on the **S**mall or **L**arge option button. The pattern in the dialog box changes to show you what it looks like with the new colors and density.

If a dialog box pops up stating that nothing is currently selected, that's because you clicked one of the Fill tools without having selected an object first. Close the dialog box, select your object, and then try, try again!

Fill Your Objects with This

Those Corel geniuses never stop: here are yet some more ways to fill your objects! Uniform and fountain fills may be less fun than textures and patterns, but they still get the job done. A uniform fill is a single, solid color fill including black, white and shades of gray. You can add this type of fill to your objects using the color palette at the bottom of your screen, or via the Uniform Fill dialog box, which you can access by clicking on the **Uniform Fill Dialog Box** tool (the tool that looks like a wheel of colors) on the Fill tool flyout menu.

Fountain fills mix two colors or tints of colors gradually (for example, a blend of green and white). They are also sometimes called gradient fills because of their gradual blend of colors. There are three types of fountain fills: linear, radial, and conical. Generally, you'll use linear fills when filling rectangular and square objects, and you'll use radial or conical fills to fill rounded objects. Think of linear fills as fills that change color from start to finish. Radial fills change color in circular increments, and conical fills are similar to pie wedges that change in color. Does that help? Hope so—if not, play around with them in the Fountain Fill dialog box, and you'll get the idea pretty quickly. But we didn't tell you about the Fountain fill dialog box yet, did we? Shameful. To open it, display the Fill tool flyout menu, and click on the **Fountain Fill Dialog Box** tool, which is the third tool from the left on the top row.

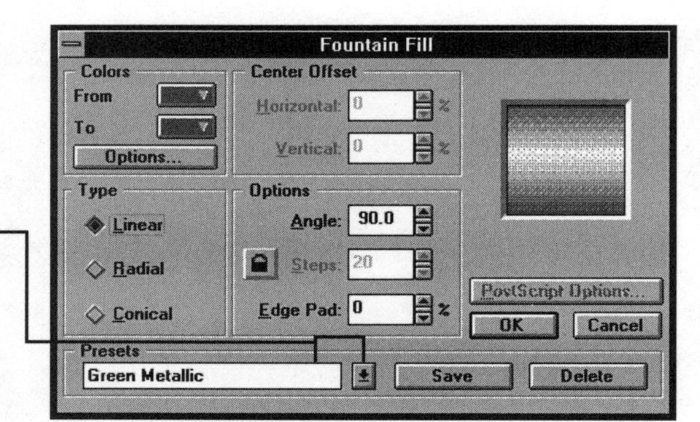

Using the Fountain Fill dialog box.

First decide which colors you want to use for the fill and select them in the Colors area of the dialog box. Then choose the type of fill you want: **L**inear, **R**adial, or **C**onical. You may want to take the time to check out some of CorelDRAW!'s preset fills; some of them are pretty cool! Just open the Presets drop-down list, and then click on a fill. The sample at the right side of the dialog box shows you what the fill looks like.

The Least You Need to Know

When it comes to color, the possibilities are endless. Have fun experimenting!

- Use the Fill tool flyout menu, the Fill roll-up, or the color palette to fill your objects.
- Click on the **Fill** tool to display the flyout menu. You can access the Fill roll-up, Fill dialog boxes, and certain shades from the Fill tool flyout menu.
- Use the Texture Fill dialog box to change the color of texture fills.
- Use the Two-Color Pattern dialog box to add color to and change the density of two-color pattern fills.
- Uniform fills are your basic black, white, and gray fills.
- Fountain fills blend two colors, and come in linear, radial, and conical designs.

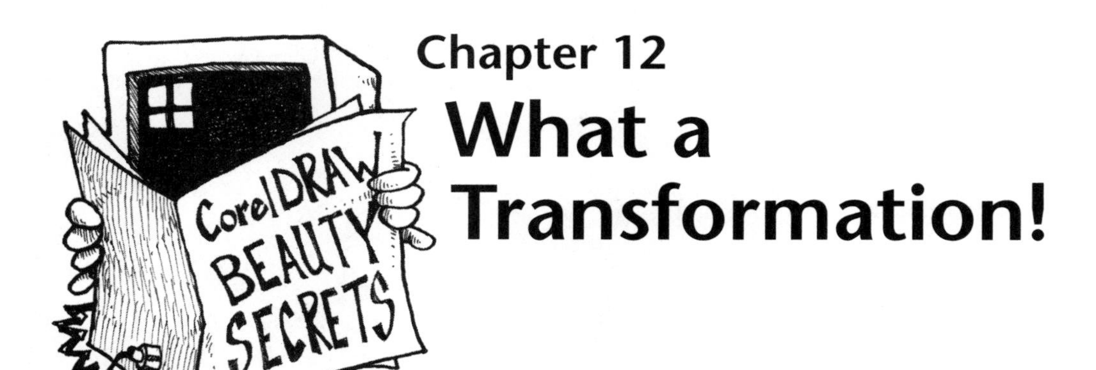

Chapter 12
What a Transformation!

In This Chapter

- Looking at your object in a mirror
- Turning about face
- Skewing up royally
- Seeing a different perspective

Women's magazines often feature articles on makeovers, because we love to see how someone can be transformed from a plain, ordinary human being into a creature that knocks people off their feet. Well, you can be a veritable makeup artist with your CorelDRAW! objects, transforming them from ordinary objects to eye-catching, unique graphics (kind of like going from an ugly duckling to Christie Brinkley). By rotating, flipping, skewing, and changing the perspective of a drawing, you can give it a whole new look.

Mirror, Mirror on the Wall

Who's the fairest of them all? Well, your drawings, of course. But let's say you don't like the direction your drawing is facing. You have a palm tree that clearly shows an easterly wind is blowing by the way its leaves are bent, but everything else in your drawing is being blown to the west. Flip the palm tree, or in CorelDRAW! terms, mirror the object. *Mirroring* an object flips it to the left or to the right horizontally, or upside down or right side up vertically.

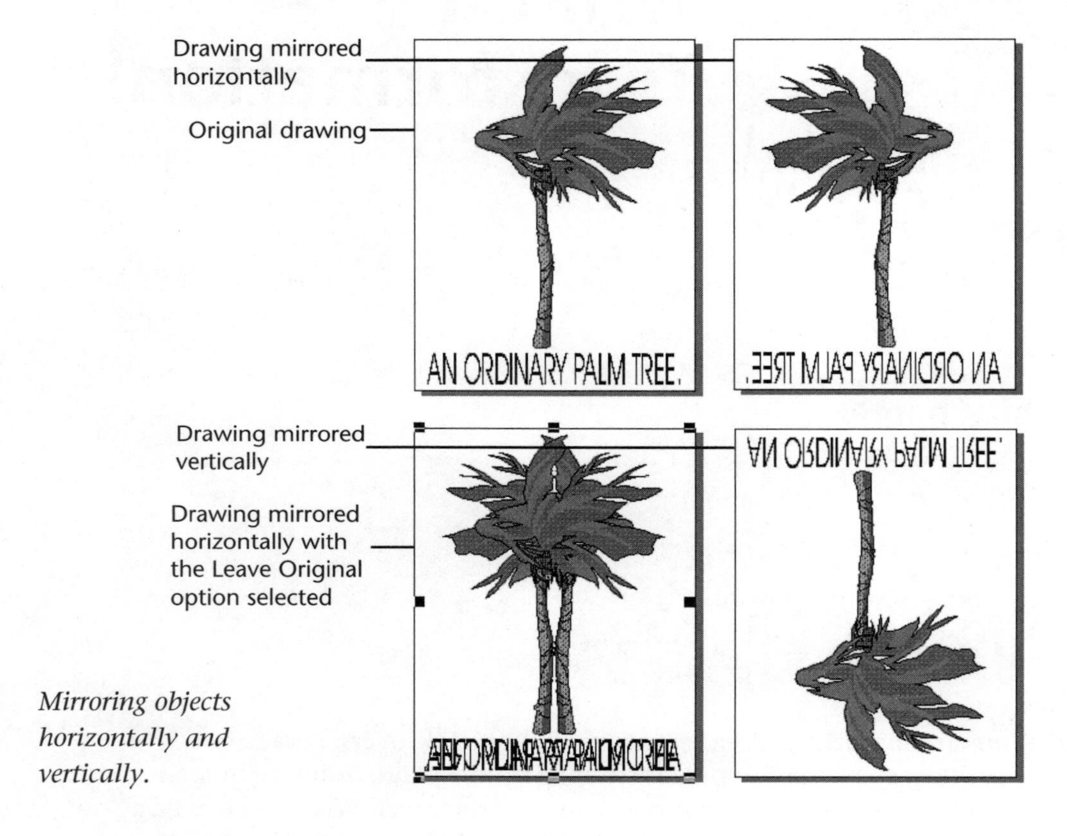

Mirroring objects horizontally and vertically.

5.0

To mirror, flip, rotate, and skew objects, you'll use the new Transform roll-up instead of individual commands on the Effects menu. To display the Transform roll-up, select the **T**ransform Roll-Up command from the Effe**c**ts menu.

To mirror an object, select the **S**tretch and Mirro**r** command from the Effe**c**ts menu. In the Stretch and Mirror dialog box, click on the H**o**rz Mirror button to flip an object from side to side, or the **V**ert Mirror button to flip an object upside down or right side up. If you want to leave a copy of the original object (which can produce an interesting effect with the right object), mark the **L**eave Original check box.

Around and Around We Go

BY THE WAY

You can use the Stretch and Mirror dialog box to enlarge or shrink an object, but it's usually easier to resize an object by dragging its selection handles. However, if you prefer to enter exact scaling measurements, enter values in the Stretch **H**orizontally and/or Stretch **V**ertically option boxes. To keep the proportion of the object intact, enter equal values in the option boxes.

Okay, so your object is facing the right direction, but now you want it tilted to one side. Picky you! But CorelDRAW! will gladly comply with your desire. You can rotate an object any number of degrees you want. Just click on an already selected object, and *rotation handles* appear instead of the regular selection handles. In fact, if you're like most humanoids, you already discovered these rotation handles earlier by accident, and wondered what the heck they were. Well now you know.

When the rotation handles are showing, drag one of the handles in the direction you want the object to tilt. For example, to tilt the palm tree to the right, you'll drag the rotation handles in the upper right corner to the right and downward. If you want to rotate an object in increments of 15 degrees, hold down the **Ctrl** key as you drag.

CorelDRAW! - PALM.CDR
File Edit Layout Effects Text Arrange Display Special Help
(10.14, -2.83)
angle: -33.5 degrees
inches
As you drag, the cursor changes to a rounded horseshoe.
A dotted box shows you how much you're tilting the object.

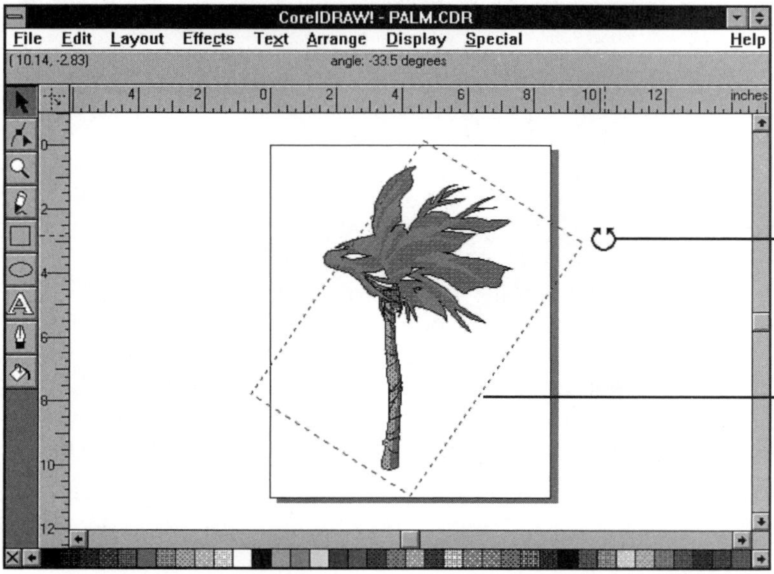

Rotating the palm tree to the right.

Notice when you drag, a dotted box appears showing you how much of a rotation you're producing. Also, the cursor changes to what looks like a rounded horseshoe. When your object is tilted to the correct degree, release the mouse button. Bingo! Your object that once stood proper and erect now sloppily tilts to the side.

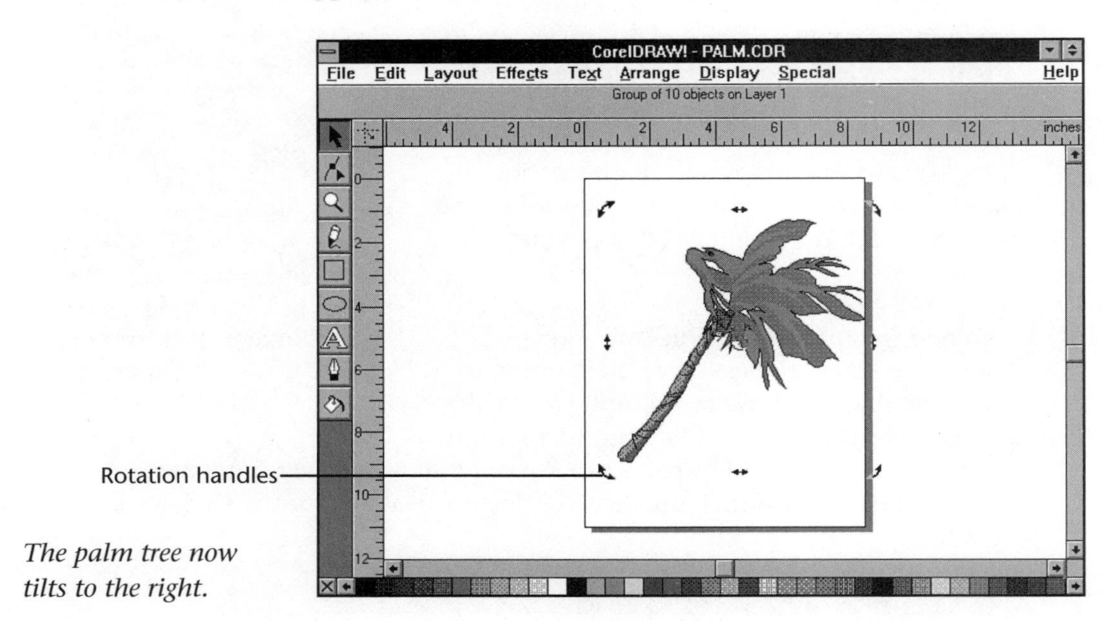

The palm tree now tilts to the right.

TECHNO NERD TEACHES...

If you want to rotate an object an exact amount in degrees, we recommend that you use the Rotate & Skew dialog box instead of skewing around trying to achieve the degree of rotation manually—especially if you're dealing in fractions. To open the Rotate & Skew dialog box, choose **R**otate & Skew from the Eff**e**cts menu. Type the degree to which you want the object rotated in the Rotation Angle box. Check out the degree scale in the dialog box if you need help entering the correct degree for the angle you're looking for.

A Skewed Understanding

You've flipped, tilted, rotated . . . what's left? How about skewing? You may have been taught all about skewing at a young age, but just in case you weren't, *skewing* is slanting an object. To skew an object, select the **R**otate & Skew command from the Effects menu. The Rotate & Skew dialog box appears.

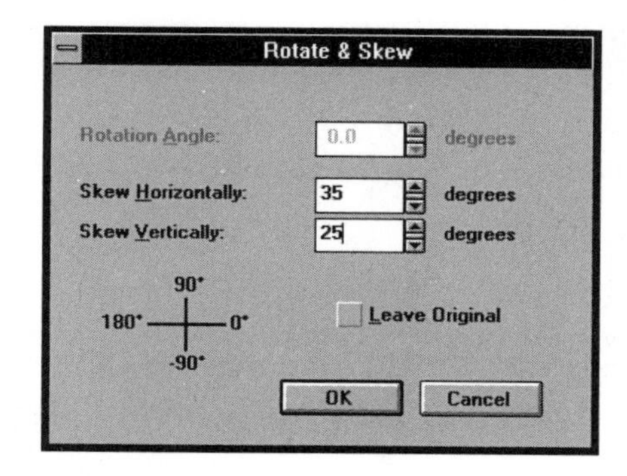

The Rotate & Skew dialog box.

In the Skew **H**orizontally box, enter the number of degrees you want to slant the object to the right or left, following the degree scale provided for you in the dialog box. For example, if you wanted to slant an object a little to the left, you would enter a number between 105–115 degrees. If you want to slant the object up or down as well, enter an amount in the Skew **V**ertically box.

To leave a copy of the original, mark the **L**eave Original box. Click on **OK**, and presto—your object is now skewed.

A skewed object slants.

After you've rotated or skewed an object, its selection box changes to a new size to fit the new object. It is difficult to return a rotated or skewed object to its original orientation. Leaving the original intact doesn't even help because it becomes part of the rotated or skewed object. So be sure to select the Edit Undo command right away if your object's new tilt or slant isn't just right.

That's Your Perspective

Haven't we just been changing an object's perspective by slanting, tilting, rotating, and mirroring it? Not really. At least, not the way the Add Perspective command does. Changing an object's *perspective* gives depth to the object. You can create one-point and two-point perspectives. If you shorten one side of an object, you're creating a one-point perspective; you can achieve a two-point perspective by changing two sides of an object. A two-point perspective adds the greatest depth.

To begin, select your object and then choose Add Perspective from the Effects menu. A dotted box with four corner handles appears around the object. Move the mouse over one of the corner handles until the cursor changes to a cross. You're ready to change the object's perspective.

For a one-point perspective, drag the handle vertically or horizontally. Holding down the **Ctrl** key as you drag causes the handle to move perfectly up or down or to the right or left. For a two-point perspective, drag diagonally towards the center of the object or away from the object. If you drag toward the center of the object, this pushes the object further away from you, while dragging away from the

Hold down the **Ctrl** and **Shift** keys while dragging, and the opposite handle mysteriously moves the same distance in the opposite direction as you're dragging.

Changing an object's perspective.

object pulls the object out towards you. When you release the mouse button, CorelDRAW! redraws the object in its new perspective.

The Least You Need to Know

Now you know how to do all sorts of weird stuff to your drawings. You can tilt 'em, slant 'em, skew 'em, flip 'em. . . . By the sound of it, you'd think we were talking about burgers!

- To flip an object to the left, right, or upside down, use the **S**tretch and Mirror command on the Effe**c**ts menu.
- To rotate an object, click on it to select it, and then click on it again to display its rotation handles. Then drag one of the handles to the degree you want the object tilted.
- To slant an object, use the Rotate & Skew dialog box. Select the Effe**c**ts **R**otate & Skew command to open the dialog box, and then enter an amount for the degree you want the object slanted.
- To give an object depth, choose the **A**dd Perspective command on the Effe**c**ts menu. Then drag the corner handles towards the center of the object or away from it to create a 3-D look.

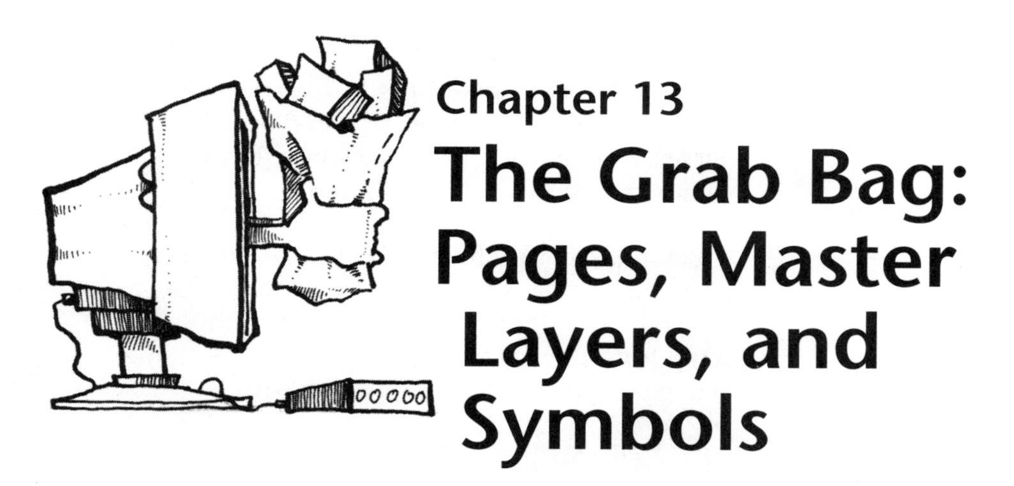

Chapter 13
The Grab Bag: Pages, Master Layers, and Symbols

In This Chapter

- Decorating your page
- Pages and pages and pages, oh my!
- Introducing the friendly Master Layer
- More symbols than you can grasp

As a kid, I always thought grab bags were fun to explore—full of surprises and all sorts of different knick knacks. Well grab bags are not just a part of your past. This chapter is just like one of those grab bags: it's packed full of various tidbits on CorelDRAW!—knick knacks that didn't really fit in the other chapters. Though pages, layers, and symbols, might not be as exciting as the treasures you found in your childhood grab bags, they are definitely more valuable (a quality we, as adults, are supposed to be more concerned with anyway—right?).

What a Page!

"What *type*, what *size*, what *orientation*?" Does this sound like questions you ask yourself when looking for a date? But what if we throw in "display facing pages?" Then you'd know we were talking about CorelDRAW! page

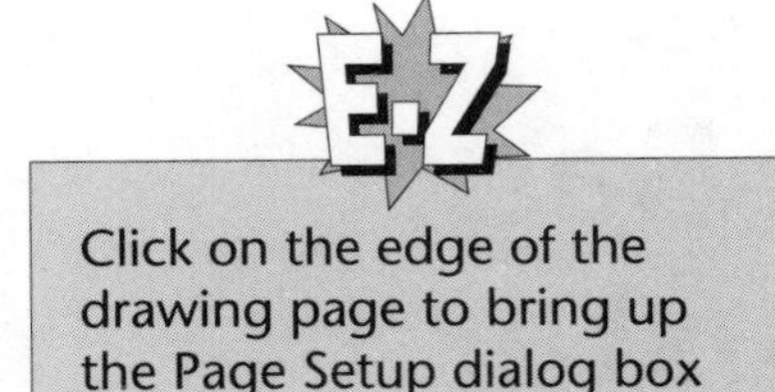

Click on the edge of the drawing page to bring up the Page Setup dialog box in a jiffy.

setup, and not characteristics of your next romance. You'll soon learn that type, size, and orientation impact your drawing just as much as they impact your love life.

Before we go any further, let's open the Page Setup dialog box. This will make it much easier for you to follow along as we talk details. Choose **Page Setup** from the **Layout** menu. Watch out—another dialog box you have to get familiar with.

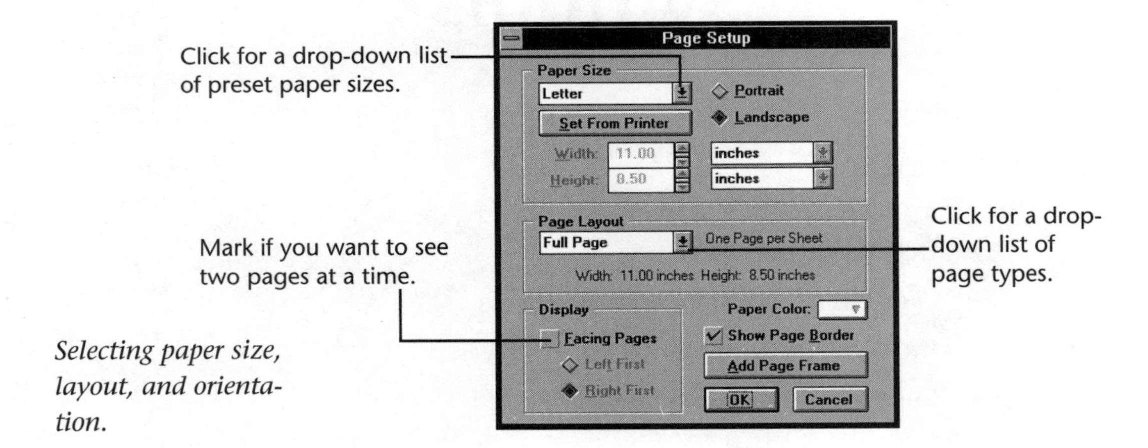

Selecting paper size, layout, and orientation.

It's just like CorelDRAW! to have done most of the work for you already, isn't it? Choosing a paper size is a true joy, because there are preset dimensions in the Paper Size drop-down list box for almost any page setup you could possibly think of. For instance, those amazing CorelDRAW! designers included dimensions for envelopes (in all sorts of different styles), a slide, a tabloid, and the old traditional—legal, to name a few.

All you have to do is click on the drop-down list arrow next to the Paper Size box, and then pick the size you want. Picky people can choose Custom and put their own unique, special paper sizes in the Width and Height boxes, if it helps their sense of worth.

You may already know all about orientation, but just in case: *portrait orientation* is long and narrow, and *landscape orientation* is short and wide. It all depends on your preference. To make it clearer: portrait is a regular piece of paper, and landscape is a regular piece of paper turned sideways.

Page Layout is where you can have some fun. Click on the drop-down arrow button next to the Page Layout box and you'll see what we mean. You've got options for booklets and books, folds from the top, folds from the side—even tents. Now we could put all these goodies into a spiffy little table here and give you a brief spiel on what each style is about, but there's a quicker way for you to find out. Just click on a style, and a descriptive sentence pops up to the right. If this still doesn't satisfy your curiosity, you could always slam that OK button and see how the drawing page changes. To reap the full benefit of this process, though, make sure you have marked Facing Pages in the Display box and have inserted a couple pages (see the next section for details on the latter).

About the Facing Pages option: This is great if you're creating a book or booklet, because you can see the pages as they'll appear when printed. Choose Right First if the first page in your drawing will fall on the right side of the book or booklet.

I Want More of Those

You're the type that's never satisfied with just one—and we're not talking potato chips here . . . we're talking pages. Okay then, all

CorelDRAW! 5's Page Setup dialog box looks completely different from the Page Setup dialog box in CorelDRAW! 4. However, most of the options are the same, they're just located in different places. One thing that's new is that version 5's Page Setup dialog box has tabs. To change page size and orientation, click on the **Size** tab. To change the page layout, click on the **Layout** tab. If you want to view facing pages of a multi-page document, click on the **Display** tab.

You do things your own way, don't you. To access the Page Setup dialog box, you'll choose Page Setup from the File menu instead of from the Layout menu. You then get a special dialog box that has fewer options than the Page Setup dialog box in CorelDRAW! 4 has. But be happy: in this case, less is more, because the options you don't get, you probably wouldn't have used anyway.

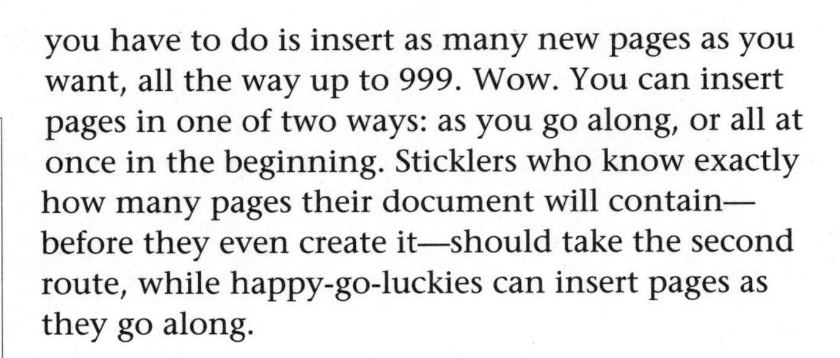

"Bummer." "What a drag." "Should have spent the extra $300." These might be some of your sentiments after we unload this bad news on you: CorelDRAW! 3 ends at one page. That's right: one page per file. Sorry, you can look for the Insert Page command all you want, but you won't find it! Also, go ahead and skip the next section on the master layer, 'cuz version 3 don't have that neither. But hey, look on the bright side, you're $300 richer.

you have to do is insert as many new pages as you want, all the way up to 999. Wow. You can insert pages in one of two ways: as you go along, or all at once in the beginning. Sticklers who know exactly how many pages their document will contain—before they even create it—should take the second route, while happy-go-luckies can insert pages as they go along.

Choose **I**nsert Page from the **L**ayout menu. A runty-looking dialog box appears with, believe it or not, only four or five options. To insert a specific number of pages, type in the amount in the Insert Pages box. To insert pages as you go along, next to Page, enter the page number before or after which you want to insert a new page (or pages). Then select **Before** to insert the new page in front of the page you specified in the Page box, or **After** to . . . well, you know where CorelDRAW! will put it.

Use the **L**ayout **G**o To command to sprint to a far-away page. Press **Ctrl** and click the page forward icon to go to the last page, or press **Ctrl** and click on the page back icon to go to the first page. One more thing: to go forward or back five pages at a time, click the page forward or page back icon with the right mouse button.

Whenever you work with a multi-page document, a nifty page icon mysteriously appears in the lower left corner of the screen. This icon does more than you'd think. Sure it tells you what page you're currently on, as well as the total number of pages in the document, but you can also use the left and right arrows to move back and forth between the pages. And, when you're on the first or last page, a small plus sign appears. Click on the plus sign, and bang! The Insert Page dialog box appears.

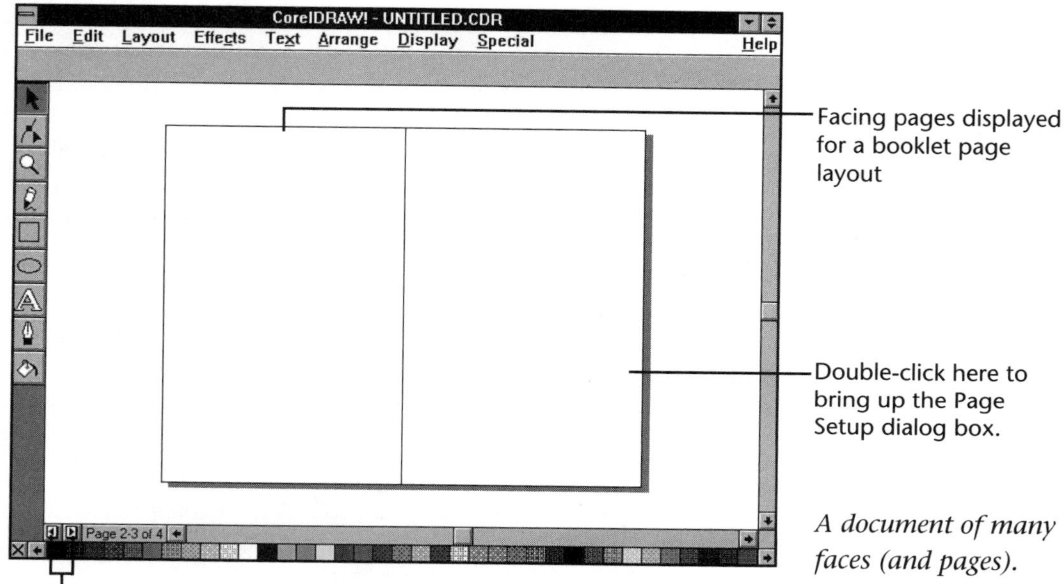

A document of many faces (and pages).

The Master Layer Does Graphics

Let's get one thing straight: this section is not about bricks, eggs, or promiscuity. If you're familiar with headers and footers, a master layer works kind of like headers and footers. A header is something like a page number or a title that automatically appears in the same place on every page in a document (sometimes not the first page). Like, you know, type it once and have it show up on every page—like maybe a company logo. If that's what you want, that's what a master layer will do for you (see, no bricks or eggs involved). And if you haven't put two and two together yet, we'll spell this out for you very clearly: you'll only use master layers for multi-page documents.

If you want to keep the master layer from showing up on a page, move to that page and double-click on **Master Layer** in the Layers roll-up. The Layer Options dialog box appears. Unmark the Set Options for **A**ll Page check box and the **Visible** check box, and click on **OK**.

Talk of layers can kind of frighten people. But keep it simple. Unless you're some pro-joe artist who plans to use layers upon layers, all you ever need to have is two layers in your document: 1) the regular layer that holds all the stuff you draw and 2) the master layer that holds repeating text and graphics (stuff you want to show up on all your pages). Okay, Layer 1 is already done for you whenever you start a drawing. That leaves Layer 2: the master layer. When you select **L**ayers Roll-Up from the **L**ayout menu, the Layers roll-up appears on your screen.

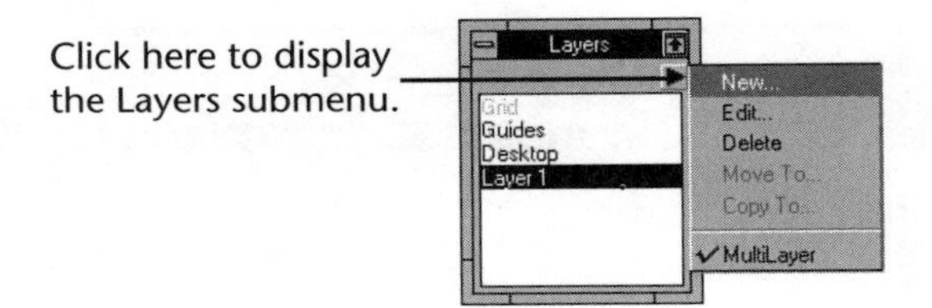

Creating a master layer.

Click on the right arrow to display the Layers submenu, and then choose **New**. The Layers Options dialog box appears. To make your life easier, type **Master Layer** where it says "Layer 2." This will help you remember that this layer is indeed the master layer. Mark the **M**aster Layer check box, and then click on **OK**. The term Master Layer now appears in the Layers roll-up list, and is highlighted because you're currently working on the master layer.

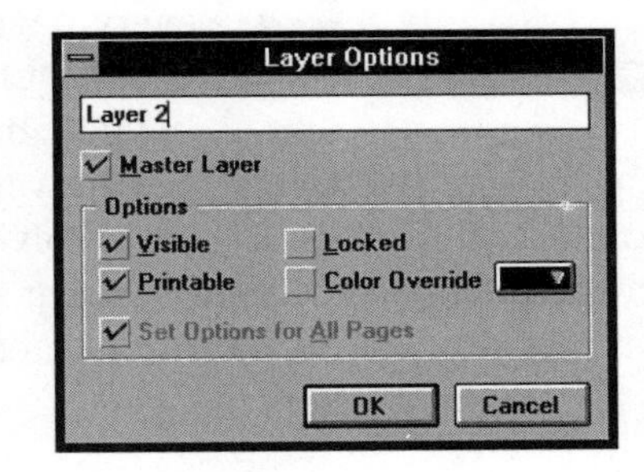

The Layers Options dialog box.

So go ahead and draw your graphics and add the text you want to show on all your pages. When you're done, select **Layer 1** in the Layers roll-up to return to the original layer, where you can continue creating your document. Notice the objects from the master layer show up here, and when you switch to other pages, you'll see those same objects again.

The Symbolism of It All

Before you spend hours drawing your version of a cat, check out CorelDRAW!'s cat symbol. Instead of wasting a good day's work concocting a pair of scissors, take a look at CorelDRAW!'s scissors symbol. Get the point yet? There are bunches and dozens and hoards of symbols that CorelDRAW! has already drawn for you, so why spend time drawing your own if one of CorelDRAW!'s will suffice?

Where Do Symbols Fit In?

Symbols can be used as symbols in text, or they can be enlarged and modified to be used as full-scale drawings. But keep in mind, these are simple, one-dimensional sketches—nothing at all like CorelDRAW!'s sample drawings or clip art (discussed in Chapter 4, "Juggling Drawings").

We had a hard time deciding where symbols fit in relation to CorelDRAW!'s other features. Graphics? Text? But we weren't the only ones. You probably wouldn't guess where Corel stuck the Symbol tool. Symbols are not created with a drawing tool (like circles and squares); you don't have to go through any fancy importing to use them (like sample drawings); and although you can include them as part of text, we have a hard time considering symbols text because they are graphics, after all (and because the Western world dropped pictures for an alphabet, oh, ages ago). But guess where CorelDRAW! put the Symbols tool? Yep, as a subtool on the Text tool flyout menu. And the real reason we're even making an issue out of this is so that you might remember to look for the Symbol tool on the Text tool flyout menu!

And to prove our point . . . guess what? Corel woke up with version 5. The Symbol tool is no longer on the Text flyout menu. (They finally discovered that that was a weird place for the Symbol tool to be, we suppose.) To display symbols in CorelDRAW! 5, click on the **Symbol** icon on the Icon palette at the top of the window. The Symbol icon is still a star, and it's located near the right edge of the palette.

Sticking a Symbol onto the Page

Now to the logistics. Click on the **Text** tool (the A) and hold the button until the Text flyout menu appears. Now click on the star, which is the **Symbols** tool. The Symbols roll-up appears.

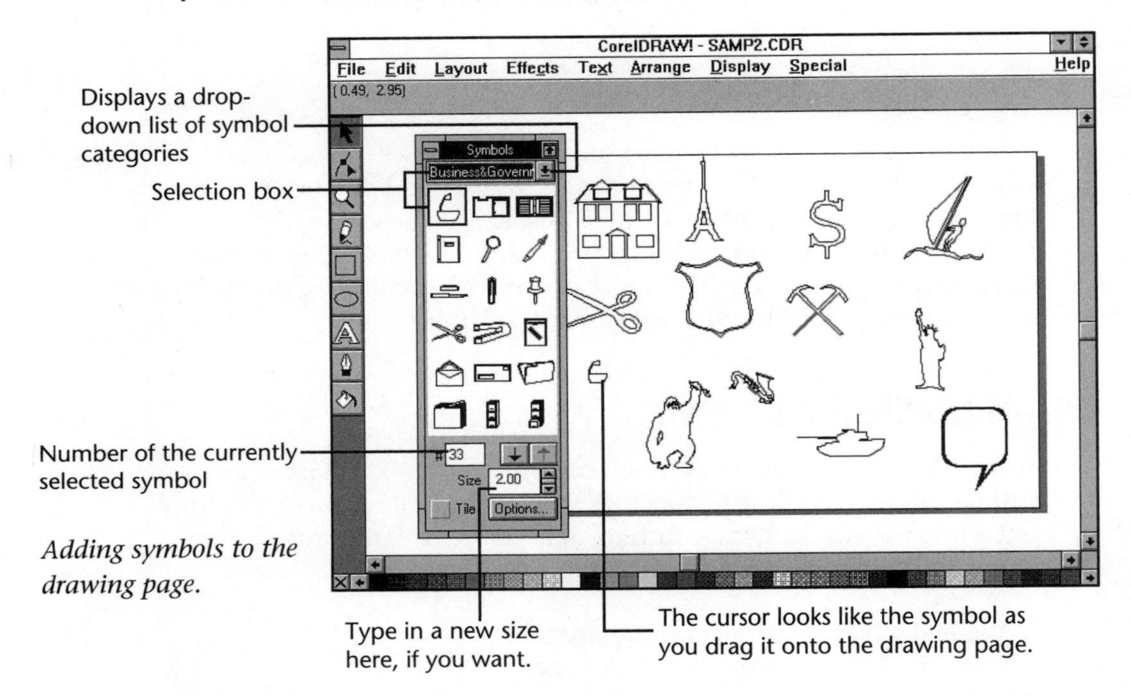
CorelDRAW! - SAMP2.CDR
File Edit Layout Effects Text Arrange Display Special Help
Symbols
Business&Governm
Size 2.00
Tile
Options...
Displays a drop-down list of symbol categories
Selection box
Number of the currently selected symbol
Type in a new size here, if you want.
The cursor looks like the symbol as you drag it onto the drawing page.

Adding symbols to the drawing page.

When you select the **Symbols** tool, you get a dialog box instead of a roll-up. Fortunately, most of the options are the same as those in the Symbols roll-up. The main difference is that when you click on a symbol in the left side of the dialog box, a submenu appears, from which you can select the symbol you want.

CorelDRAW!'s symbols are organized in categories. For example, the Business & Government category houses symbols like scissors, overhead projectors, business envelopes, file holders, and cabinets, whereas the Animals category contains symbols from gorillas and monkeys to cats and dogs. All you need to do now is decide what type of symbol you want and select the category it is likely to fall under.

To access a category's symbols, you need to select the category from the drop-down list at the top of the Symbols roll-up. Click on the drop-down list button near the top right corner of the roll-up to display the drop-down list. When you click on the category you want, CorelDRAW! displays the symbols belonging to that category in the roll-up. Most categories contain pages and pages of symbols. So use the up and down arrows in the roll-up to display additional symbols for a category (there's no way CorelDRAW! could squeeze all those symbols into that little area!).

To use one of the symbols, simply drag it onto the drawing page. From there, you can size, transform, and edit the symbol as you would any object. You can enter a size in the Size box before you drag the symbol to the drawing page, but it's probably easier to just resize the symbol after it's on the page.

Tile It Away

One more tidbit. CorelDRAW! lets you tile the symbol across the page, if it suits your fancy. If you mark the Tile check box and then drag a symbol onto the drawing page, you get several copies of the symbol lined up and down across the page. You can modify the tile format by clicking on the Options button in the Symbols roll-up.

The Least You Need to Know

Well, now you know something about page setup, layers, and symbols. Have fun exploring!

- Set up the page size, orientation, and layout in the Page Setup dialog box, accessed through the Layout Page Setup command.
- Use the **L**ayout **I**nsert Page command to add additional pages to your document (not applicable in CorelDRAW! 3).
- For multi-page documents, use a master layer when you want to show the same graphics and text on all or several pages (not applicable in CorelDRAW! 3).
- Use symbols to dress up text, or use them as drawings. Drag a symbol from the Symbols roll-up (accessed by the Symbols tool) to the drawing page.

Recycling tip:
tear this page out and photocopy it.

Chapter 14
Rolling the Presses

In This Chapter

- Getting your printer ready
- A veritable publisher
- If you don't like the position and size . . .
- Printing only portions of drawings

We all want it published, don't we? Oh, our ideas don't have to be put into print by a major printing house, a popular magazine, or a prestigious journal—just to get our creative thoughts onto paper is usually enough. After all, mankind has sought to make his ideas concrete since the dawn of time; whether it was drawing on rocks, etching on parchment, engraving wood, or typing on a manual typewriter, the goal has always been the same: something tangible! Somehow a job just doesn't seem to be done until you can hold it in your hands. In this chapter, you'll learn how to transfer those lofty pieces d'art you create in CorelDRAW! into something you can hold in your hands.

Before You Do Anything Else

I'm not sure I've seen anyone quite so frustrated as a former co-worker who, try as she might, couldn't seem to get her stuff to print. And she didn't want anyone's help. This battle was between her, CorelDRAW!, and her printer, she said. She fooled around with settings in CorelDRAW!'s Print dialog box, and adjusted and readjusted the menu settings on her LaserJet. She even opened "the darn thing" up a couple of times. But no luck. After toiling a couple of hours, she finally crawled into my office, looking mighty defeated. And after I told her where the problem lay, she looked stumped. "You mean I was looking in the wrong place the whole time?" she grumbled. Uh-huh.

You version 5 users will need to take an additional step to access the Print Setup dialog box. When you choose the File Print Setup command, a dialog box appears that lets you select your printer and printer quality only. To access the dialog box you see in the figure, click on the Setup button.

Sometimes we get so caught up with the Print dialog box that we completely ignore a more critical dialog box: the Print Setup dialog box. If the settings in the Print Setup dialog box do not match the settings of your drawing and printer, you'll never get the copy you want. People often overlook the Print Setup dialog box because it is a Windows dialog box, which usually needs no adjusting for other software programs. But when you work in CorelDRAW!, it's likely you're not using the standard page orientation and size. You may even be printing to a printer you haven't used before, such as a PostScript printer. This is when the settings in the Print Setup dialog box become relevant.

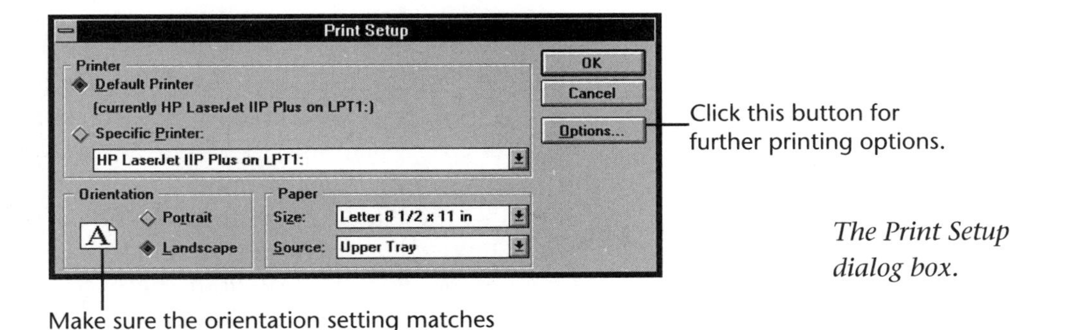

The Print Setup dialog box.

To open the Print Setup dialog box, select **Print Setup** from the **File** menu. Three crucial options in this dialog box can really trip you up if they aren't set right:

- Make sure the orientation setting in the dialog box matches the orientation of your drawing. Remember, **Portrait** prints across the short side of a piece of paper, and **Landscape** prints across the long side of a piece of paper.
- CorelDRAW! has several preset page sizes and layouts. Many of those use nonstandard paper. If you're using anything other than 8 1/2 x 11-inch paper, select the correct paper size in the Size box.
- If you work on a network at the office, or if you have more than one printer, make sure you've selected the correct printer in the Printer box. If you're trying to use a printer other than your default printer, but you haven't selected that printer in the Specific **Printer** box, you'll be pulling your hair out trying to get anything to print.

If you work on a network and have multiple printers to choose from, you have one more something to worry about. Your system probably runs a batch file when it starts up to capture the printer you use most often. If you intend to use a different printer, make sure you've captured that printer. Otherwise, even if you choose the right printer in the Specific Printer box, CorelDRAW! will still download the file to the default printer, and you'll get a bunch of jibberish when you print.

The CorelDRAW! 3 Print Options dialog box doesn't show a picture of your drawing. That means you can skip the next section, because you won't be able to change the size and position of your drawing manually if you haven't got a picture to work with!

Put It on Paper

Okay, you've done the dirty work and made sure your printer is set up correctly. Now to print. Number One: Make sure the drawing you want to print is currently open. Number Two: press **Ctrl+P** or choose the **F**ile **P**rint command. Voilà, the dialog box you've been waiting for all your life suddenly presents itself on your computer screen. And what a dialog box. It takes up almost the entire screen.

And . . . what's this? A miniature picture of your drawing is stuffed into the dialog box so you don't have to feel like you're working blindfolded. But this is more than just a pretty picture. You can resize and reposition this picture with your own bare hands to print your drawing just the way you want it. Read the next section for the lowdown on this feature.

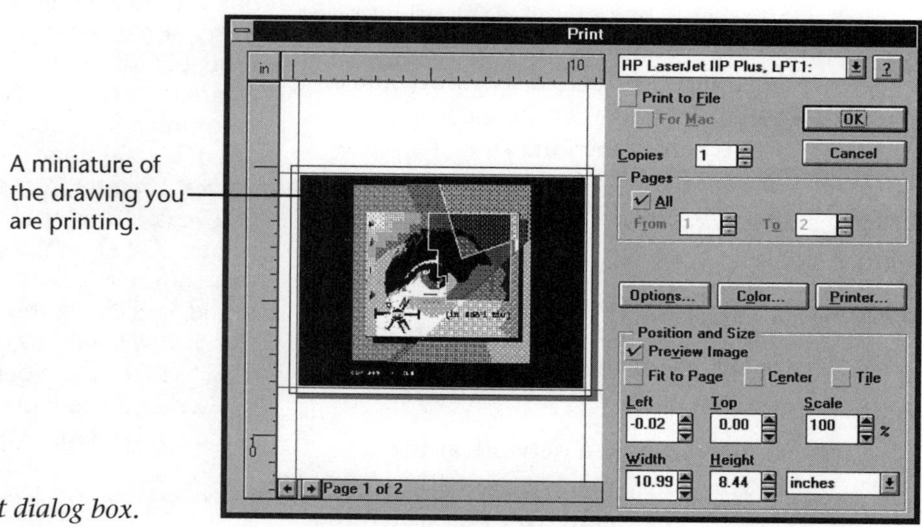

The Print dialog box.

If you only want one copy of your entire drawing just as is it, go ahead and click on **OK**. Your drawing will be sent to your printer and printed out. If you want more than one copy of your drawing, type or scroll to the number of copies you want in the Copies box (and unless you want a thousand copies, don't worry about any limit). If you're printing a multi-page document and want to print out only certain pages, unmark the **All** check box under Pages, and then enter the first page you want to print in the **From** box, and the last page in the **To** box. For example, if you only wanted to print page 7, you'd type 7 in the **From** box and in the **To** box. However, if you wanted to print pages 7, 8, and 9, you would type 7 in the **From** box and 9 in the **To** box.

The version 5 Print dialog box has two parts. The first dialog box you see after you choose the **File Print** command contains the options you're likely to change most frequently, like the print range and the number of copies you need. To view your drawing within the dialog box and access further options, click on the **Options** button.

Life's All About Position and Size

Your position in life. Your position at work. The size of your car. The size of your house. Well, here's another size and position that concerns you: the size and position of your drawing relative to the printed page. As you saw in the last section, the Print dialog box shows a sample of your drawing. Without a lot of hassle, you can make your drawing the size you want and position it on the page where you want just by dragging selection handles and the drawing itself.

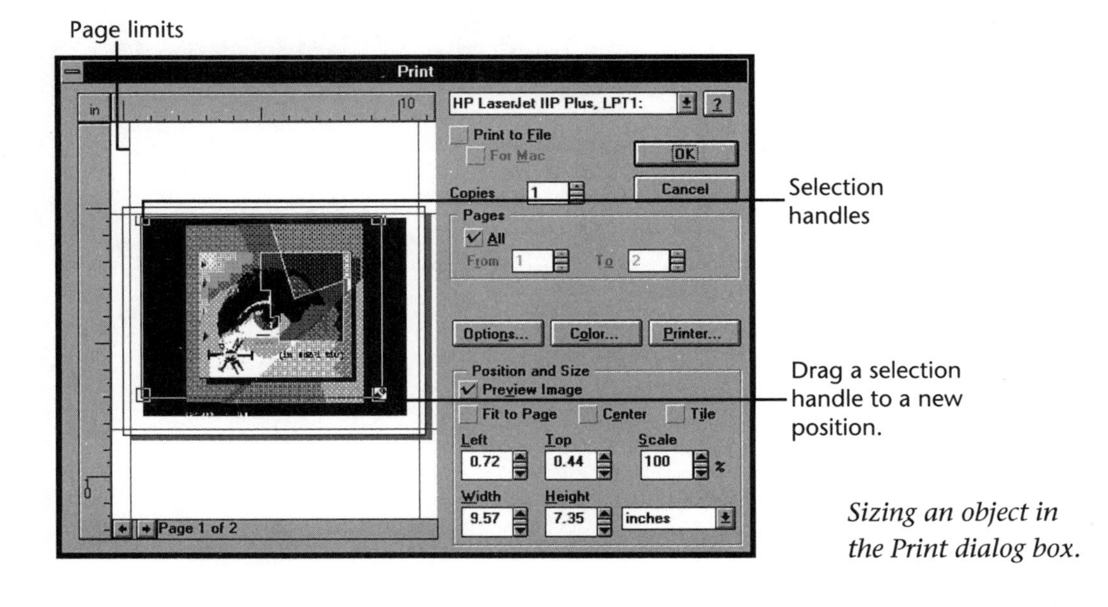

Sizing an object in the Print dialog box.

Just click on the drawing to select it. When you do, four *selection handles* appear on the corners. Drag a selection handle to resize the drawing. Then, when the drawing is the size you want, drag the entire drawing to the desired position on the page.

Picky Printers Print These

Sometimes you may not want to print an entire drawing. Therefore, CorelDRAW! gives you the option of printing only certain drawings on the page. Select the objects you want to print before you open the Print dialog box. Then, when you open the dialog box, you'll notice a new option appears, called Selected O**b**jects Only. Mark this box to print only the objects you select. Also notice that only the selected options appear in the sample page, so you can see what the printout will look like and make sure that you selected all the objects you wanted.

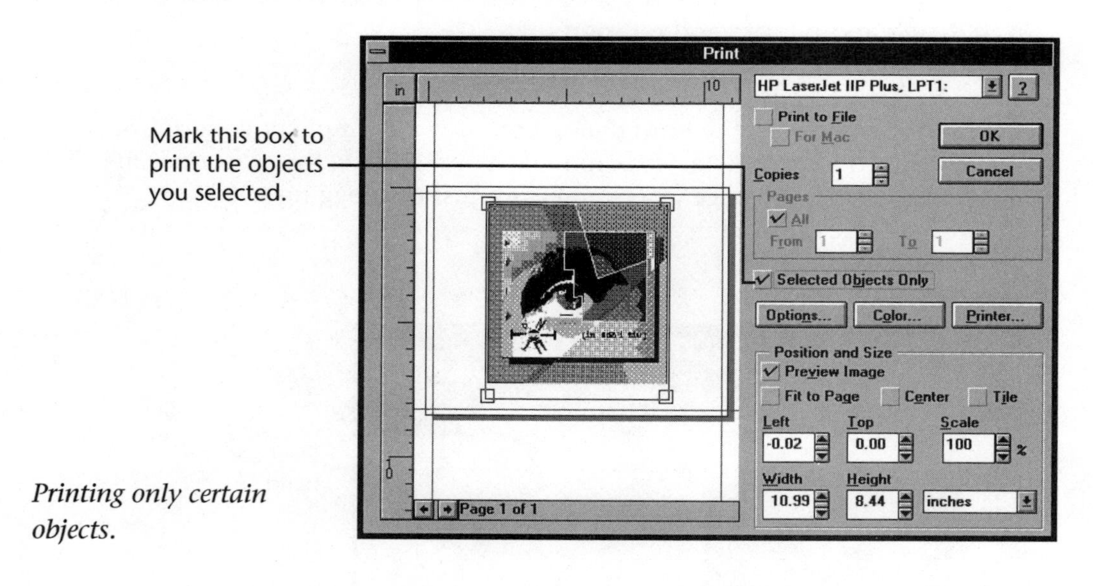

Printing only certain objects.

The Least You Need to Know

Printing is painless and easy with CorelDRAW!.

- Make sure everything is set up correctly in the Print Setup dialog box.
- Open the Print dialog box by pressing **Ctrl+P**.
- Resize or reposition the drawing on the printed page using the sample in the Print dialog box.
- To print only certain objects, select the objects before you bring up the Print dialog box, and then mark the Selected O**b**jects Only check box.

A page is a terrible thing to waste.

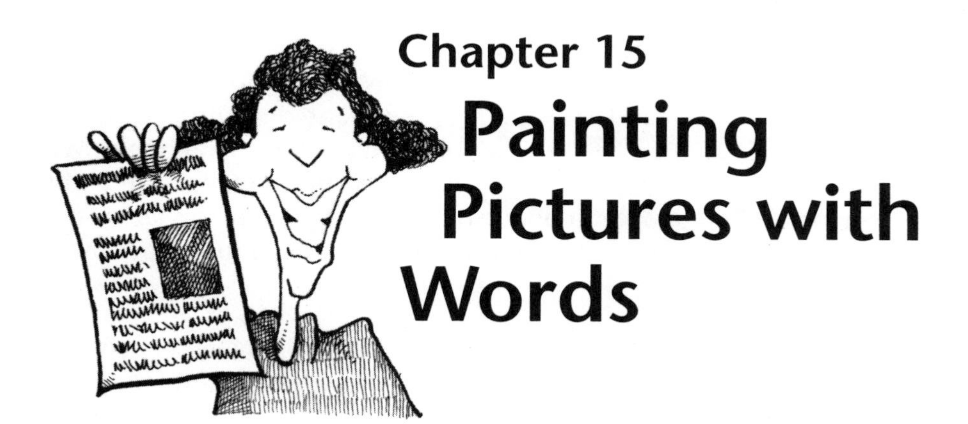

Chapter 15 Painting Pictures with Words

In This Chapter

- Check out that artsy text
- A blob of paragraph text
- Going with the flow

Some of the loveliest pictures in the world have been painted with words instead of brushes—like the openings of *The Tale of Two Cities*, *A Farewell to Arms*, or the book of Genesis. Your CorelDRAW! text may not be as dramatic or poignant, but it is still considered an object d'art (well, at least an object). All text you enter in CorelDRAW! works just like your graphic objects. You can reshape text, move text, copy text, apply special effects to most types of text, and more. You'll learn all about CorelDRAW! text in this and the next few chapters.

The Tale of Two Types of Text

You've certainly figured out by now that CorelDRAW! seldom, if ever, gives you only one way to do something, right? Well, you can enter text into your drawings in—you guessed it—two different ways. There's artistic text for you creative types, and paragraph text for you, well, dull types; no, actually the two types of text have nothing to do with your level of

Artistic text Strings of text you enter using the Artistic text tool. You can apply special effects to artistic text.

Paragraph text Blocks of text you enter using the Paragraph text tool. Paragraph text automatically wraps to the next line.

original thought. You'll use *artistic text* for short text entries, and *paragraph text* for long stretches of text. For example, if you wanted to add a pageful of text, 99% of the time you'd use paragraph text. We don't say 100% because, technically, you can write a pageful of artistic text (and you know there are always those of us who want to do things our own way).

There are two more things to remember about the two types of text. One, when it comes to special effects (such as blending, extruding, and twisting text), artistic text is always the one for you. You can fit artistic text to a path (which you'll find out how to do in Chapter 17), or blend or extrude artistic text. You can do nothing of the sort to paragraph text. So if you want to do all sorts of weird, twisted things to your text, think artistic. Two, if you want text to automatically wrap to the next line when it reaches the right margin (as it does in any respectable word processor), think paragraph text. You have to manually return text to the next line when you use artistic text.

The Text of a True Artist

Remember, choose artistic text when you want to add only a small amount of text to your drawing, or when you want to apply special effects to the text. (See Chapters 17, 18, and 19 for details on how to apply special effects to text.) To enter artistic text, click on the **Artistic text** tool on the toolbox (the capital "A"). The cursor changes to a cross. Click on the drawing page where you want to begin the text, and then type the text. If you want to constrain the text to a certain width, press **Enter** to move to the line below, and then continue typing. If you don't press Enter and your text reaches the end of the drawing page, the text will flow outside the page margin. It keeps going, and going . . .

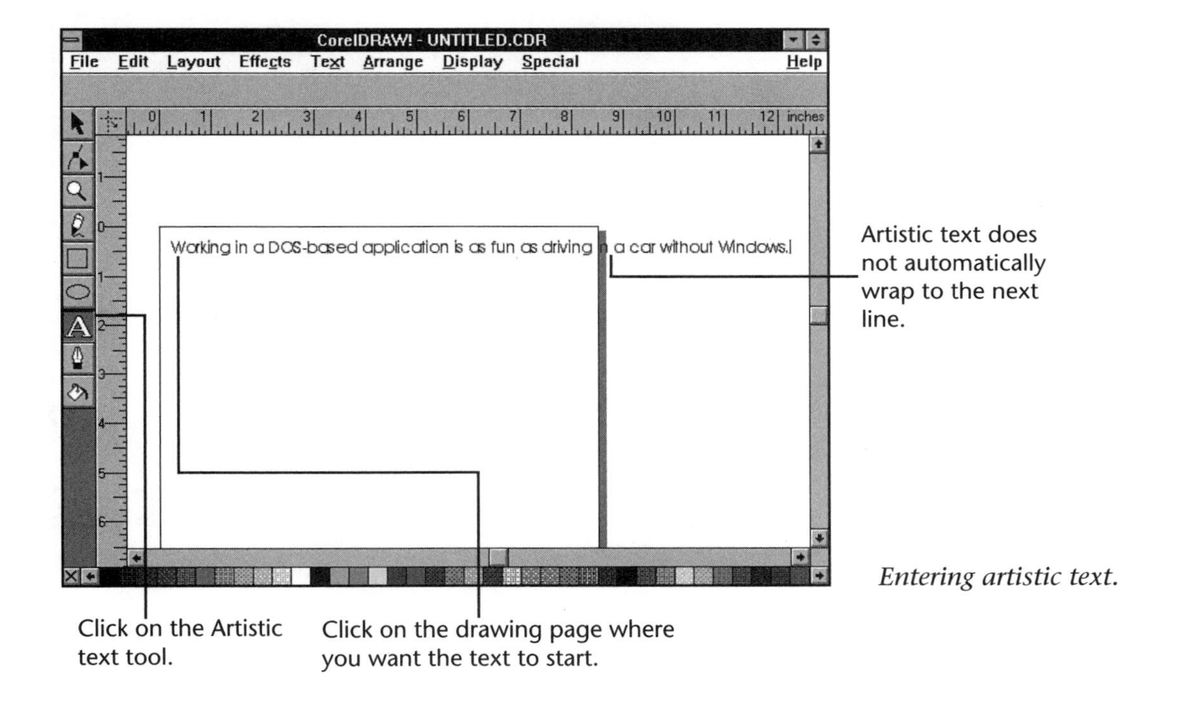

Entering artistic text.

Don't worry if the text isn't exactly the right size or in exactly the right spot. You can move it the way you move any other CorelDRAW! object: just select it with the Selection tool, and then drag it (making sure to touch the text itself) to a new location. If you want to resize the text, select it and drag one or more of the selection handles until you get the right look. You can really transform the appearance of your text just by stretching it or lengthening it.

You can also give text a new look by changing its text attributes in the Text roll-up. For details, read the section "Giving Your Text a Makeover" later in this chapter.

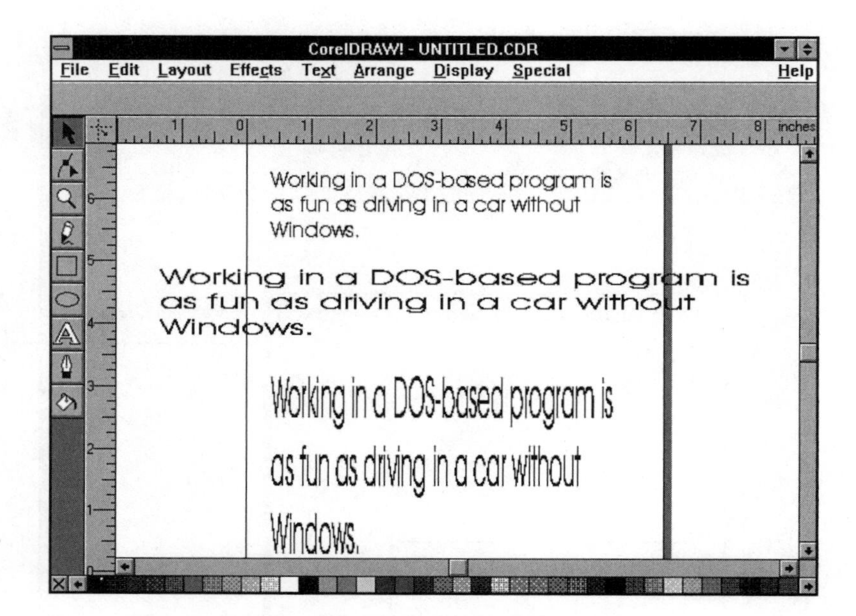

Changing the look of artistic text.

A Paragraph-Type of Guy

When you have blocks of text to enter, paragraph text is the way to go. This type of text automatically wraps to the next line when it reaches the right margin. Although you can move or resize paragraph text the way you do artistic text, you can't apply special effects to paragraph text. However, you can format paragraph text into columns and use tabs and bullets, which you can't do with artistic text.

To access the Paragraph text tool, hold down the mouse button on the Artistic text tool (the capital "A"). A flyout menu appears. Click on the **Paragraph text** tool, which looks like a miniature page of text, or press **Shift+F8**.

If you want to fill the whole page up with text, just click on the drawing page and start typing. CorelDRAW! adds a frame to the page, and places the text you enter inside this frame. You can later select the frame with the Selection tool and resize it or move it.

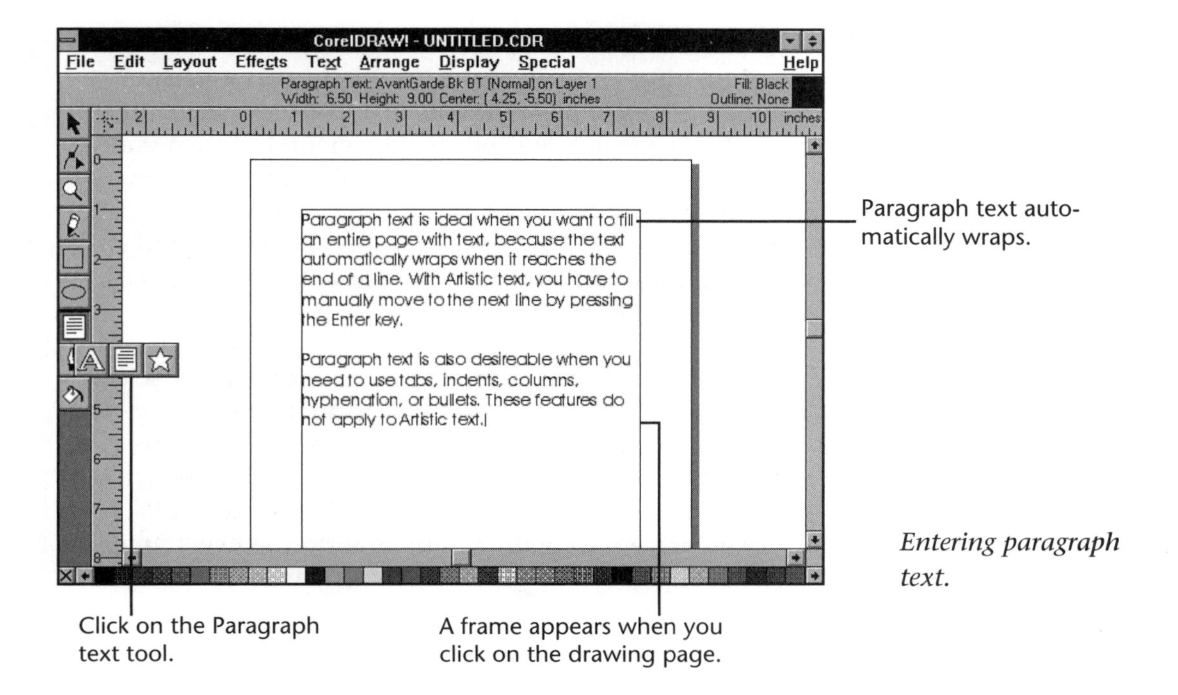

Entering paragraph text.

If you want to confine text to a certain area, select the **Paragraph text** tool but, instead of clicking on the drawing page, drag the mouse to form a box the size you want. As you drag, a dotted marquee box appears showing you the dimension of the text frame. Now when you type, your text will be confined to this frame.

There is no Paragraph text tool for you. Instead, click on the **Artistic text** tool, and then drag the mouse to form a dotted marquee box on the drawing page. Text you enter inside this box is handled like text you enter using the Paragraph text tool in CorelDRAW! 4.

Going with the Flow

If you're creating a brochure or newsletter, you don't have to worry about where your text ends at the bottom of the page or column. Because CorelDRAW!'s paragraph text rests inside a *frame* (a holding box for paragraph text), you can flow text from one frame into another—which can be very important if you edit text at a later time. Say, for example, that you have your text frame just the right size on the first page of a multi-page document, but then you have to add or delete some of the frame's contents. If you had the text flowing into other frames on subsequent pages, adding or deleting words would not affect the appearance of the first page, because text would either be pushed back or up to make way for the new material, or to compensate for the removal of certain text.

To have text flow from one frame to another on the same page, select the paragraph frame. The selection box that appears looks a bit different from a regular selection box: the top and bottom middle handles are hollow instead of filled boxes. Click on the hollow box on the bottom of the frame. The cursor changes to a unique design (it looks like the Paragraph text tool with an arrow). Drag the mouse to form the new frame. As you drag, the dotted marquee shows you the dimensions of the frame.

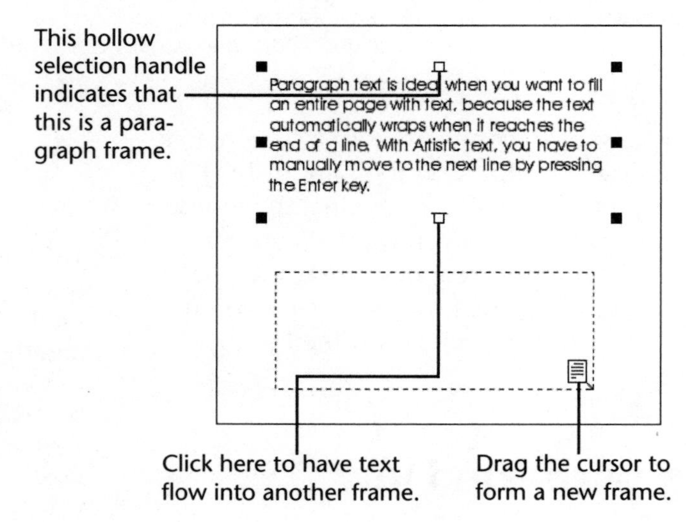

Creating a new paragraph frame.

When you release the mouse button, the text that didn't fit into the first paragraph frame now appears in the new frame. A plus sign appears inside the top middle selection handle to let you know that the text inside this frame is actually the overflow from a previous frame.

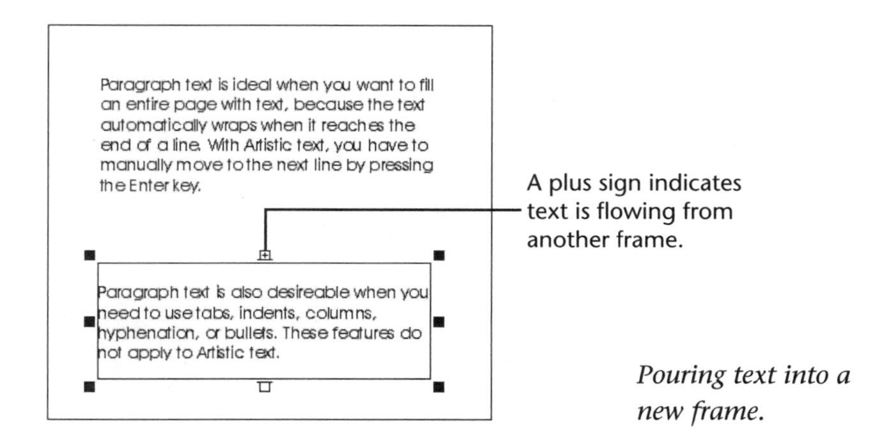

Pouring text into a new frame.

Giving Your Text a Makeover

You can beautify your text in CorelDRAW! just as easily as you can get a beauty makeover. (Watch out Cindy Crawford.) CorelDRAW! offers the same type of formatting options that all decent desktop publishers do: you can change the font style and size, change the text alignment, and apply boldface or italic to the text. The easiest way to reset these text attributes is by using the Text roll-up. To display the roll-up, select the Text **R**oll-Up command from the Text menu.

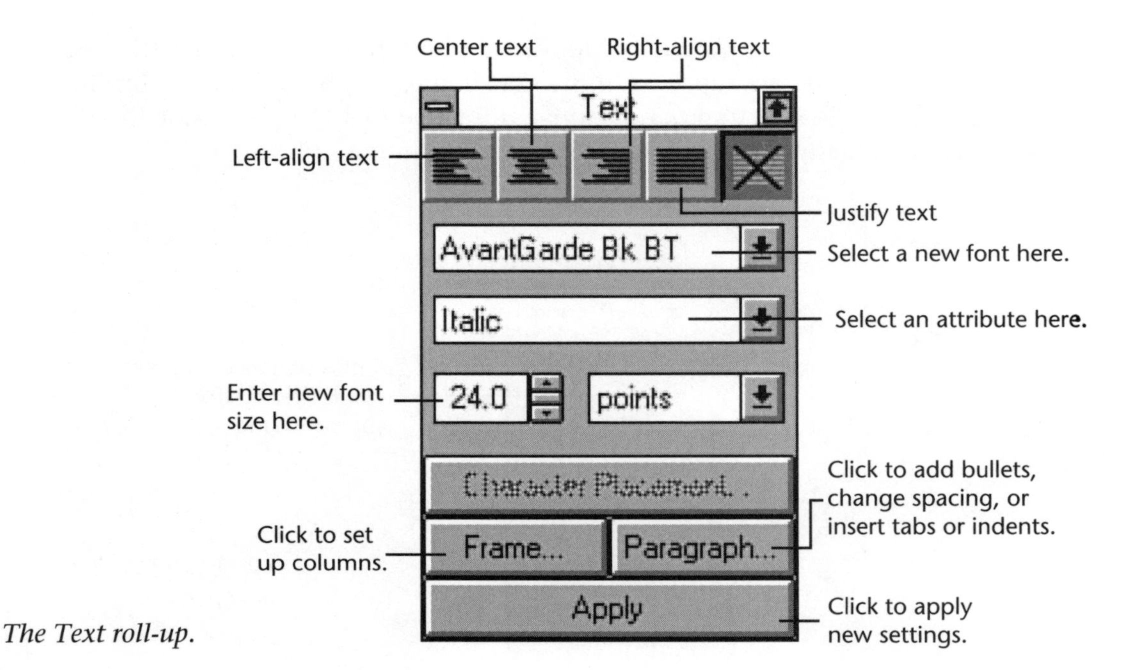

The Text roll-up.

You can use the Text roll-up to format both artistic and paragraph text. First select the text, and then change the settings in the Text roll-up. Finally, click on **Apply** to apply the new attributes to the text.

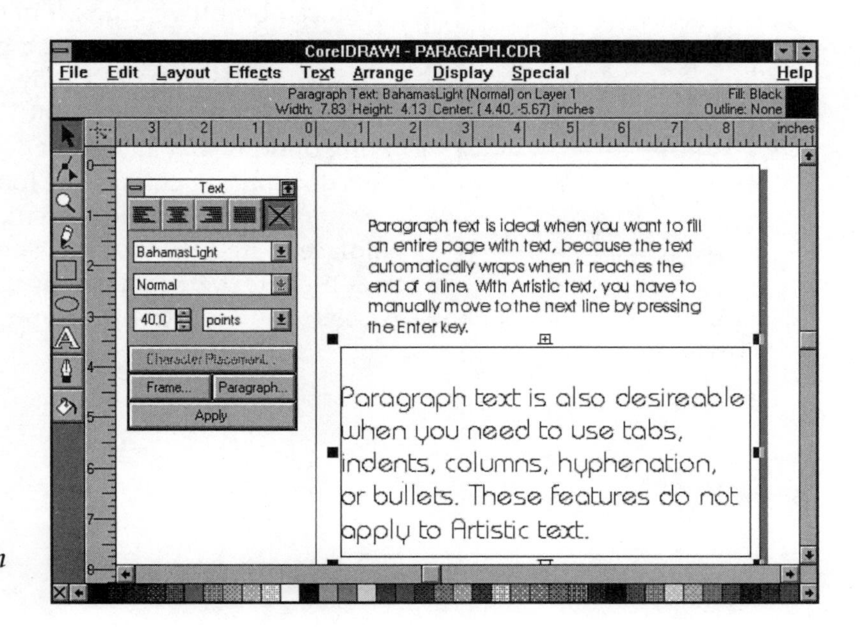

Formatting paragraph text.

If you want to change the spacing of characters, words, or lines, or to add more or less space before or after the paragraph, click on the **Paragraph** button in the Text roll-up. The Paragraph dialog box appears. Change the settings in the Spacing box options. For example, to assign double-spacing, enter **200%** in the Line box.

The Text roll-up looks different in CorelDRAW! 3 than it does in CorelDRAW! 4 and 5. Instead of a drop-down list for boldface and italic attributes, the roll-up has buttons for boldface and italic. To apply either of these attributes, simply click on the appropriate button. Additionally, the roll-up has Superscript and Subscript buttons.

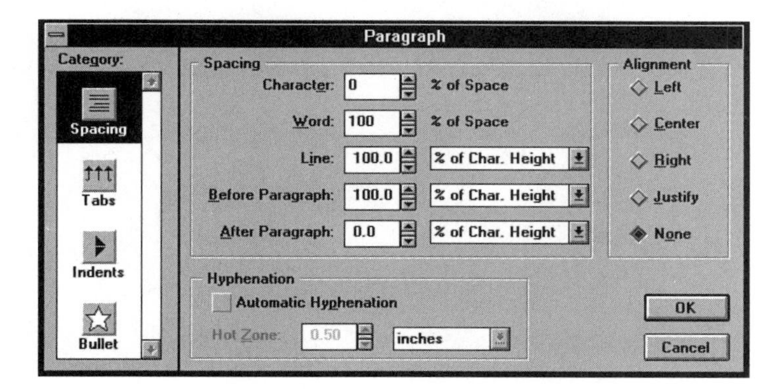

The Paragraph dialog box.

The Paragraph dialog box has several other options, which you'll learn more about in the next chapter, "Revamping Text."

The Least You Need to Know

Use artistic text for short strings of text; use paragraph text for blocks of text. The Text roll-up lets you format both types of text.

- Click on the **Artistic text** tool and then on the drawing page. Now type your text!
- Select the **Paragraph text** tool from the Text flyout menu, and then click on the page to enter a full page of text, or drag the mouse to create a frame of any size for your text.
- To flow paragraph text from one frame to another, use the hollow selection handles on the frame.
- To display the Text roll-up, select the Te**x**t Text **R**oll-Up command.
- Click on the **Paragraph** button to adjust character, word, and line spacing.

Part III

Textual Matters

My mother told me that of all of her six children, I definitely had the most unusual first word. It wasn't "mama" or "dada," or even "sis" or "bro." It went something like . . ."t-w-i-n-k-i-e." It's amazing how a first word can characterize the rest of a person's life.

What is a CorelDRAW! document without words? Even one little word like "Sadam" can give a whole new meaning to a picture of a donkey. Or, imagine how the title "White Water" would add special meaning to a simple drawing of a crystal clear lake. To aid you in your effort to enliven your drawings with words, this part is devoted to telling you all you need to know about CorelDRAW! and text.

Chapter 16
Revamping Text

In This Chapter

- Squeezing text into columns
- Putting a bullet to your text
- Being on the lookout for a word
- Corel's exchange policy
- Getting your a-b-c's straight

Here it is: the part you always dread—but that has to be done—editing your writing. Your teachers did it for you when you were young. But now the responsibility lies on your shoulders. Fortunately, CorelDRAW! gives you some tools that make it easy to format your text into columns or bulleted points, to search for text and replace it with something new, and to spell-check your text. Thank goodness for the last feature, huh?

Pillars and Columns

If you're creating brochures or newsletters, you may want to break your paragraph text into columns. Like most other text formatting, the easiest way to add columns to your text is via the Text roll-up. To display the roll-up, select the Text **R**oll-Up command from the Text menu. Select the paragraph frame containing the text you want to place in columns, and then click on the **Frame** button in the Text roll-up. The Frame dialog box appears.

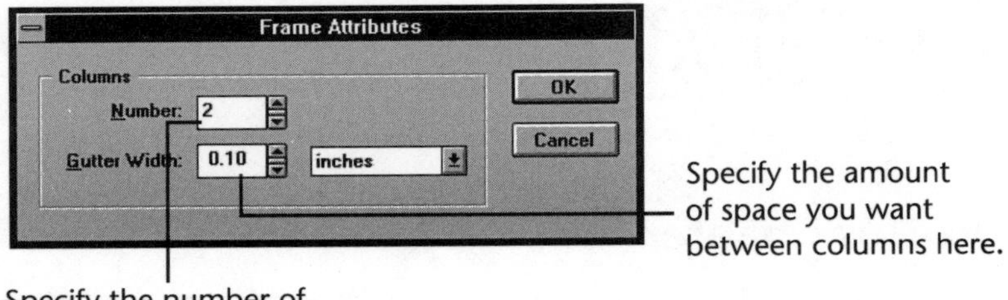

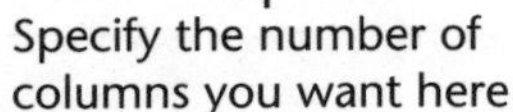

The Frame dialog box.

If you want to format more than one paragraph frame into columns at the same time, hold down **Shift** as you click on each frame, and then click on the **Frame** button.

Decide how many columns you want to break the text into and enter that amount in the **N**umber box in the Frame dialog box. The amount of space between columns is called *gutter width*. You should enter a .10 inch gutter width as a minimum; otherwise your text may look too crowded. To get just the right size, you may have to use the trusty old trial and error method. Enter a gutter width amount in the **G**utter Width box, and then click on **OK**. Click on **Apply** in the Text roll-up. CorelDRAW! formats the paragraph text inside the frame according to your specifications.

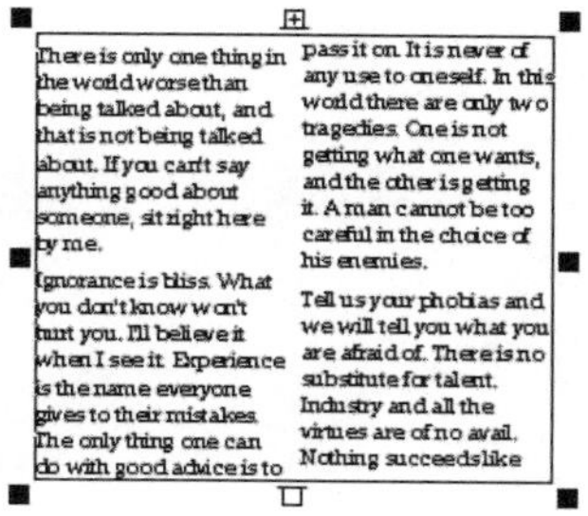

Formatting paragraph text into two columns.

Wacko Bullets

I never would have thought I'd be condoning the use of bullets, let alone encouraging them. But since CorelDRAW! bullets have nothing to do with guns, and certainly can't hurt anybody, I guess it's okay. So when you're breaking text apart into separate points for emphasis, try adding a bullet at the beginning of each point. Bullets help draw a reader's eye to the text, as well as spruce up the overall appearance of a document.

You're probably used to boring bullets—you know, small circles or squares, or at best, little diamonds or triangles. Well, CorelDRAW!'s bullets will blow you away (figuratively speaking). Not only does the Symbols library include a few bullet categories, but you can use any CorelDRAW! symbol as a bullet. Yep, that means you can use little anteaters, miniature houses, or tiny watches as bullets, among a zillion other things.

To add bullets to paragraph text, display the Text roll-up. Then either select the paragraph frame that surrounds all of the paragraphs in front of which you want to place a bullet, or select only certain text in the paragraph frame by dragging the mouse over the text. In the latter case, CorelDRAW! only adds bullets to paragraphs included in the selected text. Then click on the **Paragraph** button in the Text roll-up to display the Paragraph dialog box.

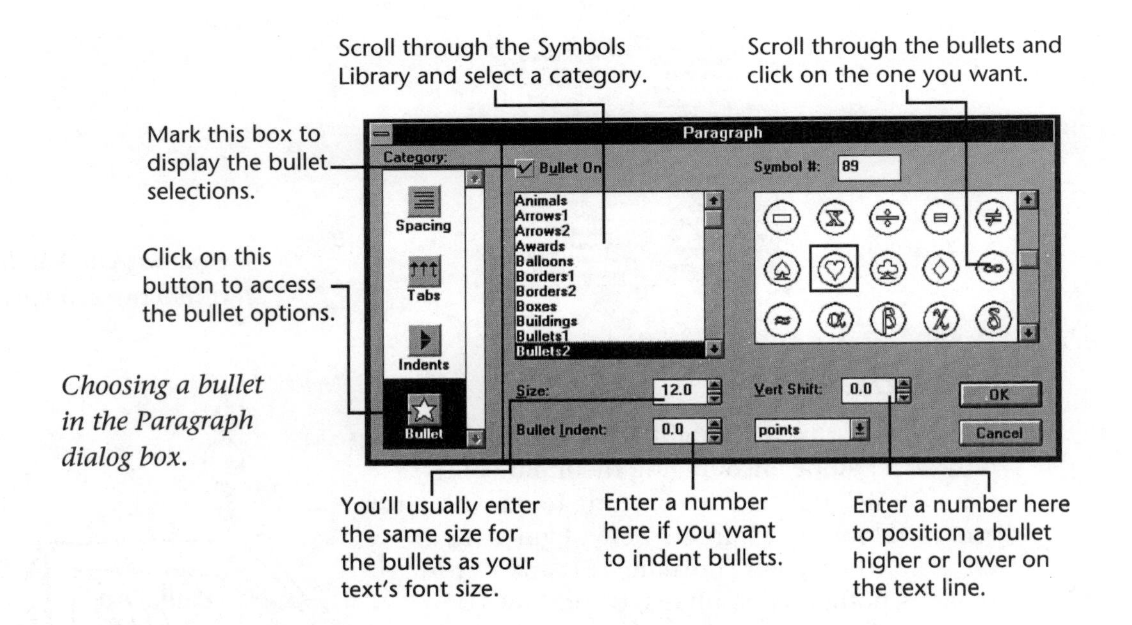
Scroll through the Symbols Library and select a category.
Scroll through the bullets and click on the one you want.
Mark this box to display the bullet selections.
Click on this button to access the bullet options.
Paragraph
Category:
Spacing
Tabs
Indents
Bullet
Bullet On
Symbol #: 89
Animals
Arrows1
Arrows2
Awards
Balloons
Borders1
Borders2
Boxes
Buildings
Bullets1
Bullets2
Size: 12.0
Vert Shift: 0.0
OK
Bullet Indent: 0.0
points
Cancel
You'll usually enter the same size for the bullets as your text's font size.
Enter a number here if you want to indent bullets.
Enter a number here to position a bullet higher or lower on the text line.

Choosing a bullet in the Paragraph dialog box.

5.0

Version 5's Paragraph dialog box looks a bit different from that in version 4. However, the options still work the same. Instead of clicking on a Bullet button, click on the Bullet tab in the Paragraph dialog box to access the bullet options.

The first thing you need to do after the Paragraph dialog box opens is to click on the **Bullet** button at the left side of the dialog box. This displays the bullet options in the dialog box. Everything in the dialog box will be dimmed until you mark the Bullet On check box. Next, scroll the Symbol categories and select the one you want. The symbols of the category you select appear in the right hand box of the dialog box. Scroll through the bullets and click on the one you want. Now make sure that the number in the Size box matches the font size of your text; otherwise the bullets may distort the spacing of your text. If you want to indent bullets, add a number in the Bullet Indent box. At last, click on the **OK** button, and then click **Apply** in the Text roll-up.

You can change the overall spacing of paragraph text using the **Spacing** button in the Paragraph dialog box.

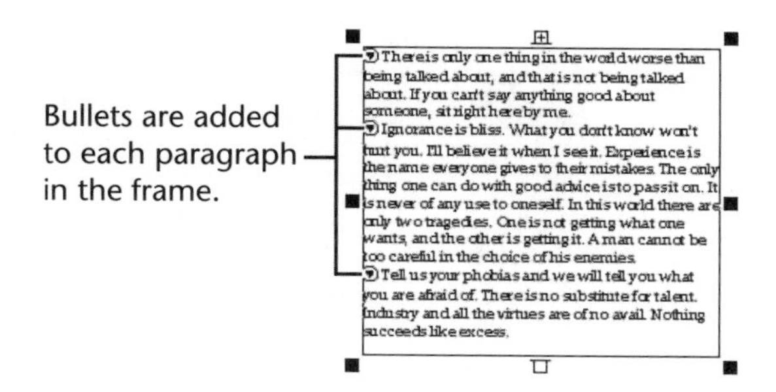

Bulleted text.

Is there a big gap between the first and second lines of the bulleted text? . . . Did you forget to make the size of the bullet the same size as your text? Probably so, because having a bullet size larger than the text it precedes will create a rift in the text. So be sure you enter the correct font size in the Size box of the Paragraph dialog box.

I've Finally Found What I Was Looking For!

One of the nice things about CorelDRAW! is that it has features that you don't expect graphics programs to have. CorelDRAW!'s Find feature lets you locate certain words or phrases in your text. This can be helpful if you want to review information on a certain term, person, function, or any element of your text. For example, you might search for Guarantee if searching a product brochure.

Select the text in the paragraph frame containing the text you want to search. Then select Find from the Text menu, and the Find dialog box appears. In the Find What box, type the word or phrase you're looking for. If you're searching for a capitalized word, or want to find only capitalized occurrences of a word, mark the Match Case box.

Click on the **Find** button to begin the search. CorelDRAW! searches the text for the specified word or phrase, and highlights the first occurrence of the word, if there is indeed a match. When you're ready to find the next occurrence of the word, click on the Find Next button. When you're done searching, click on the **Close** button to close the dialog box.

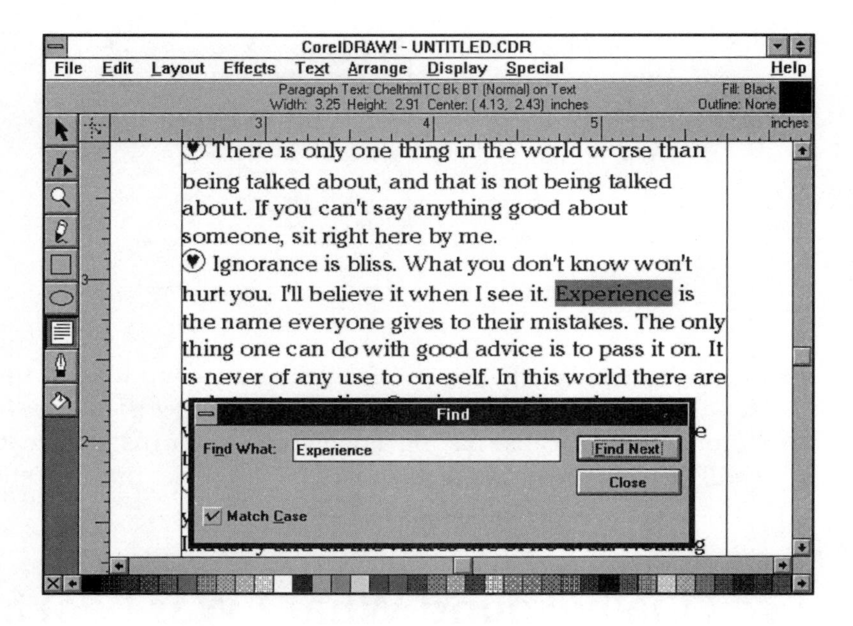

Searching for a word.

I'd Like to Exchange This, Please...

In addition to finding text, you can replace each occurrence of a word or phrase with a new word or phrase. It sure beats having to scrutinize the text yourself and replace the word manually.

To open the Replace dialog box, choose the Replace command from the Text menu. This dialog box should look familiar to you; it looks just like the Find dialog box, except there is a Replace With box and a couple of new buttons. Type the word or phrase you're looking for in the Find What box and type the new word or phrase in the Replace With box. If you want to replace every occurrence of the word or phrase, click the Replace All button, and you're done. If you want to replace only certain occurrences of the word or phrase, you'll have to use the Find and Replace buttons to review each occurrence. If you don't want to replace this occurrence of the word, click Find Next. If you do want to replace the text, click Replace and then Find Next. Keep doing this until you've scanned all of the text. Click on **Close** to close the dialog box and return to your text.

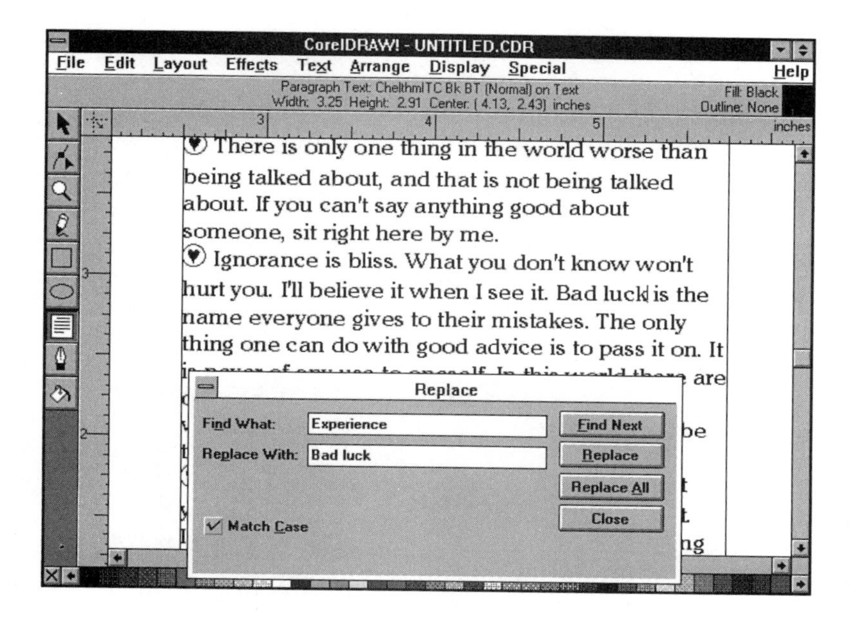

Replacing "Experience" with "Bad luck."

Only Goobers and National Spelling Bee Champions Don't Spell-Check

Imagine how all of our stoic English teachers would have cringed if we could have told them we really didn't need to learn how to spell because the future would hold little boxed machine that would check spelling for us. But we didn't know about computers then—unfortunately. We know some of you are master spellers (thanks to those same stoic English teachers), but no matter how sure you are that you did not misspell any words, spell-check your text before printing your documents—or you may cringe when you review the printed material. There's really no good excuse not to spell-check. It's simple and takes such a short amount of time, there's no reason to pass it up. ("I forgot" may be the only acceptable excuse.)

To spell-check text, first select the artistic text, paragraph frame, or specific text in a paragraph frame. Choose the Spell Checker command from the Text menu, and the Spelling Checker dialog box opens. Click the **Check Text** button; notice that the contents of the dialog box change. CorelDRAW! searches the text for words it doesn't recognize. When CorelDRAW! comes across a word that is not in the CorelDRAW!

dictionary, CorelDRAW! displays the word in the Word not found box. It's helpful to let CorelDRAW! give you suggestions of what the word was intended to be, because then you can just double-click on the correct suggestion instead of typing in the correct spelling. More than that—you might not even know the proper spelling since you missed it the first time! So mark the Always suggest box, and CorelDRAW! will give you a list of suggested words for each unrecognized word it finds.

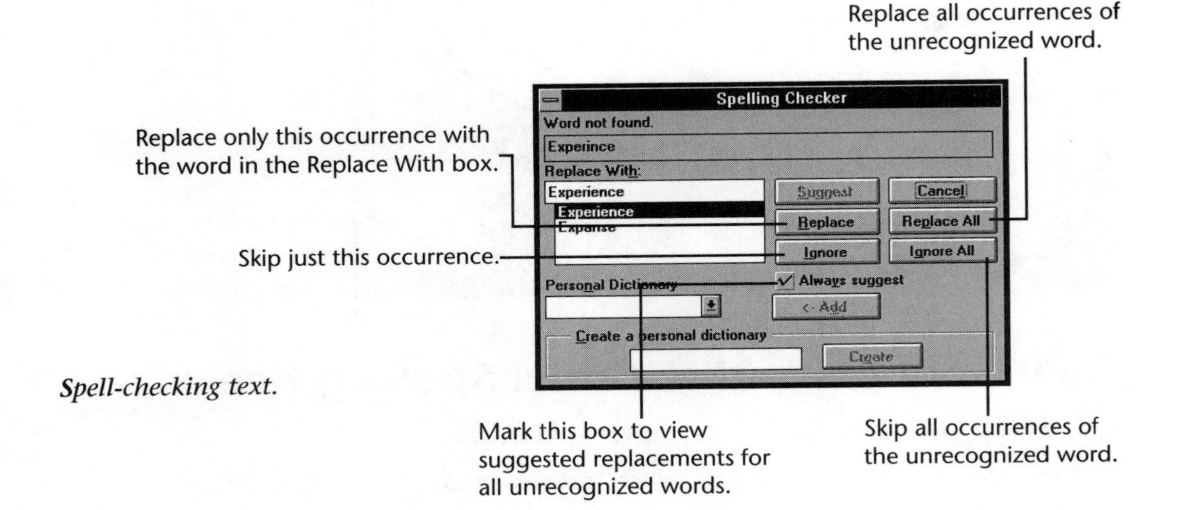

Spell-checking text.

There are several possible actions you can take for each unrecognized word.

- If you want to replace a misspelled word with the correctly spelled word, choose the **R**eplace button. This option replaces only this occurrence of the misspelled word.
- To replace all occurrences of a misspelled word, click on the Replace A**l**l button. Unless you're nervous about changing a whole slew of words at once, we suggest you choose this option rather than **R**eplace because you won't have to review the same misspelled word over and over.

- If the word is spelled correctly, but CorelDRAW! just didn't know that, click on Ignore.
- To ignore all occurrences of a word, choose Ignore All. You'll probably find yourself using this button a lot for proper names you use in your documents.

If you use certain proper names or technical terms consistently in your documents, you should create a personal dictionary to which you can add these names and terms so that CorelDRAW! doesn't stop at them all the time. To create a personal dictionary, click on the Create button and then type a name for the dictionary, up to eight characters. From now on when CorelDRAW! stops on a word that you know is spelled correctly, select your dictionary in the Personal Dictionary box, and then click on the Add button. CorelDRAW! will no longer consider this word misspelled.

When all words in the selected text have been checked, CorelDRAW! displays a dialog box telling you so. Click on the **OK** box to conclude the spell-checking process and return to your text.

I clicked on the wrong button! I wanted to ignore a word, but I clicked on Replace **A**ll instead. Here's what you need to do: click on **Cancel** to close the Spelling Checker dialog box. However, this in itself does not affect any of the spelling corrections made while you were working in the dialog box. You must use the **E**dit **U**ndo command to reverse the last spelling correction you made.

Some Editing Tidbits

We all need to go back to our text and make changes sometimes. (I have yet to meet someone who typed and constructed text perfectly the first time around.) So here are some editing tidbits that might help you as you play around with your text trying to get it just the way you want it.

To accomplish this . . .	Do this . . .
Select an entire artistic text string	Click on the **Pick** tool, and then click on any character in the artistic text sting.
Select entire an paragraph text block	Click on the **Pick** tool, and then click anywhere on the paragraph text. This selects the paragraph text frame.
Select individual characters or words in an artistic text string	Select the artistic text with the Pick tool. Click on the **Artistic text** tool, and then point to where you want to begin the selection. Drag the mouse over the text you want to select. The text becomes highlighted.
Select individual characters or words inside a paragraph text frame	Select the paragraph text frame with the Pick tool. Click on the **Paragraph text** tool, and then point to where you want to begin the selection. Drag the mouse over the text you want to select. The text becomes highlighted.
Edit artistic text	Select the artistic text with the Pick tool. Click on the **Artistic text** tool, and then point to where you want to insert text, edit text, or delete text. Click to place the cursor inside the artistic text string. Add, change, or delete characters and words as needed.
Edit paragraph text	Select the paragraph text frame with the Pick tool. Click on the **Paragraph text** tool, and then point to where you want to insert text, edit text, or delete text. Click to place the cursor inside the paragraph text frame. Add, change, or delete characters and words as needed.

To accomplish this . . .	Do this . . .
Move text	Select the artistic text or the paragraph text frame with the Pick tool, and then drag the selection to a new location.
Change the shape of an individual character in artistic text	Select the artistic text with the Pick tool. Click on the **Shape** tool. Nodes appear along the artistic text characters. Drag a node to a new location to change the shape of a particular character.

The Least You Need to Know

CorelDRAW! has many of the same text editing features that the most powerful word processors have.

- To put text into columns, choose the **Frame** button in the Text roll-up.
- To add bullets to text, click on the **Paragraph** button in the Text roll-up, and then on the **Bullets** icon in the Paragraph dialog box. Mark the Bullet on check box, and then select the bullet you want to use.
- To search for a word or phrase, select the text you want to look through, and then choose the Text Find command.
- To find and replace text, select the text you want to look through, and then choose the Replace command from the Text menu.
- To spell-check text, select the text you want to check, and then choose the Text Spell-Checker command.

Here's that blank page thing again.

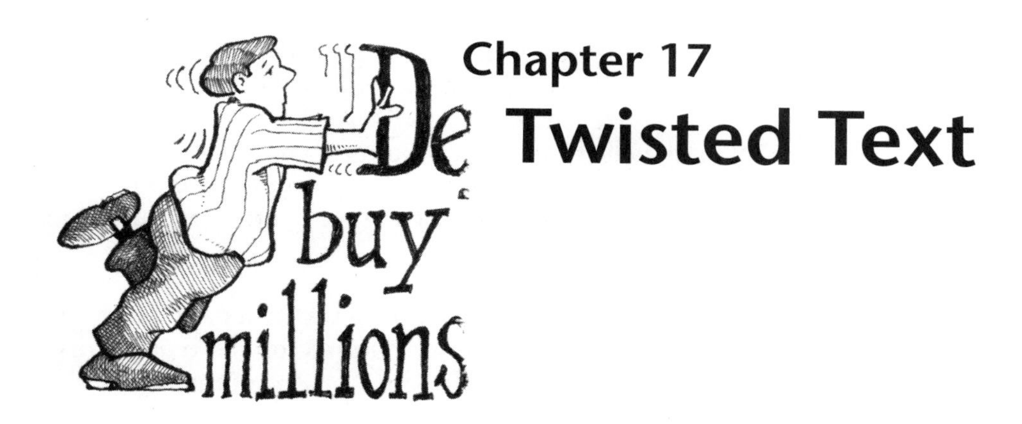

Chapter 17
Twisted Text

In This Chapter

- Winding text around shapes
- Learning the tricks of the Fit Text To Path roll-up
- How the text stands up
- An interesting position
- Strewing text along curves and lines

Would it seem strange to you if Dorothy left a trail of words behind her on her way down the yellow brick road? It may now, but after you wind your CorelDRAW! text up and around and through and over all sorts of paths, the combination of words and roads will no longer seem weird. You can create any object, shape, or line and have text wrap around it, inside it, underneath it, or on top of it. It's called fitting text to a path, and it's what this chapter is all about.

Shapely Text

When we were kids, my little brother used to hang around me religiously. Wherever I went, he was on the path just behind me. You've got the same big sister/little brother relationship in CorelDRAW!: The big sister is any

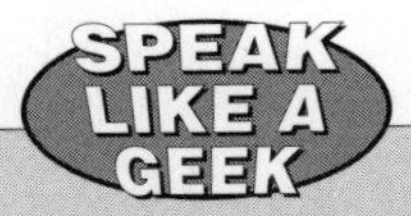

Fitting text to a path Wrapping text around an object, curve, line, or CorelDRAW! symbol.

Path Essentially, the outline of any object, or a line or curve.

shape or curve, and the little brother is text. You can hang any artistic text you've written around a rectangle, circle, or CorelDRAW! symbol—basically any object. And if you want to, you can delete the object and leave the text still hanging, as if the object were still there (good thing little brothers can't do that in real life).

CorelDRAW! calls it fitting text to a path, and you use the Fit Text To Path roll-up to do it. To open this roll-up, choose the Fit Text To Path command on the Text menu. Before you have any objects selected, most of the options in this roll-up are dimmed.

First of all, you need an object to wrap your text around. So draw an arc, circle, square, or ellipse, or choose a CorelDRAW! symbol from the Symbol roll-up. Then type the text you want to wrap around the object. Important note: you can fit only artistic text around an object; paragraph text won't budge. If you want to add cool shapes to paragraph text, use an envelope instead (read the next chapter, "The Envelope Please," for details).

So you have your object, and you have your artistic text. Select both by holding down the **Shift** key and clicking each with the Selection tool (but you should already know all about that). Wow—did you see the Fit Text To Path roll-up do all sorts of funky things? Certain options popped up that weren't there before, and others brightened up. That's because you're all set to fit your text to your object's path. We won't get into any of the objects now, so go ahead and click on the **Apply** button. Voilà! Your text now wraps along the outline of your object.

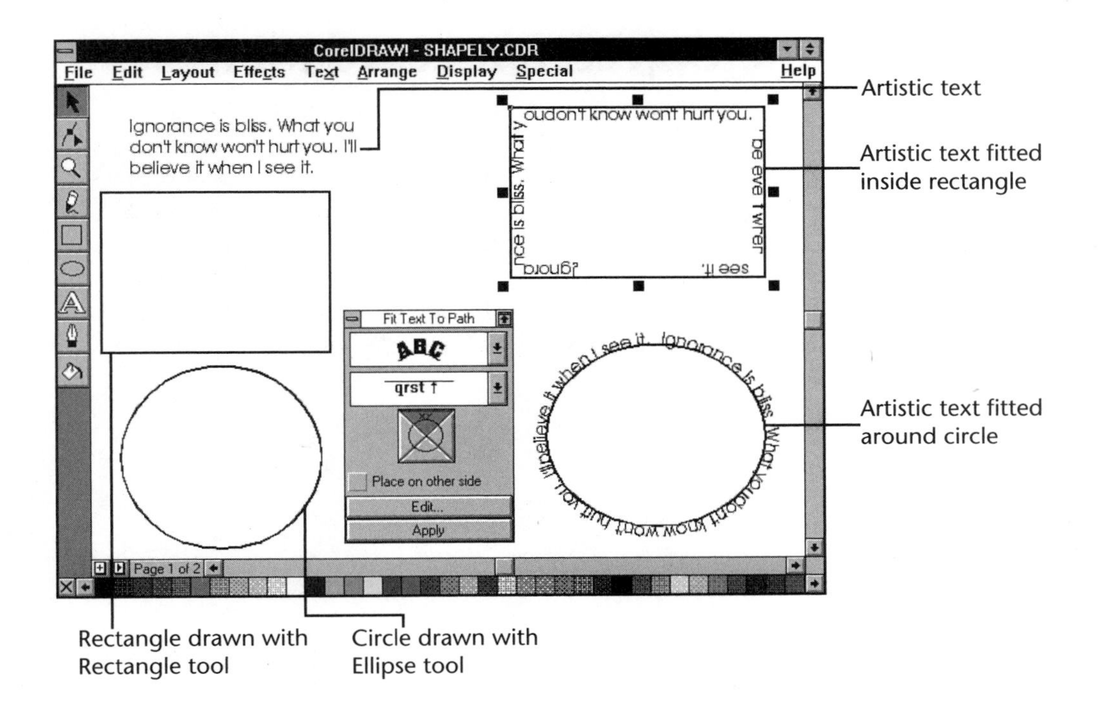

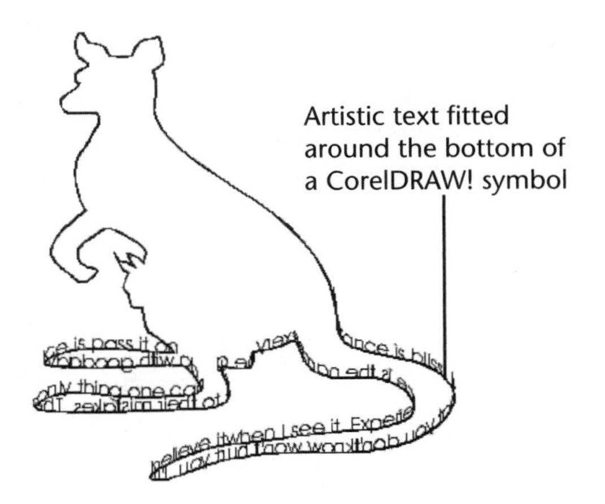

Fitting text around a rectangle, circle, and CorelDRAW! symbol.

Get Rid of the Object, But Leave the Text

Sometimes you might want to wrap text around an object, without the object showing. This is a can-do with CorelDRAW!. At any time after you've fitted the text to the object's path, just select the object and press the **Delete** key. Understandably, this might make you a bit nervous. But don't worry, CorelDRAW! just deletes the object, leaving the text intact. To delete the whole object and the text, you'd either have to press **Delete** twice, or click on the object twice with the Selection tool (so that the rotation handles appear), and then select **Delete**.

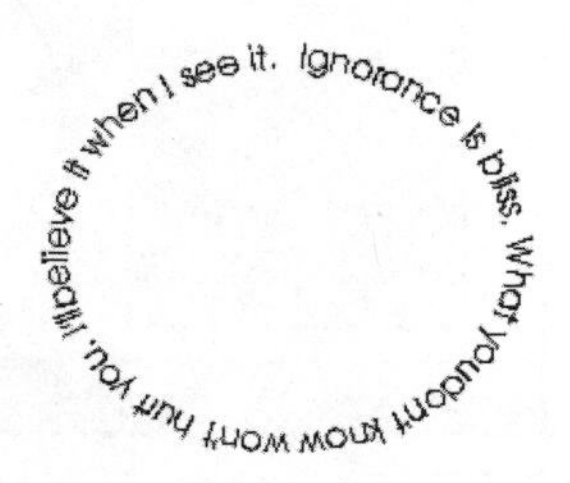

Deleting the object after you've fitted the text can create cool looks.

Unveiling the Fit Text To Path Roll-Up

In the last section, we brushed off any special effects you could apply to the text you fit to an object. That's because there are so many different ways you can have the text wrap that this subject deserves its own section. And we're not even to that subject yet. We're just going to tease you and give an overview of the different options in the Fit Text To Path roll-up.

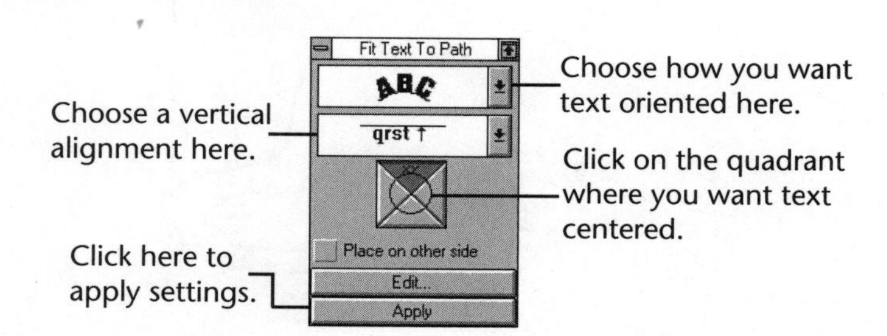

The Fit Text To Path roll-up.

Be happy that this roll-up has fewer options than most. When you're fitting text to an ellipse or rectangle (versus a straight or curved line), a little box shows up in the middle of the roll-up with a circle divided into fours. This box lets you determine which quadrant your text will be centered in. Just click on the quadrant you want.

You can change the setting of fitted text at any time. Just select the text in the object, choose new settings in the Fit Text To Path roll-up, and click on the **Apply** button. Make sure that you select the text in the object. If the options in the roll-up do not become available, that means you selected the object as a whole, and not the text. Try again by clicking directly on a letter.

The first drop-down list in the roll-up determines how your text will be oriented. You have a choice of rotated letters, vertical skew, horizontal skew, or upright letters. But those titles won't help you half as much as seeing exactly what the options do to the text. To see samples, read the next section, "Your Text's Orientation."

If CorelDRAW! were a human being, it'd definitely be pro-choice. With the amount of choices CorelDRAW! gives you to perform functions, you can sometimes get lost. Well there's another decision you need to make. How do you want your text to be wrapped: on the inside, outside, right on top of the object's outline, so that the bottom tips of the text sit on the object's outline, completely detached from the outline, or drawn right through the outline? By the way, this is called vertical alignment, and if you want to see some examples, review the upcoming section, "How It's All Wrapped Up."

Your Text's Orientation

Life's not as simple as just fitting your text to a path. How you want the text fitted is an issue to be reckoned with, since the overall look of the fitted text changes drastically depending on the selection you make in the first drop-down list of the Fit Text To Path roll-up. If we tried to describe the different choices to you, you'd probably just end up with a mish-mash of ideas that might or might not have any accuracy. So instead, we drew some objects and formatted each with a different type of text orientation. So just look at the figures below to decide what look you want.

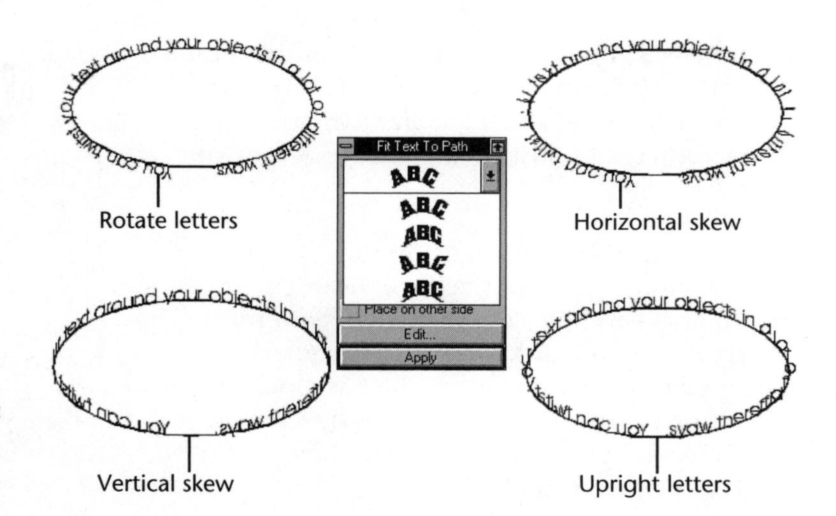

These text orientation choices are available to you.

It's a Wrap!

This is another choice CorelDRAW! gives you—that changes the look of your text wrap, just as much as the text orientation. This time you get to choose how you want the text wrapped on the outline (officially called the *vertical alignment*). You can have the object's outline go right through the middle of the text, not even touching the text, aligned at the top or bottom of your text, or aligned so that the bottom points of your text touch the outline. For example, check out the figure below.

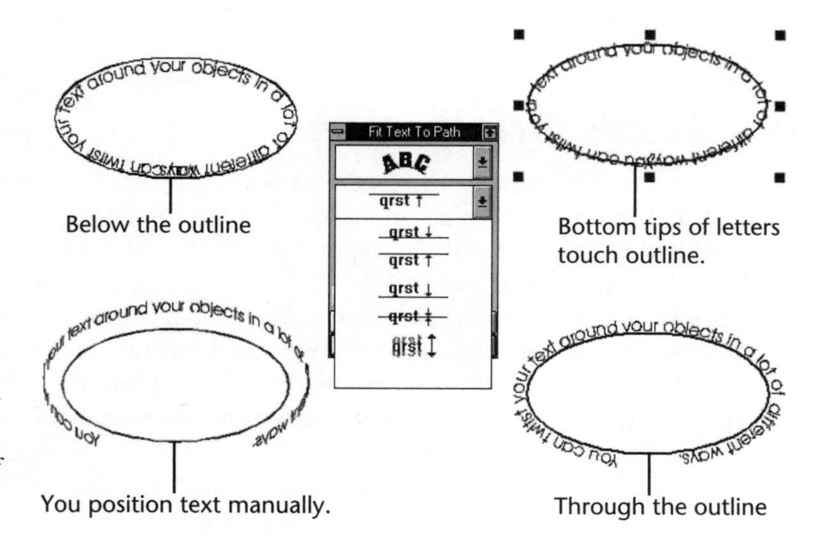

Determining the vertical alignment of the fitted text.

Making your choice is as simple as selecting the appropriate style in the roll-up's drop-down list. However, one of the styles needs further instruction. When you choose the option that lets you manually adjust where the text is positioned, you must first apply the selection as you would any of the other options. Then you drag the fitted text either away from or towards the object to where you want it, and release the mouse button.

The Long and Winding Road

Remember, you can fit text to any path—including straight or curved lines. Just draw the line, write the artistic text, select both, display the Fit Text To Path roll-up, choose your settings, click on **Apply**, and presto: your text hangs around the line. But you should know all these steps by now; they're the same as for fitting text around shapes. The only reason we even broke this into its own section is because you get an added option when you fit text to a line.

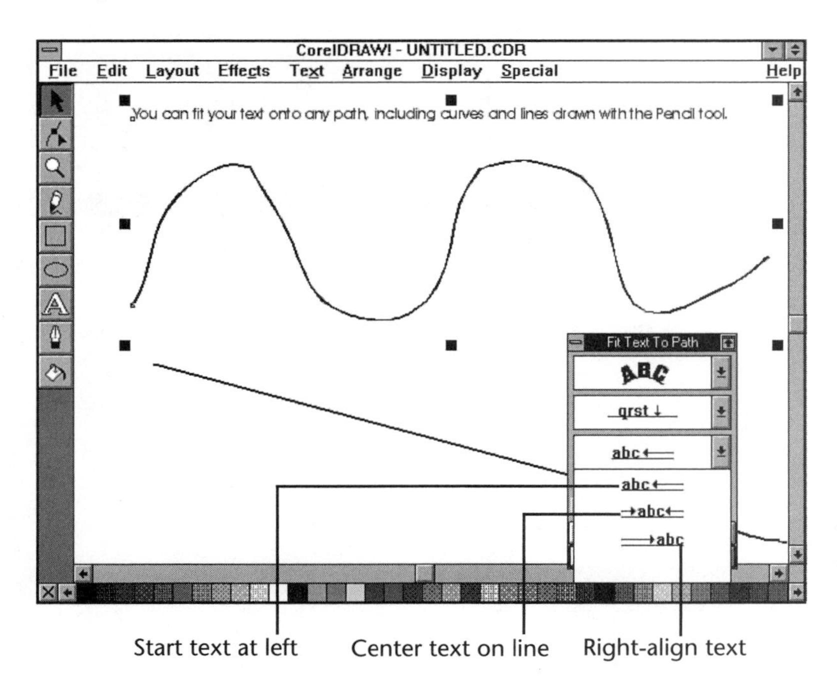

Fitting text to a straight and curved line.

You'll notice a third drop-down list appears in the roll-up now. This list lets you determine where you want to place the text on the line. You can have the text start at the start of the line, appear centered on the line, or appear right-aligned (so that the end of the text lines up with the end of the line).

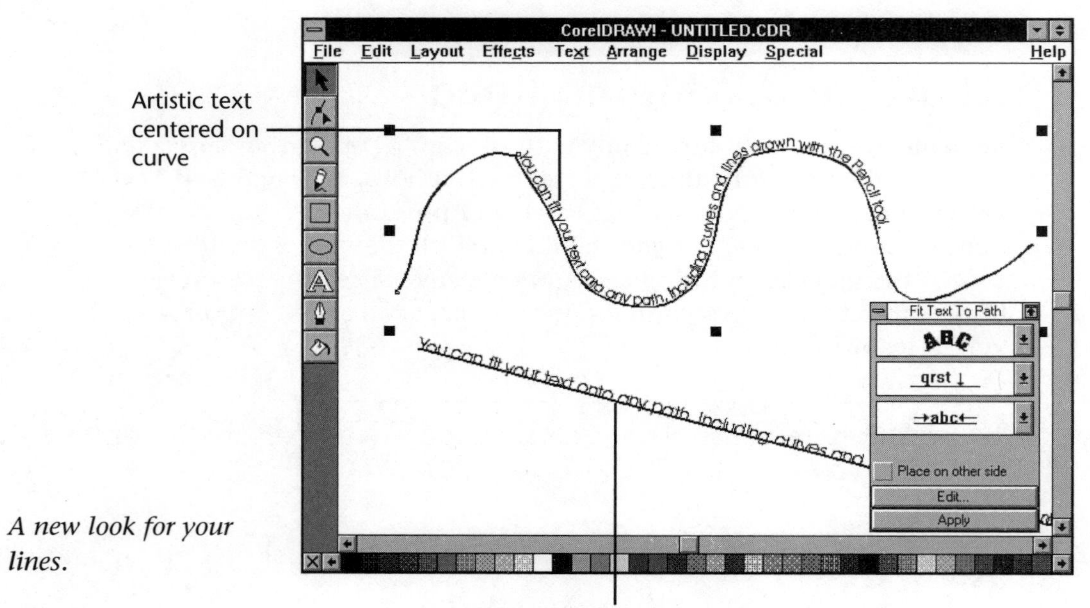

A new look for your lines.

The Least You Need to Know

You can do all sorts of cool and unusual things with text, shapes, and lines, thanks to CorelDRAW!'s Fit Text To Path feature.

- To display the Fit Text To Path roll-up, choose the Text Fit Text To Path command.
- Only artistic text can be fitted to a path.
- Select the text and the object you want to wrap around the object. Choose your settings in the roll-up, and then click on **Apply**.
- You can remove the object by selecting the object and then pressing **Delete**.

Geez, another &*!@ blank page!

Chapter 18
The Envelope, Please

In This Chapter

- A peek at the Envelope roll-up
- Picking the envelope you want
- Stuffing the envelope
- Pushing the contents this way and that way
- Artistic text inside an envelope
- An envelope overflowing with paragraph text

CorelDRAW! envelopes are not at all like the envelopes you know. Normally, when you try to squeeze too many papers into an envelope, the envelope bulges as a result. In other words, the envelope's shape changes to fit its contents. Well just the opposite is true with CorelDRAW! envelopes. When you place text or a graphic into a CorelDRAW! envelope, the text or graphic changes shape to fit inside the envelope. You'll learn all about this new type of envelope in this chapter.

Opening the Envelope

When you think of a CorelDRAW! envelope, think of the word "shaper." The latter probably would have been a more accurate term, but Corel most likely wanted to use a name you're more familiar with, like "envelope." The only problem with the term "envelope" is that it gives you the false notion that the object inside the envelope will affect the envelope's shape (like a bulk of papers would make a regular envelope bulge), when in fact just the reverse is true: a CorelDRAW! envelope changes the shape of your object. As long as you remember this last point, you'll be fine. You might also want to keep in mind that you can apply an envelope shape to any single or grouped object, to artistic text, and to paragraph text.

Before we go any further, let's look at some enveloped objects and at the Envelope roll-up, from whence you will perform all of your envelope operations. To display this roll-up, select the Envelope Roll-Up command from the Effects menu. That's easy enough.

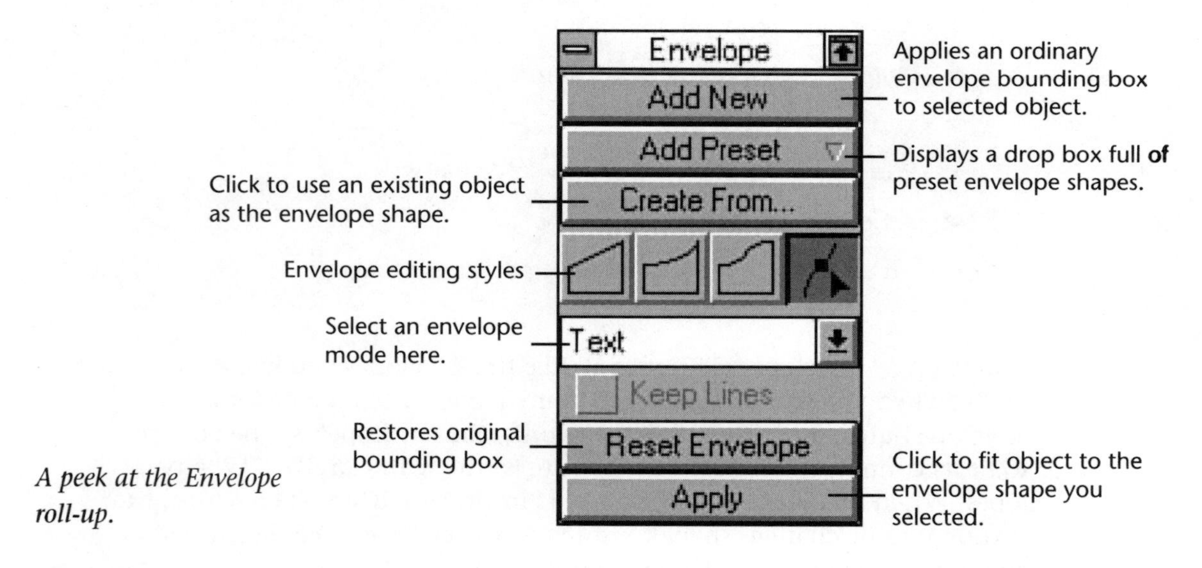

A peek at the Envelope roll-up.

CorelDRAW! envelopes come in all sorts of sizes and shapes—certainly a better selection than your local post office offers. There are three sources of envelope shapes you can choose from:

- The default envelope shape (Add New command on the Envelope roll-up). This is a regular box shape that you can change into whatever shape you want by dragging its nodes and control points (see Chapter 10, "Nodes, Nerds, and Segments: Reshaping Your Objects," if you need help manipulating nodes).
- Preset envelope shapes (Add Preset command on the Envelope roll-up). These are predesigned shapes provided for you by CorelDRAW!. Check out these shapes before trying to design your own. What you want may already be here.
- An existing object as the envelope shape (Create From command on the Envelope roll-up). If there's an object in your drawing that you want to use as the envelope shape, choose this command, and then click on that object.

Good news and bad news. First the good: no roll-up for you to worry about (don't those version 4 and 5 users look green with envy!). Instead, the first four commands on the Effects menu deal with shaping objects with envelopes. Now the bad: no preset envelope shapes for you either. You'll have to rely on your node editing abilities to get the envelope shapes you want. (Hint: skip the next section on preset envelopes.)

You'll go through the entire process of applying an envelope shape to an object in the next section, but just to give you an overview, here's the basic procedure. Select an object, choose an envelope shape from one of the three sources listed above, and then click on the **Apply** button. I know you wish that was all there was to it and that we could just end this chapter right here, but you're not so lucky. Still, using envelopes really isn't that hard. So read on!

Stuff Your Object into an Envelope

Okay, you've got the Envelope roll-up displayed, and you've selected an object (well, if you haven't done so, do please!). In this section we'll show you how to use a preset envelope to shape your object. Two sections later in this chapter ("An Artist's Envelope" and "Blob-like Paragraphs") show you how to use the default envelope shape.

Click on the **Add Preset** button in the Envelope roll-up. A list box appears with lots of little strange-looking objects. Scroll through the selection until your eyes rest on a shape that appeals to you. I'm going to choose the raindrop. You don't like this shape you say? Well, choose your own.

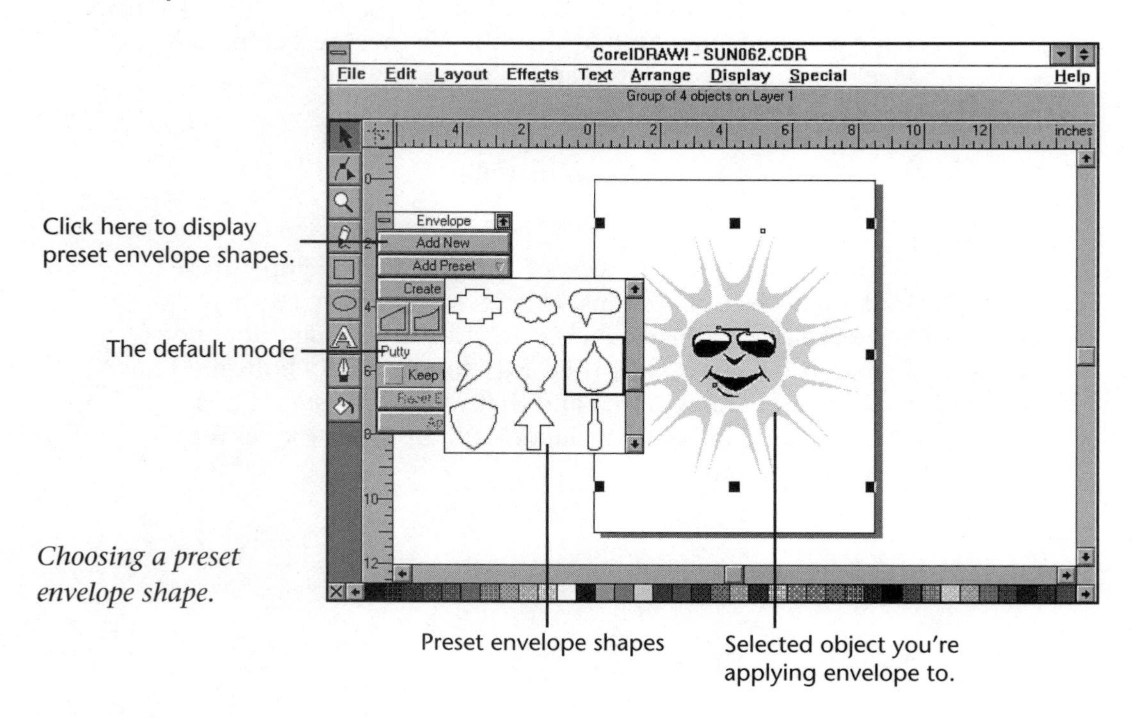

Choosing a preset envelope shape.

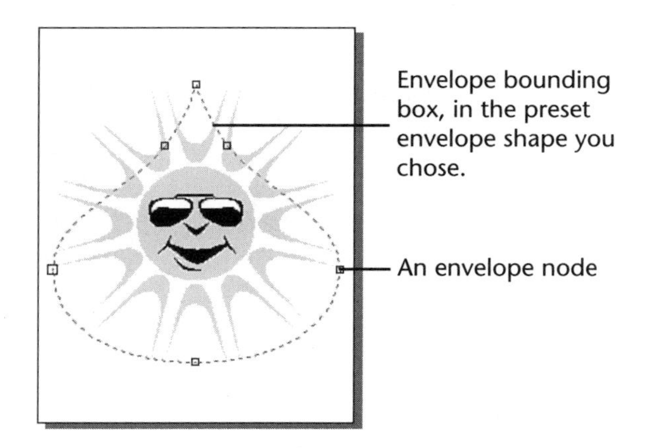

The envelope shape sits on top of the object.

After you choose the preset envelope shape, a dotted outline in that shape appears on top of the selected object. This dotted outline, called the *envelope bounding box*, contains nodes. Remember those from Chapter 10, "Nodes, Nerds, and Segments: Reshaping Your Objects"? If not, you might want to flip through that chapter again—but not right now, because we're going to leave the envelope shape intact (that is why we chose a preset design after all). The section "An Artist's Envelope" (later in this chapter) discusses how you can reshape an envelope using the bounding box nodes.

SPEAK LIKE A GEEK

Bounding box A dotted line with nodes that appear when you select an envelope shape. You can change the shape of the envelope by moving the nodes on the bounding box.

Envelope editing style Determines how the envelope's shape moves when you drag its nodes and control points.

Notice that nothing has happened to the selected object yet. Now why can that be? Remember that old Apply button? (It comes back to haunt you again.) Actually, CorelDRAW! doesn't immediately shape the object for a good reason. If you want to edit the shape of the envelope before applying it to the object, you can. And then when you're done, you can click on the **Apply** button in the dialog box, and . . . presto! bingo! bango! (we're running out of exclamations): your object's shape changes to fit the envelope.

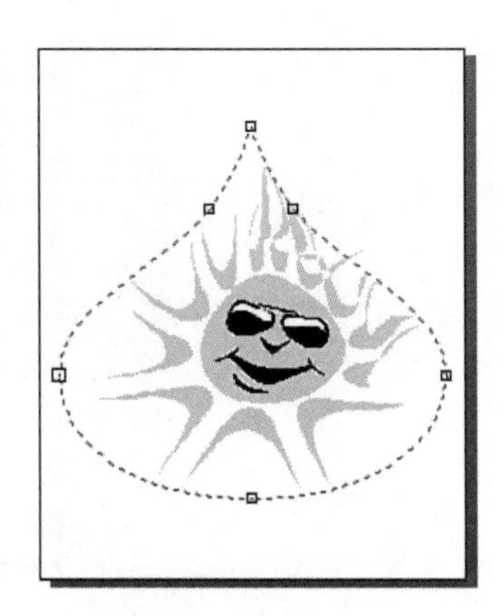

The object now fits inside the envelope.

Modes Really Shape Up Your Objects

There's one thing we didn't mention yet that we need to mention: the *envelope mode*. This term can be best described visually, but we'll try with words too. The envelope mode you choose determines just how badly your object is going to be distorted. The default mode, Putty (maybe for "molds like putty in your hands?"), usually distorts an object drastically. If you want to get some really funky looking distortions, try the Vertical or Horizontal modes (actually, Vertical is primarily for shaping text, but there's no law saying you can't use it on your graphics). The Original mode is CorelDRAW! 3's one and only envelope mode. If you want to open a CorelDRAW! 3 envelope-shaped object in CorelDRAW! 4, choose the Original mode.

The Original mode is the only mode available in CorelDRAW! 3, so you must select Original mode to edit a CorelDRAW! 3 envelope.

To change a mode, display the drop-down list in the Envelope roll-up (where it currently says "Putty"). Select a new mode from the drop-down list. Click on the **Apply** button in the Envelope roll-up. CorelDRAW! changes the selected envelope object to the new mode.

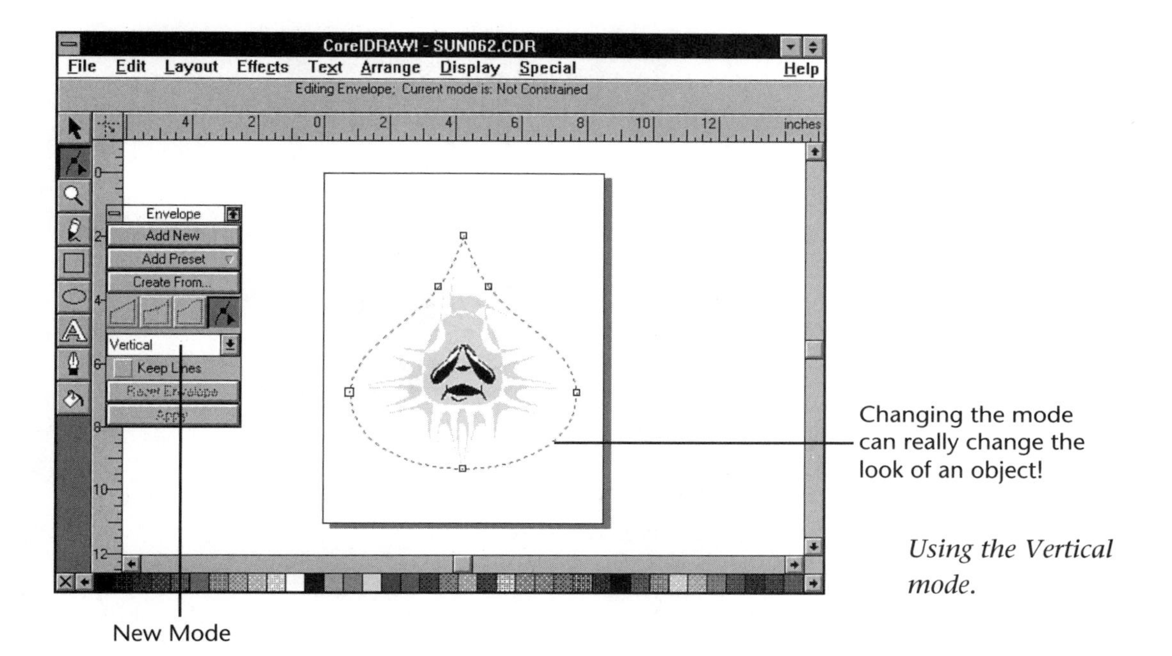

Using the Vertical mode.

An Artist's Envelope

Envelope mode Determines the severity of the distortion created by an envelope.

Let's move on from graphics to text. You can shape both artistic and paragraph text with an envelope. You'll find the Envelope roll-up to be indispensable when you want to combine text and graphics. By shaping with an envelope, you can squeeze artistic text into any tight spot within one of your graphics. For example, if you want to create a clock or watch and place the word "time" inside its face so that the word fills the entire face, you can do so with the Envelope roll-up. Follow the steps outlined below.

1. Type the text you want using the Artistic text tool (refer to Chapter 15, "Painting with Words," if you need help).
2. Draw the object to which you want to add the text, or use one of the CorelDRAW! symbols or clip art images.

3. Move the text object over the graphic so that it sits where you want it. Leave the text object selected.
4. Display the Envelope roll-up (select Effects Envelope Roll-Up).

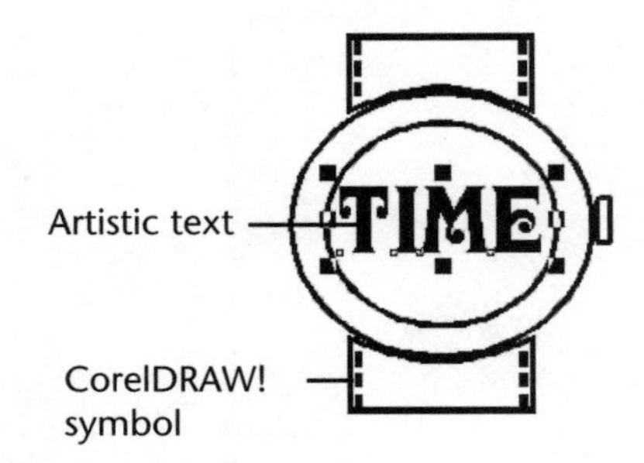

Combining text and graphics.

5. Click on the **Add New** button in the Envelope roll-up. A bounding box appears over the text.
6. Drag the nodes and control points on the bounding box so that the bounding box forms a rough circle. If you need help manipulating nodes and control points, read Chapter 10, "Nodes, Nerds, and Segments: Reshaping Your Objects."

TECHNO NERD TEACHES...

When you edit an envelope's nodes, you get four different editing styles to choose from. These four styles are represented by the icons just below the Create From button in the Envelope roll-up. By default, the far right icon is selected, which allows you edit the nodes just as you do with the Shape tool. You can select multiple nodes and move them all at once. With the other styles, you can only move one node at a time. The easiest way to see how the other styles affect the envelope shape is to try each one out on an envelope.

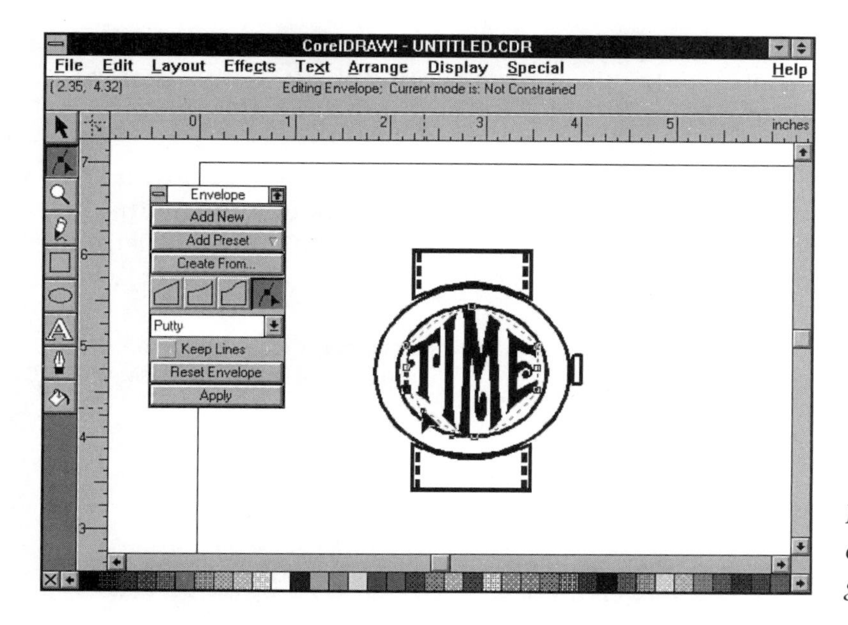

Manipulating the envelope's nodes to get the desired shape.

7. Click on the **Apply** button in the Envelope roll-up. CorelDRAW! distorts the text so that it fits the envelope.

Time now fills the face of the watch.

Blob-Like Paragraphs

In addition to artistic text, you can apply envelope shapes to paragraph text. Just select the paragraph text frame, and then click on the **Add New** button to add an envelope to the text frame. Manipulate the nodes into the shape you want, and then click on the **Apply** button. The paragraph text frame actually changes into the shape of the envelope; but keep in mind that this frame only shows when the paragraph text is selected. So if any of the text shows outside of the frame, you don't have to worry.

You can also shape the paragraph text using a preset envelope, a CorelDRAW! symbol, or the shape of an existing object (in the latter case, use the Create From button).

You can also come up with some neat looking effects by removing the shape after you distort the text with an envelope. For instance, in the example above, you would be left with a circular "TIME" if you deleted the watch after shaping the word with the envelope.

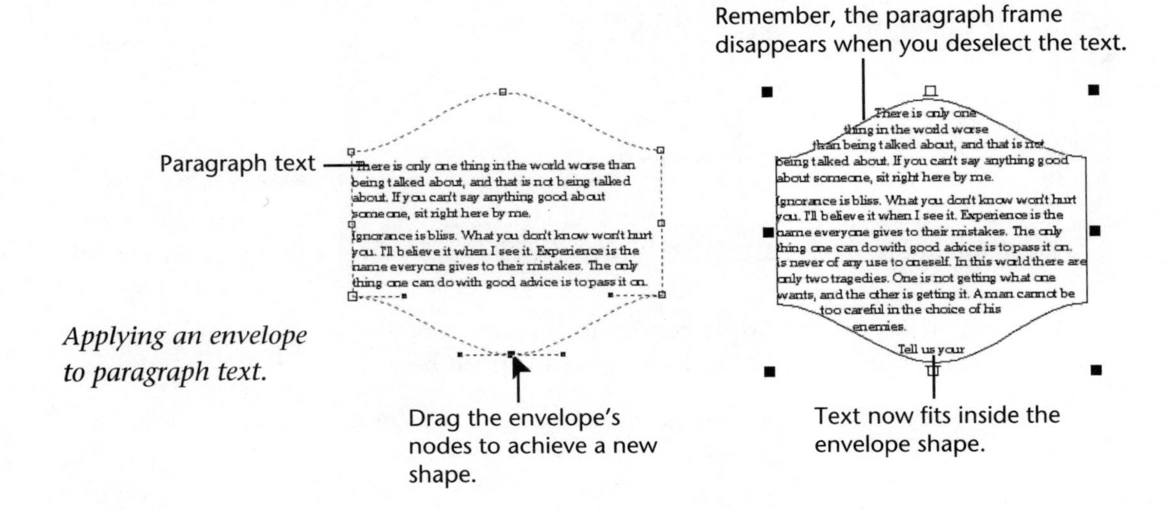

Applying an envelope to paragraph text.

The Least You Need to Know

You can create all sorts of neat shapes for your graphics and text using CorelDRAW!'s Envelope roll-up.

- To display the Envelope roll-up, select the Effects Envelope Roll-Up command.
- To use the default envelope style so you can shape it into your own desired shape, click the **Add New** button in the Envelope roll-up.
- To use one of CorelDRAW!'s preset envelope shapes, choose the **Add Preset** button in the Envelope roll-up, and then click on the preset shape you want.
- Click on the **Apply** button in the Envelope roll-up to fit your object into the envelope shape.
- Try using different envelope modes to achieve unique effects.

This page suitable for doodling.

Part IV

Some Way Cool Stuff

I was only a kid when "Star Wars" came out. I remember sitting dazzled in front of the movie screen as my brain sucked in all the visual special effects. The bar scene (remember—with all the funky-looking characters) was the ultimate high. What an experience! The only bad thing was that when I got home, turned on the tube, and sat down to watch Star Trek, I could no longer be awed by any of the futuristic props. This was a great tragedy since it had always been my favorite show (thank goodness "The Next Generation" arrived on the scene a bit later). Still, seeing the special effects in "Star Wars" was worth it.

Prepare to be dazzled by some of CorelDRAW!'s special effects. You can transform two-dimensional objects into full-bodied 3D objects instantly. You can squeeze your text into the shape of a banana. And you can blend two objects together so that you can't tell where one begins and the other ends. Have fun playing around with CorelDRAW!'s special effects (and don't worry, this won't make "Deep Space Nine" seem out of date).

Chapter 19
Blend for 30 Seconds

In This Chapter

- Taking a glimpse at the Blend roll-up
- Ooh! Your first blend
- See how many different blends there are?
- Man meets woman in a sample blend

A shape here and a shape there, a pinch of patience, a lot of tries, several steps, 180 degree rotation, and a loop—throw all this into your CorelDRAW! blender. Then blend; and what do you have? One really cool blended object in CorelDRAW!. If you have any "DRAW-pro" friends, you may have glimpsed some of their snazzy graphics. Ask one of them: chances are he or she used the Blend feature to achieve some of those sophisticated effects. You'll understand just how to blend your objects after you go through this chapter. And who knows, maybe then you'll feel like a "DRAW-pro" yourself.

Your New Blender

CorelDRAW!'s Blend feature works differently than your kitchen blender does. Instead of mixing everything up so that you can't detect what the original ingredients were (as your kitchen blender does), CorelDRAW!'s blender blends two objects together with intermediate shapes, called steps. The two original objects remain distinct, but are blended together by the shapes in between them. You can blend any type of objects, including objects you draw, CorelDRAW! symbols, clip art, and text. The figure below shows two CorelDRAW! symbols (from the first page of the Animals category) blended together; it comprises a man and a woman, with one hundred intermediate shapes. Now if it were only this easy in real life for a man and woman to blend together and live harmoniously. Well, I guess it would be boring—but I could sure give it a try for a little while.

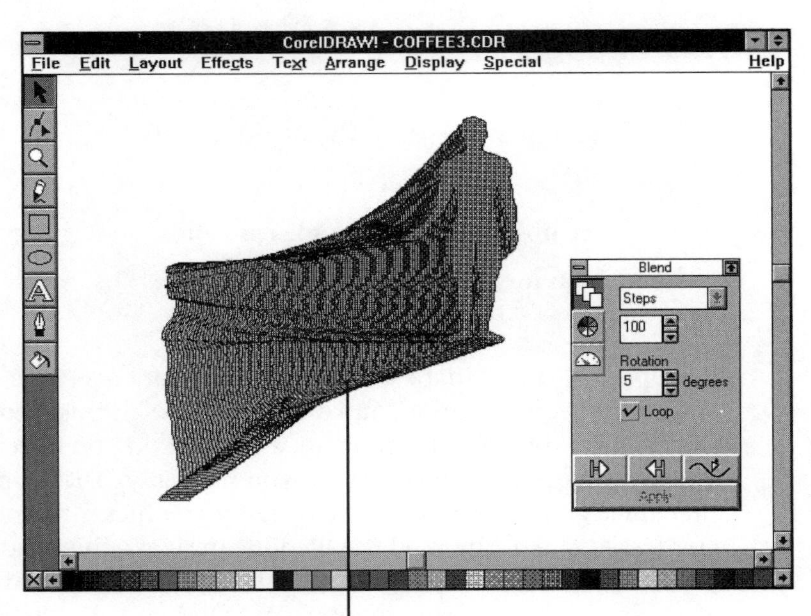

You can create interesting designs with CorelDRAW!'s Blend feature.

A Peek at the Blend Roll-up

The Blend roll-up has lots of options, but you'll probably only need to use a few to get the blends you want. Rather than cover all of the options, we'll review the essential ones, and let you explore the more advanced features later on, after you've used the Blend feature enough to feel comfortable with it.

The Blend roll-up has three different displays. You'll use the default display to create your blends. If you ever click the wrong button and lose this display, you can always return to it by clicking the upper left button in the roll-up, which looks like three stacked pages. To display the Blend roll-up, select the Effects Blend Roll-Up command or press **Ctrl+B**.

SPEAK LIKE A GEEK

Steps Intermediate shapes between two objects blended together with CorelDRAW!'s Blend roll-up. You determine the number of steps in a blend.

Intermediate shapes The same as steps, these are the objects between two blended objects.

Start and end objects The two original objects that you blended together.

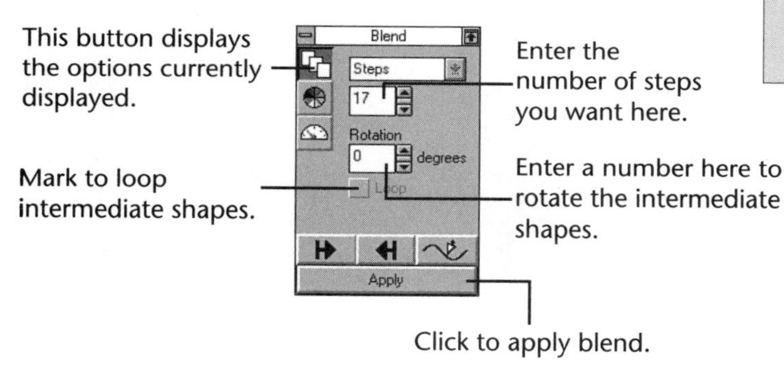
This button displays the options currently displayed.
Mark to loop intermediate shapes.
Blend
Steps
17
Rotation
0
degrees
Loop
Apply
Enter the number of steps you want here.
Enter a number here to rotate the intermediate shapes.
Click to apply blend.

The default display in the Blend roll-up.

A Blend of This and That

Enough talk—let's get down to some action. We'll show you about the different ways you can blend objects by using two simple objects: a red square and a blue rectangle. You'll achieve very different looks depending on the type and color of objects you blend. But at least by seeing how to blend these two objects you'll learn about the blending process so that you can experiment with blending other objects on your own later.

5.0

Though the version 5 Blend roll-up looks a bit different, it functions the same as the one in version 4. The difference is that the icons that change the roll-up display are positioned at the top of the Blend roll-up in version 5, instead of at the left side of the roll-up (as in version 4).

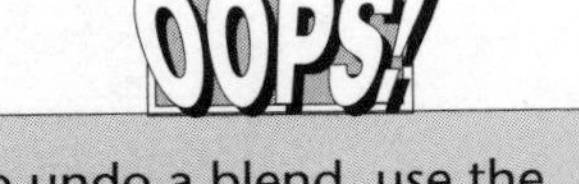

To undo a blend, use the **E**dit **U**ndo Blend command.

After you've displayed the Blend roll-up, select both objects to be blended. Until you select the objects, most of the options in the Blend roll-up will be dimmed and inaccessible. When you have selected the objects, the options should become available.

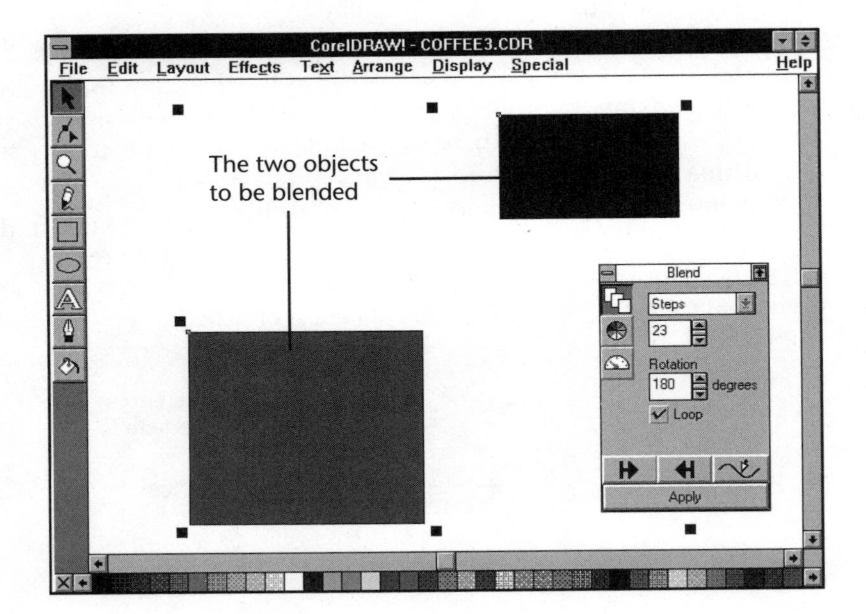

The first step: select both objects.

Enter the number of intermediate shapes you want to use in the box just below the word Steps. (For instance, you'd type 23 for the sample blend we're creating.) Usually, you'll just have to experiment with different numbers of steps until you achieve the effect you're looking for. Click on the **Apply** button, and you're the proud new owner of a CorelDRAW! blended object.

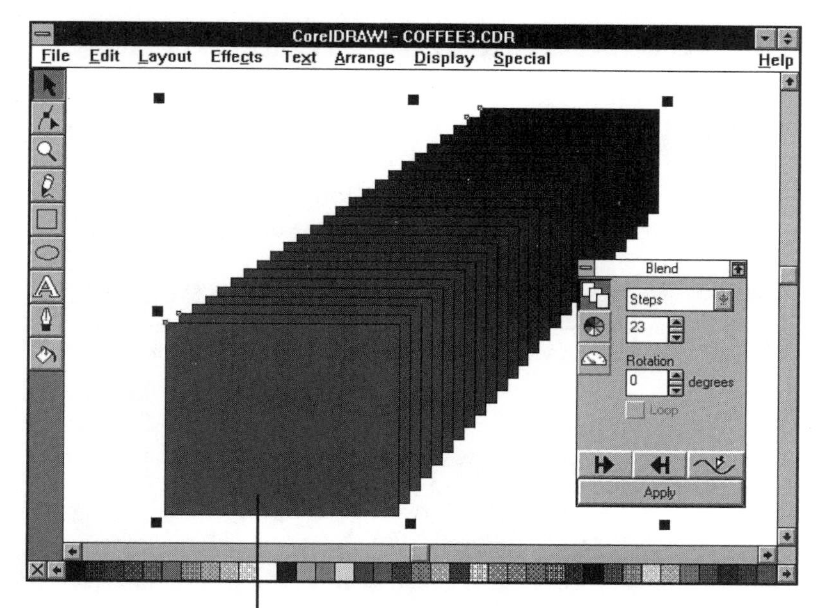

A simple blend.

TECHNO NERD TEACHES...

The two objects that you blend are a dynamically related group. This means that if you edit the start object or end object (the two original objects), the blend changes accordingly. For example, if you moved the start object up on the drawing page, the intermediate shapes would move up accordingly. Or, if you changed the end object's color, the color of the intermediate shapes would change also. You can edit the start or end objects by clicking on them.

What a Little Rotate Will Do

You can twist the intermediate shapes between two blended objects to achieve quite a different look. The Blend roll-up contains a Rotation box that controls the degree to which the intermediate shapes are rotated. You can enter any number from 0 to 360 in the Rotation box. Notice how different the blended objects look when we apply 180 degree rotation.

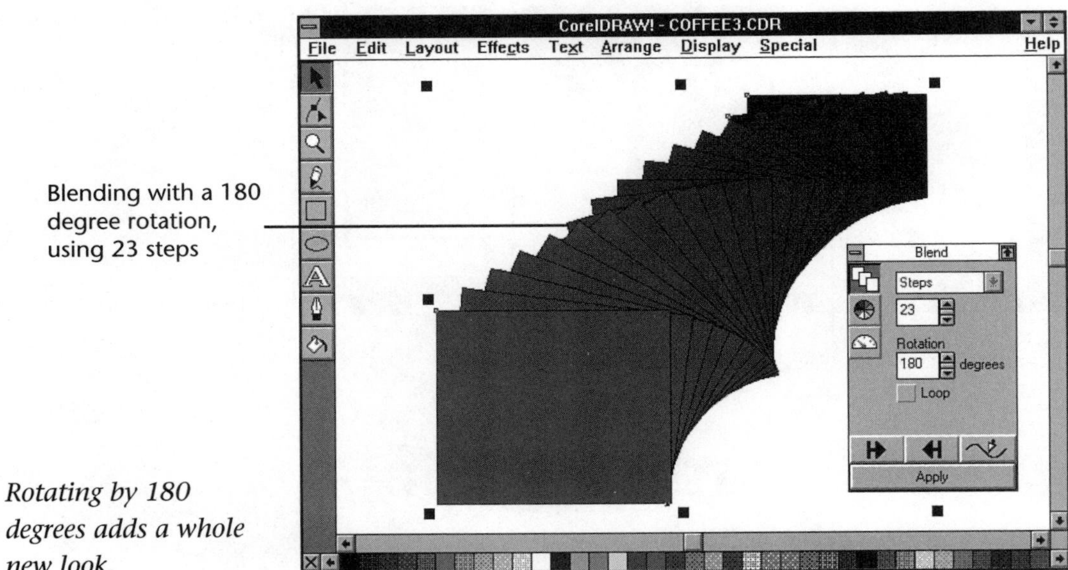

Rotating by 180 degrees adds a whole new look.

Loop To cause intermediate shapes in a blend to rotate around the halfway point of the start and end objects' centers.

Loop de Loop

Looping intermediate shapes between blended objects gives yet another unique look. When you mark the Loop box in the Blend roll-up, it gives the illusion that the intermediate shapes follow the path of a boomerang: they phase out, and back in again. Technically speaking, the shapes rotate around the halfway point of the start and end objects' centers.

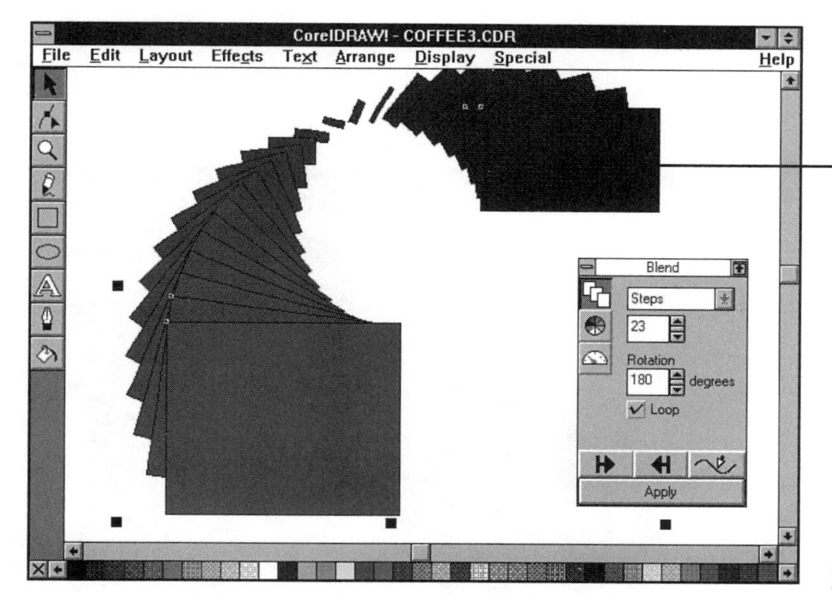

Using the Loop feature in a blend.

Twisting Until You're Dizzy

The mother of all rotations: try a 360 degree rotation for fun, and see what you come up with. But when you use such a high degree of rotation, you should use a greater amount of steps to gain a full, twisted effect. You don't have to use more steps, of course, but you'll get a completely different look with only a few intermediate shapes when you're using a 360 degree rotation. We'll say it again, the best way to learn about the Blend feature is to experiment with all sorts of different settings.

For another interesting effect, mark the **Loop** check box in the Blend roll-up, and see how you like a looped, 360-degree rotated blend with 100 steps.

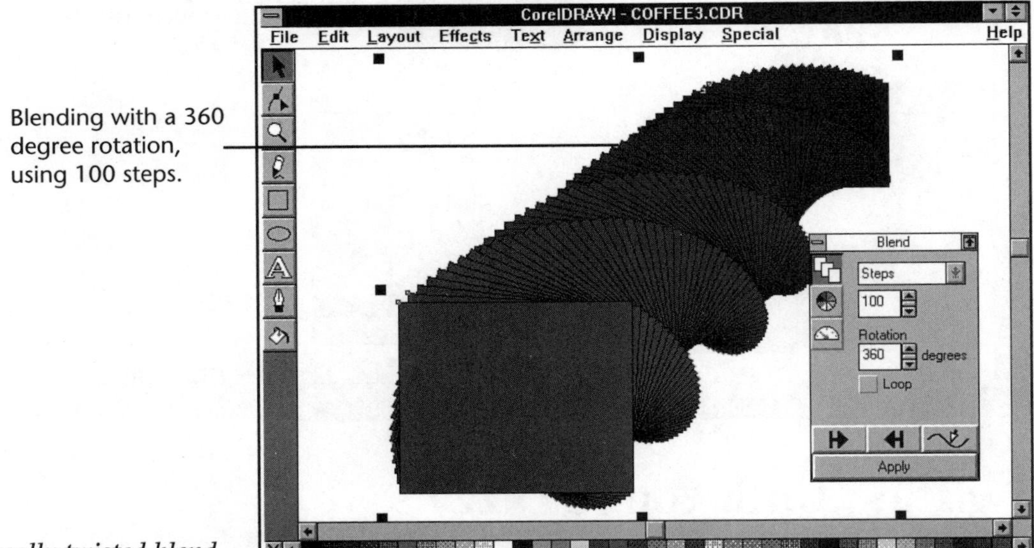

A really twisted blend.

The Least You Need to Know

You can create some really twisted objects using CorelDRAW!'s Blend feature.

- To blend two objects together, you must first select both objects.
- You'll use the Blend roll-up to create a blend. To display this roll-up, select the Effects **B**lend Roll-Up command or press **Ctrl+B**.

- You can rotate the intermediate shapes in a blend by entering a number of degrees in the Rotation box in the Blend roll-up.
- Mark the **Loop** box if you want to have the intermediate shapes look like they're following the path of a boomerang.
- For a truly twisted look, enter a 360 degree rotation with a high number of steps.

Blank (this) Subliminal (book) text (is) page (great)

Chapter 20
Getting Deep: Extruding Objects

In This Chapter

- Step one in extruding
- Your first extrude
- A shaded extrusion
- From circular to tubular
- A light at the end of the tunnel

Now here is a feature that will take your drawings from amateur status to professional. The extrude feature. We're really not sure why the Corel wizards up in Canada decided upon "extrude." Perhaps they did it because it was the weirdest term they could come up with for the CorelDRAW! process that gives your objects depth (or because they have to keep up their reputation for using bizarre terms). But for us plain folk, extruding sounds a bit kinky, and it may even make us think of something it has nothing to do with at all. Simply speaking, *extruding* in CorelDRAW! is adding depth to an object. Apply a simple mixture of new dimension lines, light effects, and shading to your objects, and you can make them look truly 3D. This chapter tells you all about how to extrude (as if you didn't already know how).

Though the version 5 Extrude roll-up looks a bit different, it functions the same as the one in version 4. The difference is that the icons that change the roll-up display are positioned at the top of the Extrude roll-up in version 5, instead of at the left side of the roll-up (as in version 4).

I've Never Extruded Before

Neither had I, until I used CorelDRAW!'s Extrude roll-up. Extruding an object adds an illusion of depth to the object, either by distorting the object's shape, by adding light and shadows, or by shading the object.

Like CorelDRAW!'s other special effect features, the first step in extruding is opening the Extrude roll-up. Choose the Extrude Roll-Up command from the Effects menu or press **Ctrl+E** to display the roll-up.

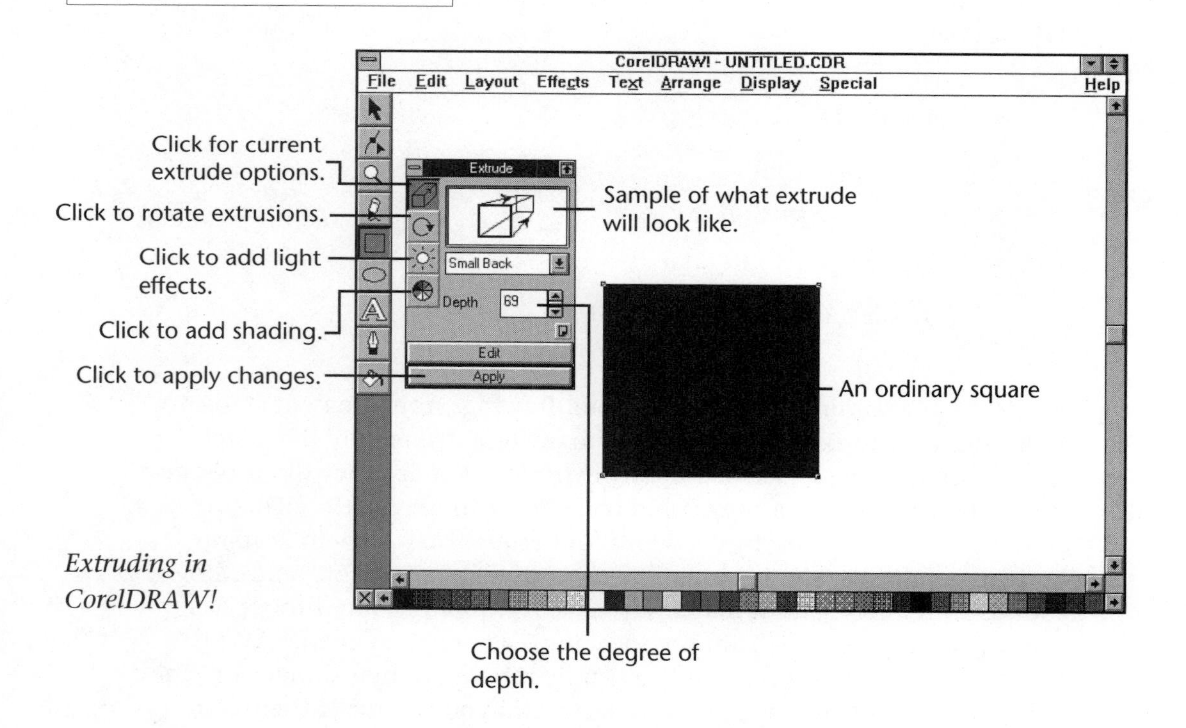

Extruding in CorelDRAW!

How Deep Do You Want to Go?

Perspective extrusion An unevenly shaped object with a 3D appearance.

Parallel extrusion An evenly shaped object with a 3D appearance.

The easiest and fastest way to learn how to extrude an object is to just do it. First, as always, select the object. Then select the options you want from the Extrude roll-up. You can choose from several different extrusion styles in the drop-down list below the sample extrusion in the roll-up.

There are two types of extrusions: parallel and perspective. *Parallel extrusions* have parallel extrusion lines (gee, that makes sense) that create an evenly dimensioned cubed look, while *perspective extrusions* have uneven extrusion lines, giving the appearance that the object is either smaller or larger in the front or back. Speaking of front or back, the extrusion styles in the drop-down list usually include "front" or "back" in the name: for instance, Small Back and Small Front. The terms small and back refer to the placement of the extrusion lines in relation to the object. The extrusion lines will appear in front of the object if you select a style with "front" in its name, and will appear in the back if you select a style with "back" in it. When you select an extrusion style, the sample picture in the roll-up shows you what the extrusion will look like.

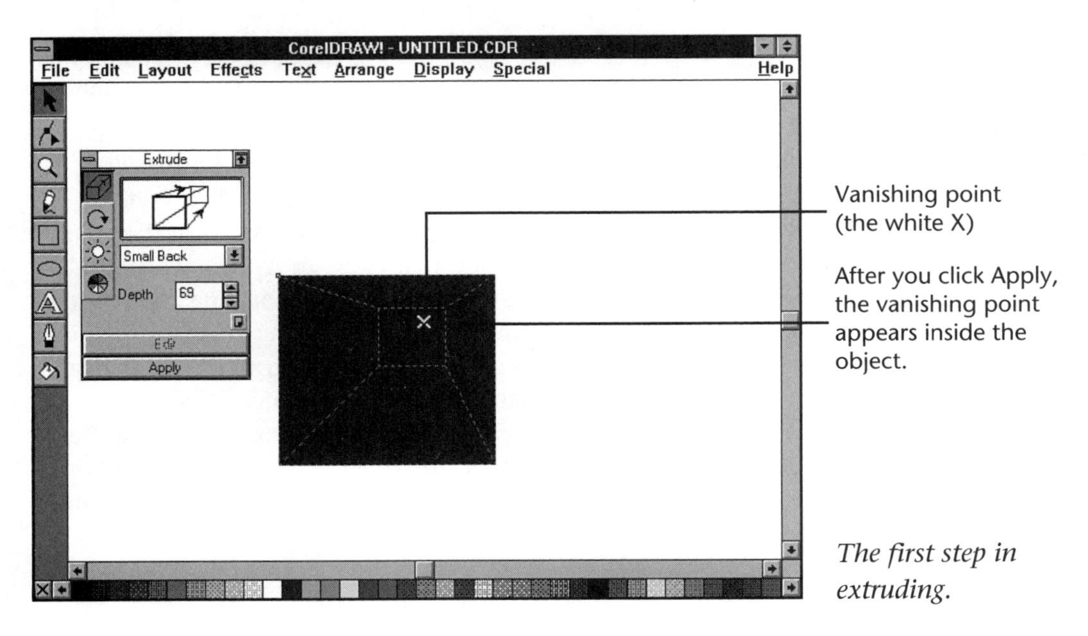

The first step in extruding.

For perspective extrusions, you can enter a number in the degree box (up to 99) that determines the extent of the extrusion. The higher the number, the deeper the object appears to be. However, you can also increase or decrease an object's depth by dragging its vanishing point, which appears after you click on the **Apply** button. In fact, why don't you do that now, so you can see what I mean.

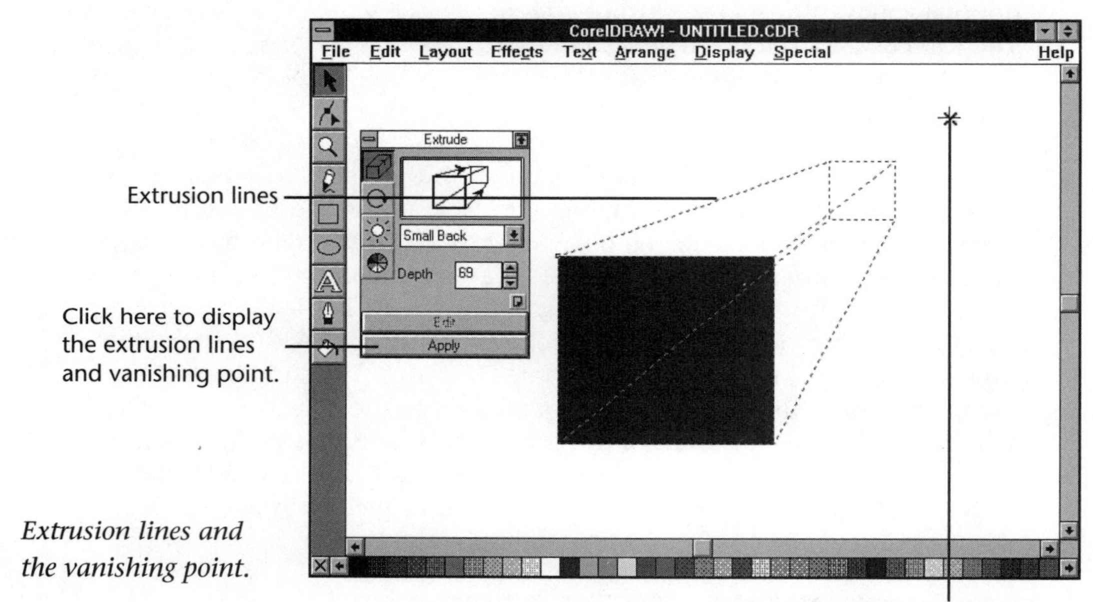

CorelDRAW! - UNTITLED.CDR
File Edit Layout Effects Text Arrange Display Special Help
Extrude
Small Back
Depth 69
Edit
Apply
Extrusion lines
Click here to display the extrusion lines and vanishing point.
Drag the vanishing point away from the object to add greater depth.

Extrusion lines and the vanishing point.

After you click on the Apply button the first time, extrusion lines (dotted lines) and a vanishing point (an "x") appear, according to the extrusion style you chose in the Extrude roll-up. If the extrusion lines are the correct size you want for your object, just click on **Apply** again, and CorelDRAW! fills up the blank space within the extrusion lines with the color of the object. Or, if the object is too deep or not deep enough, you can control the position of the extrusion lines by dragging the vanishing point to a new position. To add greater depth to the object, drag the vanishing point away from the object. To condense the object, drag the vanishing point in towards the center of the object. When the extrusion lines are where you want them, click on the **Apply** button again to complete the Extrude process.

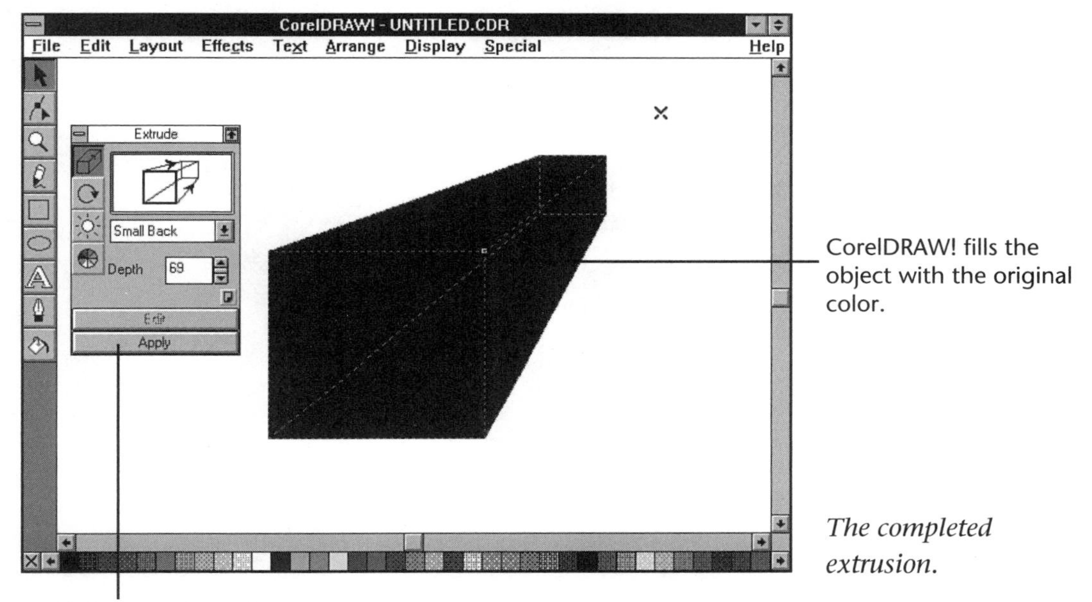

CorelDRAW! fills the object with the original color.

Click on Apply again to complete the extrude process.

The completed extrusion.

Extrusion Alert! A Shady-Looking Object!

You can have even cooler effects by applying shading to an extruded object. Click on the bottom icon on the left side of the Extrude roll-up to display the color options. You can either fill in the space created by the extrusion lines with the original color (which is the default), with a solid fill of a different color, or with a shaded fill which starts with one color and gradually changes to a different color. In the sample below, we chose a shaded fill with two different hues of blue. After you choose the color setting you want, click on **Apply** to change the extruded object's fill accordingly. The object will become filled with the new colors you choose, instead of the original color of the object.

Extrusion lines Dotted lines that appear when you're extruding an object. These lines show you what the dimensions of the extruded object will be.

Vanishing point The point to which the extended surfaces of prospective extrusions approach. The vanishing point is represented by an X. By dragging the vanishing point, you can change the depth of the extrusion.

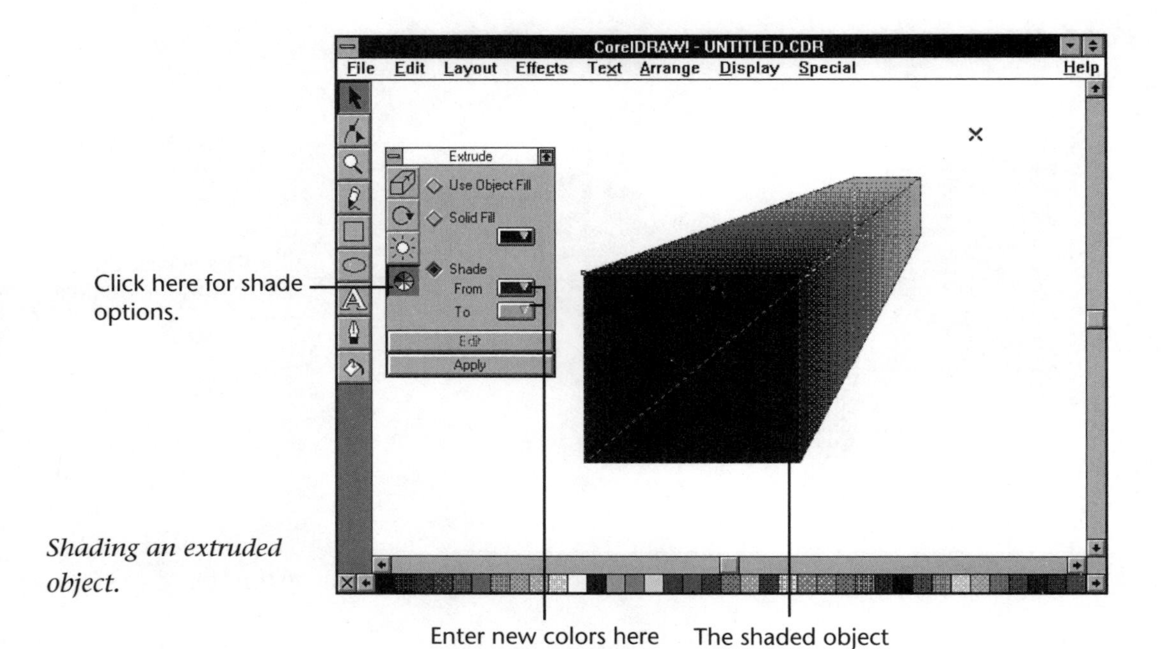

Shading an extruded object.

OOPS!

The rotation arrows in the Extrude roll-up are unavailable for parallel extrusions because parallel extrusions cannot be rotated.

A Tubular Extrusion

When extruding a circular object, you can use the rotating display in the Extrude roll-up to control the appearance of the extrusion. (To access the rotation options, click on the second icon from the top on the left side of the Extrude roll-up.) For example, draw an ordinary circle, and then click on the **Apply** button in the Extrude roll-up to show the extrusion lines and vanishing point.

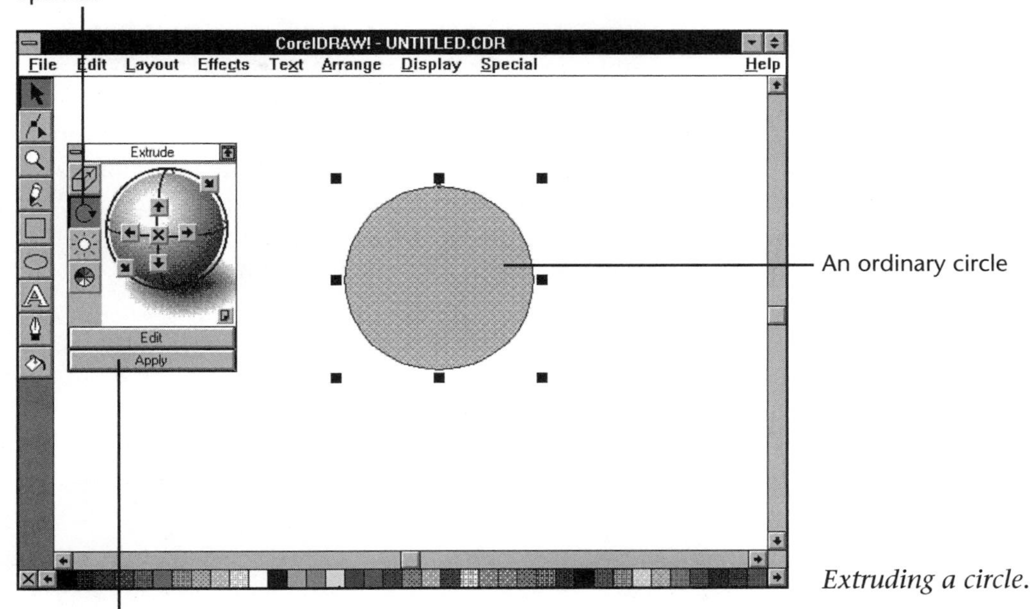

Extruding a circle.

Click on Apply to add extrusion lines and vanishing point.

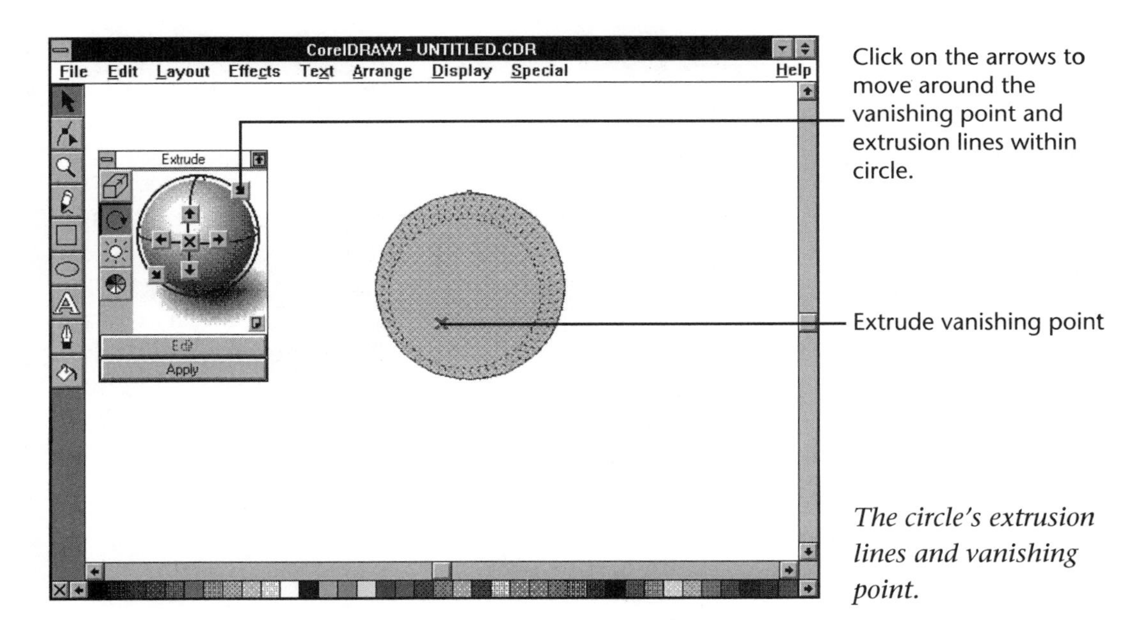

The circle's extrusion lines and vanishing point.

If you want to remove any rotation you may have created by clicking on the arrows in the Extrude roll-up, click on the "X" in the middle of the sphere. This removes all rotation and resets the extrusion lines and vanishing point to their original positions.

Clicking on the rotation arrows in the Extrude roll-up (on top of the sphere) moves the vanishing point and extrusion lines in the direction of the arrow. If you want to make the circle look like a sphere, use the arrow to get the extrusion lines where you want them within the circle, and then click on **Apply**. If you want to create a tubular effect, you have to drag the vanishing point away from the circle so that it sits anywhere outside the circle's parameter. Then click on **Apply**.

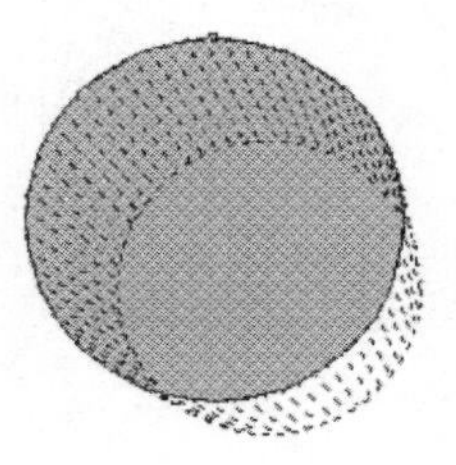

Drag the vanishing point outside the circle to create a tube.

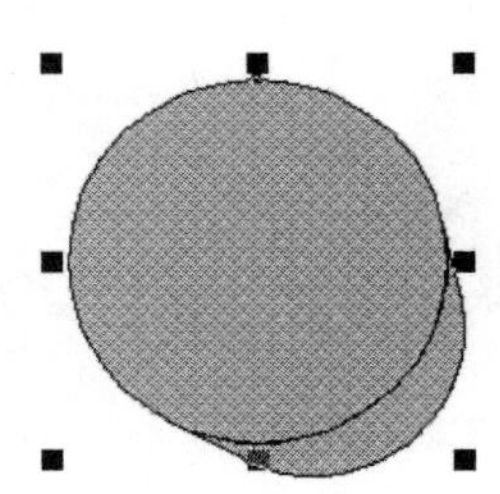

The circle is now a tube.

Come On Baby, Light My Fire

You can add a light source to your object, giving the appearance that the sun is shining on the object sitting on your computer screen. This type of light effect tends to heighten the 3D effect of your object, and can really change the object's look depending on the direction the light is coming from. To access the light source display options, click on the third icon on the left side of the Extrude roll-up (the icon that looks like a bright sun).

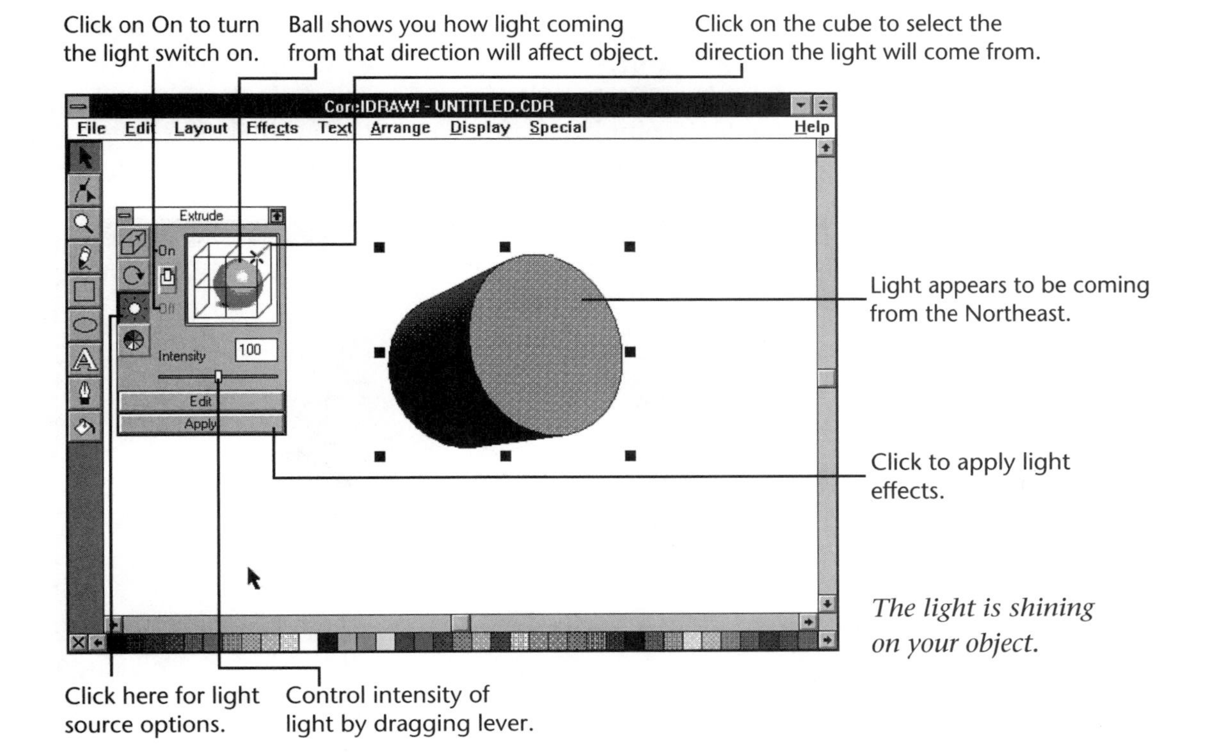

The light is shining on your object.

Before you can do anything with lighting, you have to turn on the light switch (just like it works at home, duh!). Click on **On**, and the light switch will appear to move up, enabling the light source options. If you want a brighter light, drag the lever below Intensity to the right; to dim the light source, drag the lever to the left. Click on **Apply** to apply the light effects.

What affects the light effects the most is the direction the light is coming from. The cube in the Extrude roll-up controls the light source direction. The ball inside the cube shows what effect the light source creates coming from the current direction. For example, if you click on the

Lower right corner of the cube, the sphere's look changes, with the light now originating in the southeast. When you apply this new light source direction to the extruded object by clicking on Apply again, the object's size and orientation look completely different than they did before, just because you changed the direction the light comes from.

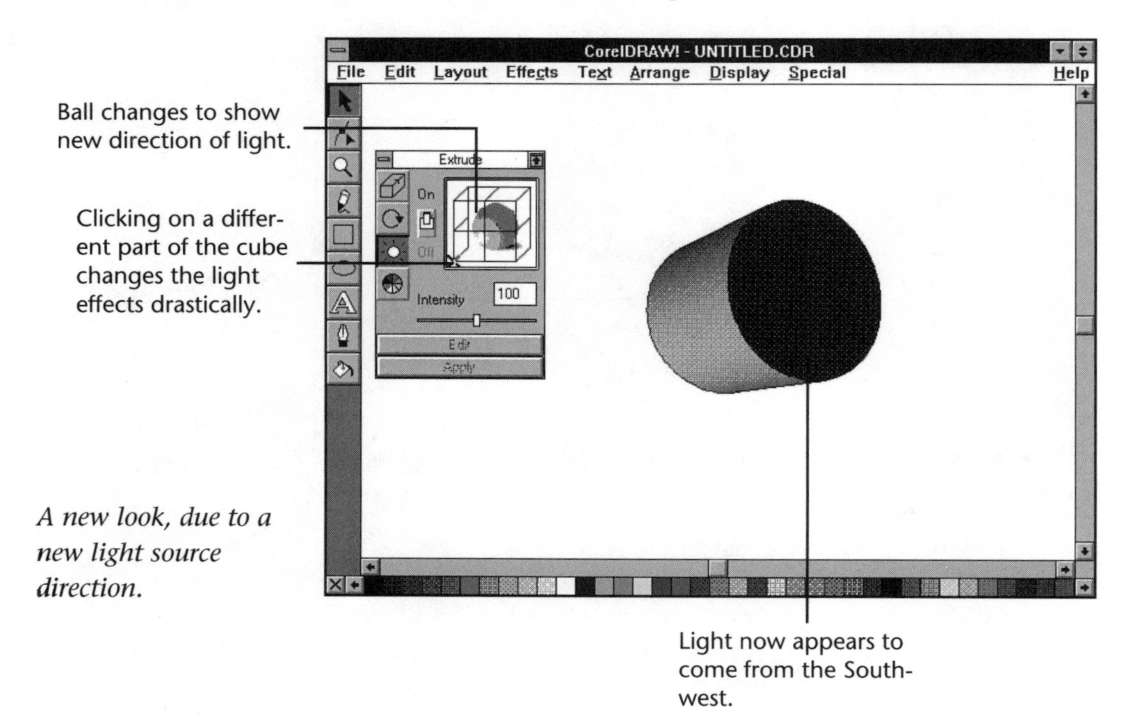
CorelDRAW! - UNTITLED.CDR
File Edit Layout Effects Text Arrange Display Special Help
Extrude
On
Off
Intensity
100
Edit
Apply
Ball changes to show new direction of light.
Clicking on a different part of the cube changes the light effects drastically.
Light now appears to come from the Southwest.

A new look, due to a new light source direction.

The Least You Need to Know

Give your objects depth by using the Extrude feature. (Put on some 3D glasses to reap the full benefit of the new look.)

- To display the Extrude roll-up, select the Extrude Roll-Up command from the Effects menu or press **Ctrl+E**.
- Select an object, select the settings you want in the Extrude roll-up, and then click on **Apply** to add extrusion lines and a vanishing point.

- If you want to increase or decrease the depth, drag the vanishing point to a new position. Either way, click on **Apply**.
- To add a shade to the object, click on the bottom icon in the left side of the roll-up, select the settings you want, and then click on **Apply**.
- To create a tube, start with a circle, and then drag the vanishing point off of the circle.
- To rotate an extruded object, use the rotation arrows on top of the sphere (select the second icon from the top on the left side of the roll-up for the rotation options).
- To add a light source to an extruded object, click on the third icon on the left side of the Extrude roll-up, select the light source direction you want, and then click on **Apply**.

What a waste it is to lose a page.

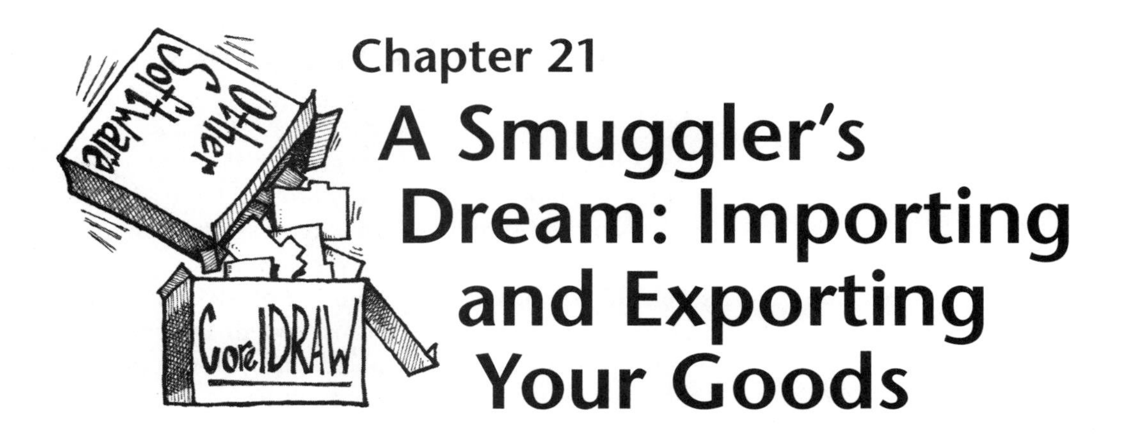

Chapter 21
A Smuggler's Dream: Importing and Exporting Your Goods

In This Chapter

- Importing a foreign file
- Exporting your files in other formats

What good would it do you to have a diamond that you could never take out of the box it came in, or that you could never give to another person? Some good, but not much. Along the same lines, what good would it do you to be able to create beautiful CorelDRAW! graphics if you could never use them outside of CorelDRAW!? For a minority of users, this wouldn't pose any problem since they don't use their CorelDRAW! graphics in other applications. But most people use CorelDRAW! primarily as a graphics program; they create drawings in CorelDRAW! for use in other documents they create in their word processor or desktop publisher, (such as Ami Pro, Word, or PageMaker). Also, users often want to use CorelDRAW! to edit or polish up graphics they've created in other, less powerful programs.

Fortunately, CorelDRAW! has extensive capabilities to exchange information between different file formats. You'll learn about CorelDRAW!'s importing and exporting processes in this chapter.

File format The format in which an application stores its software code.

File filter A code that can translate code formats from another program so that you can import files created in other programs into CorelDRAW!.

Import To open in CorelDRAW! a file that was created in a different program.

Snagging Other People's Stuff

Get ready for lots of new terms in this chapter (sorry, but there's really no way around it). Each graphics program has its own way of storing its software code. This is called a *file format*. CorelDRAW! comes with a number of file filters that translate other applications' code into code CorelDRAW! can understand, so that you can open a file created in a different program in CorelDRAW!. This process is called *importing*. You can import graphics or text from the most popular programs into CorelDRAW!, using the Import dialog box. To open this dialog box, select the **Import** command from the **File** menu. (This dialog box may look familiar to you; it's similar to CorelDRAW!'s Open dialog box.)

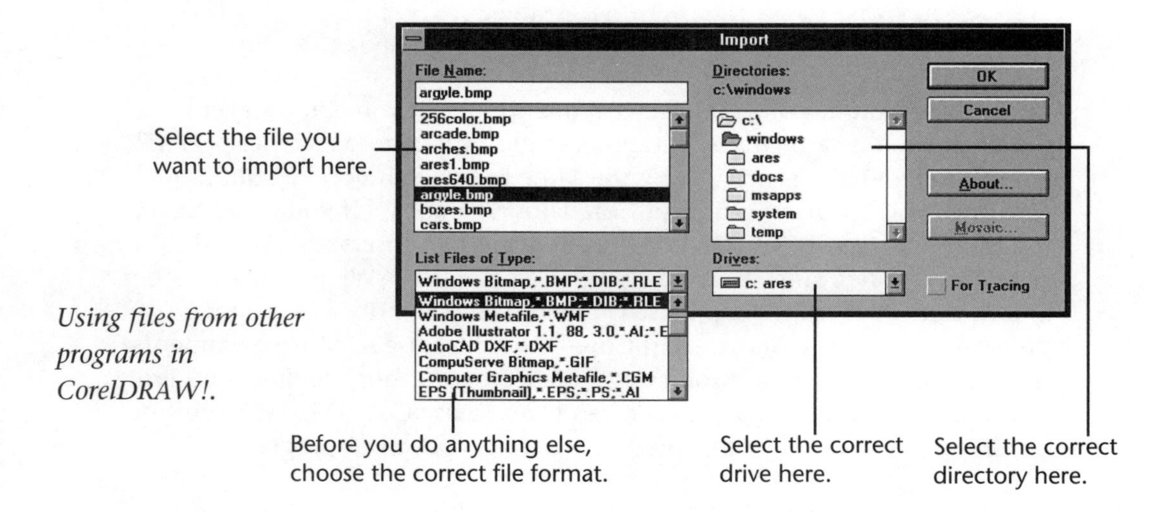

Using files from other programs in CorelDRAW!.

To import a file into CorelDRAW!, you need to know two things: the file format (which is detectable by the last three letters in the file name), and the path of the file (in other words, the directory in which the file is located either on your hard drive or on a diskette). Knowing the format and the path, select the appropriate file format in the List Files of **T**ype box in the Import dialog box. To do this, click on the scroll arrow to display a drop-down list of file types, and then scroll through the list until you find the correct file format. Click on that file format.

You may need to open a file that CorelDRAW! does not have a filter for. If so, try opening the file in the original program and then copying the graphic or text to the Windows Clipboard. Then switch to CorelDRAW! and select **E**dit **P**aste. Or, you can try saving the file under a different format (one that CorelDRAW! supports) in the original application, and then choosing that format in CorelDRAW!'s Import dialog box.

If you don't see many or any file filters in the List Files of **T**ype box, you might not have installed all or any of CorelDRAW's filters when you installed CorelDRAW!. During installation, CorelDRAW! lets you choose which file filters you want to install. You may need to reinstall CorelDRAW! if you didn't install the filters originally and you now need to import files from other applications. Read Chapter 28, "Installing CorelDRAW!," for installation guidelines.

After you've selected the file type, choose the drive and directory where the file is located. Find your file in the list of files at the left side of the dialog box. Either double-click on the file, or click on it once and click on **OK**. Alternatively, you can type the name of the file in the File **N**ame box. CorelDRAW! closes the dialog box and imports the file. You may have to wait a few moments while CorelDRAW! converts the program.

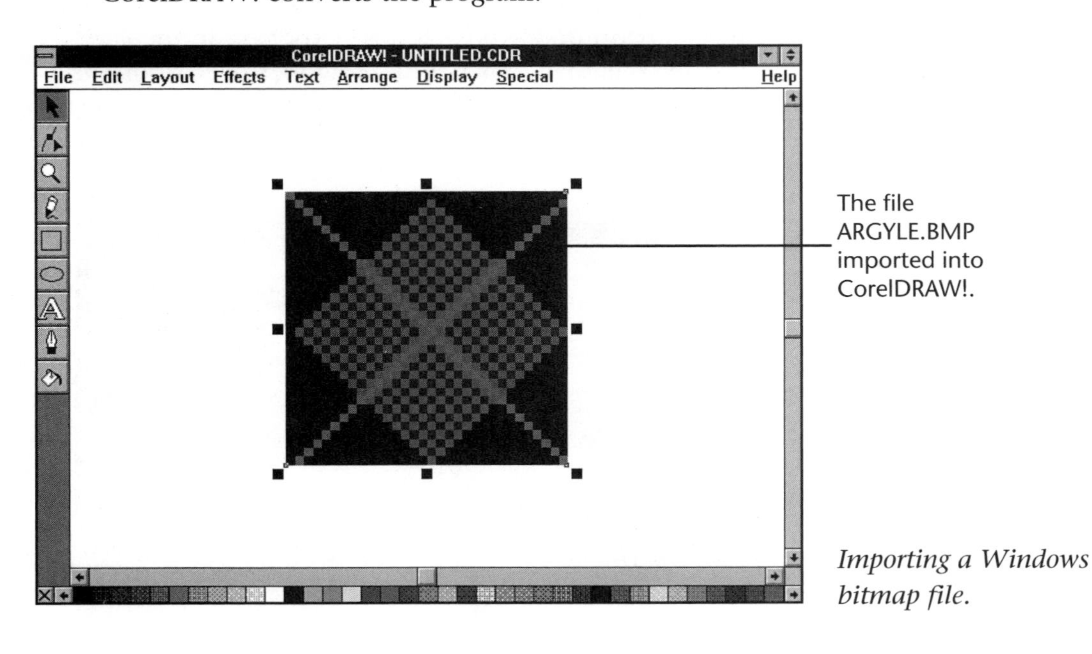

Importing a Windows bitmap file.

The Robbers Want to Take Our Stuff Now!

If you can import, then you should be able to export too, right? Well, right or wrong, you can in CorelDRAW!. *Exporting* is just the opposite of importing: instead of opening a file from another program in CorelDRAW!, you save your CorelDRAW! file in a different format so that you can open the file in a different application.

Few programs have as many import filters as CorelDRAW! does; and of those, few support CorelDRAW!'s .CDR format. Perhaps the designers of the other programs were aware that you could save CorelDRAW! files in their own file formats—so why would they need an import filter for CorelDRAW!'s format? For example, if you wanted to use a graphic you created in CorelDRAW! in Microsoft Word for Windows, you couldn't import the CorelDRAW! file in Word as it is, because Word doesn't have a CorelDRAW! file filter. What you would do instead is save the file in a .WMF or .BMP format in CorelDRAW! first. Then you could import it into Word since Word supports those types of file formats.

You must access the Export dialog box to export your drawings to other programs. To open the Export dialog box, choose Export from the File menu. The Export dialog box may look familiar to you, since it looks similar to the Save dialog box. In fact, the main difference between the two is that the Export dialog box has lots of file formats in the List Files of Type box, unlike the Save dialog box.

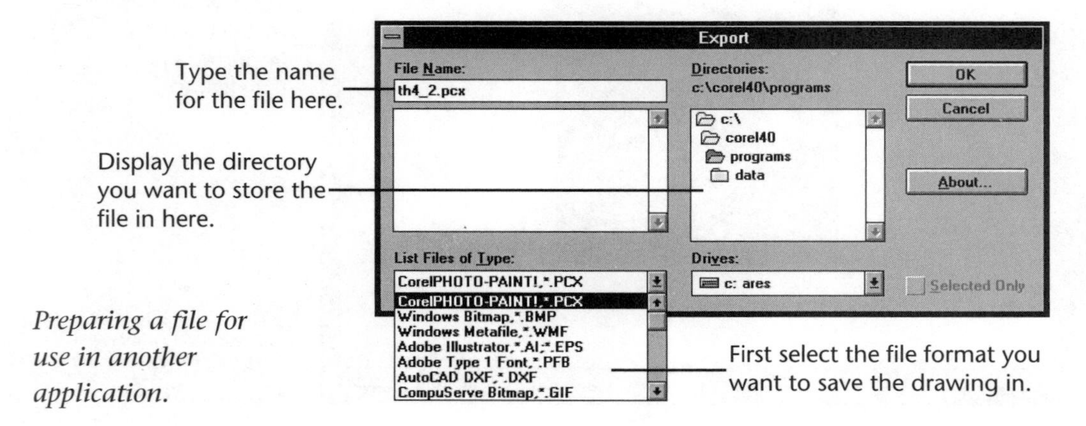

Preparing a file for use in another application.

Before you export a drawing, you really need to know the type of format you want to convert it to. To find out what file formats another application will import, either read the application's user documentation, or open the application's Open or Import dialog box, and check to see which file formats are listed. But the inspection doesn't end there. Now you have to make sure CorelDRAW! will export in the file format you need. Check the List Files of Type drop-down list in the Export dialog box in CorelDRAW! to see if the file format you need is included there. If it's on the list, you're safe. If not, go back to the other program and see if there's a third party format you can use (one that's available in both programs).

Export To save a CorelDRAW! file in a different file format so that you can use the file in another application.

Just like when you import, if you want to export but don't see many or any filters in the List Files of Type drop-down list, you did not install all of the filters during installation. You'll need to reinstall CorelDRAW! to load those filters. Read Chapter 28, "Installing CorelDRAW!" for help.

After you've selected the file format in the List Files of Type box, the rest is easy. Just select the drive and directory where you want to store the exported file, and then type a name for the file. Or, if you want to leave the name the same, just click on **OK**. CorelDRAW! will automatically add the correct extension (last three letters) to the file name.

E-Z

If you want to edit a CorelDRAW! drawing in CorelPHOTO-PAINT!, choose the .PCX format in the Export dialog box.

The Least You Need to Know

In this day and age of worldwide data exchange, CorelDRAW! fits right in. By using CorelDRAW!'s Import and Export features, you can exchange files between CorelDRAW! and most other software programs.

- To import a file, use the Import dialog box. Select the **I**mport command on the **F**ile menu to open the dialog box.
- Select the format of the file you want to open in the List Files of **T**ype drop-down list in the Import dialog box.
- To export a file, use the Export dialog box. Select the **E**xport command on the **F**ile menu to open the dialog box.
- If you don't already know what file format to export a file in, check the application's Open or Import dialog box to see what formats it supports, and then select one of those formats from the List Files of **T**ype drop-down list in the Export dialog box.

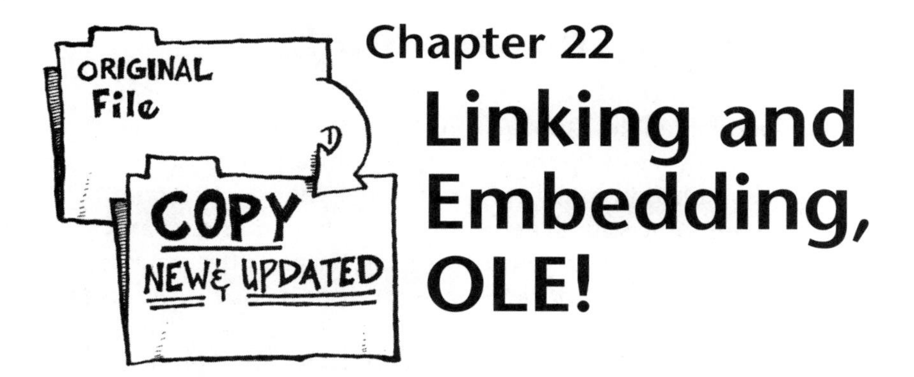

Chapter 22
Linking and Embedding, OLE!

In This Chapter

- Of links and beds
- The proud owner of a link
- Bullying your links
- Open CorelDRAW!, insert file

Normally when you see the expression "ole!," you think "yeah!," "bravo!," or "all right, dude!," right? For some reason, I think of frozen strawberry margaritas. Not so lucky with CorelDRAW!. Though you may jump for joy over CorelDRAW!'s ability to link and embed objects, "OLE" does not stand for an exhilarating feeling in CorelDRAW!. It stands for Object Linking and Embedding—true techno nerd stuff, I know. Here's a description in layman's terms (ole for the layman, huh?): You have an object . . . the object is from another software program . . . you stick the object inside your CorelDRAW! drawing—this is *embedding*. Or, you have an object . . . the object is from another application . . . you connect the object to your CorelDRAW! drawing—this is *linking*. Anytime the first object changes, the linked object in CorelDRAW! automatically changes accordingly. For a more sophisticated discussion, read the rest of this chapter.

Tell Us More About This OLE Stuff

Don't make the mistake of thinking OLE is some kind of techno monster that you'll never understand, let alone use. Just because the term Object Linking and Embedding sounds foreboding, doesn't mean the process is. Actually, linking and embedding objects is not only easy to understand, it's easy to do. Here are some simple definitions to help you understand the process better. We stay away from the technical terms in the process descriptions so you can understand the process without having new words thrown at you. But, just to cover all paths, we also define some of those techy terms you should be familiar with if you plan to link and embed objects.

Linking	You can link an object created in another application to your CorelDRAW! drawing. When you link an object, you insert a clone of the original object into CorelDRAW!. Whenever you change the original object (whether it is in CorelDRAW! or another program), the cloned object in CorelDRAW! changes accordingly. You can link an object to as many CorelDRAW! drawings as you want.
Embedding	When you embed an object created in another application into your CorelDRAW! drawing, you place a copy of the object inside the drawing. Changes you make to the embedded object in CorelDRAW! do not change the original object (unlike linked objects). When you edit an embedded object, the original application starts up so you can edit the object. For example, if you wanted to change an embedded picture from CorelPHOTO-PAINT!, CorelPHOTO-PAINT! would start so you could edit the embedded object. Basically all embedding does is give you an easy way to edit an object you created in another application, using that application. (By the way, you can embed from CorelDRAW! into another program as well.)

The difference between linking and embedding	It's this simple: when you change a linked object in CorelDRAW!, it changes the original object in the other application. When you change an embedded object in CorelDRAW!, nothing happens to the original object.
The difference between using the Copy and Paste commands, and linking and embedding	When you copy an object to the Windows Clipboard, and then paste it into a different application, there are no ties between the original and copy. Changes to the original will not be reflected in the copy, as they are with linked objects. Also, you cannot edit the copy in its original application by double-clicking on the object, as you can with embedded objects.
Object	Any piece of information created in a Windows, OLE-capable application. An object can be a single word, an area of a picture, or an entire file.
Source application (the server)	The application in which you created the object.
Destination application (the client)	The application you're linking the object to or embedding the object in (in this chapter, the destination application is always CorelDRAW!).
Server	A Windows application that you can link or embed from. Some CorelDRAW! applications are both clients and servers; others are only clients or servers.
Client	A Windows application that you can link or embed to. Some CorelDRAW! applications are both clients and servers; others are only clients or servers.

Enough discussion—let's get on to the process. We'll start with linking.

A Linky-Dink Operation

Let's say you've imported a scanned photograph into CorelPHOTO-PAINT!, and you want to use it in several of your CorelDRAW! drawings. But you plan to make changes in the future to the photo image. If you used the Copy and Paste commands, you'd have to edit each copy of the image in each of the drawings as well as the original image in PHOTO-PAINT. On the other hand, if you link the image to the CorelDRAW! drawings, all you'd have to do is edit the original image in PHOTO-PAINT, and all the linked copies in CorelDRAW! would change also.

We'll be nice and give you some orderly steps to follow that you can refer back to later whenever you want to link an object in CorelDRAW!. We'll show you how to link part of a CorelPHOTO-PAINT! image to CorelDRAW!, but you can use similar steps to link from other applications as well. One thing to always keep in mind when linking: you can only link from a file that has been saved. So if you want to link a portion or all of the file, save and name the file first!

To link part of a CorelPHOTO-PAINT! image to your CorelDRAW! drawing:

1. If CorelDRAW! is running, reduce it to an icon by clicking the **Minimize** button in the upper right corner of the CorelDRAW! window (the single down arrow).
2. Start CorelPHOTO-PAINT!.
3. Open the file containing the object you want to link.
4. Select the area of the image you want to link to CorelDRAW!. (Drag over the area with the Rectangle Selection tool.)
5. Copy the object into the Clipboard using the **C**opy **E**dit command.

To link an entire file, use the **E**dit Insert Ob**j**ect command instead. Click on the **Create from File** option, mark the **Link** check box, and specify the name of the file.

First open the image you want to link.

6. Switch to CorelDRAW!. To do so, press **Ctrl+Esc** to display the Windows Task list. Then double-click on the CorelDRAW! task.

Switching back to CorelDRAW!.

7. Select the Paste **S**pecial command from the CorelDRAW! **E**dit menu. The Paste Special dialog box appears.

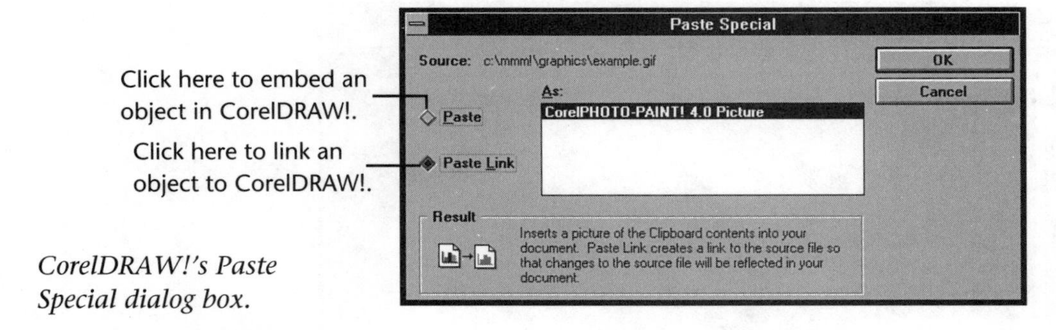

CorelDRAW!'s Paste Special dialog box.

You can also embed an object using the Paste Special dialog box. Instead of choosing Paste Link, choose the **P**aste option in the dialog box. Remember that if you embed the object, it won't change when you edit the original in PHOTO-PAINT later.

8. Click on the Paste **L**ink option.
9. Click on **OK**. CorelDRAW! places a copy of the object inside the drawing. From now on, whenever you change the original object in PHOTO-PAINT, the copy of that object in CorelDRAW! will change also because it is linked to the original (unless you specify manual updating, which you'll read about in the next section).

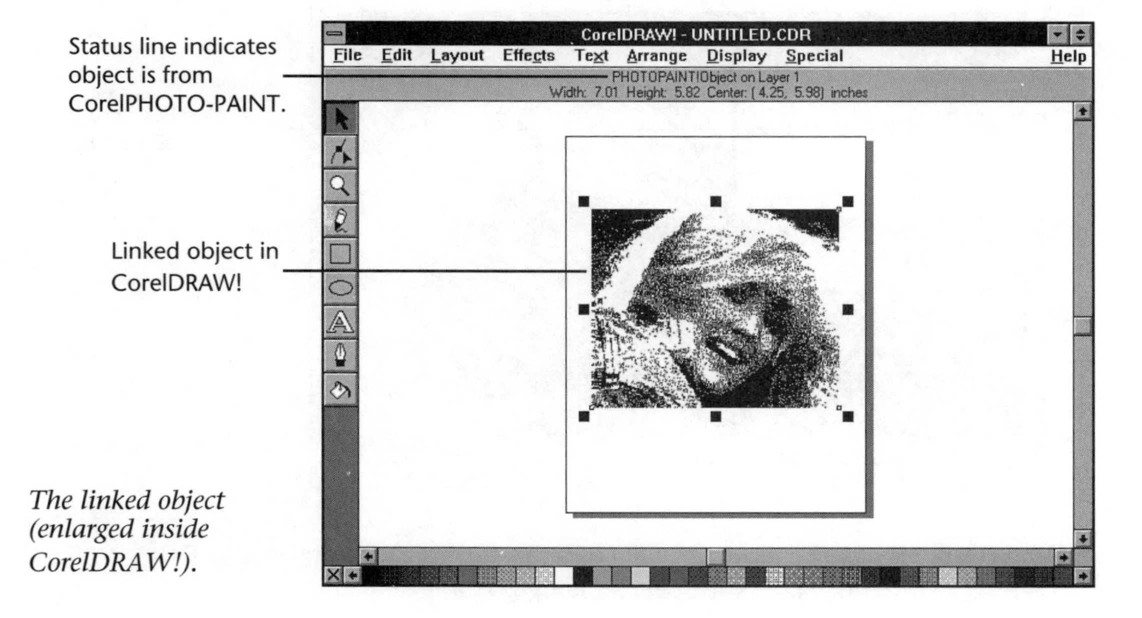

The linked object (enlarged inside CorelDRAW!).

Stop! In the Name of Links

Sometimes you link an object and then leave it alone. However, the day will probably come when you want to tell CorelDRAW! not to update the object just yet (yes, you can do this); when you want to sever the link altogether (but not delete the object) or to replace the linked object with another (Sound like we're talking about marriage here?)

You can do all these things in the Links dialog box. To get your hands on this dialog box, select the Links command from the Edit menu. (This command will not be available unless the current drawing contains linked objects.) If your drawing has multiple links, select the link you want to modify by clicking on it in the Links box. The next few sections explain the options in this dialog box in detail.

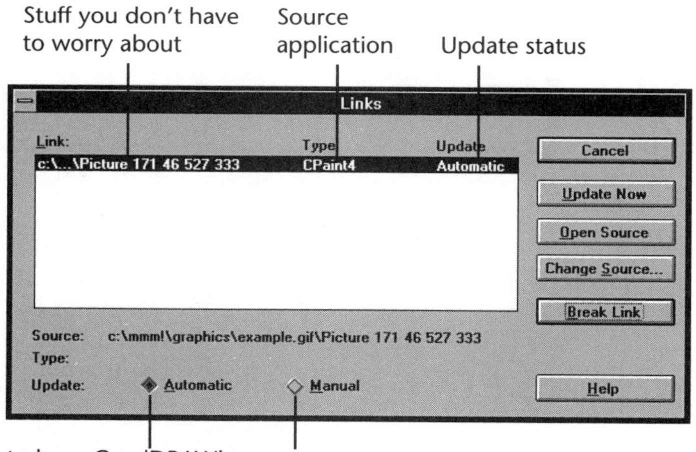

Manage your links in the Links dialog box.

No Outdated Links Here

To make it easier for you to understand things, we left out an important point about linking earlier. We'll tell you about it here, now that you're a little more familiar with linking, and you've actually created a link. Here it is: linked objects don't have to change whenever the original object changes. You get to decide. If you want the linked object to change automatically, leave well enough alone, because by default CorelDRAW! will automatically update the linked object. But, let's say that you don't want the linked object to change just now. You'll want to update it later, but not at the moment. Well, CorelDRAW! lets you do this (it's borderline-sick how this program caters to your every need, isn't it?).

The Update category in the Links dialog box contains two options: **A**utomatic and **M**anual. Look at the Update column in the Link box to see what the current Update status is for the linked object. When the **A**utomatic update option is selected, it means that each time you open the drawing containing the linked object, the linked object will change if any changes have been made to the original application since the last time you worked on the linked object. If you want to have more control over when CorelDRAW! updates the linked object, mark the **M**anual box. Now the linked object will only be updated when you click on the **U**pdate Now button in the Links dialog box.

Cutting the Umbilical Cord

If you're positive you'll never want to update the linked object again, go ahead and break all ties to the original object by clicking on the **B**reak Link button in the Links dialog box. This does not delete the linked object—it deletes the link. You'll still see the linked object in your drawing, but it can no longer be updated to reflect the current state of the original object, because all connections have been severed.

To delete the linked object itself (instead of just breaking the link), select the linked object and press the **Delete** key.

If there's any chance at all that you'll want to update the linked object at a later time, choose the **M**anual update option instead of actually breaking the link. That way, if you need the updated linked object later, it's a lot easier to tell CorelDRAW! to update a linked object than is it to relink the object. Read the previous section if you choose to update manually.

Don't Trust That Source

If you want to replace one linked object with another, you can do so without having to go through the process of deleting the first linked object and linking the new one. Instead, you can just change the source specifications in the Link dialog box. To do so, click on the Change Source button. An additional dialog box opens on top of the Links dialog box. Keep in mind that you can only replace a linked object with a complete file (not a portion of it).

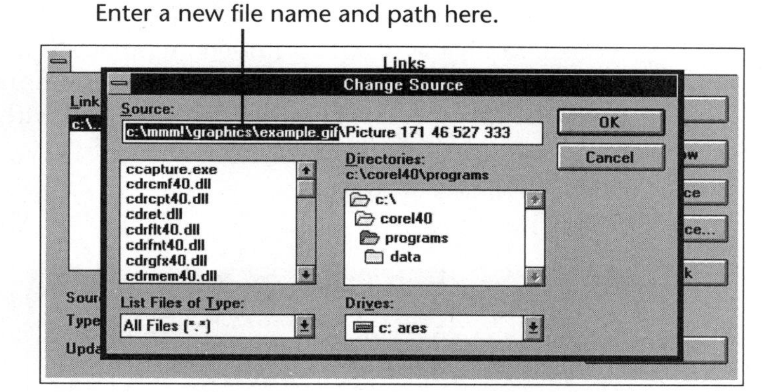

Replacing a linked object with a new one.

When the Change Source dialog box opens, the path and file name of the currently linked object are highlighted. If you already know the file name of the new object you want to link, as well as the directory it's located in, just type the file name and path (directory information). CorelDRAW! will enter this information in the Source text box. If you're not sure about the name of the file you want to link or where it's located, use the Drives and Directories boxes to find the file.

Stick It There

Now let's turn the tables. You have an object in CorelPHOTO-PAINT! that you'd like to use in your CorelDRAW! drawing, but you're going to want to make changes to the object once it's in CorelDRAW! without having those changes affect the original object in PHOTO-PAINT. If this is what you want, embed the object instead of linking it. This way you can edit the embedded object using PHOTO-PAINT tools, but from within CorelDRAW!. How so? When you double-click on the embedded object,

Here's another scenario. You intend to include an object in your CorelDRAW! drawing that you need PHOTO-PAINT tools to create. CorelDRAW! can do this (as always). You can use the Insert **Ob**ject command on the **E**dit menu to embed an application in your CorelDRAW! drawing—even if you haven't created

CorelPHOTO-PAINT! starts up. Now you can edit the object with CorelPHOTO-PAINT!'s tools. You can't do this when you simply copy and paste an object from PHOTO-PAINT to CorelDRAW!. Hint to all of you that don't get it: this is why embedding is beneficial.

When you choose the File **Insert Object** command, the Insert Object dialog box appears. By default, the Create **N**ew option is selected, which is the option you want if you haven't created the object yet. If you want to embed an object that already exists, select the Create from **F**ile command instead, and then specify the file name and path.

Click here to embed an object you haven't created yet.

Click here to embed an existing file.

Click on the application you want to embed.

Embedding an application in CorelDRAW!.

The Object **T**ype box will display the names of all the applications on your hard drive that are OLE servers. Remember, an OLE server is a Windows application that you can embed objects from. Almost all of the popular word processing, desktop publishing, presentation and spreadsheet applications are OLE servers; so if you have Word or WordPerfect for Windows, Excel for Windows, or PowerPoint loaded on your system, you'll see their names here. Additionally, you'll find many of the CorelDRAW! applications in the Object **T**ype box.

Just click on the application you want to use to create the object in CorelDRAW!. Then click on **OK**, and the application you specified will open. Create the object you want to embed. When you finish, display the **F**ile menu in the source application (i.e., not CorelDRAW!'s menu). Different software programs have different commands that will close the application and return you to CorelDRAW!. You'll either see an Exit and Return, Update, or Update and Return command at the bottom of the **F**ile menu (if all else fails, use the Exit command). Select this command to close the application, update the embedded object, and return you to CorelDRAW!.

From now on just double-click on the embedded object when you want to edit the object. (There's also a command on CorelDRAW!'s **E**dit menu to open the embedded object.) Either way, the application starts. Edit the object as desired, and then select the **F**ile Update (or whatever . . .) command to return to CorelDRAW!.

The Least You Need to Know

Linking and embedding are two great ways to reap the benefits of other programs for your drawings in CorelDRAW!.

- Link if you want the linked object in CorelDRAW! to change when you change the original object.
- Embed if you want to create an object using another application's tools.
- You can link an entire file or only portion of a file to CorelDRAW!.
- Use the Paste **S**pecial command on the **E**dit menu to link an object.
- Use the Insert O**b**ject command on the **E**dit menu to embed an object in CorelDRAW!.

It's ok to write on this page—you paid for it.

Part V

Less But Not Least: The Secondary Applications

I've always wondered what it would have been like to grow up as an only child. Instead, I grew up with three older sisters and two older brothers, a younger half-sister (I've always hated that term—what, is she cut in half?), and three older step-brothers. Like I said, I always wondered, because there was no way in this world I could ever come close to imagining it. Somehow though, I always felt I wouldn't be as complete without having my siblings around as an intrinsic part of my life.

CorelDRAW! has lots of brothers and sisters too. In fact, DRAW has a total of seven siblings: four older sisters and three younger brothers. We concentrate on the sisters in this Part because they are full-grown, full-featured applications that could stand on their own. In addition to CorelDRAW!, you get a presentation program, an animation program, a charting program, and a painting program. Somehow CorelDRAW! just wouldn't be the same without these sister applications.

Chapter 23
Charting Your Way to the Top

In This Chapter

- Good morning, CorelCHART!
- The proud owner of a baby chart
- Getting a grip on chart elements
- Your perspective needs work
- Strolling through the art gallery
- Would you look at the size of that!

Here's a little tip on how to get that raise you've been clamoring for: CorelCHART!. The next time your boss asks you for a report on anything that involves numbers, run to your computer, start CorelCHART!, and take about fifteen minutes to produce a chart that your boss will think took you a few days to develop. (Unless, of course, your boss is privy to CorelCHART!'s magical capabilities. Then you're out of luck.) Because CHART has done most of the work for you by designing sophisticated, beautiful data chart templates, all you have to do is plug in some numbers—and then take ownership of a true work of art. Don't be shy about taking the credit for your work (little that it is), the Corel team would want it that way.

Okay, from the Top

With CorelCHART! you can create spectacular-looking data charts in a matter of minutes. In case you need some brushing up on terms, a *data chart* is a graph that maps the progress, value, or cost of an item or items over time (usually). If we're talking gibberish to you here, flip the page and take a look at some of the figures up ahead; a picture of a chart can tell you what would take us, well, at least a couple hundred words.

Chart Shows the progress, cost, or value of items over time.

Chart styles Various formats that determine the layout of graphed data (for example, pie, bar, line, area, etc.). A chart style comes in several different types.

Chart types Variations of a chart style. There can be several different chart types associated with a chart style.

Speaking of words, we're certainly not going to repeat ourselves by telling you how to start and exit CorelCHART!, because Chapter 3, "Taking Your First Step," tells it all. Wherever you see the word "CorelDRAW!" in that chapter, replace it with the word "CorelCHART!," and you'll do just fine. And do the same for the first two sections in Chapter 4, "Juggling Drawings," if you need help on saving or opening files in CorelCHART!.

Ride of Your Life

We could start this chapter off with a lot of abstract information about charts, templates, numbers, rows, columns, titles, footnotes, labels, axes, and more. But the best and fastest way you can learn the basics about CorelCHART! is to go through the process of creating a chart from start to finish. So fasten your seat belt and hold on tight to the steering wheel because you're going to experience a lot of fast moves and curves in the road ahead. (At least on this ride you don't have to worry about falling asleep at the wheel.)

1. Start CorelCHART!. Whenever you start CorelCHART!, almost everything on the screen is dimmed. All you can access is the File menu.

2. Choose **F**ile **N**ew. (Later on down the road when you want to open a chart you've already created, choose **O**pen instead.) Lo and behold, your first CorelCHART! dialog box appears.

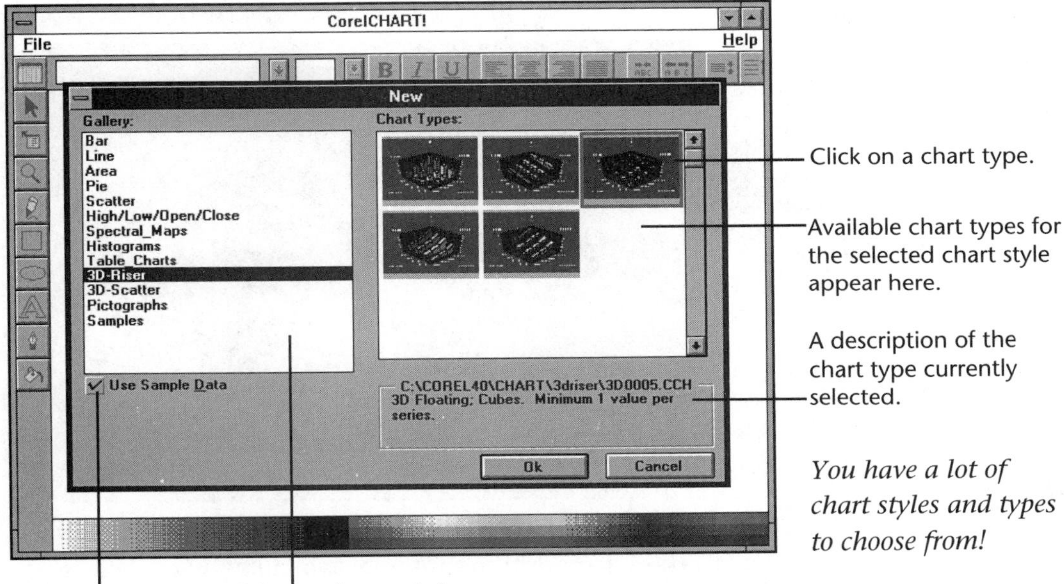

You have a lot of chart styles and types to choose from!

3. Select a chart style in the Gallery box. (For this demo, select **3D-Riser**.) When you click on a style, the available designs, called *types*, for that style appear in the right side of the dialog box.

4. Click on a chart type in the Chart Types box. (Click on the third type, for our example.) A description of the chart type you chose appears at the bottom of the dialog box.

> **3.0**
>
> By default, the Use Sample **D**ata check box is not marked. Make sure you mark this box before clicking on **OK**, or you'll be lost when you try to follow the rest of the steps!

5. If you're ready to go, click on **OK**. You might have to sit and wait for CHART to open the new file. But it's worth the wait.

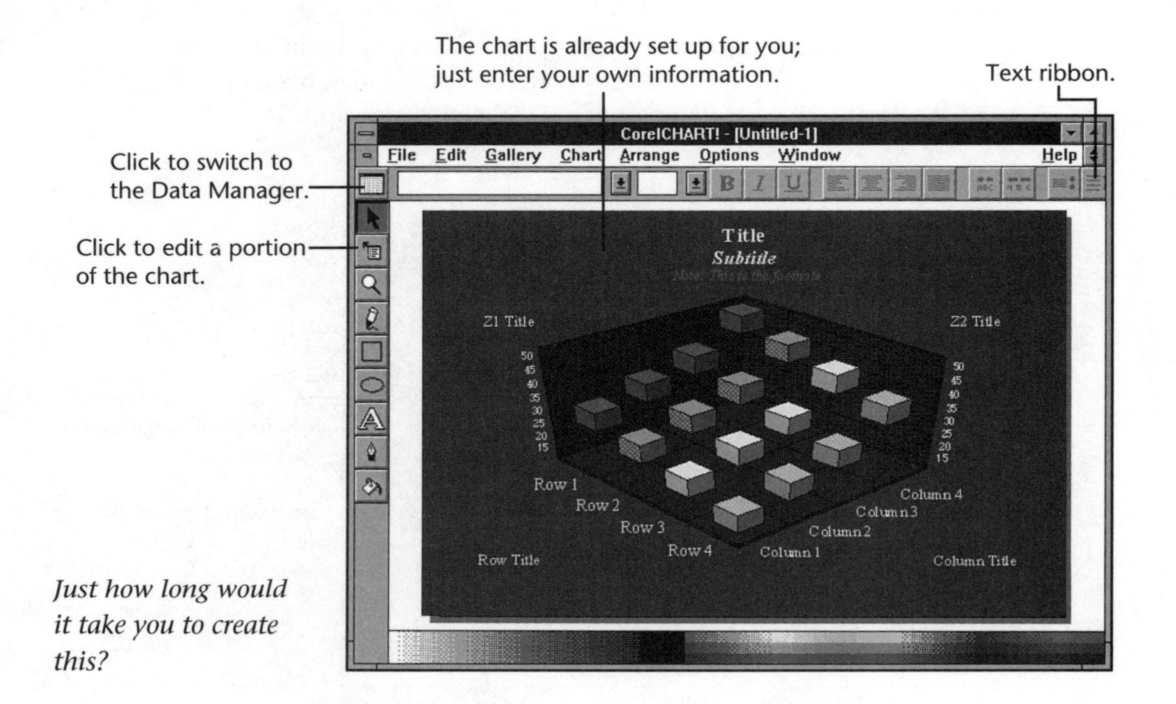

Just how long would it take you to create this?

Notice that all of the screen options are now available to you. The design sure looks similar to CorelDRAW!, doesn't it? I think the Corel designers planned it that way—but I could be wrong (not). Look at the toolbox: virtually the same tools as DRAW's toolbox, with one exception. The third tool lets you edit your chart quickly. Just click on the tool, and then click on the portion of the chart you want to modify. The Text ribbon at the top of the screen below the menu bar is new. You can use the buttons and drop-down lists on this bar to quickly format selected text. (Select text the same way you select it in CorelDRAW!.)

You've probably figured this out by now, but the words you see in the chart serve as placeholders for text you will enter. All the hard work of designing the chart, creating graphics, and selecting fonts is done, thanks to CorelCHART!. Now all you have to do is enter your own title and subtitle, column and row labels, other titles, and most importantly, values. There's a special window where you can enter all of this information, and it's called the *Data Manager*.

5.0

The CorelCHART! 5 window is similar to the CorelDRAW! 5 window. The new window has a tear-off toolbar and a Ribbon bar full of commands you can access via icons.

Switch to the Data Manager by clicking on the **Chart View** icon at the top left of the window (just above the Pick tool). This icon works as a toggle between the two views (Data Manager and Chart View). Data Manager, you say again? What exactly is the Data Manager? Hold your horses, and take a look at the next picture. That's the Data Manager. If you want a written description, the Data Manager is a spreadsheet program that holds all the data to be included in your chart. It works like 1-2-3 and Excel, yet on a less grandiose scale (much less, quite frankly).

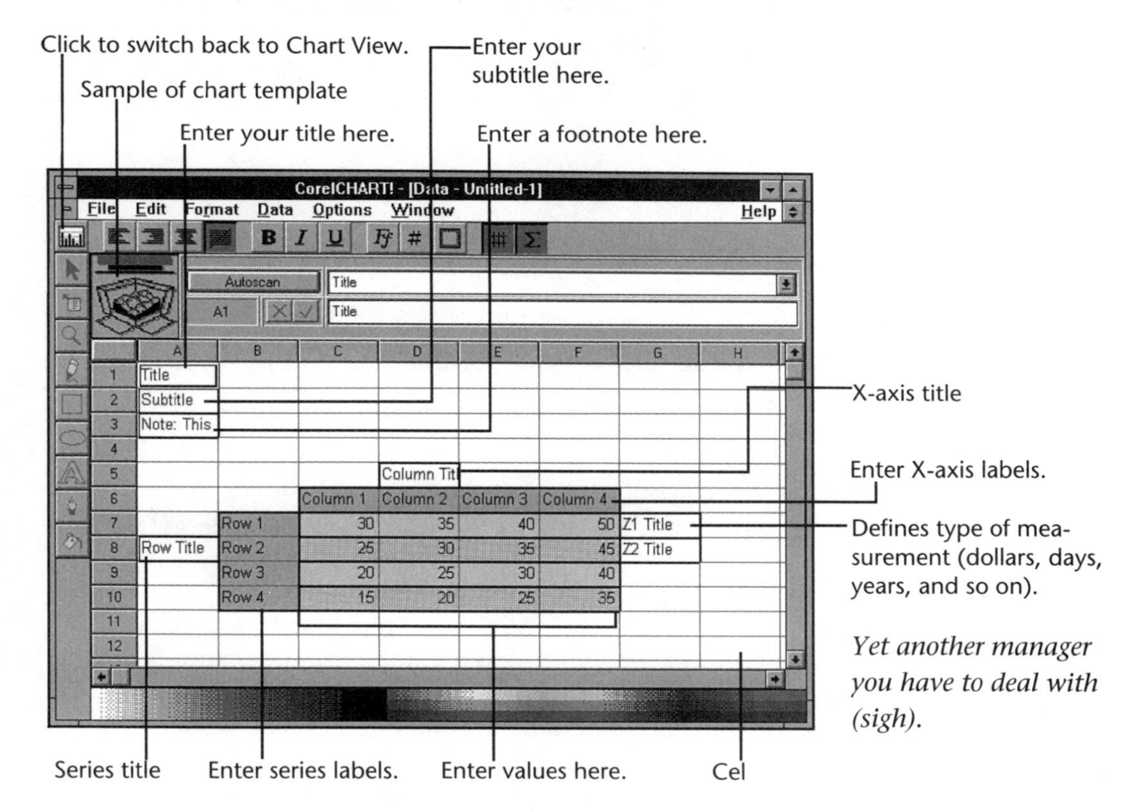

Yet another manager you have to deal with (sigh).

Replace the template words and numbers with your own by clicking on a cell and typing the word or number. (For this example, refer to the next two figures and enter the words and numbers you see there.) If you're new to charting, just enter the titles and numbers you see in the two figures below, and don't worry about words like "series" and "z-axis." We'll tell you about these and other charting elements in the next section.

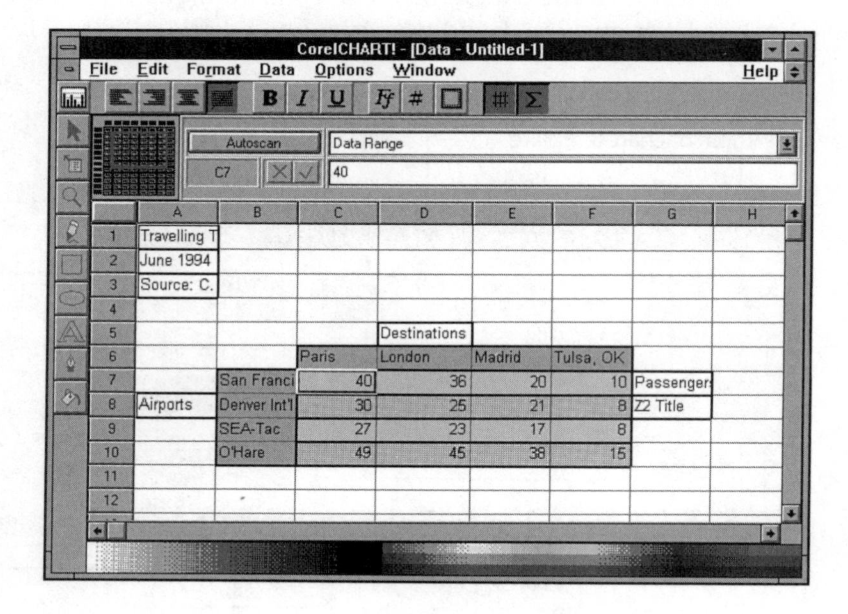

Replacing Chart's values and titles with your own.

Data Manager A spreadsheet program where you enter all of your chart data, including titles, labels, and numbers.

Cell A spreadsheet is made up of cells. A cell holds a piece of information, such as a number or title.

Series Any data item that is being charted; series items are row labels.

Click on the **Chart View** icon to switch back to Chart View. Unless you have a super speedy computer, you'll have to watch CHART draw each component of the chart one by one (which can be fun at first, but gets old after about the first time). But there it is . . . your first chart. Give yourself a big pat on the back.

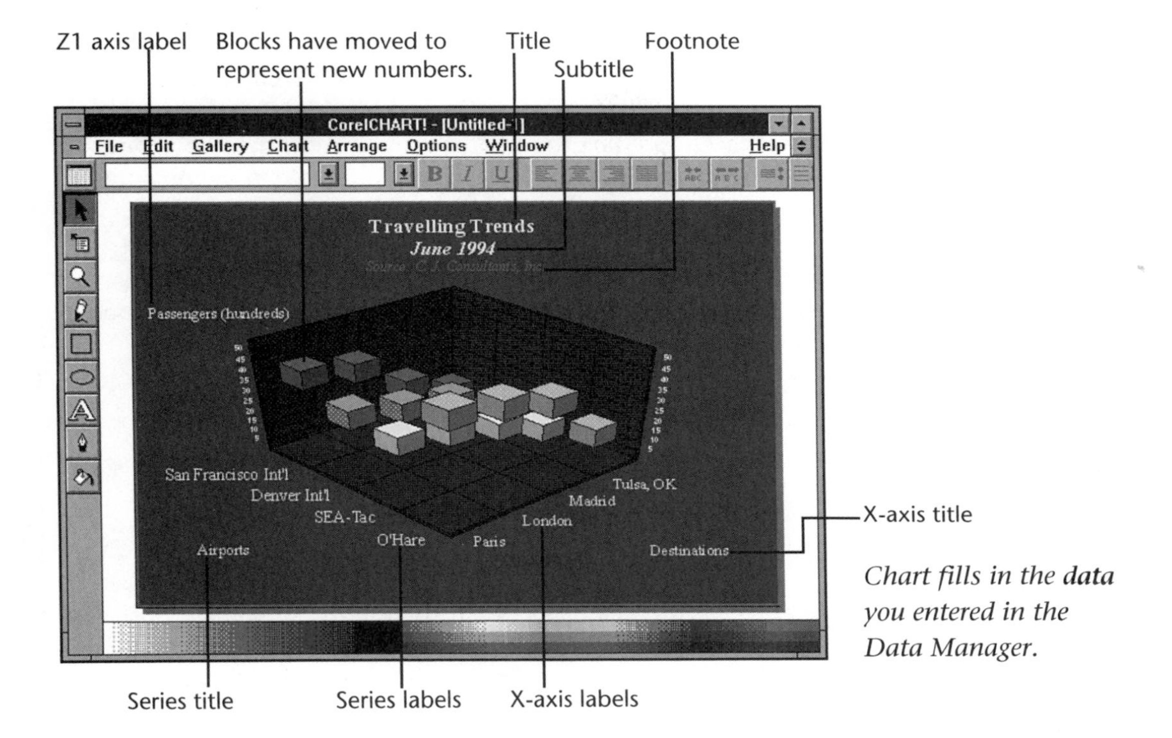

*Chart fills in the **data** you entered in the Data Manager.*

Select the Display Status command from the Chart menu. The 3D Graph Display Status dialog box appears. Use this dialog box to determine which components of the graph you want to display. For example, you can turn off walls, axis titles, and even the floor if you want. Since you didn't enter a title for the Z2 axis, click on that box so that it is no longer marked. You no longer see the Z2 Title placeholder on the chart.

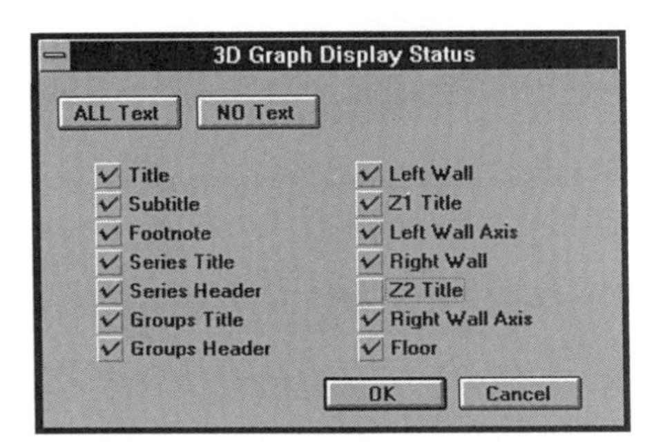

Deciding what to show and what to hide.

Well, there you go: a chart to be proud of. And you created it in such a short time—and with so few steps. Of course we only touched on CHART's capabilities, but we got the important stuff covered. You already know enough to create a dazzling chart the next time your boss asks you for a report! But, in case you want to edit your chart a bit, we'll give you some tips on how to quickly change the look of your chart.

The procedure for printing your chart is almost identical to printing drawings in CorelDRAW!. Read Chapter 14, "Rolling the Presses," if you need help printing your chart.

Why Do Chart Elements Have to Be So Confusing?

Series. X-axis. Columns. Legend. Blah. Blah. Blah. The only terms you may be familiar with are the last three. Well, we'll spell it all out for you. Charting doesn't have to be confusing, and it won't be once you get a grip on the elements of a chart.

What makes life more complicated is that the elements in two and three dimensional charts work differently. For example, you don't need a legend in a 3D chart, because the y-axis identifies series items. And in 3D charts, you actually have three axes: the regular x and y, and the z (the y in two dimensional charts), which shows the units of measure. The best way to show you the differences is visually. Take a look at the two charts below.

X-axis The bottom, outside edge of the chart frame; x-axis items are column heads.

Z-axis In 3D charts, the Z-axis title describes the type of measurement used to chart the data (for example, days, dollars, number of people); it depends on what you're measuring.

Legend Used in two-dimensional charts, a legend identifies which series items are represented by the various types of bars or lines (depending on the chart style).

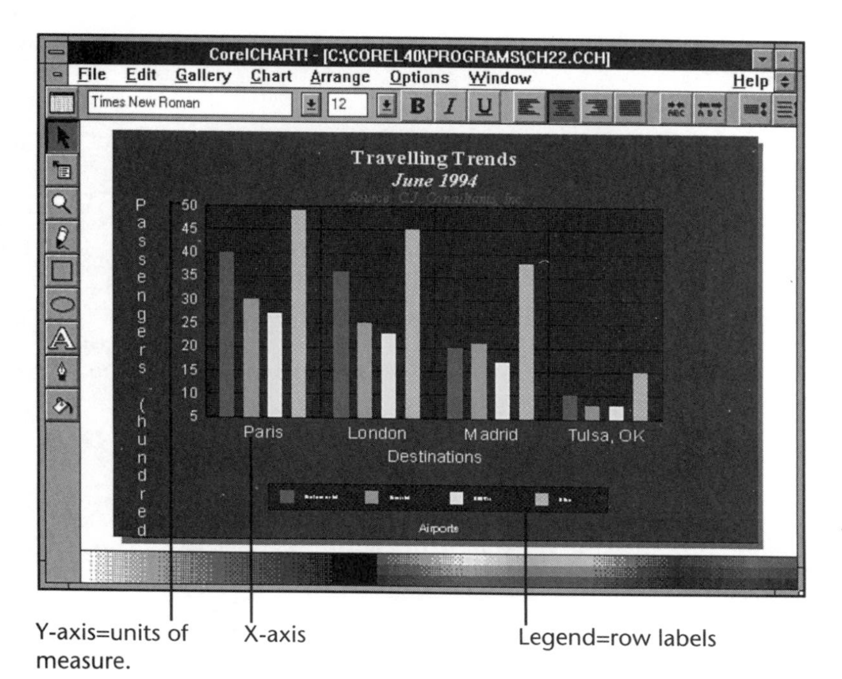

Elements of a two dimensional bar chart.

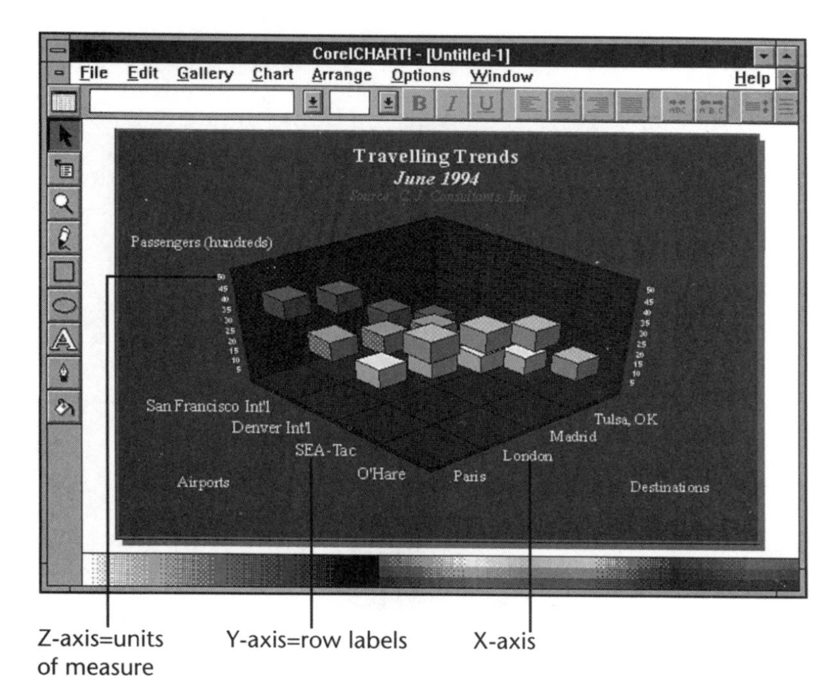

Elements of a three dimensional bar chart.

If that's not enough to clear things up for you, take a look at the table below.

Element	Description
Axis	The axis consists of a vertical (up) and horizontal (down) line, forming an L-shape. The vertical line is called the y-axis and the horizontal line is called the x-axis. You chart data by placing items up the y-axis and along the x-axis. The y-axis often contains numbers or dates. Line or bars are usually used to show where the items intersect on the axis.
X-axis	The horizontal axis line. In the Data Manager, the column heads are x-axis items.
Y-axis	The vertical axis line. In CorelCHART!, the y-axis (z-axis in 3D charts) is automatically generated based on the numbers you enter in the data range.
Z-axis title	In 3D charts, the z-axis title describes the unit of measurement; for example, months, years, days of the week, dollars (in thousands), passengers, employees—whatever you're tracking the amount of.
Series	The data that you're tracking. For example, if you're showing how many passengers have travelled from certain airports to certain destinations, the series could be either the group of airports you're monitoring, or the destinations. In the Data Manager, you enter series items as row labels. The bars and lines in charts represent the series, and a specific bar or line color represents an individual series item. In 3D charts, the series usually runs along the y-axis.
Legend	Not used in 3D charts. The legend shows which series item a bar or line color represents.

A Fresh New Perspective

Geez, enough about the elements! Let's turn to something more pleasant, and certainly easier to understand! Though you might be stumped when it comes to arranging your data in the Data Manager, you won't have any trouble at all changing how your chart looks once you've created it.

For instance, if you want to add a different perspective to your 3D chart, you can do so using a preset design. Select the Chart menu, and then choose the Preset Viewing Angles command. Something new appears: a submenu of selections with a blank space at the top. Well, drag your mouse down the selection list and see what happens to that blank space. As the mouse moves over a selection item, a small picture of that selection's style appears. So selecting a new perspective for your chart is as simple as selecting a new style on the Preset Viewing Angles menu. For example, select **California Special** and see how your chart changes.

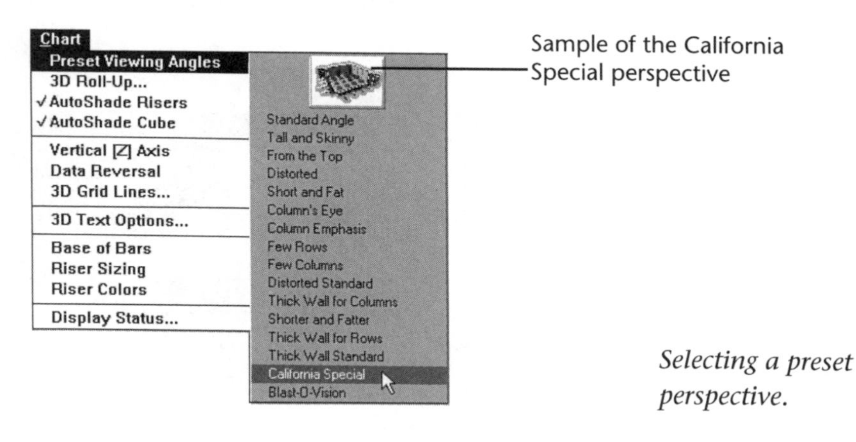

Selecting a preset perspective.

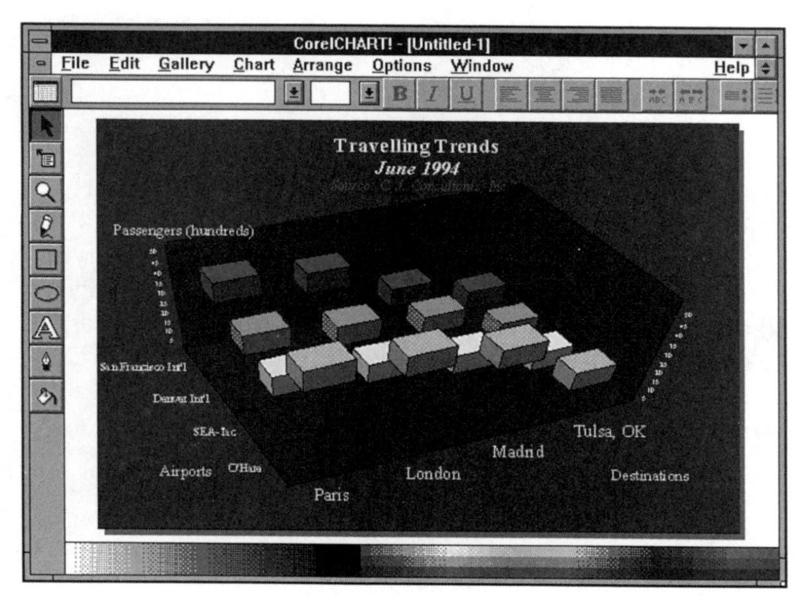

The chart as seen from a new angle.

Sorry, You're Not My Type

The chart style and type you selected when you first created your chart is certainly not set in stone. You can change to a new type or to a whole new style quite easily using the Gallery menu, which contains a list of all the chart styles available in the New Chart dialog box. The Chart menu is one of those cool CorelCHART! menus that shows you pictures: as you drag the mouse over a style on the Gallery menu, a submenu appears containing the chart types available for the style. When you move the mouse over a chart type, a picture appears in the submenu showing you what that chart type looks like.

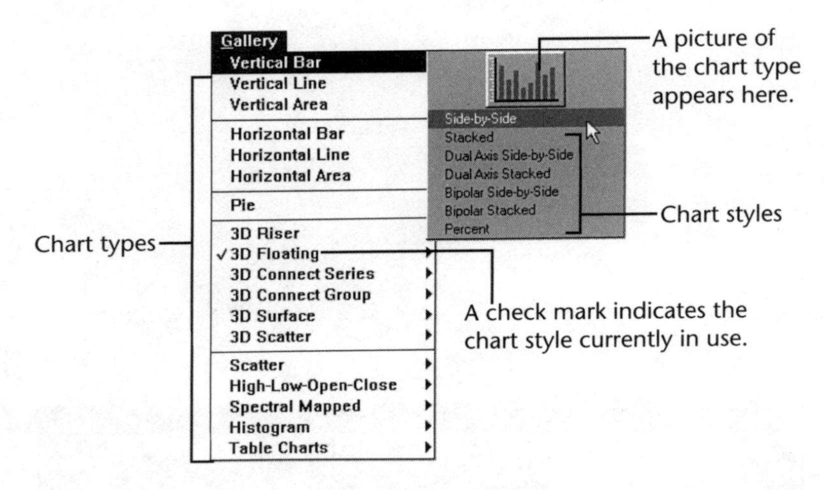

Selecting a new chart style and type.

You need to keep one thing in mind: Not all chart styles convert smoothly. For example, singular pie charts don't convert well to other types of charts because pie charts are so simplistic they can't present all of the information that needs to be charted. Even if the data from one style converts well to another style, sometimes the titles don't fit on the page or are too small to read. For example, look at how the information converted from a 3D Riser chart style to a Side-by-Side two dimensional bar chart. Though the numbers and series converted well, notice how small the legend names are and that the y-axis title was cut off.

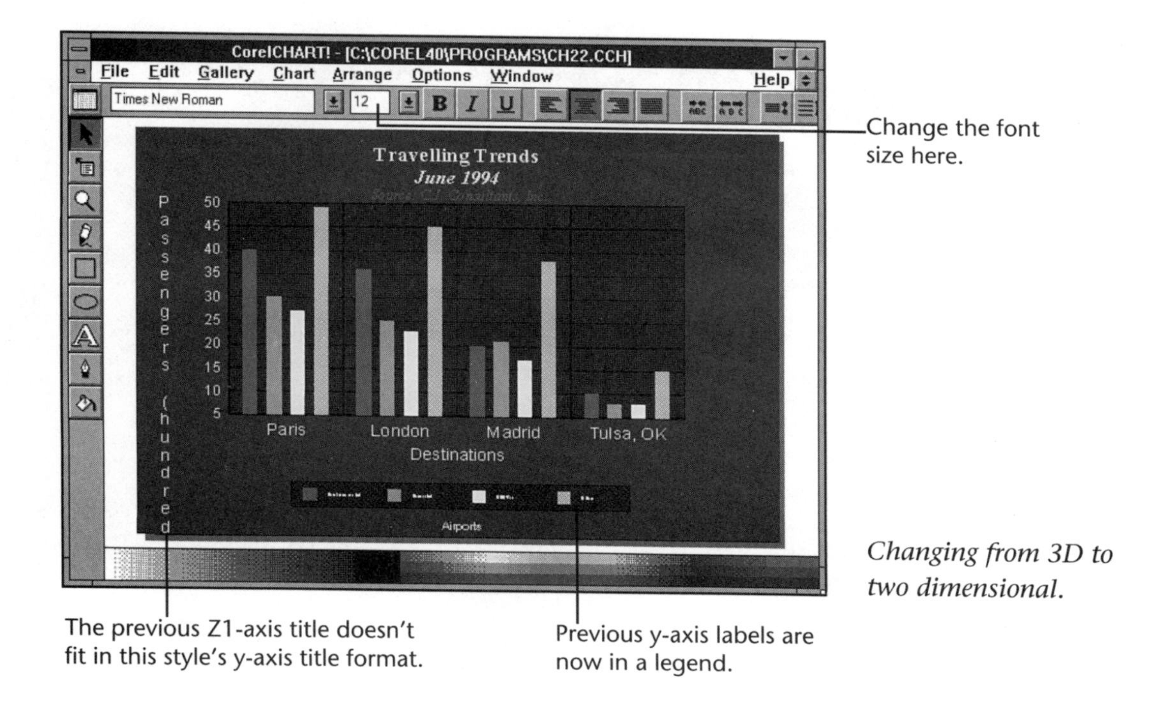

Changing from 3D to two dimensional.

It's not that big of a deal if title sizes don't convert well. You can simply shorten title names in the Data Manager if they're too long, or enlarge the font size of text that is too small to read. To increase or decrease a font size, select the text (click on it) and then change the number in the font size box on the Text ribbon. However, if you want to change the size of legend text, you first have to choose the Chart Legend command, and turn off the **Autofit Legend Text** option. Then you can resize the legend text as you desire.

Size Does Make a Difference

One of the problems with 3D charts is that the top or front bars sometimes hide other bars from view. If this happens, try using a different bar size. Display the Chart menu, and look for the word "Sizing." (The reason I can't give you the exact name of the command is that the name changes depending on the chart style you're using.) In the example, we're using a 3D Riser chart, so the command is Riser Sizing. If we were using a different style, the menu name would be different.

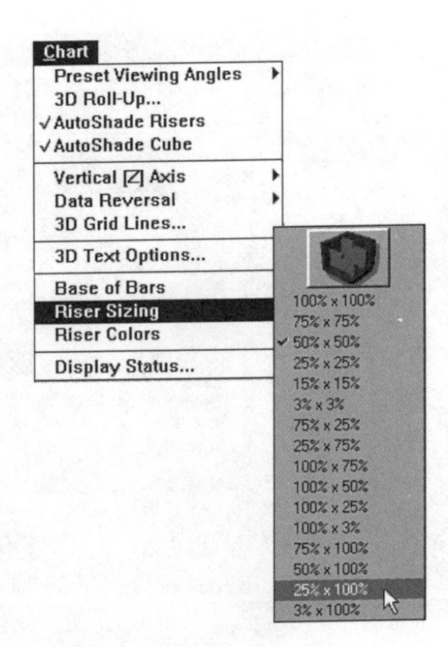

Changing the size of the bars.

The current size of the bars is checked. Remember the number in case you want to revert back to the original bar size later. You'll probably have to employ the faithful trial and error method to find which size you like the best. Although you get to see a sample of the bar size at the top of the submenu, sometimes you just don't know whether you'll like the new size until you actually see it in your chart. Look at how changing the bar size from 50% x 50% to 25% x 100% changes the look of the chart.

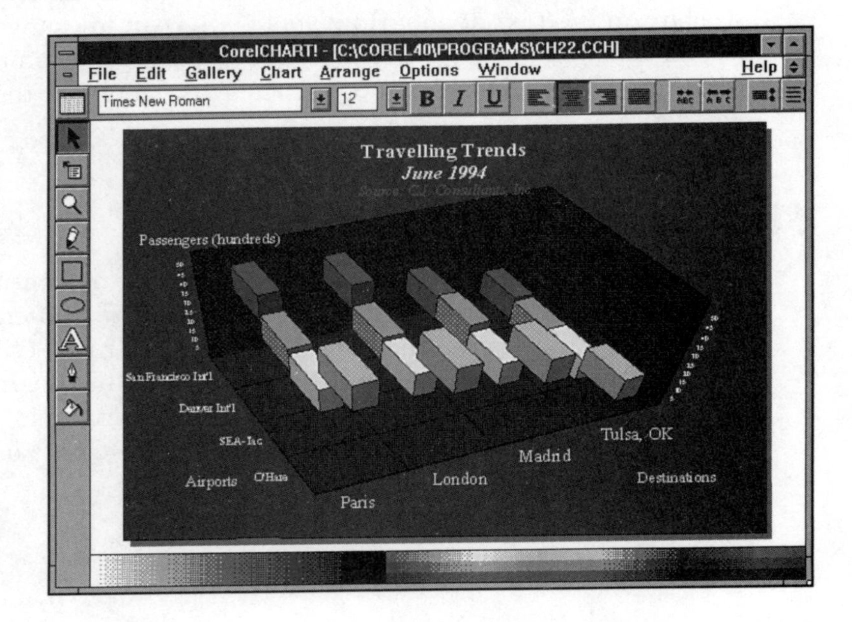

A new bar size can sometimes make the chart easier to read.

The Least You Need to Know

CorelCHART! has done all the hard work for you. Basically, all you need to do is to start a new file using sample data, and then enter your own information in the Data Manager.

- Select **F**ile **N**ew to start a new file.
- Select the chart style and chart type you want in the New dialog box.
- Switch to the Data Manager.
- Enter titles and values in the cells.
- Click on the **Chart View** icon to view the chart with its new values and titles.
- To change the perspective of the chart, select the **C**hart **P**reset Viewing Angle command.
- If you want to switch to a new chart style or chart type, select a new choice from the **G**allery menu.

No, this is not a printing error.
The page truly is blank.

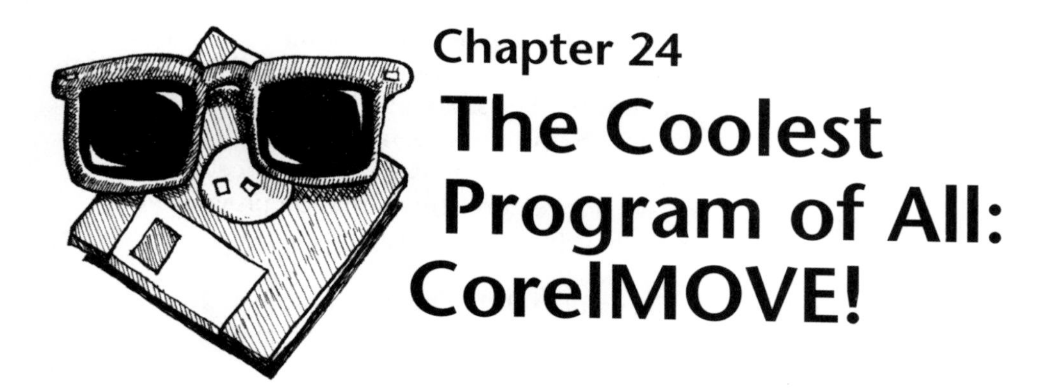

Chapter 24
The Coolest Program of All: CorelMOVE!

In This Chapter

- Yet another CorelDRAW! screen to get acquainted with
- Cartoon land
- Checking an actor out from the library
- Take a stab at interior decorating
- It moves!
- From here to there: an actor's path

When I was a kid, I couldn't wait for Saturday morning cartoons. Bugs Bunny was my hero. Little did I know then that I'd end up creating my own cartoons when I was a grown-up. I wonder if Disney has any openings?

I'm not really sure why, but for some reason, making the objects on your screen come alive and move around of their own accord is one of the most exhilarating feelings you can experience while working on your computer. Maybe it's the Frankenstein syndrome: us, mere humans, creating something that lives! Perhaps that's getting a bit carried away.

What are you doing reading this chapter?! CorelMOVE! does not come with CorelDRAW! 3. However, do feel free to look over CORELMOVE!'s features. (If you stay in this chapter long enough, you may find this feature so cool that you consider upgrading to either CorelDRAW! 4 or 5.)

You get a couple new features in CorelMOVE! 5 that you'll like. You can have multiple library roll-ups on-screen so that you can compare props and actors in different libraries. Also, you can now morph your actors, meaning you can blend from one cel to another and view all the cels in between. For example, you can change an apple into an orange and see the steps it would take for the change to take place. For more about morphing, read Chapter 27, "Blast Off with the Corel 5.0 Engine."

After all, the moment your computer shuts down, your animations cease to exist except in the sectors of your hard disk. Still, seeing objects you drew dance around the screen while you sit back and watch is a tremendously satisfying feeling. Maybe that's one reason why animation programs are rapidly gaining popularity.

Regardless, bringing your objects to life on-screen is what CorelMOVE! is all about.

The Whole Shamole in a Nutshell

This chapter describes the different steps you need to take to put an animation together. But before we go into details, here's a quick summary of the overall CorelMOVE! process. The first thing you do is create an animation file. Using props, you design a background scene that remains constant while the action takes place. Then you place actors on the stage (luckily, in CorelMOVE!, we don't have to deal with actors' personality differences). If you want the actors to move around the stage as well as moving in place, you specify paths for the actors to follow. Finally, you play the animation using the playback controls. That said, let's get into the nitty-gritties.

It All Happens Here

Like all the other CorelDRAW! programs, CorelMOVE! has its own program icon in the Corel group window. Double-click on this icon to open CorelMOVE!. (If you need more than this simple explanation, way back in Chapter 3 we talked all about program icons, group windows, and starting applications; you can turn there for further help.)

The CorelMOVE! application window looks different from all of the other CorelDRAW! program windows. It's true, none of the windows are exactly the same, but the CorelMOVE! window is the most unique. It has a grainy background, a small toolbox, a skimpy menu bar, and a bunch of stuff at the bottom that looks like the controls on a cassette tape player.

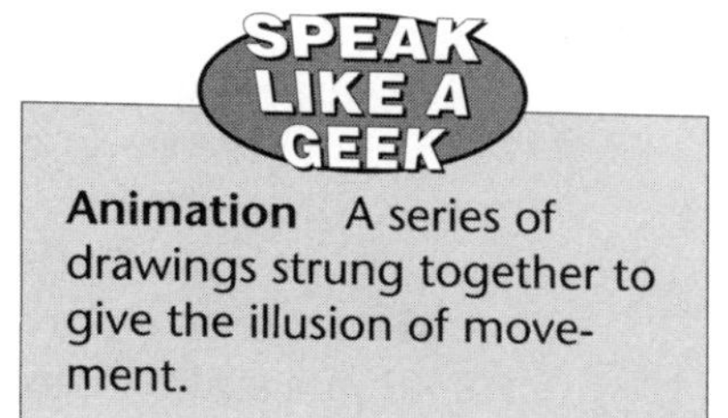

Animation A series of drawings strung together to give the illusion of movement.

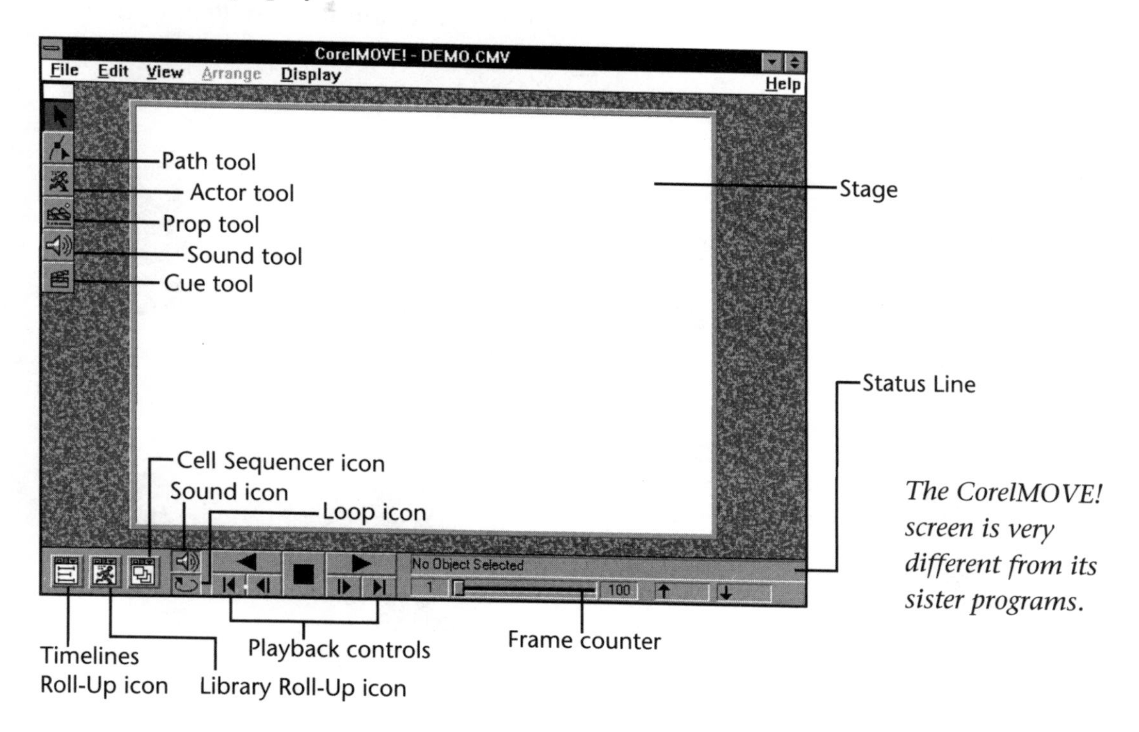

The CorelMOVE! screen is very different from its sister programs.

Is This Like Cartoons?

Did you have a graphics guru at your junior high who used to draw images of your teacher at the edges of a ton of small pieces of paper? With each page, the teacher looked a bit different. It wasn't until the guru speedily flipped through the pages with the tip of his thumb that you were able to see your teacher trip and fall to the ground, throwing her papers all over the place. That was animation. More technically, an *animation* is basically a series of drawings linked together to form the illusion of movement. Each drawing is similar, so that when rapidly displayed one after another on your screen, they appear to be a single drawing that is changing shape or moving.

The group of drawings is called an *actor*. Each individual drawing is a *cel*, so that an actor is made up of several cels (equivalent to frames in a movie). A *prop* is a graphic that sits in the background while the actor is moving. If your computer system has a sound board (such as SoundBlaster), you can include sounds in your animations. Below you will find twelve different cels of an actor.

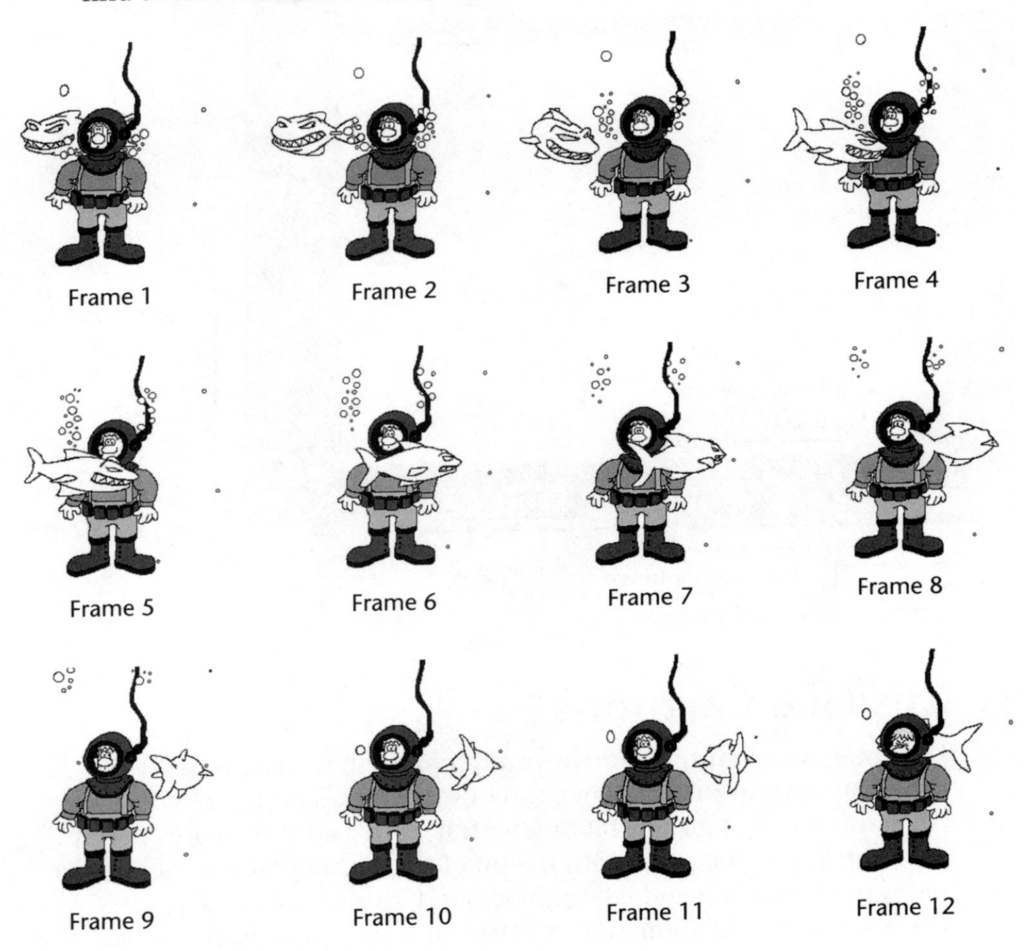

The Long and Short of It

There is a long way and a short way to set up an animation in CorelMOVE!. You can create your own actors and props (the long way), or use existing actors and props provided by CorelMOVE! or drawn in other applications like CorelPHOTO-PAINT! and CorelDRAW! (the short way). A suggestion: if you draw well, try the long way and reap the benefits of full ownership; if your drawing abilities stink, cheat and use someone else's stuff (namely, CorelMOVE!'s actors and props). There are hundreds of actors and props on the CorelDRAW! CD-ROM. You only get one library if you don't have a CD drive, so see if someone you know who does have one will download some CorelMOVE! libraries for you (MOVE actors and props are stored in libraries).

SPEAK LIKE A GEEK

Actor A drawing that moves on the screen, consisting of a series of cels.

Cels The individual frames that make an actor. Each cel is a single drawing in each stage of the movement.

Prop A drawing that remains stationary in the background as the actors move during an animation.

Sound What you hear when an animation is playing. Your computer system has to have sound capabilities (such as a Sound-Blaster card and speakers) in order for you to use sounds in your animations.

You go the high road, and we'll go the low road, and we'll come up with animations faster than you can. That's our way of saying that we'll teach you the shortcut method of creating animations in this book. If you want some bare-bones instructions on creating your own actors and props, check out the following section. If you're like me and don't have an artistic bone in your body, don't stop. Go directly past go, skip the next section. . . . You get the picture.

For All You Artsy-Fartsies

Okay, so you can draw like Norman Rockwell. Show off! Here's how to create your own actors and props in CorelMOVE! in a nutshell (for all of us non-talented types). Click on the **Actor** or **Prop** tool on the toolbox. In the New Actor dialog box (or New Prop, depending on the tool you chose), type a name for your actor or prop, and then click on **Create New**. The Object Type box displays a list of programs you can use to create your actor or prop. Click on **CorelMOVE! 4.0** and click **OK**.

A Paint window opens in which you can draw the actor or prop using the tools on the Paint Palette, which also opens on-screen. This palette contains tools from CorelDRAW! and CorelPHOTO-PAINT! that you should already be familiar with. To create additional cels, use the **I**nsert Cels command on the Paint window's **E**dit menu. You should mark the **Duplicate Contents** option in the Insert Cels dialog box so that you can edit the drawing from the previous cel to create the contents of the new cel. Choose **F**ile **S**ave when you are finished creating the actor or prop.

Shhh! We're in a Library

Whenever I think of libraries, I break out into a nervous sweat. They remind me of my college days when I would go to the library at 8:00 p.m. Thursday to get information to write a 30-page research paper due at 8:00 a.m. Monday (which, of course, I had known about for 4 weeks). Then, invariably, the crucial book or article I needed was already checked out. My heart is racing just thinking about it. Luckily, CorelMOVE!'s libraries are not like that.

Library A CorelMOVE! file that contains actors, props, and sounds to be used in an animation.

All of CorelMOVE!'s actors, props, and sounds are stored in individual libraries. Libraries can contain only actors, props, and sounds; they usually hold a combination of either actors and props or all three. The actors, props, and sounds stored in a library generally relate to a common theme so that they can be added to the stage to create a meaningful animation. For example, the sample library we'll be using in a moment contains such actors as fish, a deep-sea diver, and an octopus, and such sea-related props as a blue background for water, and an anchor.

To open a library, you can click on the **Library Roll-Up** icon located at the bottom of the screen (the middle icon of three). If you haven't created or opened a library before, you must do so now. (Later on, the last library you've used will be displayed automatically when you display the Library roll-up.) Click on the right arrow on the upper right side of the roll-up. A menu display will appear. Click on the **Open Library** command to display the Open Library dialog box.

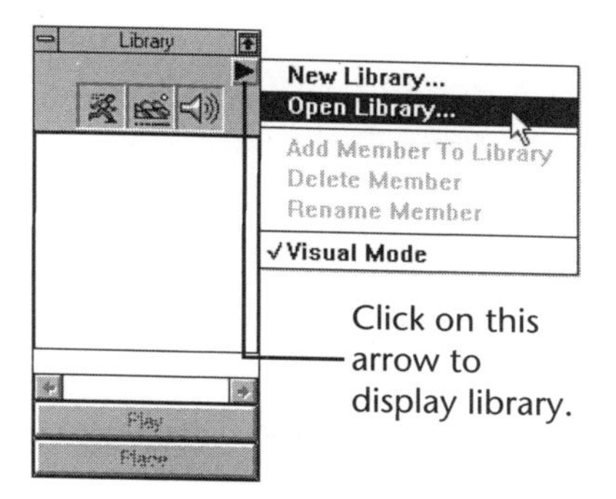

Opening a CorelMOVE! library.

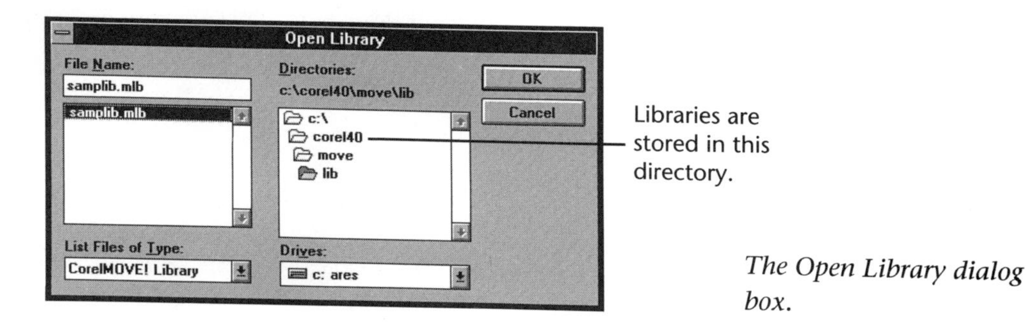

The Open Library dialog box.

For now, double-click on **samplib.mlb**. The dialog box closes, the sample library opens, and the first actor in the library is displayed in the roll-up. You can scroll through the different actors and props by clicking on the scroll arrows in the roll-up.

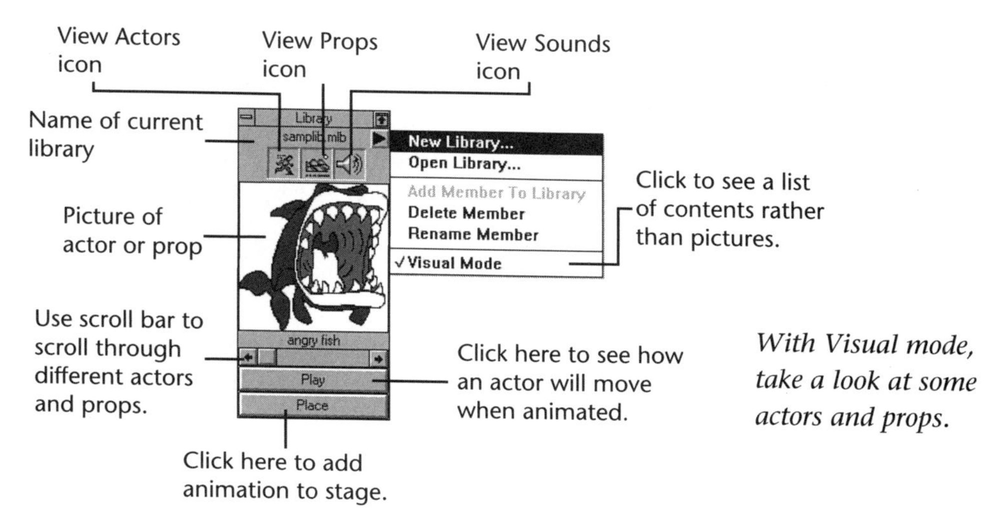

With Visual mode, take a look at some actors and props.

However, it's usually easier to view a list of the names of actors, props, and sounds. To view a list of the library's contents, click on the right arrow again to display the roll-up menu, and then click on **Visual Mode**. This turns Visual Mode off. You now see a list of contents in the Library roll-up.

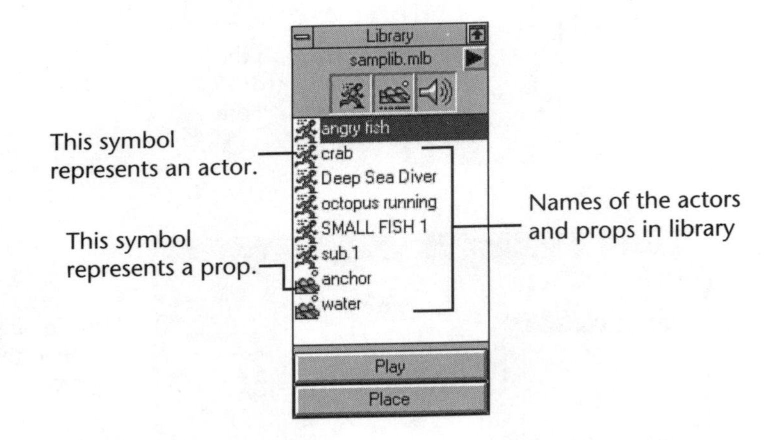

A list of the library's contents, without Visual Mode.

You can tell whether an item in the library is an actor, prop, or sound by the symbol before its name. The symbol matches one of the icons at the top of the Library roll-up. You can use the icons to display or hide certain contents in the roll-up. For example, if you're reviewing a particularly large library and want to look at props only, turn on the View Actors and View Sounds icons (the first and last icons, respectively), so that only the middle icon, the View Props icon, is depressed. To see an actor in action, click on the actor and then the **Play** button.

Watch Out, Steven Spielberg

The ultimate goal of animating is to place actors on the stage, with props positioned in the background to enhance the setting (kind of like a director has to do in a movie or play). It's usually best to place the props and then the actors, but there's no law stating you have to do it in that order. When you've decided which prop to place in the animation, click on the prop, and then click on the **Place** button. CorelMOVE! places the prop in the middle of the stage. Some props, like the water prop in the example, take up the entire stage. You may need to move other smaller props to a new position. If so, just drag the prop to a new location.

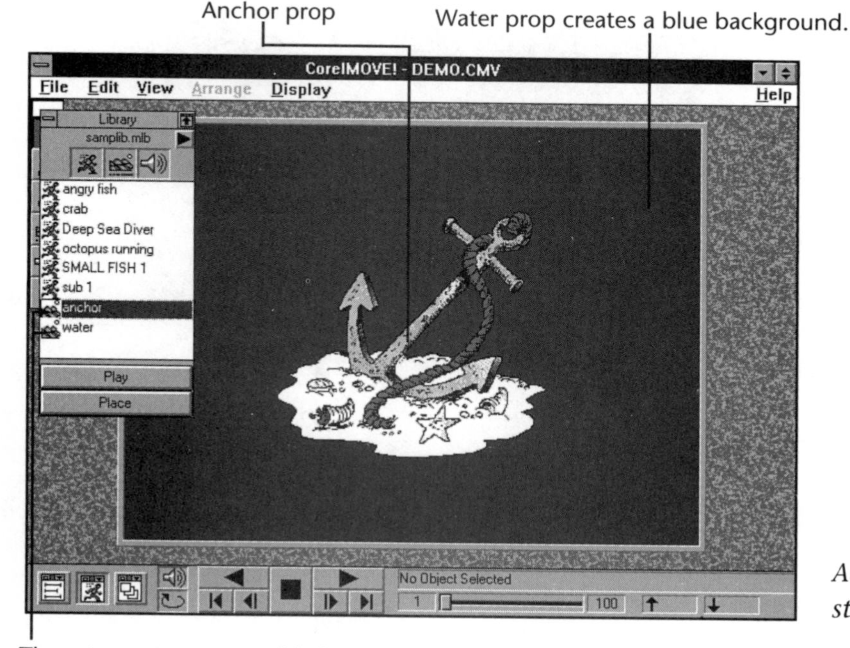

Adding props to the stage.

Add all actors to the stage in the same way: select the actor(s) and click on the **Place** button. Then arrange the actors on stage. Kind of makes you feel like a big time director—without the headache of dealing with actors' egos.

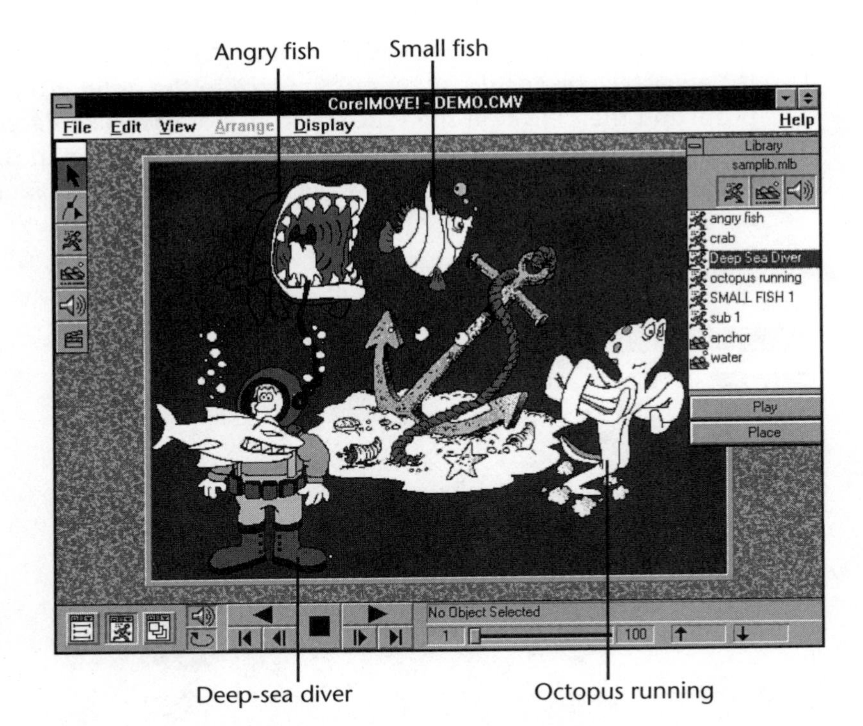

Calling all actors to the stage!

Lights, Camera, Action!

So far, for all you can tell, you've only added plain old drawings to the application window. That's nothing new. When are they going to start moving? Well, you're in the director's chair; they'll move when you tell them to. And you can tell them to move via the playback controls at the bottom of the screen. The playback control buttons let you play the animation in regular mode or in reverse, move one frame forward or backward, move to the first or last frame, and stop the animation.

CorelDRAW! 4.0 Toolbox

Use this tool	*To*
Pick	Select an object
Shape (reshape it)	Node edit an object
Zoom	Display the zoom tools
Zoom In	Zoom in on a drawing
Zoom Out	Zoom out on a drawing
Zoom Actual Size	View a drawing at size it will print
Zoom To All Objects	Zoom in on all objects
Zoom To Page	Toggle between zooming in and out on drawing page
Pencil	Draw lines and curves in Freehand mode
Bézier	Draw lines and curves in Bézier mode
Rectangle	Draw a rectangle or square
Ellipse	Draw an ellipse or circle
Artistic text	Add artistic text
Paragraph text	Add paragraph text
Symbols	Add a symbol
Outline	Display outline tools
Outline Dialog Box	Display Outline Pen dialog box
Outline Roll-Up	Display Pen roll-up
No Outline	Remove outline
Hairline Outline	Apply very thin outline: 1/4 point
Thin Outline	Apply thin outline: 2 points
Medium Outline	Apply medium outline: 8 points

Use this tool	*To*
Medium-Thick Outline	Apply thick outline: 16 points
Thick Outline	Apply very thick outline: 24 points
Outline Color	Display Outline Color roll-up
White Outline	Apply white outline
Black Outline	Apply black outline
10% Black Outline	Apply 10% black out-line
30% Black Outline	Apply 30% black outline
50% Black Outline	Apply 50% black outline
70% Black Outline	Apply 70% black outline
90% Black Outline	Apply 90% black outline
Fill	Display Fill tools
Fill Color	Display Uniform Fill dialog box
Fill Roll-Up	Display Fill roll-up
Fountain Fill	Apply a fountain fill
Two-Color Fill	Apply a two-color pattern fill
Full-Color Fill	Apply a full-color pattern fill
Texture Fill	Apply a textured fill
PostScript Fill	Apply a PostScript fill
No Fill	Remove fill
White Fill	Apply a white fill
Black Fill	Apply a black fill
10% Black Fill	Apply a 10% black fill
30% Black Fill	Apply a 30% black fill
50% Black Fill	Apply a 50% black fill
70% Black Fill	Apply a 70% black fill

alpha books

The Complete Idiot's Reference Card

A Look at the CorelDRAW! Screen

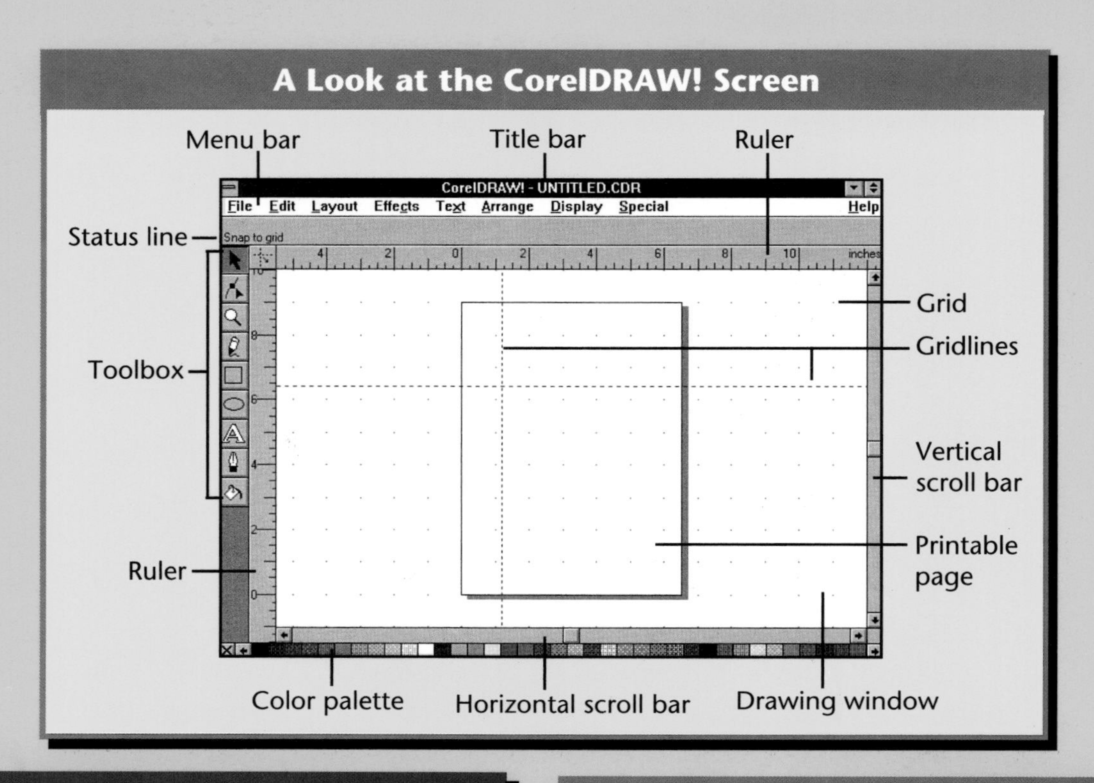

Command Shortcut Keys

Press these keys	*To do this...*
Ctrl+C	Copy object
Ctrl+X	Cut object
Ctrl+Y	Snap to grid (toggle)
Ctrl+D	Duplicate object
Alt+F4	Exit from CorelDRAW!
Ctrl+G	Group objects
Ctrl+N	Open a new drawing
Ctrl+O	Open an existing drawing
Ctrl+V or Shift+Insert	Paste
Ctrl+P	Print
Ctrl+S	Save drawing
Ctrl+Z	Undo
Ctrl+U	Ungroup object
Shift+F9	Wireframe view (toggle)

Get into the Habit of Using These Roll-Ups

Roll-Up	*Key combination or tool*
Blend roll-up	Ctrl+B
Envelope roll-up	Ctrl+F7
Extrude roll-up	Ctrl+E
Fill roll-up	Click on the Fill tool to display Fill Roll-Up. Or, click on the Fill Roll-Up tool.
Fit Text To Path roll-up	Ctrl+F
Layers roll-up	Ctrl+F3
Node Edit roll-up	Ctrl+F10
Pen roll-up	Click on the Outline tool to display flyout menu. Or, click on the Outline Roll-up tool.
Symbols roll-up	Ctrl+F11
Text roll-up	Ctrl+F2

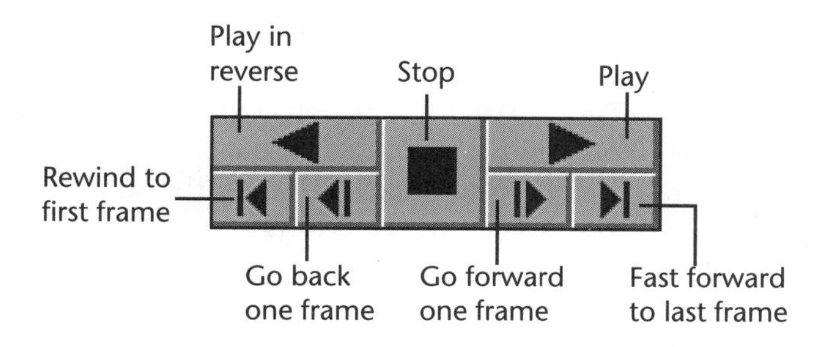

Use the playback controls to control the flow of the animation.

Though it's not nearly as exciting as seeing live action on the screen (like a dinosaur charging straight for you in *Jurassic Park*), notice how the actors in the figure below have changed from their previous positions (refer to earlier figure). That's because we started the action, and then stopped it on the eighth frame.

To go to a specific frame, use the **G**o to Frame command on the **V**iew menu, or drag the frame counter button. Also, a tip for button-shy users: you can find equivalent commands for each of the playback control buttons on the **V**iew menu.

They moved!

From Here to Eternity

Unless you specify a path for an actor, the actor will always move in place, kind of like it's walking on a lifecycle. What fun is that? By setting up paths, you can make your actors move all over the screen. You specify a path using the Path tool, which looks just like the Shape tool in CorelDRAW! (it's the second tool on the toolbox).

To specify a path, click on the **Path** tool and then on the actor. A small circle appears just above the actor. Drag the circle to where you want the actor to begin moving. This becomes the "from" point. Now click where you want the movement to end. You guessed it: this is the "to" point. A line connects the to and from points together. This line represents the path the actor will follow.

It's not enough to just create a path. If you leave it at this, the actor will jump from the "from" point to the "to" point in one frame. This won't look like animation at all. You need some intermediate steps to make the transition down the path last a while and flow smoothly. To do this, use the **Scale Path** button (the one with the +/– symbols on it) in the Path Edit roll-up, which appeared when you clicked on the Path tool. A dialog box appears, asking you how many points you want on the path. Enter the number you want and click **OK**. Make sure the **Allow Adding Points** button is on in the roll-up. Then click on the stage wherever you want the actor to move in between the to and from points. If you want the actor to follow a straight line, click on the path itself. To distribute points evenly along the path, click on the **Distribute points** button in the Path Edit roll-up. Now click on the **Play** control button, and watch your actor move along the path you specified.

The Least You Need to Know

Like a big time Hollywood director, you can make your screen come alive with CorelMOVE!.

- Double-click on the **CorelMOVE!** icon in the Corel group window to open CorelMOVE!.
- Open a library in the Library roll-up to get your hands on some actors and props. To display the Library roll-up, click on the **Library Roll-Up** icon at the bottom left portion of the screen.
- Click on the arrow in the Library roll-up to display the roll-up's menu. Click on **Visual Mode** to turn off the sample picture and display a list of the library's contents.
- Click on a prop or an actor and then click on the **Place** button, until you've added all the props and actors you want.
- Click on the **Play** control button to start the animation.
- To make an actor move around the screen, create a path for the actor to follow using the Path tool and the Path Edit roll-up.

Recycling tip:
Tear this page out and photocopy it.

Chapter 25
PHOTO-PAINT!: Watch Out, Picasso

In This Chapter

- Getting acquainted with PHOTO-PAINT!
- Opening Pandora's picture
- Sorting out the toolbox
- Really effecting your images
- Filling it to the rim with . . .

When I was a young boy, I was so impressed that human beings could be miniaturized and stuck onto the face of a photograph. I mean, how did big-boned Grandma Hite get scaled down to an image I could hold in my hand? How much more impressed would I have been if I had grown up in present times. Not only could I stick Grandma Hite onto a photograph, but I could also put her into my own computer. And with an image-editing program like CorelPHOTO-PAINT!, I could do whatever I wanted with Grandma Hite: turn her big bones into small bones, dye her hair black, put a mole on her right cheek, help her to shed a few pounds, or graft a cleft into her chin. (Geez, I'm turning her into the Wicked Witch of the West.) You name it, and CorelPHOTO-PAINT! lets you do it—to the image only though. I really was born a generation too early.

Starting 'Er Up

Same old, same old. You start CorelPHOTO-PAINT! the same way you start CorelDRAW!, except you click on a different icon (makes sense). So double-click on the CorelPHOTO-PAINT! program icon in the Corel program group window, and CorelPHOTO-PAINT! will soon grace your screen. If you need more help than that, read Chapter 3, "Taking Your First Step" (and you might also consider investing in a tutor).

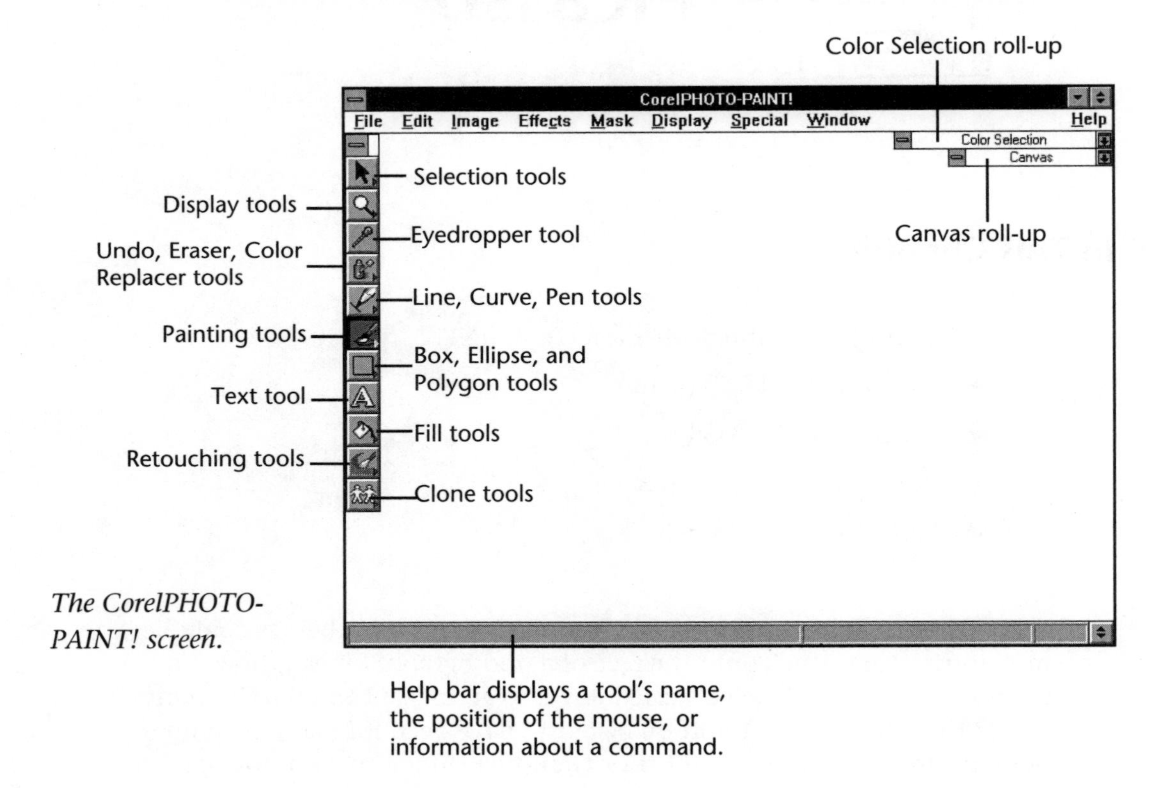

The CorelPHOTO-PAINT! screen.

Notice that most of the tools have a small arrow in their lower right corners. These arrows indicate that a flyout menu exists for that tool. A *flyout menu* contains subtools that are displayed when you hold down the mouse on the original tool. All but two of these tools, the Text and Eyedropper tools, have flyout menus.

If you have an image open, you can quickly find out a tool's name by moving the mouse over the tool. The name of the tool appears on the status line at the bottom of the screen.

The CorelPHOTO-PAINT! 3 screen looks quite a bit different than the CorelPHOTO-PAINT! 4 screen. Instead of a toolbox, the version 3 screen has a tool palette. And instead of roll-ups, you get work boxes. The Width box, color palette, and tool palette all open automatically when you start CorelPHOTO-PAINT! 3. Except for a few tools, the version 3 tool icons are the same as the version 4 tool icons. If you're not sure about a tool's function, just roll the mouse over it, and read the status line at the bottom of the screen. The function of the tool is described on the status line.

Opening a Picture

You load an image into CorelPHOTO-PAINT! the same way you open a drawing in CorelDRAW!. Choose the **O**pen command from the **F**ile menu. The Load a Picture from Disk dialog box appears. Don't let the name of this dialog box scare you: all you're doing is opening a picture (don't ask us why the Corel wizards didn't just label it the Open Picture dialog box).

In CorelPHOTO-PAINT! 5, you can open only a portion of a file if you want. This can be very helpful when you need to retouch only part of an image contained in a huge graphics file, because working with huge files can cause PHOTO-PAINT to run sluggishly.

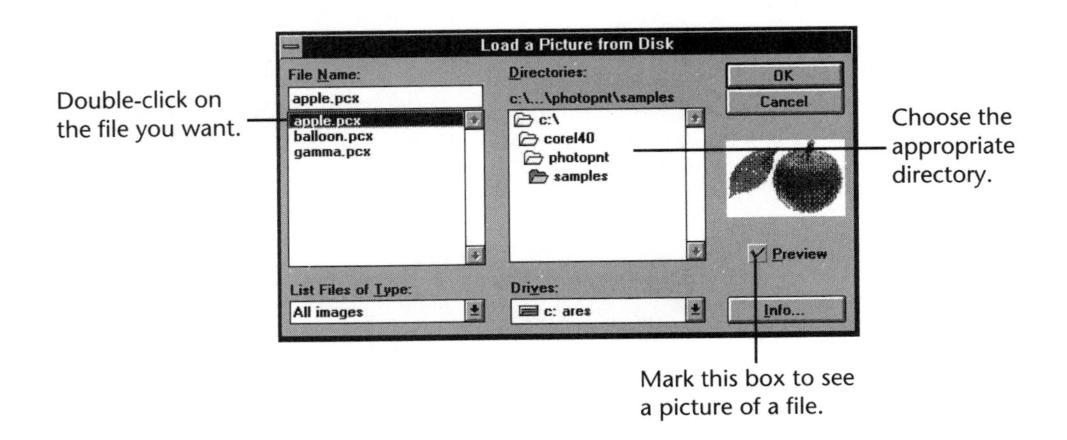

First select the directory containing the picture you want. Mark the **Preview** box if you want to view the contents of the file. When you've found the picture you're looking for, double-click on its file name. CorelPHOTO-PAINT! opens a painting window containing the picture.

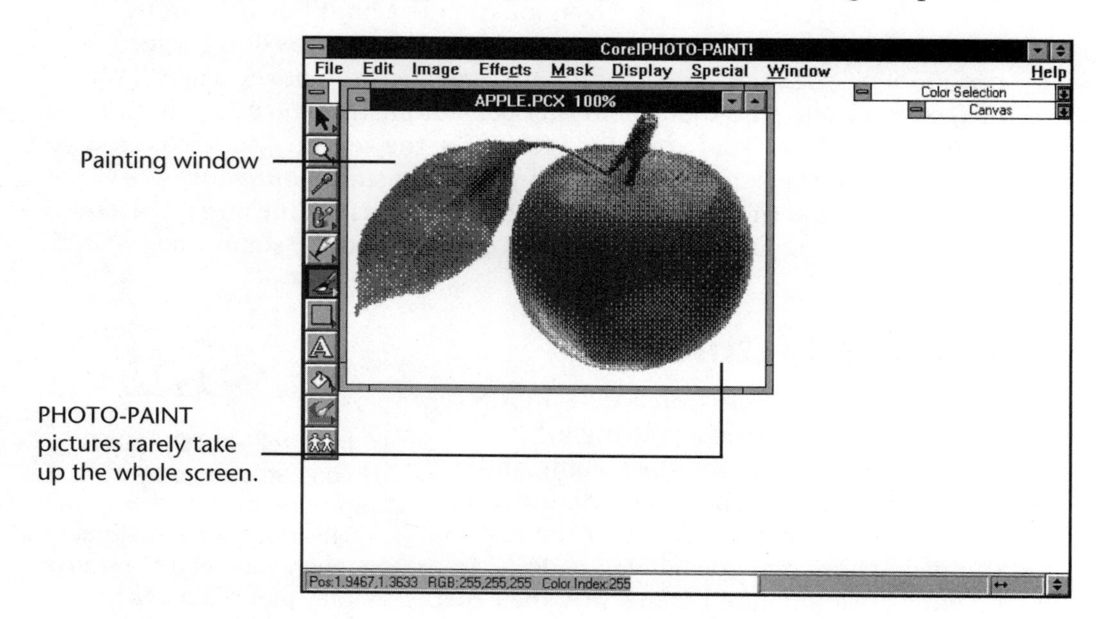

If you want to create your own picture from scratch, use the **F**ile **N**ew command instead. Specify a size and mode for your new picture in the Create a New Picture dialog box. A painting window opens; it's all yours, go ahead and begin painting!

What's Cookin' in the Toolbox

Though some of the tools' names make their functions clear, a brief description of what a tool accomplishes can sometimes work wonders for you. Keep in mind when you use CorelPHOTO-PAINT!'s tools that the outline and fill colors used will be the ones selected in the Color Selection roll-up. By default, this roll-up is displayed automatically when you start CorelPHOTO-PAINT!. Click on the **Outline/Fill/Background** button (the name changes as you click on it), and click on a color to specify the color settings for each area.

You should also select the Tool Settings Roll-Up command from the Display menu when drawing lines or shapes, or using the brush and spray tools and freehand retouching tools. If you do, you can quickly modify the line and stroke widths and sizes in the Tool Settings roll-up. Additionally, when you use some of the retouching tools, including the Freehand Smear, Sharpen, Contrast, Brighten, and Blend tools, settings for these tools appear in an extended portion of the Tool Settings roll-up. To display these extended options, click on the down arrow in the Tool Settings roll-up.

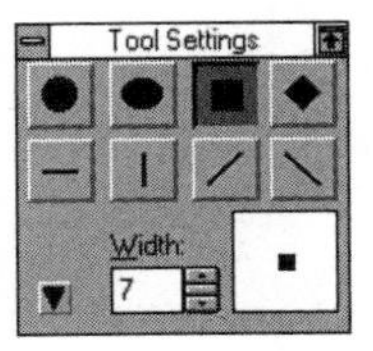

Choose tool widths and styles in the Tool Settings roll-up.

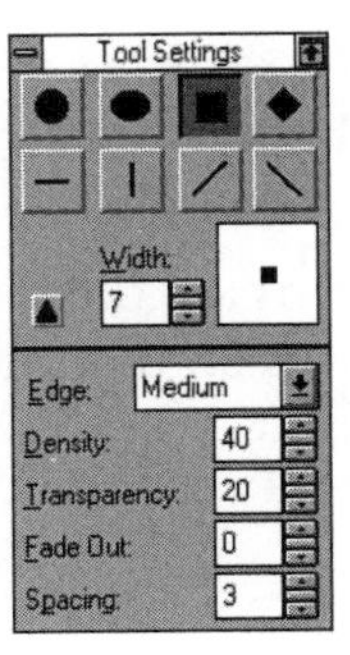

Additional settings appear for certain tools.

There's no tool for the Mask command, but you should know that you can use the **M**ask Select **A**ll command to select an entire image.

Selection Tools

Rectangle Selection tool

Select a rectangular area within an image. Drag this tool on the area you want to select.

Magic Wand Selection tool

Selects all areas in an image that are of the same hue. Click on a color to select all areas of that color.

Lasso Selection tool

Selects an irregular area. Drag this tool over the area that contains the irregular shape you want to modify.

Polygon Selection tool

Selects a polygonal area. Your actions are the same as they would be to draw a polygon: Click to start the selection, double-click to define each side, and click to end.

Display Tools

Zoom tool

Zooms in or out on an area. Click the left mouse button to zoom in; click the right mouse button to zoom out.

Locator tool

Displays a similar area in all copies of the picture you click on. For example, similar areas in all duplicates will be displayed.

Hand tool

Drags a zoomed picture to a new position so you can view another area of the picture.

Eyedropper Tool

Eyedropper tool

Picks up a color from another part of your picture, the same way a real eyedropper can suck up liquid. Click the left mouse button to pick up the outline color; click the right mouse button to pick up the fill color; hold down the **Ctrl** key and click either mouse button to pick up the background color.

Undo, Eraser, Color Replacer Tools

Local Undo tool

Returns specific portions of your picture to the way they appeared before the last command or tool you used. The Local Undo tool is no longer available once you use the **E**dit **U**ndo command.

Eraser tool

Replaces portions of your picture with the background color. Drag this tool over the areas you want to replace. If you double-click this tool, it replaces the entire image with the background color.

Color Replacer tool

Works together with the Color Selection roll-up (which opens automatically when you start CorelPHOTO-PAINT!) or with the Eyedropper tool. The Color Replacer tool removes an outline color and replaces it with a fill color. Select the outline and fill colors in the Color Selection roll-up, or use the Eyedropper tool to pick up the fill color. Drag the Color Replacer tool over the portion of the image you want to modify; the outline color changes to the fill color as you drag.

Line, Curve, Pen Tools

Line tool

Creates a line. Use this tool the same way you use CorelDRAW!'s Pencil tool. For details on how to use this tool, read the first section in Chapter 8, "Right Down to the Line."

Curve tool

Enables you to create a curve. Draw a line with this tool, and CorelPHOTO-PAINT! places circle handles on the line you create. Drag the handles to create the curve.

Pen tool

Enables you to create freehand drawings. Using this tool is similar to drawing with the Pencil tool in CorelDRAW!. (Read the first section in Chapter 8, "Right Down to the Line" for instructions.) You can adjust the width of the Pen line using the options in the Tool Settings roll-up.

Painting Tools

Paint Brush tool

Creates brush strokes in the outline color as you drag.

Impressionist Brush tool

Creates multi-colored brush strokes as you drag.

Pointillist Brush tool

Leave a path of multi-colored circles as you drag.

Artist Brush tool

Creates brush strokes wherever you drag that look like strokes in an oil painting.

Airbrush tool

Sprays the area over which you drag with the outline or fill color. Drag the left mouse button to spray the outline color, and drag the right mouse button to spray the fill color. This tool creates a mist effect, and can add shading and depth to an image.

Spraycan tool

Works like the Airbrush tool, except it leaves a splattering effect. Drag the left mouse button to spray the outline color, and drag the right mouse button to spray the fill color.

You can modify the look of any of the brushes by selecting new widths and brush types in the Tool Settings roll-up. To display this roll-up, select the Tool Settings Roll-Up command from the **D**isplay menu.

You version 3 users do not have a Tool Settings roll-up. To choose different brush styles, use the Width box that is displayed automatically when you start CorelPHOTO-PAINT! 3.

Box, Ellipse, and Polygon Tools

Hollow Box tool and

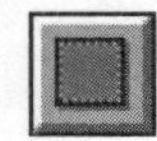

Filled Box tool

Create a rectangle or square with or without a fill color. Drag the Hollow Box tool to create a rectangle or square without any fill; drag the Filled Box tool to create a rectangle or square with a fill color. For details on drawing rectangles and squares, read "It Takes Two to Rectangle" and "What a Square" in Chapter 5, "Getting into Shape."

Hollow Rounded Box tool and

Filled Rounded Box tool

Work just like the Hollow Box and Filled Box tools, except that the rectangles and squares have rounded corners.

Ellipse tool and

Filled Ellipse tool

Create an ellipse or circle with or without a fill color. Drag the Ellipse tool to create an ellipse or circle without any fill; drag the Filled Ellipse tool to create an ellipse or circle with a fill color. For details on drawing ellipses and circles, read "An Ellipse of the Sun" and "Going in Circles" in Chapter 5, "Getting into Shape."

Hollow Polygon tool and

Filled Polygon tool

Create a polygon with or without a fill color. Drag the Hollow Polygon tool to create a polygon without any fill; drag the Filled Polygon tool to create a polygon with a fill color. For details on drawing polygons, read "Drawing Straight Lines" in Chapter 6, "Right Down to the Line."

Text Tool

Text tool

The Text tool works quite differently in PHOTO-PAINT! than it does in CorelDRAW!. When you click CorelPHOTO-PAINT!'s Text tool, the Enter Text dialog box appears. Enter your text in the text box, and then choose the font type, style, and size in the various lists in the dialog box. The color of the text will be the current outline color selected in the Color Selection roll-up. Make sure you don't click on OK until your text is just right, because when you do click on **OK**, the text is pasted into the image. Use the **Edit Undo** command if you want to redo the text.

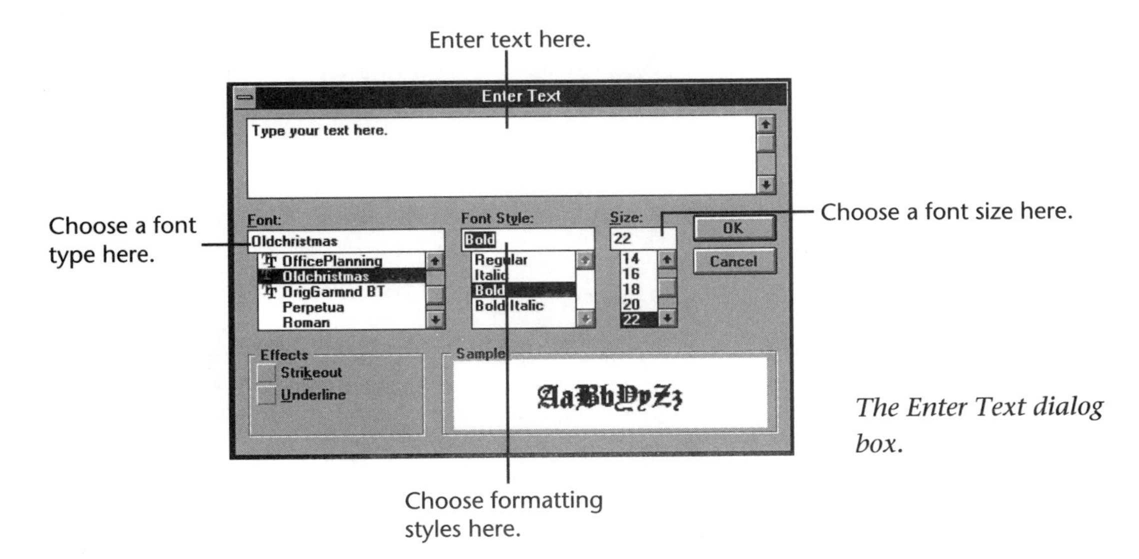

The Enter Text dialog box.

Fill Tools

Flood Fill tool

Fills an area with the outline or fill color currently selected in the Color Selection roll-up. Click the left mouse button to fill an area with the fill color; click the right mouse button to fill an area with the outline color.

Tile Fill tool

Fills an area with a repeating pattern. You'll use the Fill Settings roll-up to obtain the tile pattern you want.

Gradient Fill tool

Fills an area with a wash of colors: the fill gradually changes from one color to another. Choose the two colors you want in the To/From categories in the Color Selection roll-up.

Texture Fill tool

Fills an area with a cool-looking texture. You'll use the Fill Settings roll-up to obtain the texture style you want.

Retouching Tools

Freehand Smear tool

Smears colors in those areas of the image over which you drag this tool.

Freehand Smudge tool

Randomly mixes dots in those areas of the image over which you drag this tool.

Freehand Sharpen tool

Sharpens those areas of the image over which you drag this tool.

Freehand Contrast tool

Brightens or darkens portions of your image. A higher number in the Tool Settings box means that the light areas you drag over will get lighter, and the dark areas will get darker. So lower the number when you want less effect.

Make sure you display the Tool Settings roll-up (choose the Tool Settings Roll-Up command from the **D**isplay menu) when you use the freehand retouching tools. The Freehand Smear, Sharpen, Contrast, Brighten, and Blend tools all have control options in the Tool Settings roll-up. Click on the down arrow in the bottom left corner of the Tool Settings roll-up to display the control options for these tools.

Freehand Brighten tool

Intensifies the colors in the portions of your image over which you drag this tool. You only get one shot with this tool (only the first drag will affect the area). To intensify the effect, select a new level in the Tool Settings box.

The Brighten and Contrast tools can drastically change the look of an image, and add cool-looking effects. For example, look at how turning the contrast way up affects the lips, and how turning the brightness way down creates just the opposite effect.

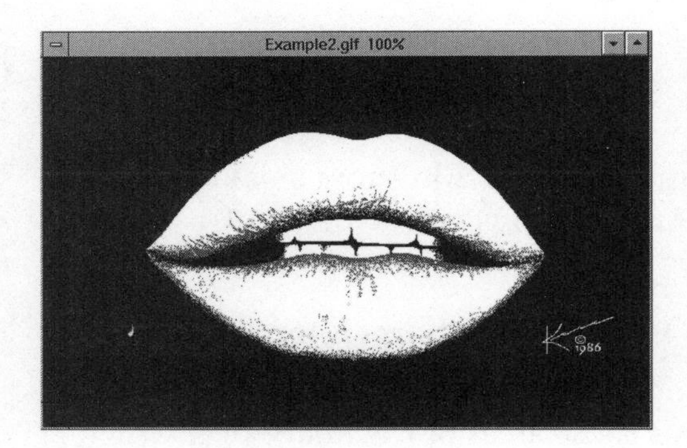

Contrast at 96%.

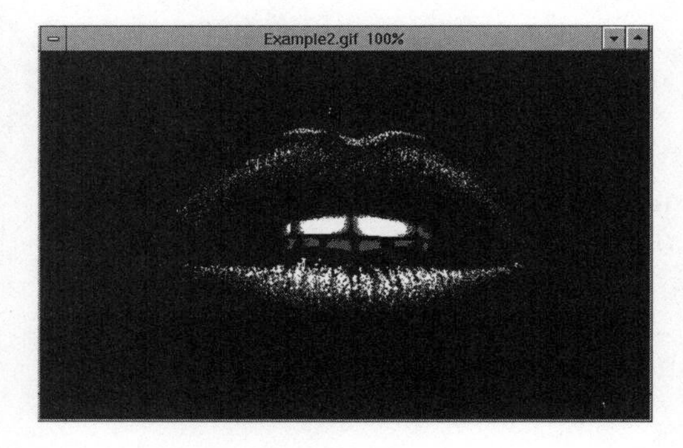

Brightness at –42%.

Freehand Blend tool

Smooths out the area of your image over which you drag the tool.

Freehand Tint tool

Adds tint to the portions of your image over which you drag. You only get one shot with this tool (only the first drag will affect the area). To intensify the effect, try a new color in the Tool Settings box.

Clone Tools

Clone tool

Copies one portion of an image to a new area in the image or to an entirely new picture. This tool is helpful to cover obstructions. For example, if you have a picture of a horse, but a gate door is blocking part of the horse's leg, you can clone from the horse's midsection to the area containing the gate, and cover the gate with the cloned portion so that the leg looks authentic (kind of like skin grafting on a human being).

Since there is no Tool Settings roll-up in CorelPHOTO-PAINT! 3, separate work boxes appear for some of the retouching tools. Adjust the controls in these work boxes to modify an image. Actually, I like some of these original work boxes better than CorelPHOTO-PAINT! 4's equivalent.

Pointillist Clone tool

Copies the color range of an area so that you can reproduce brush strokes of the same colors in the color range.

Impressionist Clone tool

Copies the color range of an area so that you can reproduce multitudinous dots of the same colors in the color range.

The Effects of Using CorelPHOTO-PAINT!

Get ready for some really funky-sounding names that you've probably not heard of before. They're all on CorelPHOTO-PAINT!'s Effects menu. Funky-sounding or not, these Effects commands will give your pictures some truly eye-opening looks. And the process is so simple. A couple of the commands will change your picture instantly when you select the command. Others access dialog boxes, where you can specify settings, preview the results, and click on **OK** when you're happy. The Preview feature is a true gem: it enables you to play around with all sort of different settings and never have to use the Undo feature. Just click on **Cancel** to restore the original image.

We'll take you on a brief tour of the Effects menu by applying each of the effects to an image and telling you a bit about the effect.

In CorelPHOTO-PAINT! 3, most but not all of the effects discussed in this section are located on the Edit Filter submenu.

CorelPHOTO-PAINT! 5 comes with some additional artistic effects. Try out the Swirl, Pinch/Punch, Spherize, 3D Perspective, and Mesh Warp effects.

Artistic Effects

You might remember Seurat, the great artist who developed the pointillism style of painting? He created fascinating paintings that were simply compilations of thousands and thousands of dots. If you stand up close to one his paintings, you can't tell what the painting depicts. It's only when you step back that you begin to see the true image portrayed by the arrangement of the dots.

The Artistic Pointillism command gives a pointillist look to your image by adding dots to it. You can adjust the size, spacing, and color of the dots in the Pointillism dialog box.

My favorite artist was always the great impressionist painter, Monet. I was so pleased to find that I could mimic his brush strokes on my computer (but I did wonder how he would feel about that). If you favor a pure impressionist look, try the Artistic Impressionism command. This commands adds impressionistic brush strokes to your image. Choose a shape and stroke direction in the Impressionism dialog box. If you want, you can modify the brush size, number of brushes, stroke length and size, and color variations, as well as the stroke spacing.

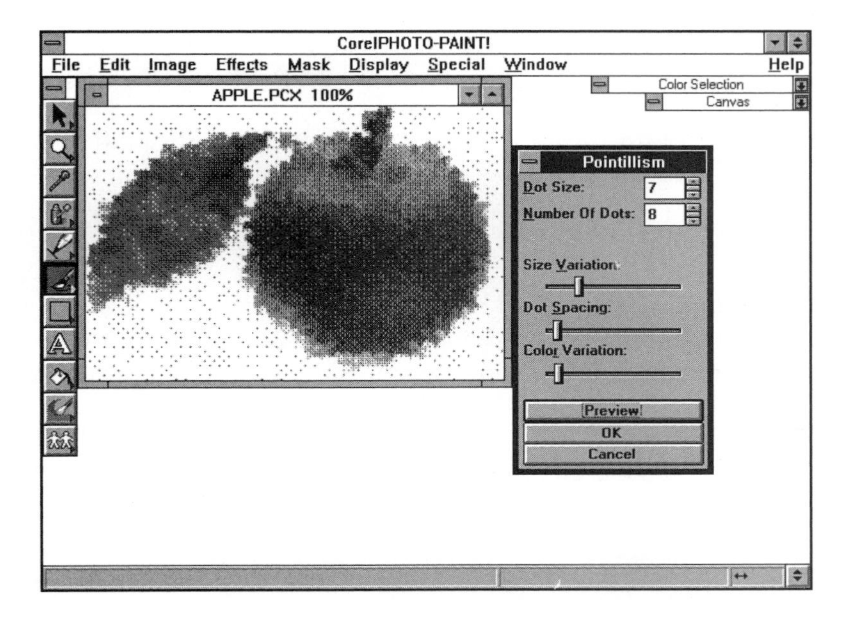

A veritable Seurat.

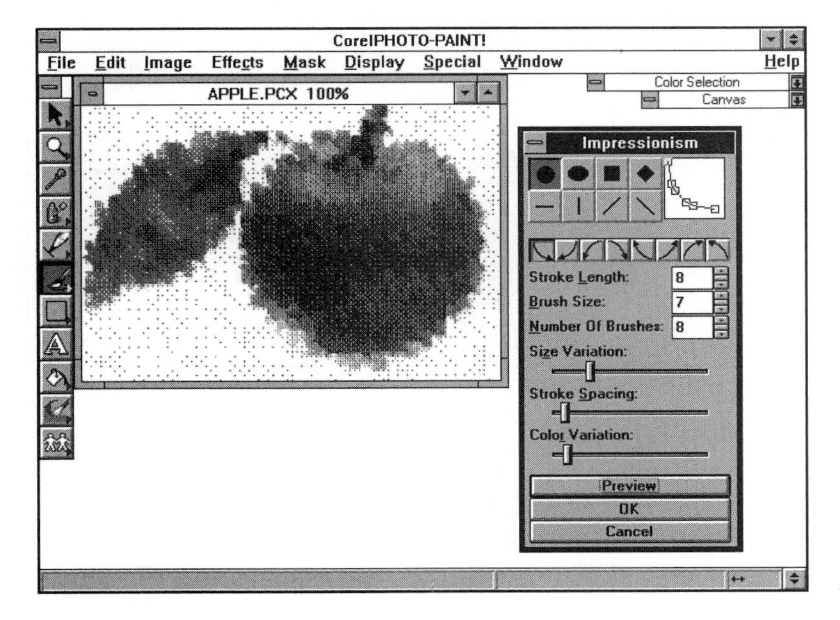

A veritable Monet.

Living on the Edge

The Edge command contains four subcommands: Edge Emphasis, Edge Detect, Contour, and Outline. Each of these gives an image quite a different look, but always involves modifying the image's edges. Because an edge is formed whenever different colors meet, the average image is likely to contain lots of edges.

The Edge Edge Emphasis command highlights edges between different colors and shapes. Depending on your image, a high Emphasis Amount setting can give your image a sharper, more defined look.

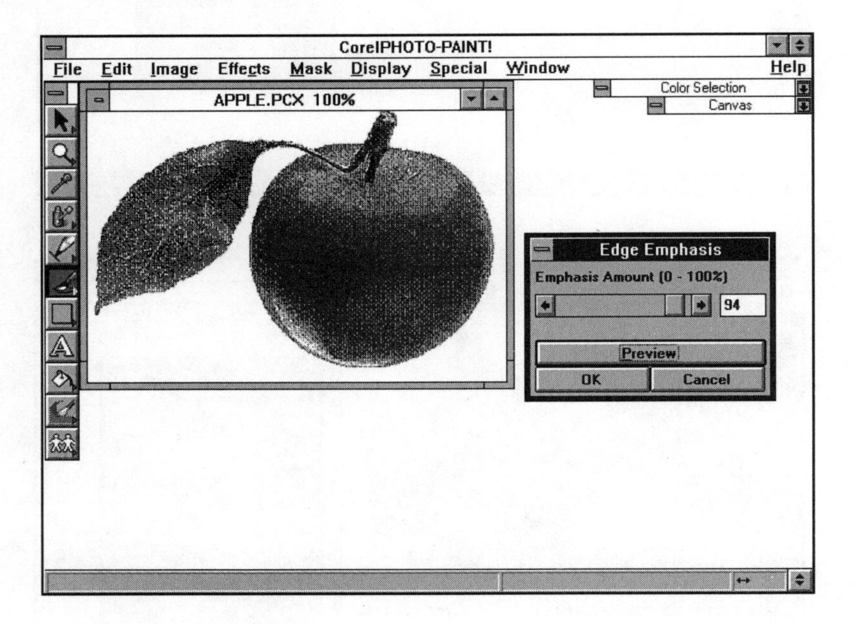

An apple, over the edge.

The Edge Edge Detect commands appear to bring your image's outline alive. The Color box in the Edge Detect dialog box determines the color of the non-outlined areas. For example, the image below now has a black background because Black is selected in the Color box. In the Sensitivity box, you can adjust how much of your image is outlined: the higher the number, the more sensitive the outline, meaning more edges of the image will be outlined. (At 10, every nook and cranny will be outlined.)

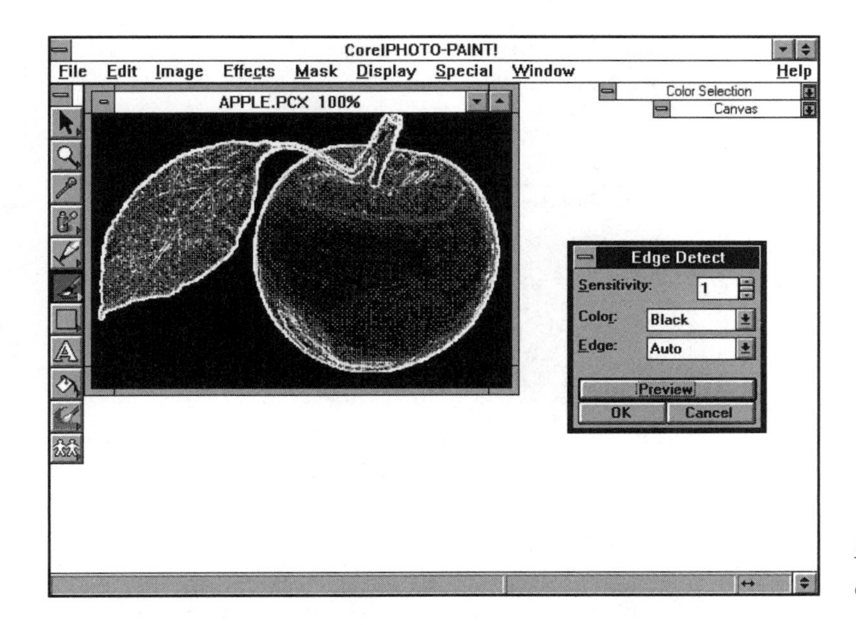

Detecting an edgy apple.

The Edge Contour command wipes out all color of an image and outlines its edges with lines. The level of Contour Threshold in the Contour dialog box determines the detail of the outline, with a higher amount generating a more detailed outline.

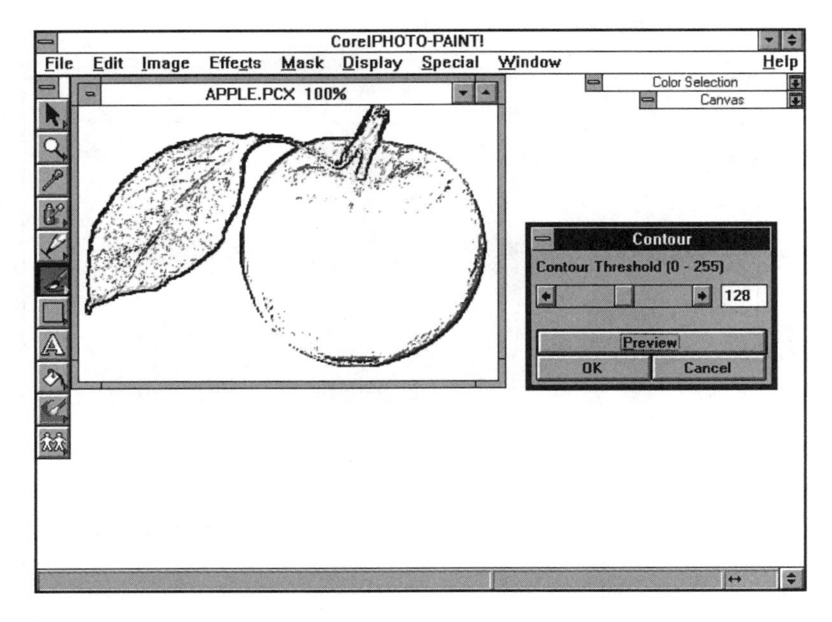

A see-through apple.

The Edge Outline command is a one-shot command. You choose this command and bingo: your image is entirely outlined, except for the barest edges.

Outline of an apple.

A Bossy Image

Something that's embossed is raised above the surface; for example, a seal can be embossed on paper (like on college transcripts). The Emboss command gives a raised feeling to your image, making it look three-dimensional, as if it were lifted off the surface. The Direction arrows in the Emboss dialog box determine which way the light source will be facing, thus affecting shadows and highlights. The Emboss Color will be the color of the entire image; gray usually works best.

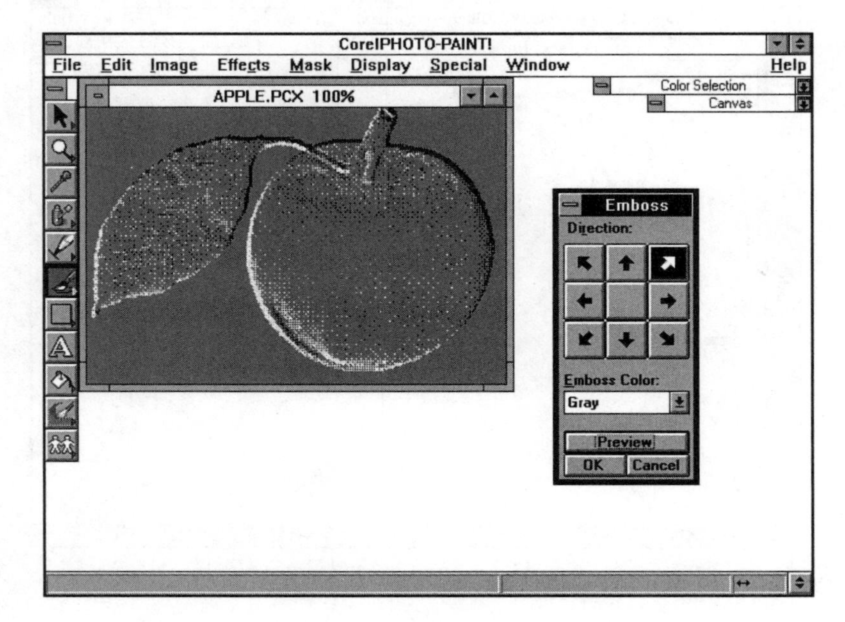

An embossed apple.

Inverting an Image

The Invert command is another one-shot command. When you choose it, your image becomes inverted instantly. This command gives a photo-negative look to a black and white image. For example, the second figure below looks just like the negative of a photograph.

An inverted apple.

A negative lady.

Jaggy de What?

For the longest time, I thought this was "jaggedy speckle," but it's "Jaggy Despeckle." This name definitely wins the contest for the weirdest Corel command name category. They should have named the command "fine fuzzy feeling" because that's the look it gives your image: a very fuzzy, blurred look. If you select the Allow Color Shift command, CorelPHOTO-PAINT! will add new colors to the image. You can determine the amount of diffusion CorelPHOTO-PAINT! applies in the Width and Height boxes. A higher number creates a more intense effect.

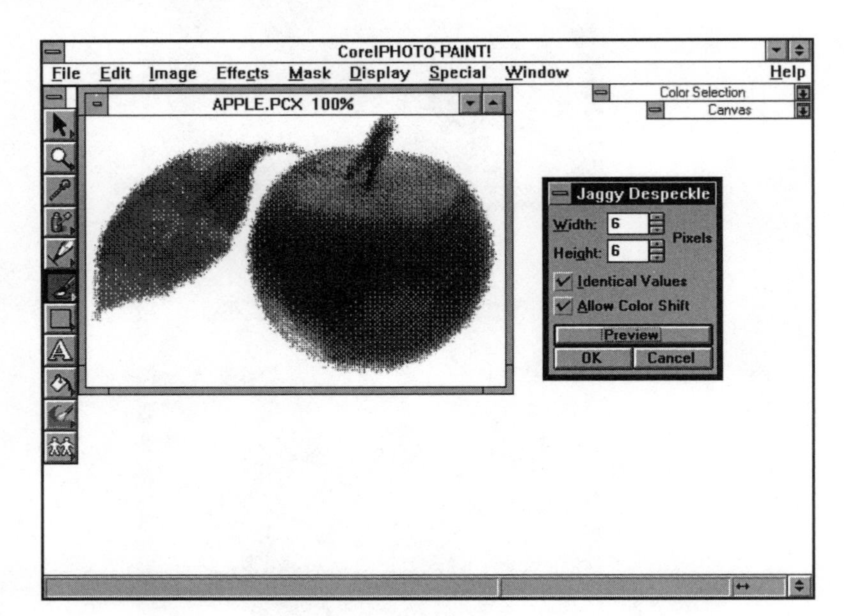

The apple, with that fine, fuzzy feeling.

In Full Motion

This is one of my favorite effects. It makes it look like your image is going places—and fast. The Motion Blur command does just what its name says: it blurs your image in one direction so it looks like the image is moving quickly in one direction. The Direction arrows in the Motion Blur dialog box determine the direction in which it looks like the image is heading. The higher the number in the Speed box, the more blurred the image becomes.

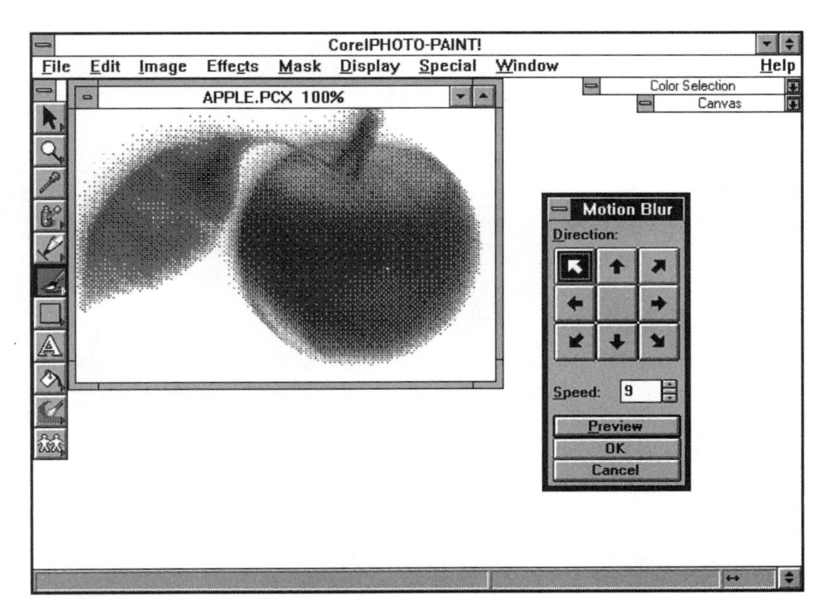

Apple on the run.

Enough Noise Already

You can add or remove noise from an image. In general, adding noise to a picture doesn't damage your ears; it just gives the image a more granular, textured look. A word of caution: when you add a high level of noise to your picture, you may have time to finish a seven-course dinner in the time it takes CorelPHOTO-PAINT! to get the image noisy. So use this command when you have time to spare. Removing noise can be useful when you have a poorly scanned image.

The Noise command contains a submenu of five commands:

- Add Noise   Adds a textured look to your image.
- Add More Noise   For you noisites who just couldn't get enough the first time, the Add More Noise command. You get three types of noises to choose from: Uniform gives an overall grainy look; Gaussian adds a heavy, larger grain size; and Spike applies a light-colored, thin grain.
- Remove Noise   Softens edges and gets rid of the jagged look that sometimes results from scanning.

- Maximum Lightens an image.
- Median Removes grainy appearance.
- Minimum Darkens an image.

Shake, Rattle, and Roll

The Pixelate command makes it look like someone is furiously shaking your picture back and forth. The command blocks the pixels of your image and creates a checkerboard appearance. You've probably seen this effect used on TV: when the media is trying to hide the identity of someone being filmed, they will pixelate the person's face so that you can not discern any features. For example, it's difficult to discern the identity of the frog in the second figure following, isn't it? The numbers in the Width and Height boxes determine the size of the blocks.

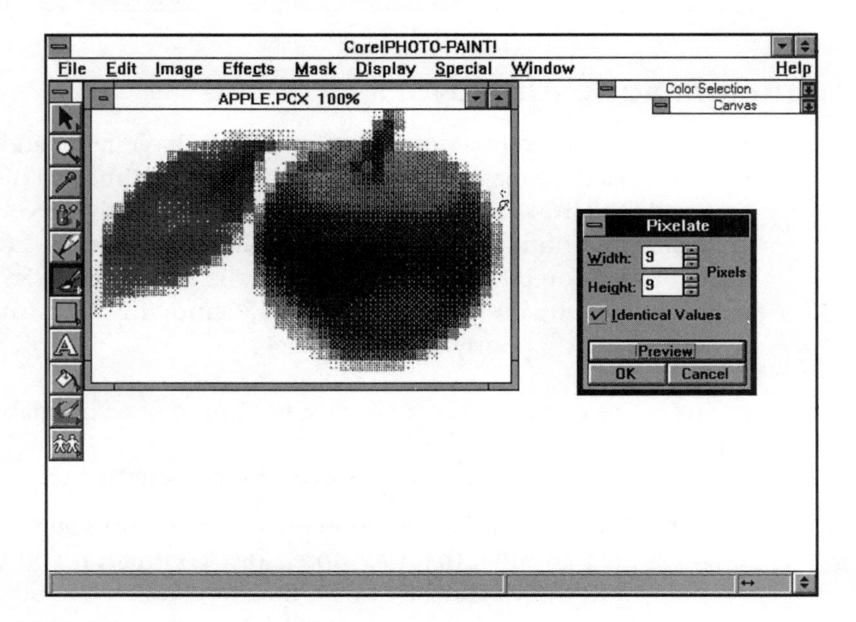

A pixelated apple.

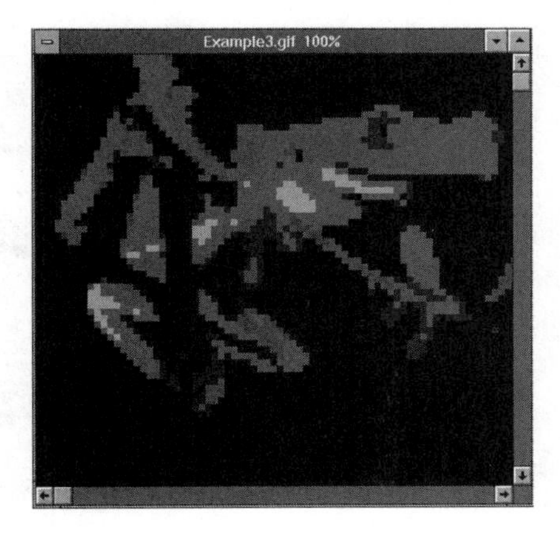

Protecting a frog's identity by pixelating.

Posterized Milk

The Posterize command switches gradient areas in your image to solid colors. It can give your image a choppy look since the gradual change of color in the gradient areas is changed to a solid block of color.

That's Psychedelic, Man

Those Corel geniuses never quit. Psychedelicize? If there were a contest category for the "most bizarre result of making an adjective a noun," the Psychedelicize command would win hands down. Still, us '60s offspring still enjoy it. You'll probably enjoy the radical effects it gives your images, too. To heighten the weirdness, increase the number in the Level box.

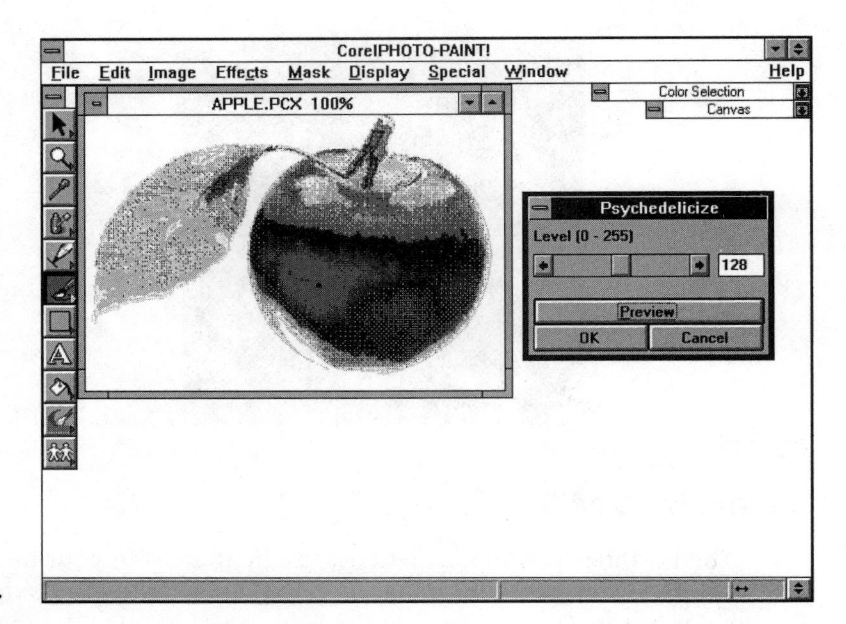

A groovy kind of apple.

Solarizing

The Solarize command reverses the look of an image, giving it a photographic negative result. This command differs from the Invert command because you can determine the level of effect you want, unlike the Invert command, which produces a complete reversal. The lower the level in the Threshold box in the Solarize dialog box, the less negative effect you'll get.

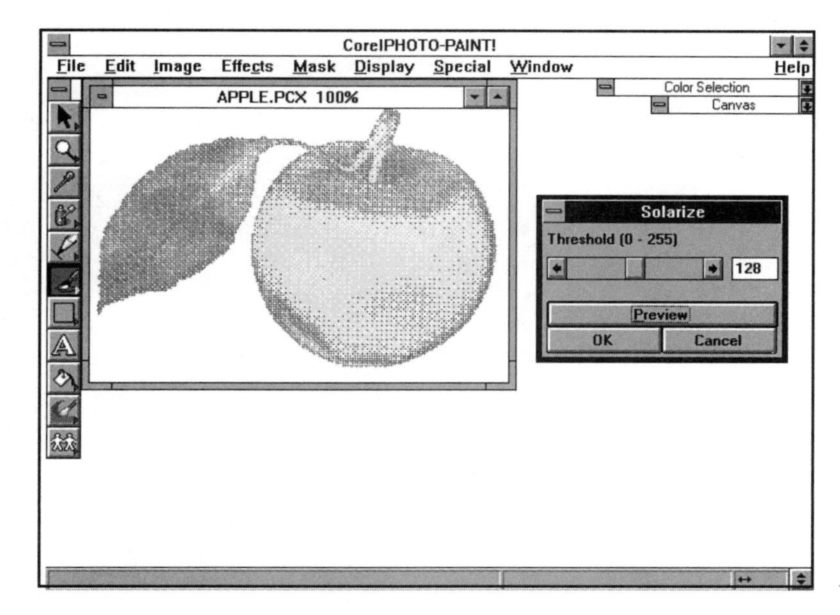

A solarized apple.

The Least You Need to Know

With CorelPHOTO-PAINT!, you can apply all sorts of fascinating and sometimes weird effects to pictures and objects.

- Double-click on the **CorelPHOTO-PAINT!** icon to start CorelPHOTO-PAINT!.
- Select the **F**ile **O**pen command to open a picture.
- Use the tools in the toolbox to select images; change the screen display; undo actions; draw lines, curves, polygons, rectangles, or ellipses; enter text; add fills; and touch up and clone images. (How's that for a mouthful?)
- Use the commands on the Effe**c**ts menu to add spectacular effects to your images.

This page isn't blank—it just looks that way.

Chapter 26
SHOW and Tell

In This Chapter

- CorelSHOW! greets you with a hug
- Hiding in the background
- SHOWing off
- Get going, SHOW
- Plopping things into your show
- Show on the run

Remember how anxious you'd get when show and tell time came around during elementary school? Some of you wanted to show your stuff so bad! Others wanted to crawl and hide inside a cabinet in the play kitchen. What happens now, when it becomes show and tell time with your boss? Do you strut your stuff with pride, or think about exactly what you'll say when you call in sick? Either way, why don't you let CorelSHOW! perform in your place. You can create screen shows that will dazzle your boss more than anything you've done before, and if you despise giving presentations before others, you can sit back and become part of the audience as CorelSHOW! does its thing. You don't even have to say a word if you don't want to. This chapter tells you about the screen shows you can create with CorelDRAW!'s presentation program called CorelSHOW!.

All Aboard the Screen Show Express

Welcome to the world of presentations. If you're not quite sure what a presentation program does, here are a few pointers. Whenever you need to present information to others, you can use a presentation program to put it all together. You compile all of your information into a presentation package. This package can include charts, drawings, diagrams, tables, text, bulleted points—what ever gets your message across to your audience. To use CorelSHOW!, you add these components to a CorelSHOW! file, and then do one of the following:

- Print the file, pass out handouts to the members of your audience, and then review the material with them.
- Create 35mm slides and present your information in the form of a slide show.
- Create overhead transparencies to display before your audience as you discuss the points of your presentation.
- Create a screen show, and either sit back and watch with the audience as the screen show plays back on a computer, or use the mouse to manually progress to the next screen, discussing the points of each screen as you go along. Or, you can give a copy of your screen show to all colleagues who need to review your material, and let them watch the screen show at their leisure on their own computers (this can be a very effective means of training others on a subject).

CorelSHOW! is a great program to use if you have a fast computer with a lot of memory. If your computer doesn't fit in that category, expect to be frustrated when you work in SHOW because it will run sluggishly while you're working, and your presentation will play back very slowly. This is because, in order to embed objects from other applications like CorelDRAW!, those applications must actually start up (so that you have CorelSHOW! and CorelDRAW! running at the same time). This isn't a problem for powerful computers, but wimpy computers will really suffer, and you will too because you'll have to wait around a lot. CorelSHOW! 5 is a bit faster, but it still won't run fast if it's on an old computer.

Since most users will want to use CorelSHOW! to create screen shows, and since this is a newer, less familiar way to administer presentations, this chapter focuses on creating a screen show. You can always print your screen show using the **File Print** command for handouts or overhead transparencies.

CorelSHOW! is the only CorelDRAW! program that greets you with a welcome message each time you start it. About starting it: just double-click on the **CorelSHOW!** program icon in the Corel group window in the Program Manager. If you need further help starting CorelSHOW!, read the beginning of Chapter 3, "Taking Your First Step."

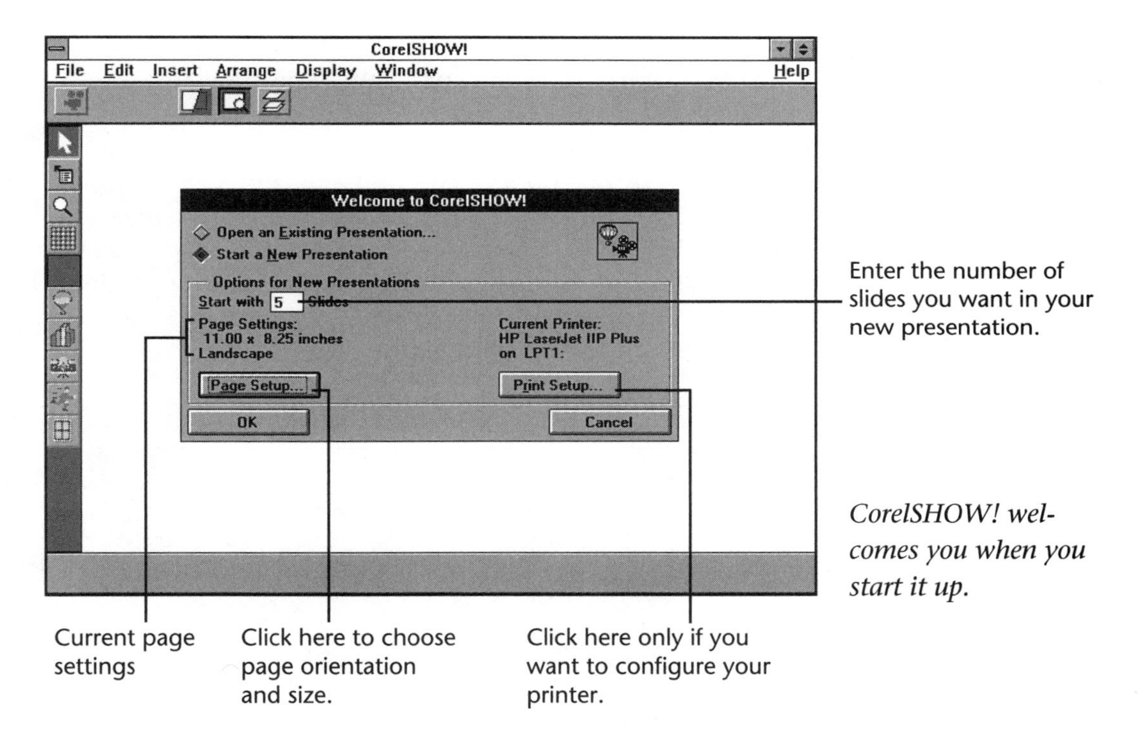

CorelSHOW! welcomes you when you start it up.

The first choice you need to make when the CorelSHOW! Welcome screen appears is whether you want to open an existing presentation or start a new one. To open a presentation you already created, click on the Open an Existing Presentation option button. Then you can specify the file you want to open in the Open Files dialog box which will appear.

Since the Open an Existing Presentation option is selected by default, the options for creating a new presentation appear dimmed in the Welcome dialog box. You must click on the Start a New Presentation option to access the options and create a new presentation. See if the current page settings in the Options for New Presentations area meet your needs. By default, CorelSHOW! designs the presentation page for a screen show as an

11" x 8 1/4" page in landscape orientation. (This your standard piece of paper turned sideways.) If you want to create handouts, overhead transparencies, or 35mm slides, or use a different page size, click on the Page Setup button. The Page Setup dialog box appears. Here you can select the type of presentation page you want and determine its orientation.

SPEAK LIKE A GEEK

Presentation A collection of slides that can be printed, made into 35mm slides, used for overhead transparencies, or run in a screen show.

Slide A presentation page.

By default, CorelSHOW! creates 5 slides for you. That's not very many, but you can always add more slides as you go along. And it's usually easier to work with a small number of slides and add more as you go along, than it is to manipulate a large twenty-slide file from the start, which can be quite cumbersome. However, the number of slides you include is entirely up to you. Just enter a number in the Start with Slides box.

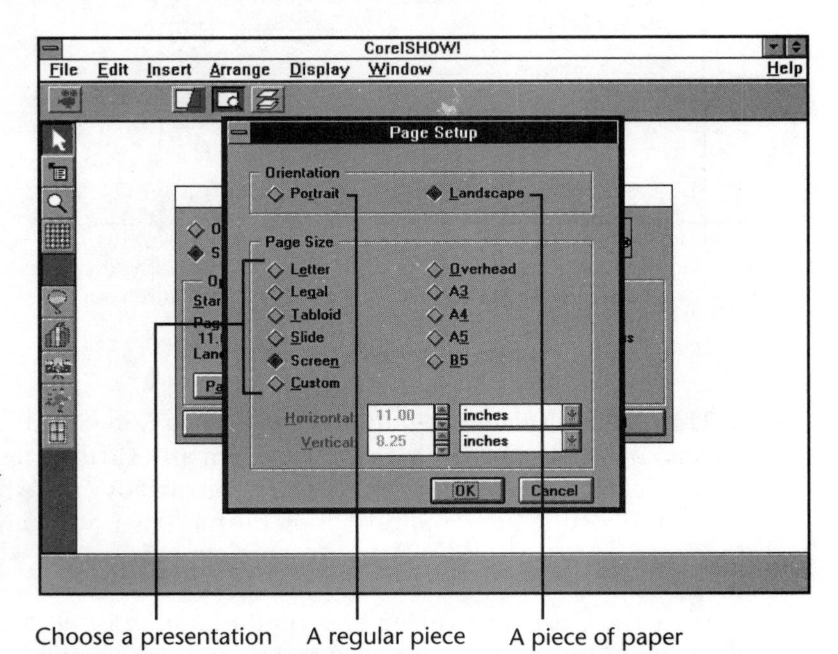

The Page Setup dialog box pops up on top of the Welcome dialog box.

After you've selected an orientation and page type, click on the **OK** button. CorelSHOW! returns you to the Welcome dialog box. Click **OK** again to close the Welcome box. CorelSHOW! sets up the page size and orientation as you specified, and creates the number of slides you specified. You're ready to go.

Paging All Tools

The CorelSHOW! 3 toolbox doesn't have nearly as many tools as the one in CorelSHOW! 4. In fact, you only get five tools: the Selection, Background Library, CorelDRAW!, CorelCHART!, and Other Applications tools.

The CorelSHOW! screen is full of fun features. At the top of the screen, you can click on a button to instantly change to the background view, to slide view, or to slide sorter view (which shows you tiny pictures of all the slides in your presentation). There's a box where you can specify how long a slide should stay on-screen, and a button to run the screen show when it's you're ready. The two clocks at the bottom of the screen help you keep track of how long a screen show has been running, and the total time it will take to run the show. But the most important part of the screen is the toolbox on the left side. You'll use these tools to perform most of the CorelSHOW! functions.

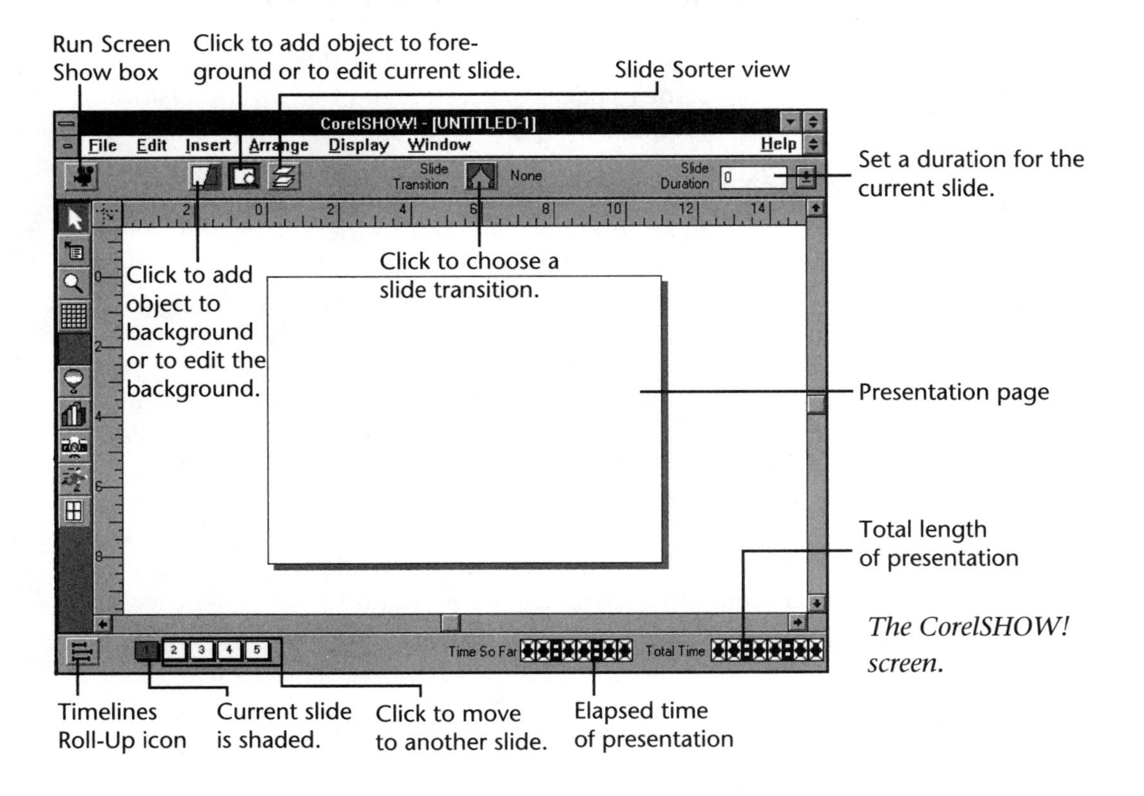

The CorelSHOW! screen.

Toolbox Icon	Function
	Selects an object.
	Displays pop-up menu for quick access to commands.
	Enables you to zoom in on the presentation page.
	Accesses the background libraries.
	Adds a CorelDRAW! drawing.
	Adds a CorelCHART! chart.
	Adds a CorelPHOTO-PAINT! image.
	Adds a CorelMOVE! animation.
	Adds objects from other Windows OLE capable applications.

Background layer The common full-page design used in all slides of the presentation.

Foreground layer The top layer of a slide where you can place individual objects that sit on top of the background layer.

A Background Search

A presentation is made up of *slides*, which are the same as pages. Each slide has two layers, a *background layer* and a *foreground layer*. The background layer holds the full-page design that usually appears on every slide in the presentation; objects you add to an individual slide sit in the foreground layer of the slide. Most presentations use the same background for each slide (for continuity), but each slide has its own unique foreground.

You can add a background to the presentation at any time, though it's often the first step you'll take. By adding the background first, you can arrange foreground objects so that they fit nicely on the background. CorelSHOW! comes with a collection of backgrounds you can add to your presentation slides. To access these backgrounds, click on the **Background Library** tool on the toolbox. When you click on the Background Library tool, the Open Files dialog box appears.

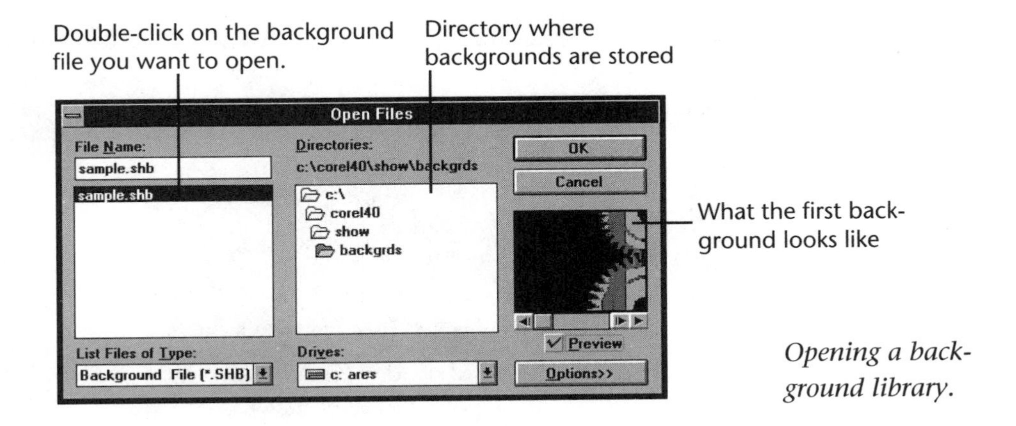
Double-click on the background file you want to open.
Directory where backgrounds are stored
Open Files
File Name:
sample.shb
Directories:
c:\corel40\show\backgrds
OK
Cancel
c:\
corel40
show
backgrds
What the first back-ground looks like
Preview
List Files of Type:
Background File (*.SHB)
Drives:
c: ares
Options>>

Opening a back-ground library.

Double-click on the background library you want. This opens the background library, which SHOW places inside a window. You can preview all of the backgrounds in the library in this window, four at a time. Use the scroll bar to see the rest of the backgrounds. When you find the back-ground design you want, click on it and then click on the **Done** button. CorelSHOW! adds the background to all the slides in your presentation.

The CorelDRAW! Install program places a sample background library in the C:\COREL40\SHOW\BACKGRDS directory. This library holds about a dozen backgrounds you can add to your presentations. Many more back-ground libraries are available on the CorelDRAW! CD-ROM. So see if you can get your hands on some of those additional libraries!

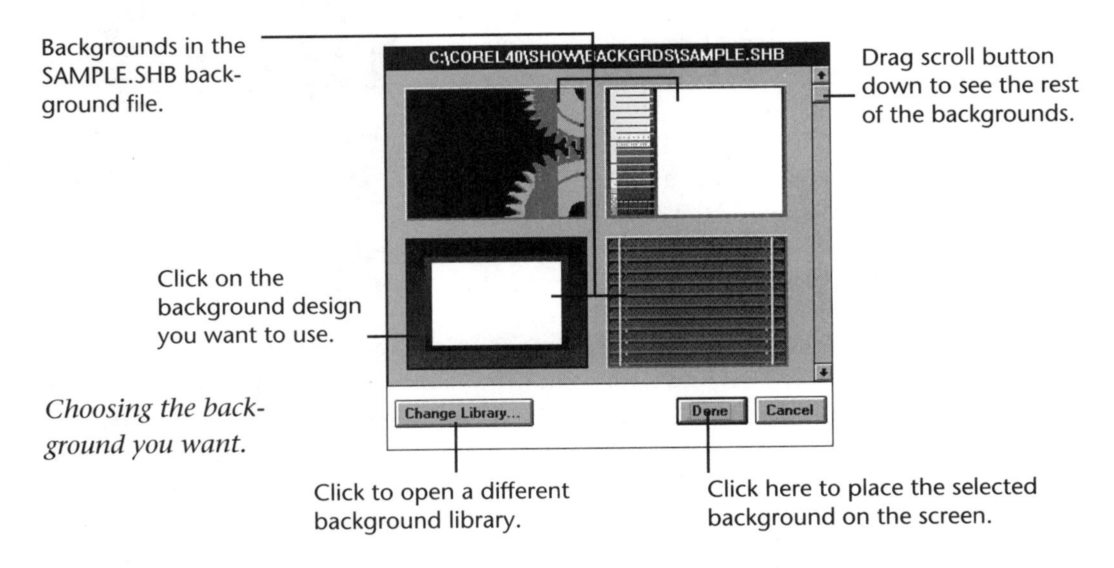
Backgrounds in the SAMPLE.SHB back-ground file.
C:\COREL40\SHOW\BACKGRDS\SAMPLE.SHB
Drag scroll button down to see the rest of the backgrounds.
Click on the background design you want to use.
Change Library...
Done
Cancel
Click to open a different background library.
Click here to place the selected background on the screen.

Choosing the back-ground you want.

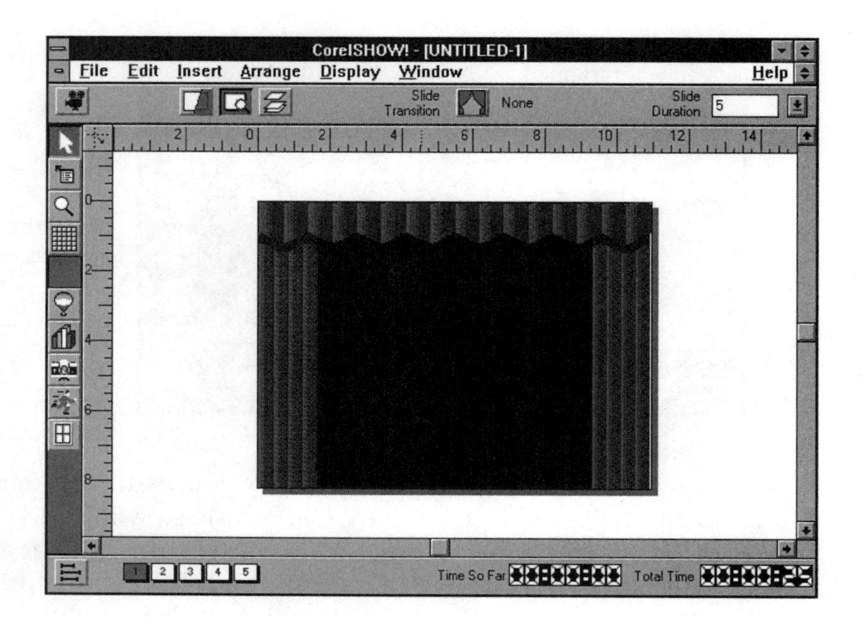

A background added to the presentation.

What's Inside

Okay, you've got the background set up, now you have to fill individual slides with charts, drawings, images, animations, etc. In order to include objects in your presentation, you'll embed them from other programs. CorelSHOW! itself doesn't have any charting, drawing, painting, or animation capabilities. After all, you have CorelCHART!, CorelDRAW!, CorelPHOTO-PAINT!, and CorelMOVE! for all of that. CorelSHOW! is just the place you stick all of your charts, drawings, images, and animations when you want to include them in a presentation.

In addition to using the other CorelDRAW! components, you can embed or link objects and files from other applications to your presentation. *Embedding* inserts a severed copy of the object or file, while *linking* inserts a copy which will change anytime you change the original object or file. In this chapter, we'll concentrate on embedding objects and files.

Chapter 22, "Linking and Embedding, OLE," tells you all about embedding and linking files. So if these snazzy terms boggle you, turn to that chapter for details.

The best way to show you how to embed an object and file from another application into CorelSHOW! is to spell it out in steps. So here we go. . . .

Embedding an Object

If you want to embed an object you haven't yet created in another application, follow the steps below.

Embed To insert a copy of an object from another application so that when you double-click on an embedded object, the application used to create the object starts up.

Link To insert an object from another application so that whenever you change the original object in the other application, the linked object changes automatically.

1. Display the slide to which you want to add the object.
2. Click on the **Foreground View** icon at the top of the CorelSHOW! screen.
3. Click on the toolbox icon that represents the program from which you want to insert the object (that program is called the source application). For example, to embed a chart, click on the CorelCHART! icon; CorelCHART! is the source application. Or, to embed an image, click on the CorelPHOTO-PAINT! icon, which makes CorelPHOTO-PAINT! the source application.
4. Drag the mouse button to create a rectangle on the page. Make the rectangle the size you want for the object. This step starts the source application. For example, if you clicked on the CorelDRAW! icon to embed a drawing, the CorelDRAW! program now starts.
5. Create the object in the source application.
6. After you've finished creating the object, display the File menu in the source application. Depending on the application, you'll see an Update, Exit and Return, or just plain Exit command. Select whichever command is available to you. If you choose just Exit, a dialog box may pop up asking you whether you want to update the object. Choose Yes or OK.

Voilà! The object is now embedded into your CorelSHOW! presentation slide. You can move, scale, or resize it as you wish.

Embedding a File

Perhaps you've already created the chart, drawing, image, or animation you want to include in your presentation. If so, follow these steps to embed the file into your CorelSHOW! presentation:

1. Display the slide to which you want to add the file.
2. Click on the **Foreground View** icon at the top of the CorelSHOW! screen.
3. Display the Insert menu and select the Object command to open the Insert Object dialog box. By default, the Create New option is selected.

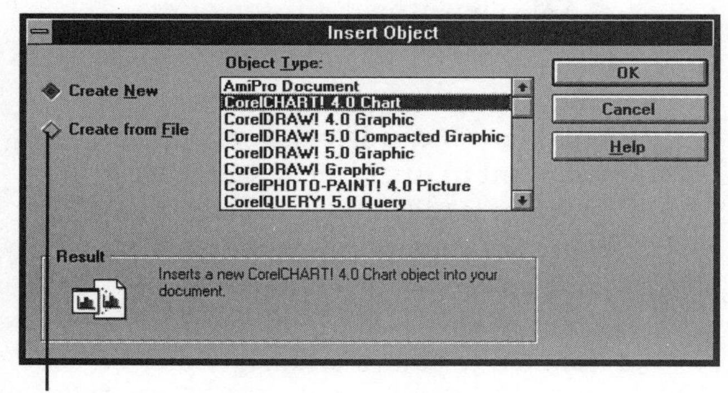

The Insert Object dialog box's default appearance.

Click here to embed an existing file.

4. Click on the Create from File option. The look of the Insert Object dialog box changes.

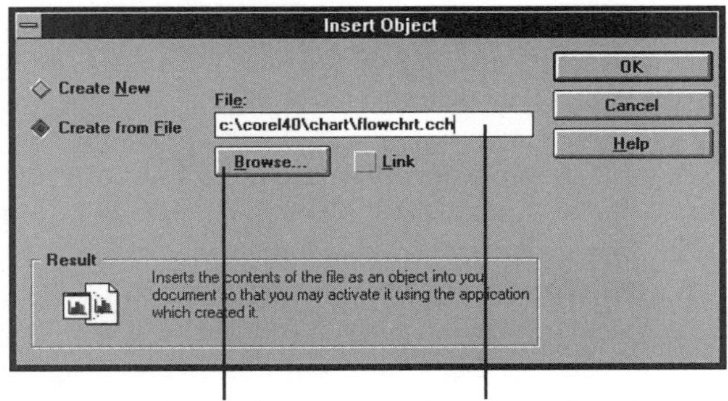

Embedding an existing file.

5. Type the name of the file you want to embed, including the directory it's located in. For example, to embed the file FLOWCHRT.CCH in the CHART directory, you'd type: C:\COREL40\CHART\FLOWCHRT.CCH. If you don't remember the exact spelling of the file, or what directory it's located in, click on the **B**rowse button and search for the file in the Browse dialog box.
6. Click on **OK**.
7. Drag the mouse button to create a rectangle on the page. Make the rectangle the size you want for the object. This step starts the source application. For example, if you clicked on the CorelDRAW! icon to embed a drawing, the CorelDRAW! program now starts.

The contents of the file now appears on your presentation page, selected. You can move, scale, or resize it as you wish.

Ladies and Gentlemen: The Show Is About to Begin!

Picture this: you go to a play, and during intermission you get in line to get a drink. You've been in line for 15 minutes, there are still 10 people in front of you, and the lights dim to indicate the show is about to begin. You start to panic. Do I wait to get the drink that I've been waiting so long for already? Or do I scrap it so I don't miss any of the show? When it comes to a Corel show, you don't want to miss a minute, so get out of line and get back to your seat. The show is about to begin.

When you get to see your presentation play back on your computer screen, it's called a *screen show*. CorelSHOW! lets you choose how you want your screen show to run. For example, you can have the screen show run on its own, or you can manually advance to the next slide by clicking the mouse button. You can have the show run over and over again until you press Esc (which is great if you're using your computer as a message center for passers-by). Or, perhaps you're one of those people who likes to be able to point things out on the screen as you discuss what's showing (you know—always talking with your hands). If so, you can display a pointer on the screen and use your mouse to point to whatever you're talking about at the moment.

To choose screen show options, select the **D**isplay **P**resentation Options command. The Presentation Options dialog box appears. Choose an automatic or manual progression, and any other options you might want.

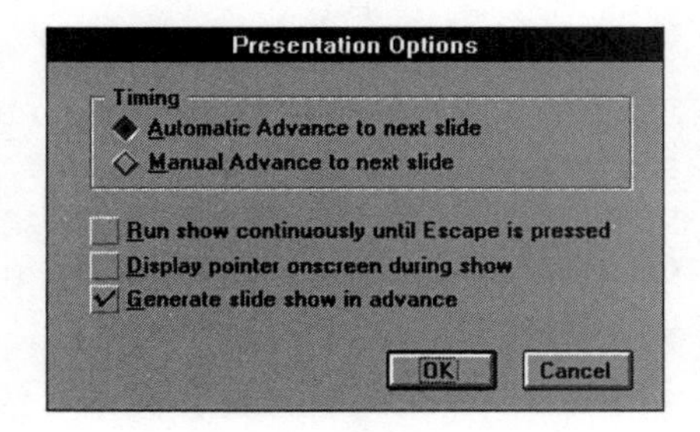

Specifying how you want the screen show to run.

If you choose an automatic screen show, you need to select a duration time for each of your slides. Display the slide, and then click on the arrow to the right of the Slide Duration box (in the upper right corner). This drops down a list of times. Click on the number of seconds you want the slide to remain on-screen during the screen show.

Now you're ready to run the show. Click on the **Run Screen Show** box just above the toolbox, or choose the File Run Screen Show command. It may take a while for CorelSHOW! to get things together. When SHOW is ready to run the presentation, it displays a dialog box asking you if want to start the presentation. Click **OK** when you're ready. And there it goes: the slides of your presentation flash before you.

Incidentally, the "start screen show" check is a helpful feature when you're running your screen show before others during a meeting. You can select the **Run Screen Show** box before the meeting to get everything ready to run, and then click on the **OK** button at any time during the meeting. This way you avoid having to wait around for a few minutes while CorelSHOW! loads the screen show.

Let's Get the Show on the Road

At times you may want to distribute a screen show to individuals on disk, so that they can run the presentation in the privacy of their offices. The only requirement is that the recipient's computer is running Microsoft Windows 3.0 or higher. The recipient does not need to have CorelSHOW! installed on his or her computer!

In order to have a screen show run on another computer, you must save the presentation as a screen show onto a floppy diskette. Don't do this until the presentation has been fully edited, because you cannot edit a screen show file. When the presentation is ready, put a diskette into a floppy drive and select the File Save As command. Mark the **Screen Show Only** check box in the Save As dialog box. Click on the Segment File for Portable Media option button to place the screen show file onto a floppy diskette. When asked, indicate which floppy drive the diskette is located in.

Now you need to copy CorelSHOW!'s SHOWRUN.EXE file onto the floppy. This file lets the recipient run the screen show (even if CorelSHOW! is not installed on his computer). To copy this file to the floppy diskette, use the Windows File Manager. If you're not familiar with the File Manager, exit Windows. At the DOS prompt, switch to the appropriate floppy drive by typing **B:** or **A:** and pressing **Enter**. Then type the following line and press **Enter**:

COPY C:\COREL40\CORELSHOW\SHOWRUN\SHOWRUN.EXE

3.0

With Corel 3, use **C:\COREL30\CORELSHOW \SHOWRUN\SHOWRUN.EXE.**

You'll see the message "1 file copied," if the file was copied properly to the diskette. Read the next section for information on what you need to do to run the screen show on another computer.

Instructions for the Recipient

5.0

With Corel 5, use **C:\COREL50\CORELSHOW \SHOWRUN\SHOWRUN.EXE.**

It would probably be a good idea for you to include instructions with your diskette on how the recipient can run the screen show on his or her computer. You're in luck! . . . We laid out the instructions for you in the "Put It to Work" section of this book. All you need to do is turn to the "Put It to Work" section and find the page with the portable screen show instructions. Make copies of the page and hand them out along with the diskettes containing your screen show.

The Least You Need to Know

Show off your CorelDRAW! stuff with CorelSHOW!. Run a screen show during a meeting, or hand out disks so colleagues can run the show on their own computers!

- Double-click on the **CorelSHOW!** icon to start CorelSHOW!.
- Set up the presentation pages for 35mm slides, overhead transparencies, handouts, or a screen show.
- Select a background for your slides.
- Embed charts, drawings, images, and animations in your presentation from CorelCHART!, CorelDRAW!, CorelPHOTO-PAINT!, CorelMOVE!, or another application.
- Indicate how you want to run the screen show in the Presentation Options dialog box, accessed through the **D**isplay **P**resentation Options command.
- Run the screen show by clicking on the **Run Screen Show** box.
- Distribute your screen show on disk so that others can run the show on their own computers.

**This page is about nothing,
absolutely nothing.**

Part VI
Always the Logistics

When I was young I ran free and wild—I didn't know the word schedule, had no idea of what bills were, had never heard of taxes, could care less about rules. A shocker came when I reached adulthood: no one had ever told me that you grow up so that you can follow schedules, pay bills, do your taxes, and make rules (for your children) ... Why is it we adults can never seem to escape the logistics?

The bad news: this Part deals with CorelDRAW! logistics (you know, the stuff you have to deal with before you can move on). The good news: you may be able to skip each one of the chapters in this Part. If you have no interest in CorelDRAW! 5, are a seasoned Windows user, and have already installed CorelDRAW! on your computer, go ahead and run wild and free. You can skip this Part entirely!

However, don't skip the chapter called "Put It to Work," You'll find some cool CorelDRAW! ideas in this chapter.

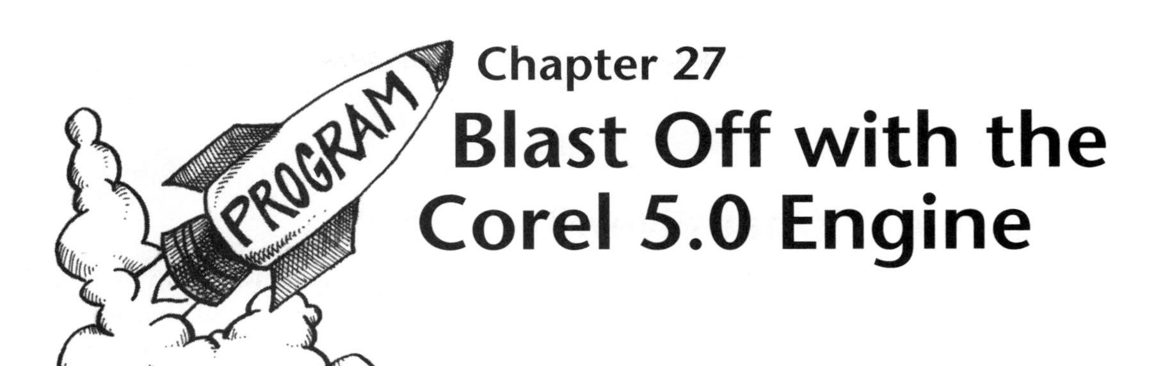

Chapter 27
Blast Off with the Corel 5.0 Engine

In This Chapter

- And now . . . even more programs
- A brand new look
- Seeing it all from a new View
- The look and feel of Presets
- Introducing Ms. Type Assist
- Things just aren't the same around here

Last year we bought a brand new Ford Mustang with 225 horsepower and a 5.0 engine. Have you ever driven a car with a 5.0 engine? It's one powerful ride. You have to have a good grip on the steering wheel and go easy on the gas pedal (unless you're like my husband and actually enjoy turning corners at 50 mph).

Using the new CorelDRAW! 5.0 is like riding in our Mustang. It's powerful, and you're in store for an exhilarating ride if you've been using CorelDRAW! 4. (You CorelDRAW! 3 users had better really hold on tight.) The first thing I noticed about CorelDRAW! 5.0 was how much faster it moved. You'll discover this during installation; what took versions 3 and 4 several minutes to compute takes version 5 only a few moments. And there are all sorts of new knick knacks you're sure to enjoy, as well as changes to existing features that you'll appreciate. This chapter will tell you all about Corel's new 5.0 engine.

Hey Dude, What's New?

All sorts of things are new. What you'll probably appreciate more than anything else is CorelDRAW!'s new speed. No more twiddling your thumbs while you wait around for a large file to open. This enhanced speed is sure to heighten your own performance since you'll spend more time working and less time waiting. Geez, I hate waiting. Like when it's Friday at noon, you're at the bank, there are about 150 people in front of you, and they all want to open a new account. Sorry. I digressed.

Each of the programs common to versions 3 and 4 have undergone changes and offer brand new features. Beyond that, you get a fully equipped, powerful desktop publisher with CorelDRAW! 5, called CorelVENTURA!. Some new utilities have been added as well, such as CorelQUERY! and the ARES Font Minder.

CorelDRAW! even looks different when you first open it, because a new feature, called the Ribbon, sits on top of the application window where the old status line used to be, and the status line is now back at the bottom of the screen. (It's just like Corel: first the status line was on at the bottom of the screen in version 3, then at the top in version 4, and now again at the bottom in version 5. What can we expect in 6?) The Ribbon holds icons for all sorts of common commands that you'll use while working in CorelDRAW! 5 (we'll tell you about these commands later).

Finally, a whole new array of features has been implemented in version 5. They range from PowerTyping and tabbed dialog boxes to preset graphic styles and morphing. You'll learn about these new features later in this chapter too.

Just When You Thought There Were Already Too Many . . .

Add a brand new program and enhanced features to the existing programs, and you've got CorelDRAW! 5. Check out the new features and benefits below.

CorelVENTURA!

There was always one thing lacking from CorelDRAW! that would make it the best and most complete business package on the market: a desktop publisher. You could do everything with graphics, but not much with heavy-duty text—until now. Welcome CorelVENTURA!, a full-featured desktop publishing application. With CorelVENTURA!, you can create full-fledged books, using CorelDRAW! graphics to spruce them up. CorelVENTURA! works similarly to other Windows desktop publishers, and supports standard Windows text editing conventions. Like its sister applications, CorelVENTURA! uses roll-ups and gives you direct access to CorelDRAW! fills and outlines.

CorelPHOTO-PAINT! 5

Handling really large files in PHOTO-PAINT has always been a long, tedious process. You've always excused it with, "Well, it is a large file." Fortunately the Corel designers weren't so generous. They came up with a way to get around the burden of manipulating huge graphics files. You can now crop or open only partial areas of files so that you don't have to deal with the whole shamole at once.

Additionally, CorelPHOTO-PAINT! 5 gives you some new funky artistic effects (as if the present ones weren't fun enough), like Swirl, Pinch/Punch, Spherize, 3D Perspective, and Mesh Warp. Not only are there some new brush styles, but now you can create your own custom brushes as well.

CorelCHART! 5

I always wanted to be able to add CorelDRAW! effects to some of my charts directly from CHART. Well, now I can. Version 5 gives you direct access to CorelDRAW! fills and outlines, not only for the charts themselves, but for the Data Manager contents as well. If that's not enough, you get ten new chart styles, and a whopping 300+ new spreadsheet functions.

To see how PHOTO-PAINT 5's new artistic effects look, choose them in turn from the Effects menu and apply them to any selected image.

The new CorelCHART! screen looks similar to the new CorelDRAW! screen. It has a tear-off toolbar and a Ribbon bar full of commands you can access by clicking on an icon.

CorelMOVE! 5

Have you ever heard of morphing? It's the latest craze in graphics programs. It is somewhat like bringing CorelDRAW! blended objects to life (that's meant to be a hint). Think about it:

Morphing . . . metamorphosis . . . cocoon . . . butterfly

That's it. *Morphing* is the act of turning one object into another with a very smooth transition. You may have seen the commercial where one man starts shaving his face, and then all of a sudden his face totally melts into the shape of another man and before you know it there's a completely different face being shaved, and so on. (You may have also seen morphing in a recent Michael Jackson video.) That's morphing. It's turning an apple into an orange before you know what's happening, and CorelMOVE! has it now.

With Move 5, you can display multiple libraries, and you get 300+ new actors, props, and sounds. But you really need a CD drive to reap the benefits of the latter.

CorelSHOW! 5

With CorelSHOW! 3 and 4, using more than 20 slides makes your screen show unstable. Not so with version 5, which boasts significant gains in performance. You can enter your text directly on-screen like you do in all the other popular presentation programs, and you can create speaker notes that help you not look like a fool when you are giving a presentation.

CorelDRAW! 5.0 Gets a Facelift

Something strikes you as different the moment the CorelDRAW! 5 screen opens. It's not immediately obvious, but it's there . . . on top of the window, under the menu bar . . . a strip with all sorts of icons. Yes, it's called a Ribbon bar. This new Ribbon bar holds a lot of icons that perform functions for you when you click on them. For example, when you click on the Open Drawing Dialog Box icon, the Open Drawing dialog box opens (who'd've guessed?). And when you click on the To Front icon, the selected object moves in front of all other objects. You'll become dependent on these icons as time goes on. The following table introduces you to each of the Ribbon's icons.

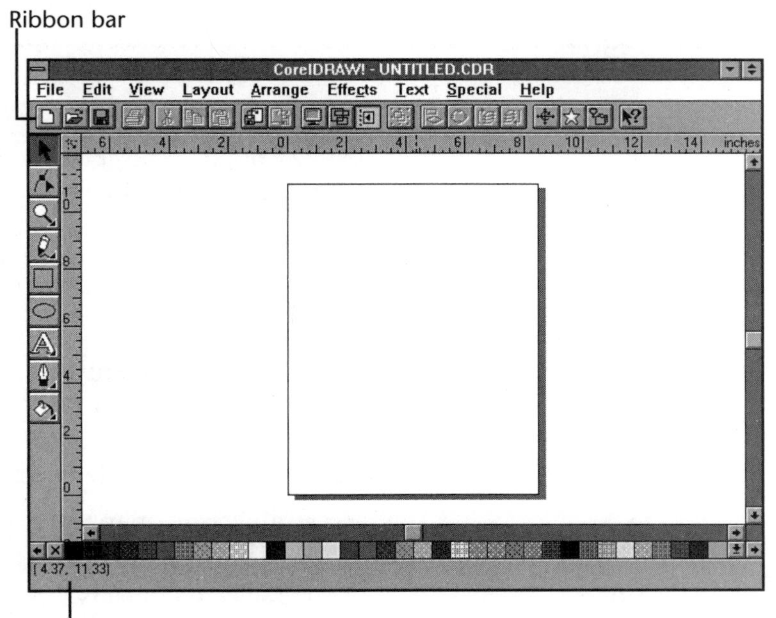

The new version 5 screen.

Icon	Name	Description
	Open Drawing	Starts a new drawing.
	Open Drawing Dialog Box	Opens an existing drawing.
	Save Drawing	Saves a drawing.
	Print Drawing	Prints a drawing.
	Cut	Cuts a selection and places it on the Windows Clipboard.
	Copy	Copies a selection to the Windows Clipboard.
	Paste	Pastes contents of the Windows Clipboard into a drawing.

continues

continued

Icon	Name	Description
	Import	Imports a drawing that was created in a program other than CorelDRAW!.
	Export	Saves a drawing in a format other than CorelDRAW!'s.
	Full-Screen Preview	Removes everything from the screen except a drawing.
	Show Wireframe	Switches to wireframe view (works as a toggle).
	Snap	Forces objects to snap to grid (works as a toggle).
	Group	Groups selected objects.
	Align	Aligns selected objects.
	Convert to Curves	Converts a line object to curves.
	To Front	Moves selected object in front of all other objects.
	To Back	Moves selected object in back of all other objects.
	Transform Roll-Up	Displays Transform roll-up.
	Symbols	Displays Symbols roll-up.
	Mosaic	Displays Mosaic roll-up.
	Context-Sensitive Help	Displays a Help cursor. (Click on the item with which you need help.)

The names of the icons tell what the icon does. There are only a few icons that may need more of a description:

- The Symbols icon displays the Symbols roll-up. There is no longer an icon on the Text menu for symbols.
- The Mosaic icon opens the Mosaic roll-up. The Mosaic roll-up can be used for displaying, organizing, and moving graphic files.
- The Snap icon snaps objects to guidelines. This icon toggles between snapping and not snapping to guidelines.
- The Full-Screen Preview icon displays a full-screen view of your drawing, without any of the CorelDRAW! interface showing. To get back to regular view, press any key.

The Open Drawing icon does not open an existing drawing. It works like the File New command, displaying a blank drawing page on the screen (after asking you if you want to save changes—if you haven't yet). Use the Open Drawing Dialog Box icon to open an existing drawing.

A Menu with a View

There's a new menu on the CorelDRAW! 5 menu bar. It's called the View menu, and it lets you determine how you want the screen display to look. Using the View menu, you get to tell CorelDRAW! to display or hide rulers, the color palette, or the entire CorelDRAW! interface—which is basically everything you see on your screen besides your drawing. The Roll-Ups command on the View menu lets you pick and choose which roll-ups to display. For more on this command, read the section "Rolling Down the River" later in this chapter.

You can also switch to a floating toolbox and to wireframe view from the View menu. If you want to redraw the screen so that it's completely updated, you can use the Refresh Window command. Finally, you can set bitmap and color correction options using commands on the View menu.

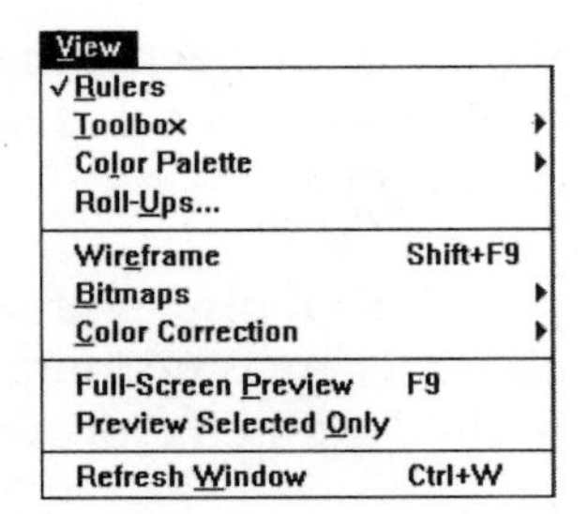

A whole new menu for you.

Preset Notions

Presets are probably the most fun of the new CorelDRAW! features. More than that, they're really helpful. A *preset* is a predesigned set of graphic styles, including coloring, shading, and special effects. You get about twenty presets to choose from. A preset can be applied to an object or to text. You simply select the object or text, display the Presets roll-up, choose the preset style you want, and click on the **Apply** button. Instantly, your object or text is transformed according to the settings of the preset you chose.

The only way you can access presets is through the Preset roll-up. To display this roll-up, select the Presets Roll-Up command from the **S**pecial menu.

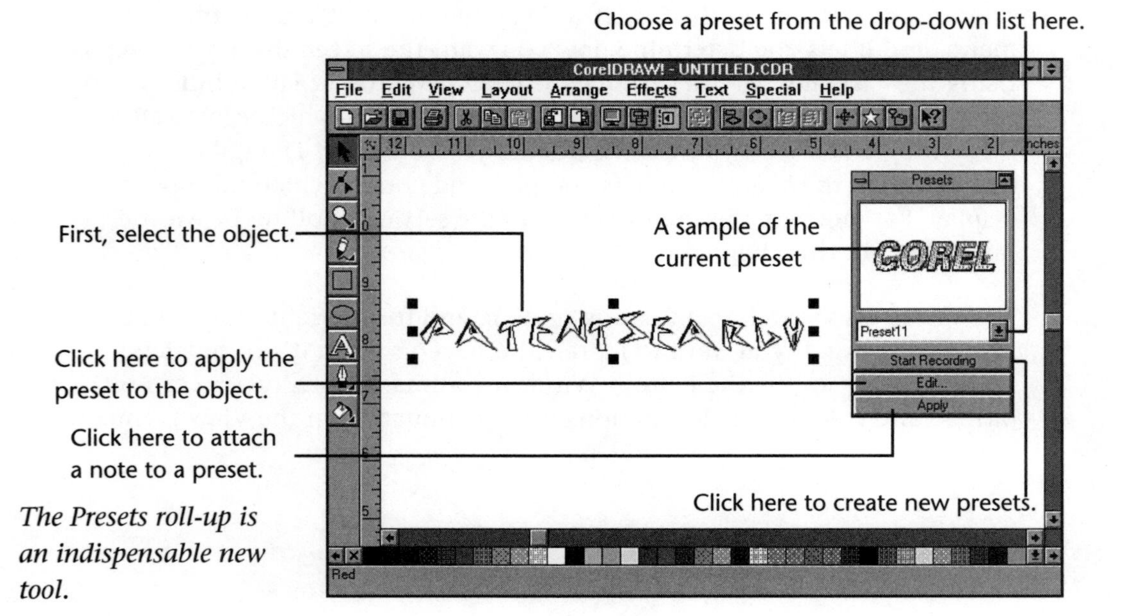

The Presets roll-up is an indispensable new tool.

CorelDRAW! makes it easy for you to choose a preset, because a sample of each preset appears in the Presets roll-up. To see how a preset will change the look of your object, click on the preset name in the drop-down list just below the sample box. When you see a style you like, leave it displayed in the roll-up, and then click on **Apply** (make sure you've already selected the object you want to modify). CorelDRAW! adds the preset's formatting to the selected object, which completely changes the look of the object.

SPEAK LIKE A GEEK

Preset A predesigned graphic style, including set colors, shades, and special effects. When you apply a preset to an object, the object's appearance changes to conform with the preset's style.

Interface Everything you see on the CorelDRAW! screen besides the drawing (for example, menus, tools, icons, status line, and so on). The interface serves as a middleman between you and your drawing.

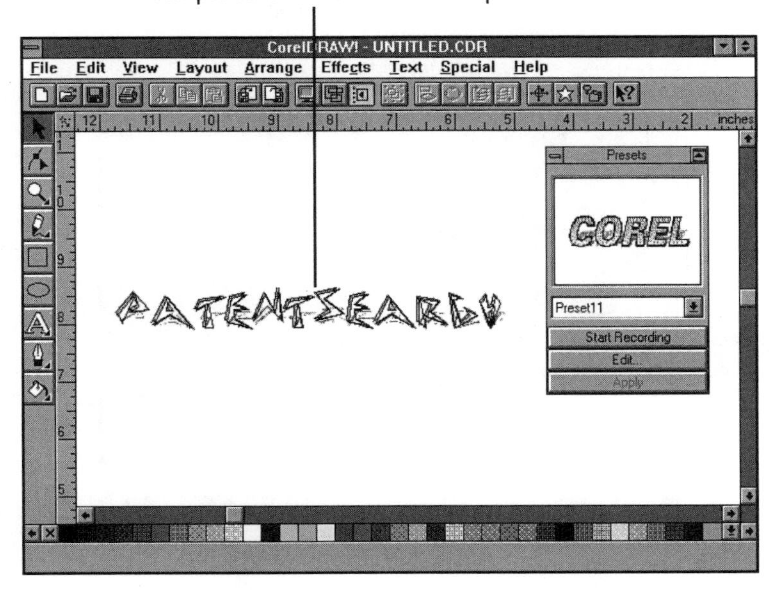

An object formatted with Preset11.

A New Way of Entering Text

Man, I wish my word processor (Ami Pro) had this new CorelDRAW! text feature; it probably will some day, considering how all the programs seem to "borrow" each other's best features (doesn't CorelDRAW!'s new Ribbon bar look mighty similar to the button bar in Word for Windows?). The feature I'm talking about is the new Type Assist command, and in particular the Replacement Text feature of the Type Assist command. With the Replacement Text feature, you type a character (like L), and CorelDRAW! automatically replaces it with a designated string of words (like "living on the edge").

Anyone who has ever written a lengthy report, manual, or book will treasure this new command, because you will no longer have to type the name of a product, person, or place over and over in your document. Instead, you can designate a letter, and have CorelDRAW! do the dirty work. For example, if you're writing a manual on a new program called "Supercalifragilisticexpialidocious," you would only have to type an "S," and CorelDRAW! would fill in the rest.

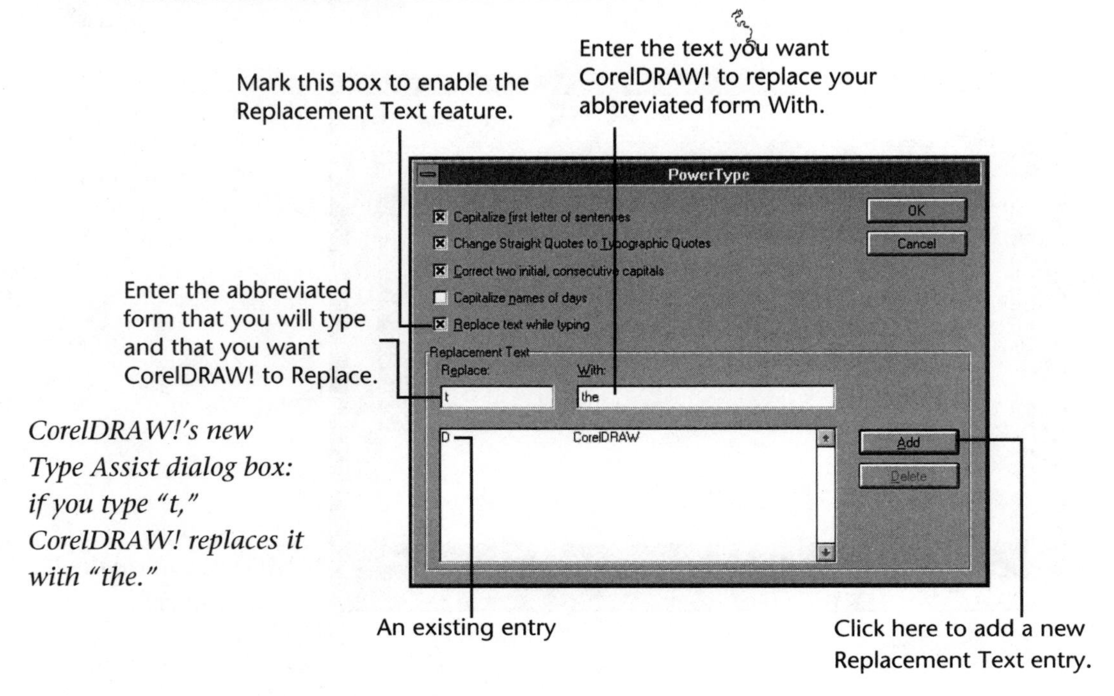

CorelDRAW!'s new Type Assist dialog box: if you type "t," CorelDRAW! replaces it with "the."

To use the Replacement Text feature, you open the Type Assist dialog box by selecting the Type Assist command from the Text menu. In order to use the Replacement Text feature, mark the **R**eplace text while typing option. Then enter a letter, character, abbreviation, or short word (basically, whatever you want) in the Replace box. (You should enter something that clearly reminds you of the text that will replace it.) In the **W**ith box, enter the text with which you want CorelDRAW! to replace the character you just entered in the Replace box. For instance, whenever I type "D," it will be replaced with the name "CorelDRAW!" as shown in the figure above. Click on the **A**dd button to add your new entry to the list at the bottom of the dialog box. Click on **OK**, and you're all set. From now on, when you type the character in the Replace text box, CorelDRAW! will replace it with the text in the **W**ith box.

The Type Assist command has some additional features you're sure to use. Remember how your crabby old elementary teachers always shouted, "Capitalize the names of days and the first word in a sentence!" Well, now you could tell them to bug off, because CorelDRAW! will do all of that for you. Just make sure you mark the appropriate boxes in the Type Assist dialog box, and CorelDRAW! will automatically capitalize the first letter of each sentence and all days of the week (which means you can type in lowercase most of the time now, without hearing echoes of school day reprimands).

I entered the Replace character, but nothing happened! Probably because you didn't do one of two things in the Type Assist dialog box: 1) mark the Replace text while typing box, or 2) click on the Add button to add the Replace character to the list of entries. So open the dialog box again and do it right this time!

Teaching an Old Dog New Tricks

In addition to all this new stuff, some very common CorelDRAW! features have changed. Hopefully you're smarter than your old dog, because you need to learn some new tricks. Some dialog boxes work differently, you can display roll-ups via a much easier method, and object transformation commands are now grouped together in one roll-up.

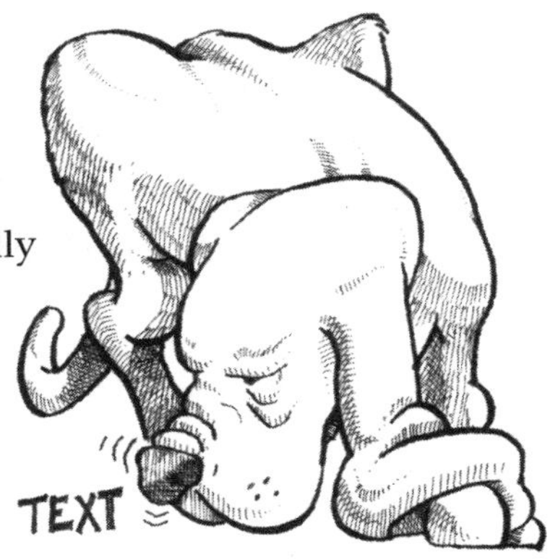

The Tabs Have It

A dialog box isn't just a dialog box anymore. At least not all of them are anyway. Several of the CorelDRAW! dialog boxes have been broken into sections and now have tabs, which you can use to access the various sections in the dialog box. It's somewhat analogous to flipping through manila folders until you find the folder you need. For example, take a look at the new Preferences dialog box.

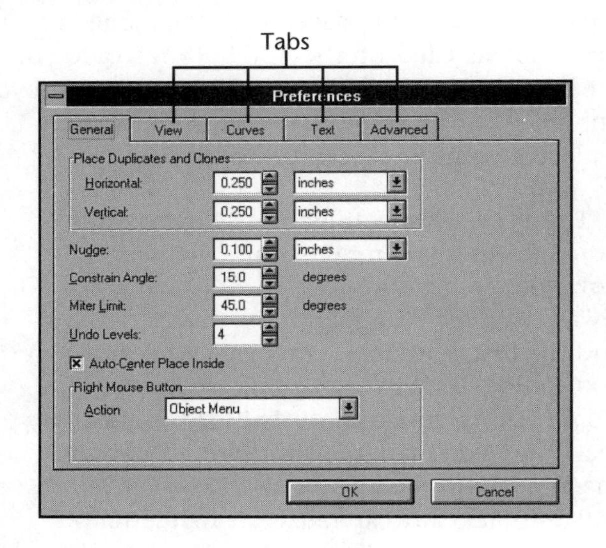

A tabbed dialog box.

CorelDRAW! has broken the Preferences dialog box into five categories. Each category has a tabbed title at the top of its box. They are General, View, Curves, Text, and Advanced. A tab represents each of these categories. When you click on a tab, the options for that category appear in the dialog box. For instance, when you click on the **View** tab in the Preferences dialog box, the View options are displayed in the dialog box, and the View tab sits on top of the rest of the tabs. You can no longer access the General options; you'd have to click on the General tab to regain access to the General options. So you can flip through different categories by clicking on different tabs.

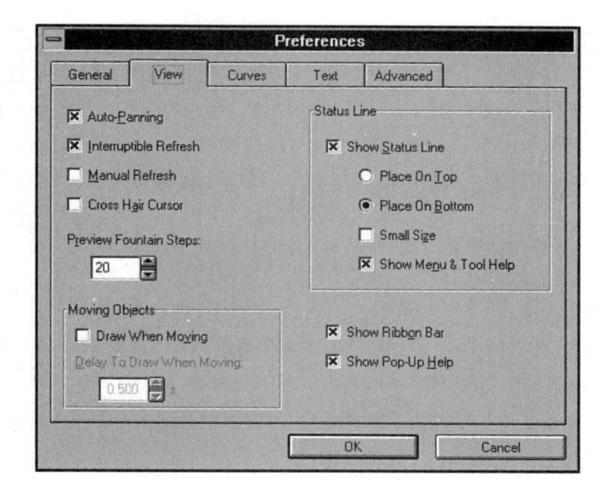

Clicking on a tab displays a new category of options.

Rolling Down the River

Specifying which roll-ups you want displayed each time you start CorelDRAW!, how they are displayed (rolled up or down), and whether they are arranged is much easier to do now. You can configure all of this in one dialog box, instead of selecting a command for each roll-up, fixing each roll-up just the way you want it, and then selecting an option in the Preferences box (as you had to do in versions 3 and 4). No more fuss. Just pick your settings in the Roll-Ups dialog box, and you're all set. To display this dialog box, select the Roll-**U**ps command from the **V**iew menu.

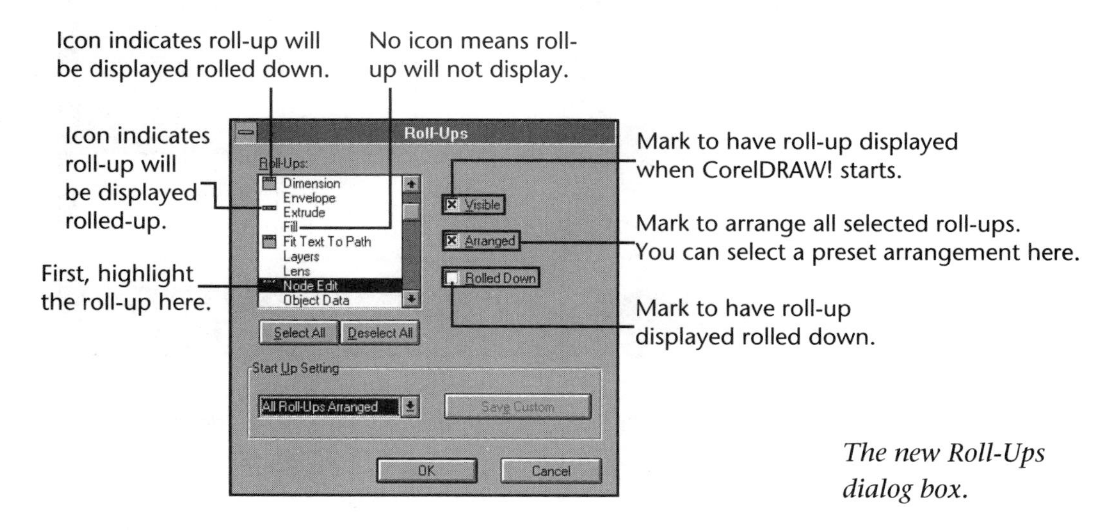

The new Roll-Ups dialog box.

Here's the process: For each roll-up you want displayed when CorelDRAW! starts up, highlight the roll-up in the Roll-Ups dialog box, and then mark the **V**isible check box. If you want the roll-up to be displayed rolled down, mark the **R**olled down check box. If you want the roll-up arranged on top of other roll-ups, mark the **A**rranged check box. Alternatively, if you want all selected roll-ups to be displayed arranged, select the **All Roll-Ups Arranged** option in the Start Up Setting box.

When you mark a roll-up to be visible but not rolled down, a small icon appears to the left of the roll-up's name, reminding you that this roll-up has been marked for display, but rolled-up. If you select the Rolled Down option for a roll-up, a larger icon is displayed, indicating that the roll-up will be displayed opened.

The figure below shows how the roll-ups will be displayed, according to the settings shown in the previous figure.

As specified, the Node Edit and Extrude roll-ups are displayed rolled up.

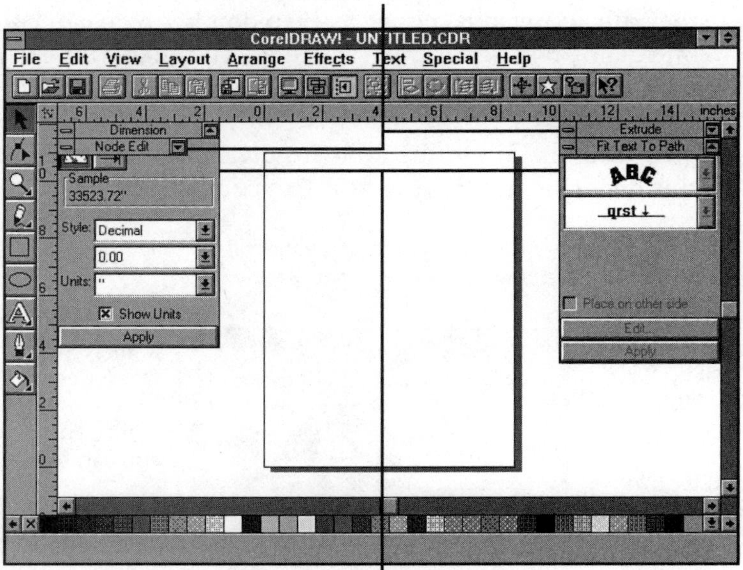

As specified, the Fit Text to Path and Dimension roll-ups are displayed rolled down.

The roll-ups arranged as specified in the Roll-Ups dialog box.

A True Transformation

If you look for an Arrange Move command, you won't find one. We'll spare you a lot of searching and tell you a secret: All move options now appear in a brand new version 5 roll-up, called the Transform roll-up. We won't even tell you how to display this roll-up via the menu, because CorelDRAW! gives you a quicker way: just click on the **Transform Roll-Up** icon on the Ribbon bar to display the Transform roll-up. (Okay, we give in. For those of you who are allergic to icons, you can also display the roll-up using the Effects Transform Roll-Up command.)

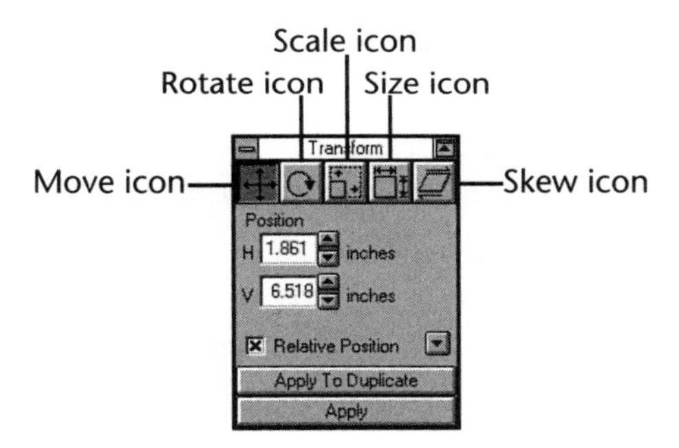

The Transform roll-up, with the Move options displayed.

You can use this new Transform Roll-Up to move, size, scale, rotate, and skew your objects. For each function, the options in the Transform roll-up differ. The five icons at the top of the roll-up move you between the different displays. It's really simple: just click on an icon to access the function you want. Take some time to select each icon and see what options are available and to experiment with the various options.

The Least You Need to Know

Wow! CorelDRAW! 5 really blows you away, doesn't it? Those Corel wizards have done well.

- Version 5 includes a full-featured desktop publisher, called CorelVENTURA!.
- The CorelDRAW! 5 screen now includes a Ribbon bar, chock-full of command icons.

continues

continued

- A new View menu makes it simpler to control the screen display.
- Presets let you format objects instantly. To display the Presets roll-up, choose the **S**pecial **P**resets Roll-Up command.
- The Type Assist feature makes typing easier, faster, and more fun. Check it out using the **T**ext Typ**e** Assist command.
- Arrange roll-ups all at once in the new Roll-Ups dialog box, accessed through the **V**iew Roll-**U**ps command.
- Move, size, scale, rotate, and skew your objects with the new Transform roll-up. Click on the **Transform Roll-Up** icon on the Ribbon bar to display the roll-up.

Chapter 28
A Crash Course on Windows

In This Chapter

- Windows you don't have to clean
- Your new pet mouse Ben
- A look at the menu
- Introducing the dialog box
- It's quitting time

We'll be frank: you won't get very far in CorelDRAW! if you don't know anything about Windows. In fact, you might not get anywhere. Fortunately, Windows is the epitome of a user-friendly program. It's visually oriented, so you don't have to store a bunch of command letters in the back of your mind as you work (unlike the old WordStar and WordPerfect programs you may have been brought up on). No, if you want to choose a command, all you have to do is drop down a list of commands and click on the one you want. And for you die-hard keyboard users, don't worry: Windows has keyboard combinations that activate most of the commands.

If the words "Bad command or filename" appear on your screen after you type **win** at the DOS prompt, try typing **cd\windows**, pressing **Enter**, and then typing **win** again.

Hey! Yo! You Trying to Start Something?

Yes. Windows. Before you can begin a CorelDRAW! session, you must start Windows. But that's easy to do. Turn on your computer and monitor (well, we have to cover everything). At the DOS prompt, which usually looks like c:\>, type **win** and press **Enter**. If you're lucky, Windows might just start automatically when you turn on your computer.

Immediately after you start Windows, the Program Manager appears on your screen. The Program Manager is like that Welcome sign on your front door: it greets you when you start Windows, and it's where you close Windows at the end of the day.

Your program group icon is called Corel Graphics.

Your program group icon is labeled Corel 5 (gee, you probably could've guessed that).

Programs are arranged by groups in what's called program group windows. These group windows hold program icons representing different programs. To close a group window, double-click on the program group's **Control-menu box** in the upper-left corner of the program group window. When you close the group window, it becomes minimized into a program group icon located at the base of the Program Manager window. Later when you want to open the group window again, just double-click on the program group icon.

You can start a program by double-clicking on its program icon. The Corel programs, including CorelDRAW!, are located in the Corel 4 group, which is represented by the Corel 4 group icon.

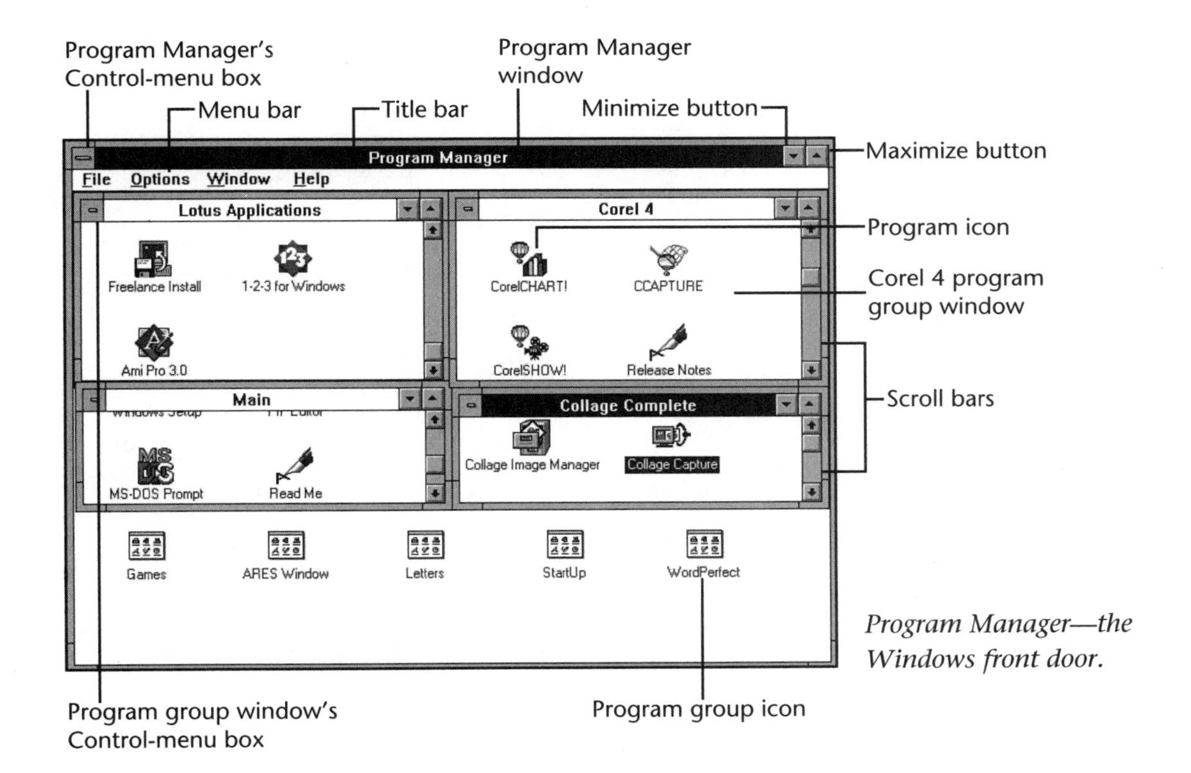

Program Manager—the Windows front door.

Title bar Displays the name of the window or program.

Program group windows Contain program icons that allow you to run programs.

Program Manager's Control-menu box Closes Windows when you double-click on it (after you confirm that you want to exit).

Menu bar Holds all of a program's drop-down menus.

Program group window's Control-menu box Closes the group window when you double-click on it.

You can push your windows around in lots of ways. To reduce a window to an icon, click on the down arrow in the upper right corner of the window. To restore the window, double-click on the icon. To move a window, drag the title bar. To resize a window, drag one of the edges of the window or art.

SPEAK LIKE A GEEK

Corel 4.0 group icon A miniature picture that represents the window holding all of Corel 4.0's program icons.

Windows The operating system that serves as the foundation for the Corel 4.0 programs. Many popular software programs run under Windows. You cannot run CorelDRAW! without Windows.

Program Manager The main window which appears on your screen when you start Windows. This is where you manage your groups, icons, programs, and files.

Program Manager window The main Windows window, which holds program groups for the Windows programs installed on your system.

Program icon Starts a program when you double-click on it.

Corel 4 program group window Contains the program icons for all of the CorelDRAW! programs you installed.

Maximize button Maximizes a window to its largest size.

Minimize button Reduces a window to an icon.

Scroll bars Move you through a window.

Program group icon Opens a program group window when you double-click on it.

Windows can run in Standard or Enhanced mode. Standard mode is designed for older, less powerful computers. Make sure you are running Windows in Enhanced mode before you attempt to start any of the Corel programs. *Enhanced mode* supplies more memory to Windows than the regular Standard mode does. The CorelDRAW! programs need this extra memory to run efficiently because they are such large, involved programs. If you're not sure whether Windows is running in Enhanced mode, type **win/3** at the DOS prompt (instead of just **win**). This will force Windows to run in Enhance mode.

Mousing Around

Getting the hang of the mouse is somewhat analogous to getting the hang of a stick shift. At first it feels funny, you may make mistakes, and you are constantly aware of what you're doing to the mouse. After a while, however, you won't even be aware of the mouse as you work, (just like driving a car with a stick shift soon becomes second nature).

The mouse is represented by a cursor on the screen. Wherever you move the mouse on your desk, the cursor moves accordingly on the screen. The cursor usually looks like a hollow arrow. However, when your computer system is working, the cursor becomes an hourglass; you cannot perform any functions with the mouse when it is an hourglass. (The hourglass is Windows' way of telling you to hold your horses.)

Also, certain programs have other shapes for the cursor when various program tools are in use. For example, when you use the Shape tool in CorelDRAW!, the cursor becomes a thick black arrow, and when you use the Rectangle tool, the cursor becomes a cross.

You'll perform four basic functions with the mouse. Becoming familiar with the terms that describe these functions will help you understand the instructions in this book (as well as other Windows books).

Point To move the mouse to a particular place on the screen. For example, to point to an object, move the mouse cursor so it rests on top of the object.

Click To press the left mouse button once. (If you have specified the right mouse button to be the primary button, you would press the right mouse button.)

Double-click To press the left mouse button twice quickly. If nothing happens, you didn't click fast enough. Try again, pressing twice in rapid succession.

Drag To hold down the mouse button continually as you move the mouse.

Here are some examples of how you'll use the various mouse actions:

- You'll point to an area to draw a line in CorelDRAW!.
- You'll click on a menu name to open a menu.
- You'll double-click on the CorelDRAW! program icon in the Windows Program Manager to start CorelDRAW!.
- You'll drag a CorelDRAW! object to move it.

Menu, Please

Just as a restaurant menu contains the available food selections for a restaurant, a Windows menu holds various commands you can use to perform functions. Menu commands are grouped by category on various menus. For example, the Help menu holds all available Help commands.

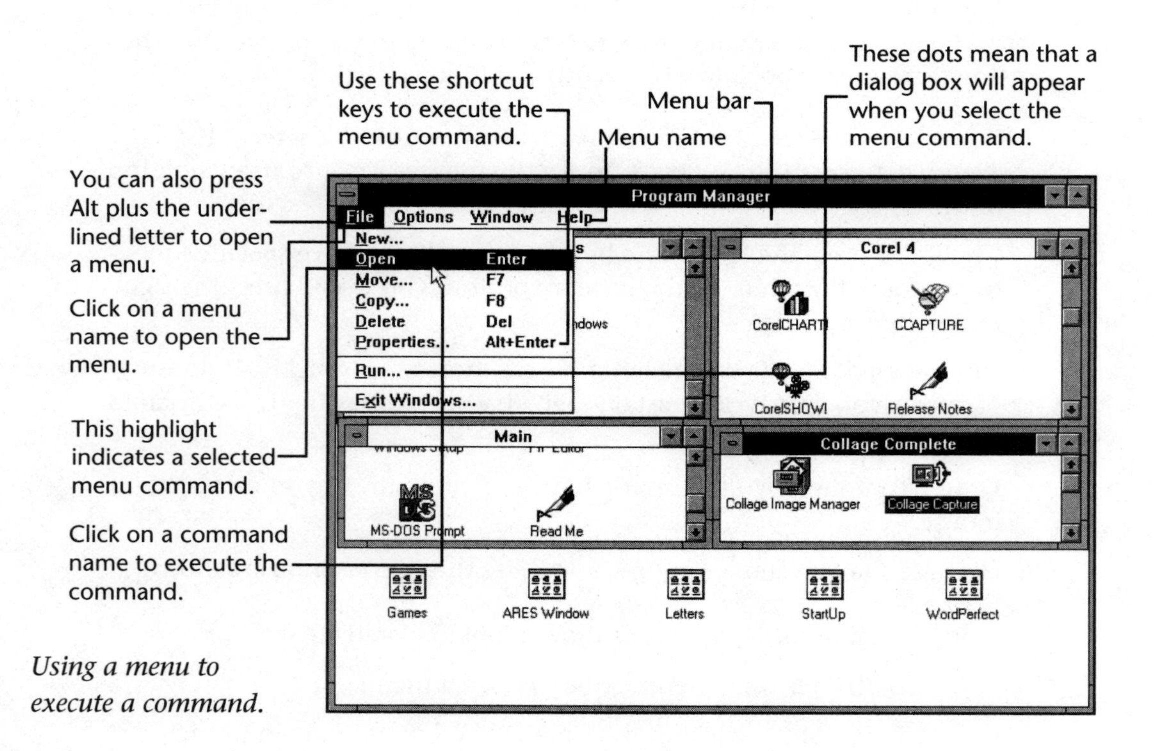

Using a menu to execute a command.

All menu names appear on the menu bar at the top of the screen. To open a menu, click on the menu name. To activate one of the menu's commands, click on the menu command name. Using the keyboard, you can press and hold the Alt key and press the menu's selection letter to open a menu. (The *selection letter* is the underlined letter in the menu's name.) Then, to activate the menu command, press the selection letter for the menu command. For example, to execute the Windows **F**ile **O**pen command using selection letters, you would press **Alt+F** and then press **O**.

Alternatively, you can hold down the mouse as you drag the highlight to the correct command and then release the mouse button, or use the up and down arrow keys to highlight the command and press **Enter**. If you change your mind about selecting a menu command, press **Esc**, click outside of the menu area, or continue to hold down the mouse button and drag the highlight back to the menu bar.

You can also activate some menu commands by using a key combination called *a shortcut key*. If a command has a shortcut key, you'll see the key combination to the right of the menu command. For example, to use Windows **F**ile **P**roperties command via a key combination, you would press **Alt+Enter**.

In CorelDRAW!, an arrow to the right of a menu command tells you that a submenu will appear when you select this menu command. Also, if a menu command appears in light gray, the command is not available under the present circumstances.

Menu command A command that appears on a menu. You can activate menu commands using the mouse or keyboard.

Key combination A combination of keys you can use to execute a command without using the menu.

Menu bar Holds a program's menus.

Dialog boxes Windows that appear to request command options, information, or the confirmation needed to carry out a particular task. Additionally, a dialog box may remind you of the consequences of the command you have just selected.

Chatting with a Dialog Box

Like every other Windows program, CorelDRAW! uses dialog boxes to let you control operations like saving files, printing, and altering text attributes. There are simpleton dialog boxes that ask you for a simple yes or no answer, and there are the demanding boxes in which you must select options from several types of formats to best suit your needs. Most dialog boxes are the latter.

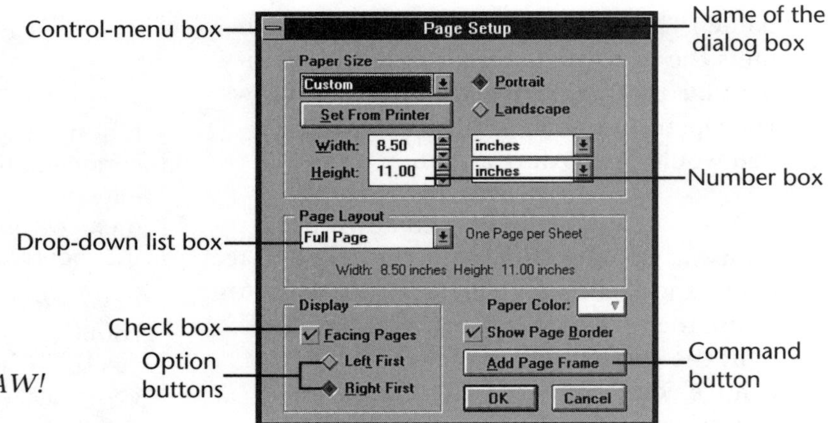

A typical CorelDRAW! dialog box.

> **SPEAK LIKE A GEEK**
>
> **Dialog box** A box in which you set options that control how an action is carried out. CorelDRAW! uses dialog boxes to ask you for more information.

To move around a dialog box, simply click on the option you want to select or change. Keyboard users can either press **Alt** plus the underlined letter in the option name, or press **Tab** until the desired option is highlighted and then press **Spacebar** to select or deselect the option. The following table shows you the elements of a dialog box and describes how to use them:

Element	Name	To use it . . .
Left First / Right First	Option Button	Click on the option button next to the option you want to select or deselect.
Facing Pages	Check box	Click on a check box to select or deselect the option next to it.
Add Page Frame	Command button	Click to access a supplementary dialog box or to execute or cancel the dialog box selections.

Element	Name	To use it . . .
Full Page	Drop-down list box	Click on the down arrow to display the list, and then click on an item.
Height: 11.00	Number box	Click to place the insertion point in the box. Type your entry. Alternatively, you can use the scroll arrow to get to the number you want.

Shut the Windows!

You can close Windows in a jiffy. But before you do, make sure you've closed all other applications.

To close Windows, double-click on the Program Manager's **Control-menu box** located in the upper-left corner of the Program Manager window, or choose the Exit command from the **F**ile menu. A dialog box appears asking you whether you really want to leave Windows. Click on **OK**. Windows closes and returns you to the DOS prompt.

If you want to run a DOS program or need to access the DOS prompt, it may not be necessary to exit Windows. You can access the DOS prompt by double-clicking on the MS-DOS prompt icon found in the Main group window. (See the first figure in this chapter to view the MS-DOS prompt icon.)

The Least You Need to Know

You'll find operating Windows and programs that run in Windows easy to do.

- Type **win** at the DOS prompt to start Windows.
- To point the mouse, move it to a particular place on the screen. Press the left mouse button once to click. Press the left mouse button twice quickly to double-click. Hold down the left mouse button continually as you move the mouse to drag it.
- Click on a menu name to display a menu, and then click on the command you want. Or, use a key combination to execute a menu command, if one is available.
- Windows dialog boxes request command options, information, or confirmation needed to carry out a particular task.
- To close Windows, double-click the Program Manager's **Control-menu box** or use the **F**ile E**x**it command. Then click on **OK** in the confirmation box that appears.

Chapter 29

Put It to Work: Ideas for Using CorelDRAW!

So you got through this entire book and know all sorts of things about CorelDRAW! tools, CorelSHOW! shows, CorelPHOTO-PAINT! images, CorelCHART! charts, and CorelMOVE! cartoons. Now what if you just left it at that: closed the book, removed CorelDRAW! from your hard drive, and went on your merry way. Sounds like a waste, doesn't it? So put some of the stuff you learned to work! We'll even help you. This chapter gives you some ideas of how you can use all the skills you learned, so that those newly acquired skills don't get tossed due to lack of use.

Another way to get your hands on some nifty tips and tricks is to send away for a copy of the Mastering of CorelDraw Journal which is published ten times a year. The Journal is in its second year of publication and is known for its real-world, tips-and-tricks focus, often covering undocumented or little-known techniques. You can call (800) 565-0815 for a free sample copy of the publication.

Way Cool Text

Don't just add ordinary text to your CorelDRAW! documents—blend the text first. You can create some interesting designs by blending two words with the Blend roll-up. First, use the **Artistic** text tool to type the word you want to blend. Then type the word again, but this time in a larger font. Place the second word below the first. Enter a high number of steps in the Blend roll-up (like 200). You'll come up with an interesting result. And for twisted text, enter a 360 degree rotation.

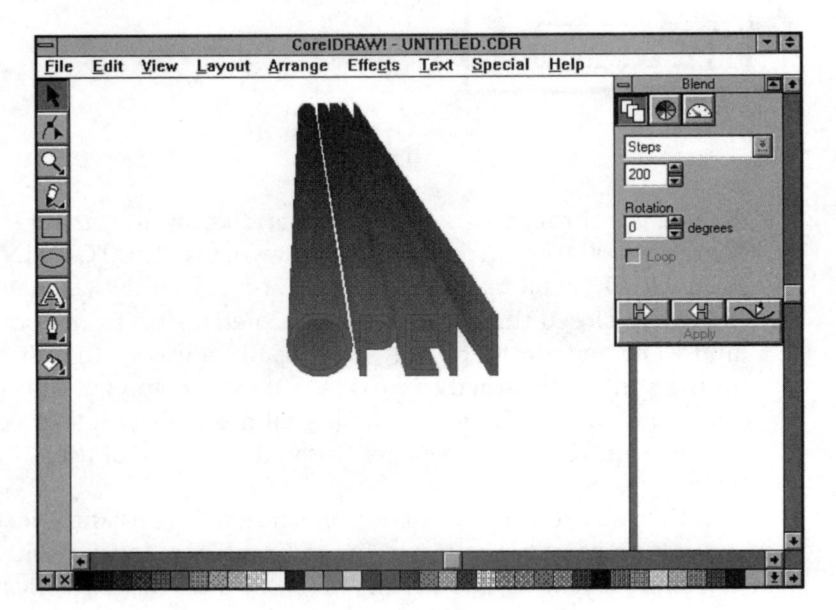

Transform ordinary text with CorelDRAW!'s Blend feature.

Don't Settle for 2D Circles

You can transform a plain old circle instantaneously into a full-bodied, three dimensional tube with CorelDRAW!'s Extrude feature. First draw a circle with the **Ellipse** tool. Then display the Extrude roll-up. Display the circular extrude settings in the roll-up, and click on **Apply**. Drag the vanishing point toward the upper right corner of the screen, and click on **Apply** again. To add more depth, use shading. Display the color settings in the Extrude roll-up, click on **Shade**, and choose the From and To colors. Click on **Apply**, and you've got yourself a dandy-looking tube.

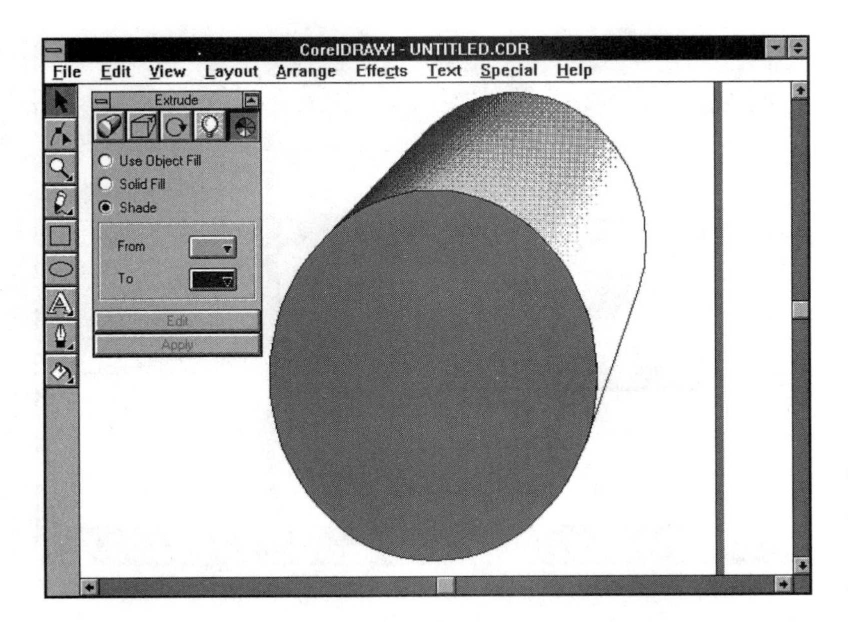

Change a circle into a 3D tube with CorelDRAW!'s Extrude feature.

A Two-Minute Logo in CorelDRAW! 5

Attention all CorelDRAW! 5 users. Want to make an interesting logo, fast? Type your company name with the **Artistic** text tool, or use a CorelDRAW! symbol (there are lots of business-like symbols in many of the Symbols categories). Display the Presets roll-up, and then choose **Preset10**. Select the text or symbol, click on the **Apply** button, and you've got a two-minute logo!

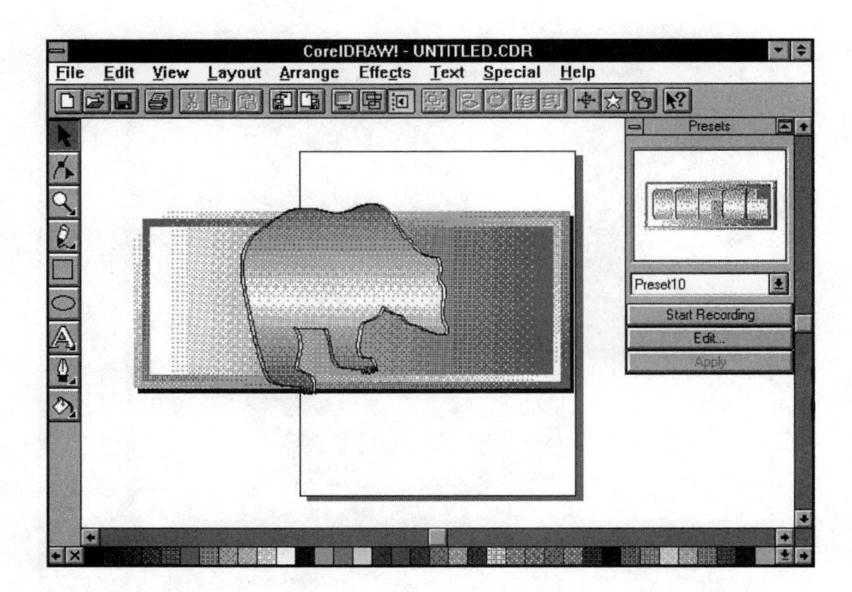

Try using Preset10 to create an interesting logo.

Let's Get the SHOW on the Road

Most people think graphics when they think of CorelDRAW!, but don't forget that you can produce nice-looking documents with CorelDRAW! as well. For instance, you can draw up a set of instructions for running a CorelSHOW! screen show on another computer in a jiffy. Just use the **Paragraph** text tool to create the body of the text. To center the title, display the Text roll-up, select the title with the **Paragraph** text tool, and then click on the **Center Justified** icon in the roll-up. To quickly add a red border, open the Page Setup dialog box, and select the page border option. Then click on the red color icon on the color palette.

Running a CorelSHOW Screen Show

1. Insert the screen show diskette in the appropriate disk drive.

2. Copy the screen show file and the SHOWRUN.EXE file from the diskette to your hard drive.

3. Select the File Run command from the Windows Program Manager.

4. In the Run dialog box, type SHOWRUN.EXE, and press Enter (or you can use the Browse feature to select the file).

5. In the Open Presentations dialog box which appears, choose the name of the screen show you want to run.

6. To start the screen show, choose the Run Screen Show command from the File menu.

Copy these instructions and include them with the screen show disk.

We've Got a Great Show Tonight

There's a great way to catch your audience's attention at the beginning of a screen show. Insert the CorelSHOW! Welcome flic into the first slide of your presentation. The Welcome flic is a short animation that gradually displays the word "Welcome" at the top of the screen. In the animation, venetian blinds with the word "Welcome" slowly close. When the blinds become fully closed, you can make out the word Welcome. Use the **Insert Animation** command to incorporate the animation into your slide show. Then play the show by clicking on the **Run Screen Show** box, and watch the animation at work.

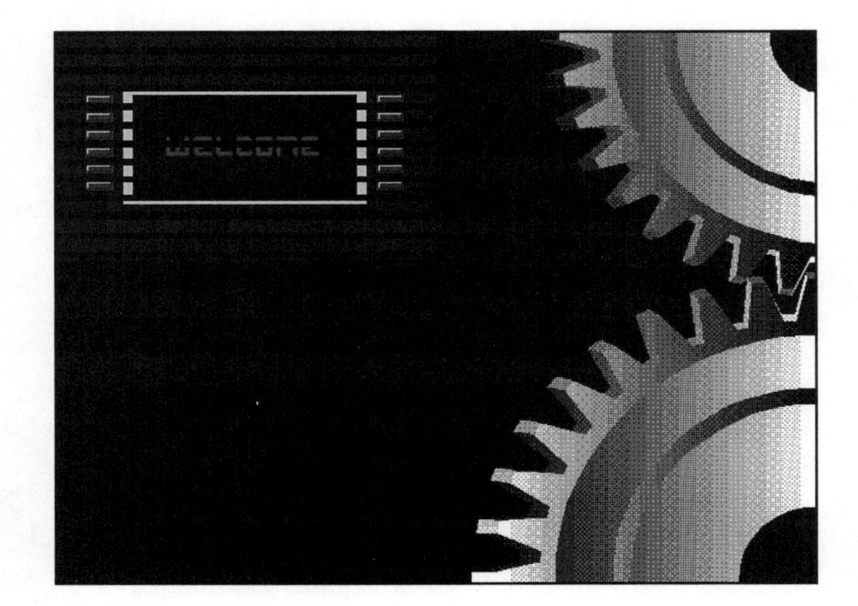

Use CorelSHOW!'s Welcome animation in one of your own screen shows.

This Is the End

There's also a neat way to add a finishing touch to your screen show. Insert the CorelSHOW! The End flic into the last slide of your presentation. The The End flic works like the Welcome flic described earlier, except the words "The End" gradually form at the top of the screen on top of a set of venetian blinds. When the blinds become fully closed, you can make out the words "The End." Use the **Insert Animation** command to incorporate the animation into your slide show. Then play the show back by clicking on the **Run Screen Show** box, and watch your screen come to life.

Close your screen show with CorelSHOW's The End animation.

In Full Motion

Don't try to add the appearance of motion to your drawings using CorelDRAW!. It's much simpler to add motion using CorelPHOTO-PAINT!'s Motion effect. Create your drawing in CorelDRAW!, and then use the File Export command to convert the drawing to a .PCX image. Now you can open the drawing in CorelPHOTO-PAINT! without any problem. Make sure the image window is active, and then select the **Motion Blur** command from the Effects Fancy menu. Presto! Your drawing looks like it's going somewhere.

Original image.

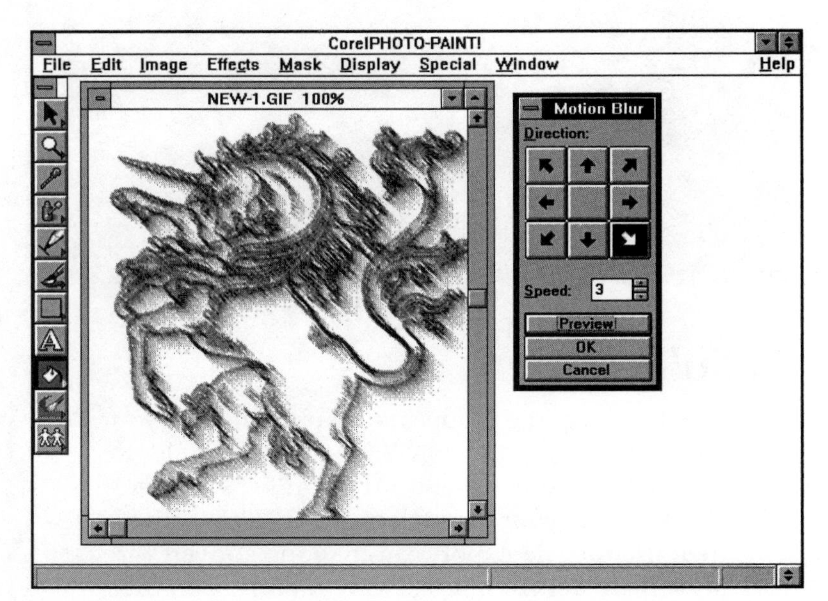

CorelPHOTO-PAINT!'s Motion Blur effect can make it look like your image is moving.

Seal of Approval

You can turn a regular CorelPHOTO-PAINT! image into a seal in just a few steps. Make sure the window containing the image is active, and then select the **Emboss** command from the Effects menu. For a seal, make sure to use a Gray emboss color. Select the upper right arrow in the Emboss dialog box. Click on **OK**, and you've got yourself an authentic-looking seal. Hint: for best results, use black and white images.

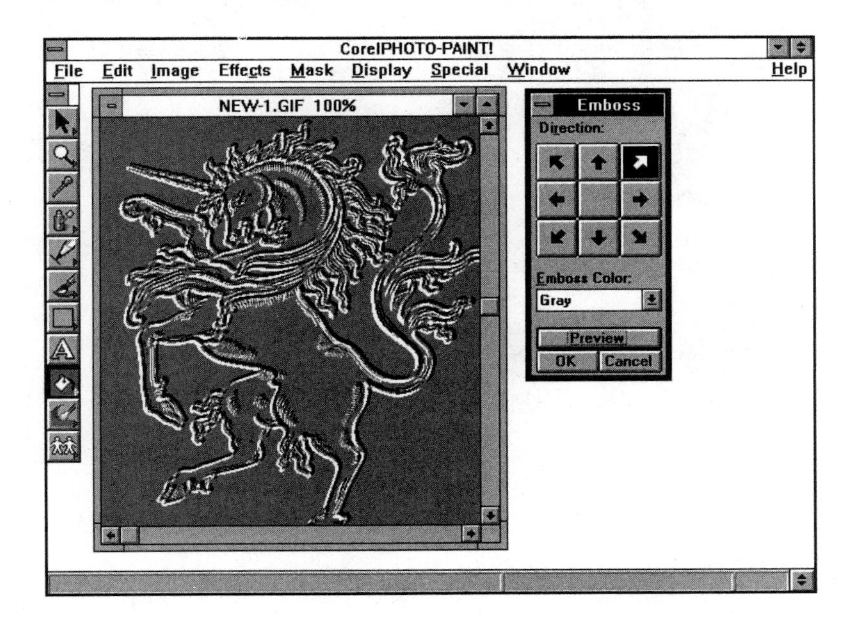

Use CorelPHOTO-PAINT!'s Emboss effect to create the look of a seal.

Bring Your Photos into CorelPHOTO-PAINT!

One of the problems with scanning photos for use in screen shows, documents, or computerized video production is that the scanned image never looks quite as polished, defined, and clear as the photo itself. But you can do something to change that. Import the scanned image into CorelPHOTO-PAINT!, and then go to work on the image with CorelPHOTO-PAINT!'s retouching tools. Use the Freehand Sharpen, Brighten, and Contrast tools, and the Edge Detect command to better define the image. Though you'll probably never get the image to look as clear as it does in the actual photograph, CorelPHOTO-PAINT!'s tools will help.

A scanned photograph image in CorelPHOTO-PAINT!.

Am I Going Blind?

We admit it has a weird name, but the Jaggy Despeckle command on the Effects Fancy menu in CorelPHOTO-PAINT! can give your image an interesting look. The command scatters colors in a picture, and can sometimes give an ethereal look to your image. If you're looking for haziness, use the **Jaggy Despeckle** command.

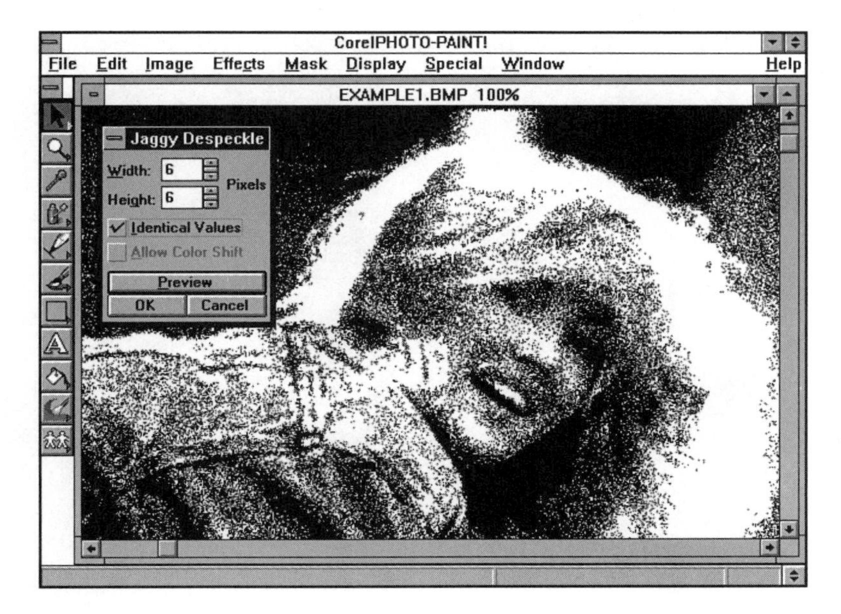

Add a fuzzy feeling to a scanned photo with CorelPHOTO-PAINT!'s Jaggy Despeckle effect.

A Photo with an Attitude

You can't import photograph negatives into CorelPHOTO-PAINT!, but you can sure make your scanned photo look like it's a negative—and in just one step. Make sure the window containing the image is active, and then select the Effects Fancy **Invert** command. Presto! You've now got a veritable negative sitting on your computer screen.

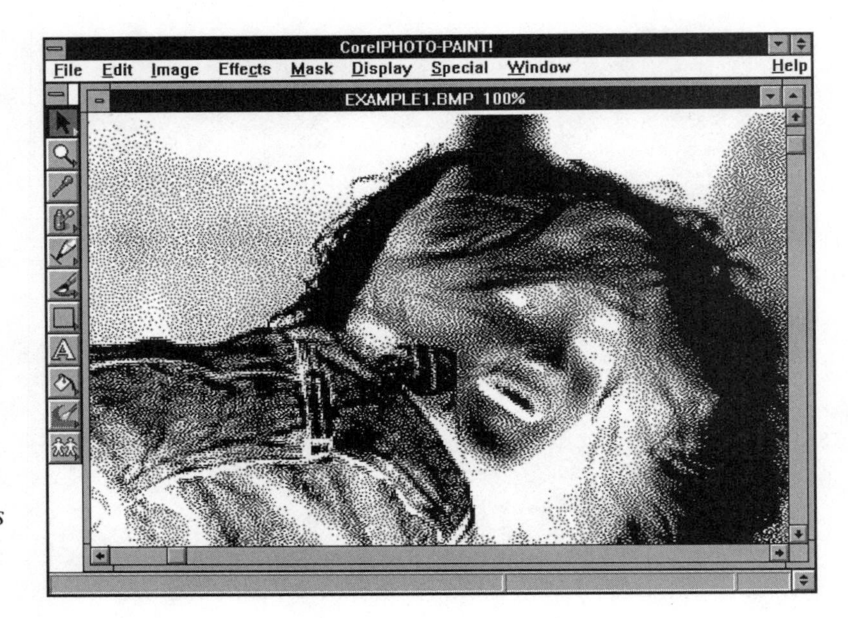

CorelPHOTO-PAINT!'s Invert feature instantly turns a scanned photo into a negative.

A Simple Bar Chart Will Always Do the Trick

Having problems trying to create a 3D chart just the way you want it? Confused by stacked bar charts, line, and area charts? The regular old bar chart is probably the simplest chart to create, as well as to read. Your boss (or whomever is going to review your chart) will appreciate being able to analyze the figures in your chart, more than trying to decipher the values in a 3D chart, albeit the 3D chart is snazzy-looking. Follow this basic rule: when numbers are most important, use a plain bar chart. If you're trying to knock the socks of the people in your audience, and specific numbers aren't as vital as an overall feel for the data's status, use a 3D riser or floating chart.

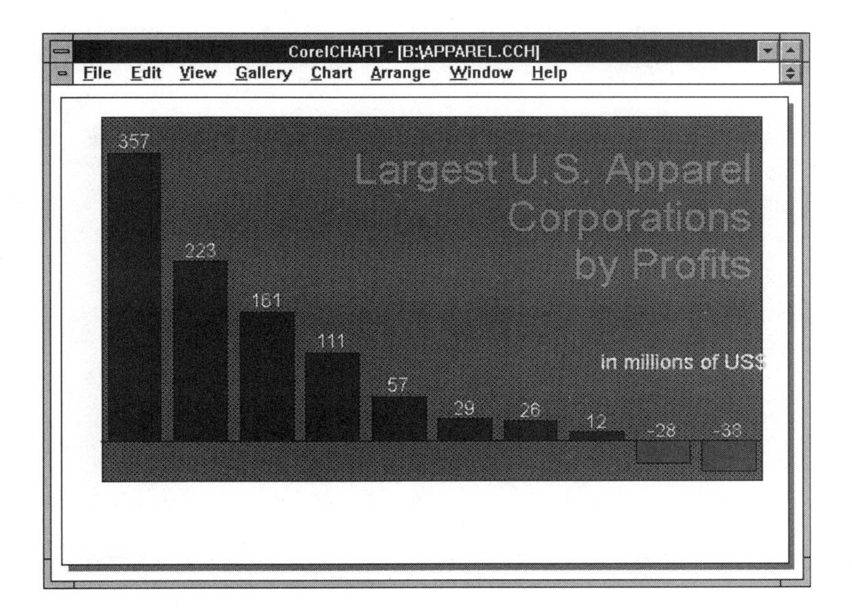

If all else fails, use a regular bar chart.

Just the Right Angle

Sometimes you don't have enough room for all of the x-axis labels in 3D charts. Because the labels are automatically sized by CorelCHART! so that they will all fit on the x-axis, the labels turn out to be so small they are almost unreadable. You can get around this problem if your chart doesn't have many columns. Use the Chart **Preset Viewing Angles** command and select the **Few Columns** angle. (This command is only available with 3D charts.) This will provide the maximum amount of space for your x-axis labels. By the way, there's also a Few Rows angle that you can use when you don't have many series.

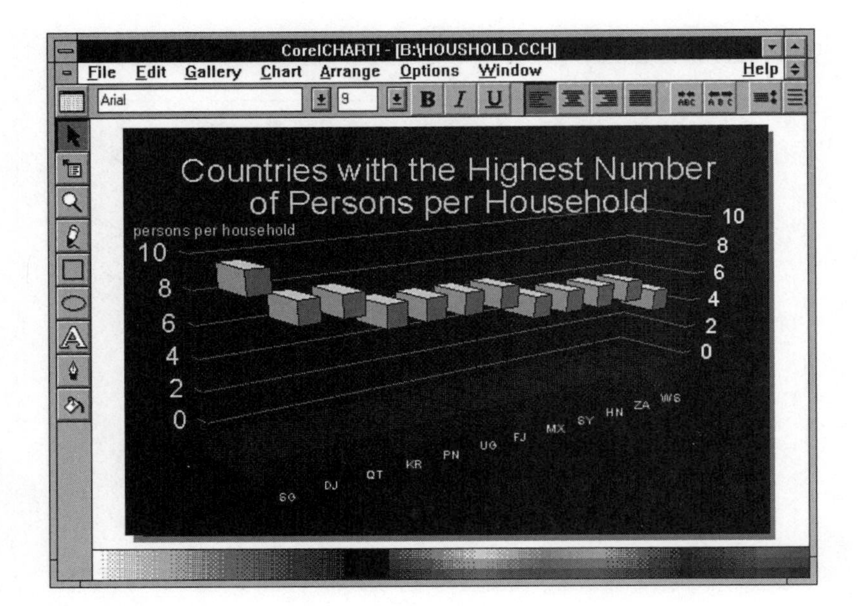

Use the Few Columns preset viewing angle when your chart has a lot of rows.

Picture Your Chart This Way

If your chart has a particular theme that can be accentuated through a graphic, use one! For example, the chart in the figure maps housing costs. You can transform the chart into a house by adding a roof. Use the **Pen** tool to create a rectangle that sits on top of the chart. Fill the rectangle using the **Fill** tool, and add a thick outline with the **Outline Pen** tool. Then create a small chimney by drawing a polygon with the **Polygon** tool, and adding a fill and thin outline to the polygon. Now the chart gets the theme across to the audience, loud and clear.

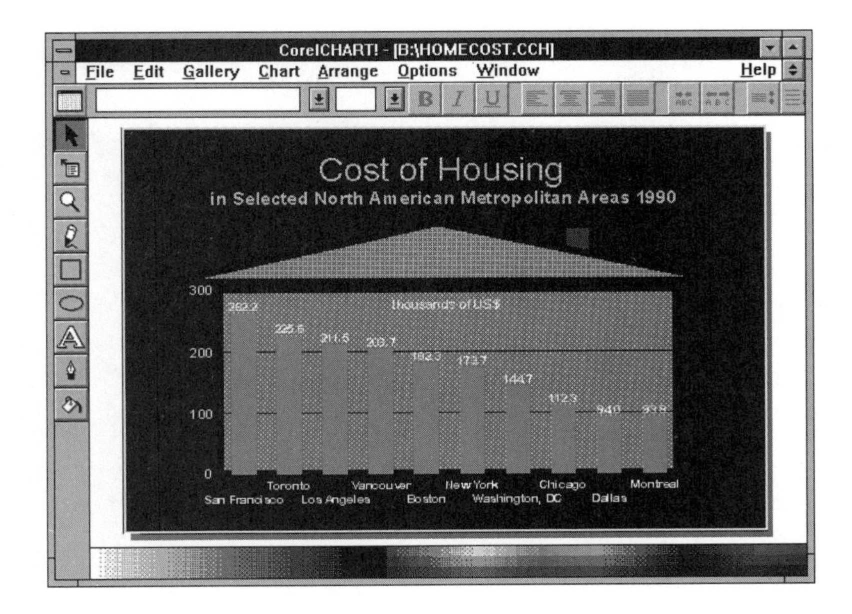

How appropriate—housing stats inside a house.

Why are you reading this page?
Can't you see it's blank?

Fill 'Er Up: Installing CorelDRAW!

Every time I take my car to a new mechanic, I get nervous he'll turn out to be of the "sleazy" sort that will rip me off and install faulty parts in my car (or the wrong parts altogether). That's why I like the fact that I'm the mechanic with CorelDRAW!. I get to decide what's installed on my computer, and I do the installation myself. So if I end up with a lousy installation job on my computer, the only sleazy mechanic I can blame is myself! But don't worry: CorelDRAW!'s installation process is pretty sleaze-proof. You just make some choices, follow some instructions, and slip some disks into your drives—and you're all set.

Installing CorelDRAW! is easy to do, but be prepared to sit around for a while—a long while. Set aside at least half an hour, and don't wander off from your computer. Every few minutes CorelDRAW! will ask you to insert a new disk, and there are plenty of them. If nothing else, your wrist gets some pretty good exercise.

All Aboard!

As you'll soon find out, you can install CorelDRAW! in two ways. The first way is to install the whole shamole with the Full Install option. You need 37 megs of free disk space to load CorelDRAW! 4.0 completely. However,

If you have a CD drive, you can install CorelDRAW! from the CD-ROM in just a few minutes. Or, if you want, you can set up CorelDRAW! to run from the CD-ROM instead. We recommend this last method if you're low on disk space since it saves about 30+ megs of your computer's disk space. However, there is a tradeoff: CorelDRAW! will run slower from the CD-ROM.

DRAW will run faster and you'll actually have room to save your files if you have more than 37 megs free. What we're getting at is that 37 megs is the bare-bones minimum. The more space you have, the better.

If your hard disk doesn't have 37 megs free, or if you don't want to install all of the programs that come with CorelDRAW!, you can use the Custom Install method. With Custom Install, you get to pick and choose which programs to install on your computer (remember, there are several to choose from). Even if you want to install all of the CorelDRAW! programs, you still may want to choose the Custom Install method, because it also enables you to pick the type of import and export filters you want and specify whether or not to install a program's sample and help files (though we strongly recommend you install both).

Regardless of which install method you choose, you'll need to complete the steps below.

1. Start Windows by typing **win** (if it isn't already running), and display the Program Manager.
2. Insert the CorelDRAW! Install Disk 1 into your floppy disk drive.
3. Select the **R**un command from the Program Manager's **F**ile menu. The Run dialog box opens.
4. In the Command Line box in the Run dialog box, type **b:setup** or **a:setup**, depending on the drive you're using.
5. Click on the **OK** button. This starts the CorelDRAW! installation process. A window pops up stating that the setup process is being initiated, so "one moment please." (Count on it being more than a moment.) When the "moment" is up, a dialog box appears, welcoming you to the CorelDRAW! setup process.

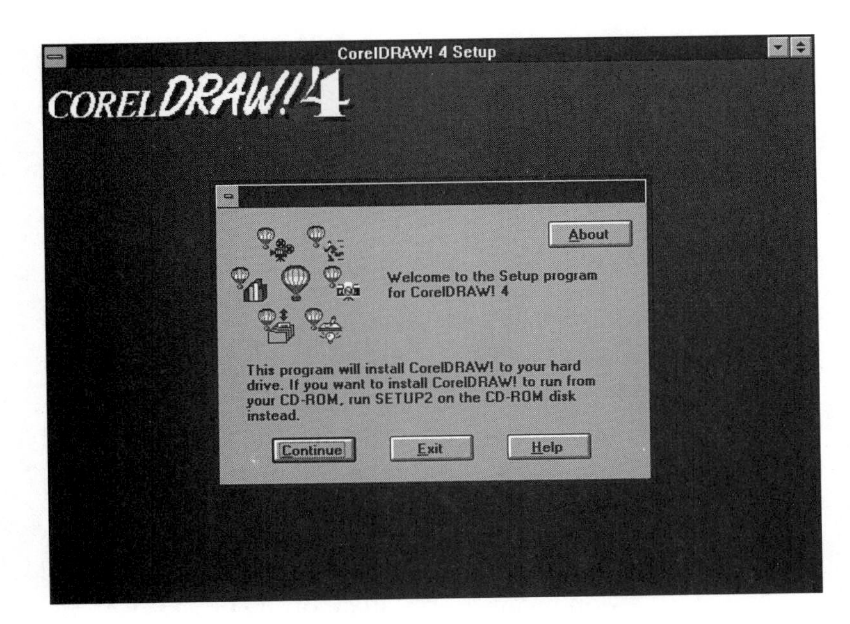

CorelDRAW! welcomes you to the install process.

6. Click on the Continue button. Corel wants to make sure you have a legitimate copy of their program, so they ask you to enter your name and the product's serial number in the CorelDRAW! 4 Setup box. You should be able to find your serial number inside the front cover of your User Manual.

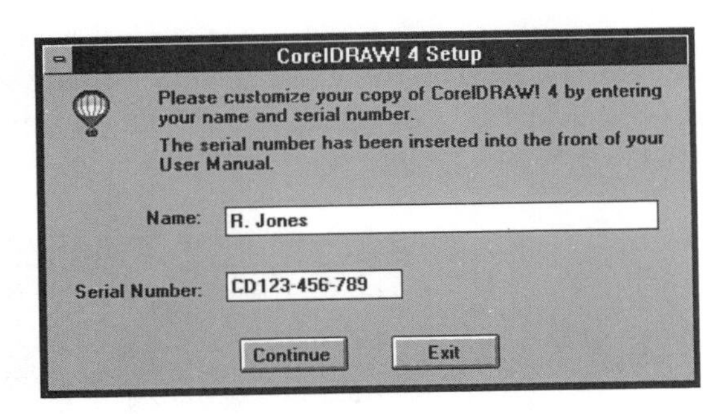

You better run if you don't have a legitimate copy!

7. Type your name (or anything you want) in the Name box, and you-know-what in the Serial Number box. Then click on the **Continue** button. The CorelDRAW! Installation Options dialog box appears.

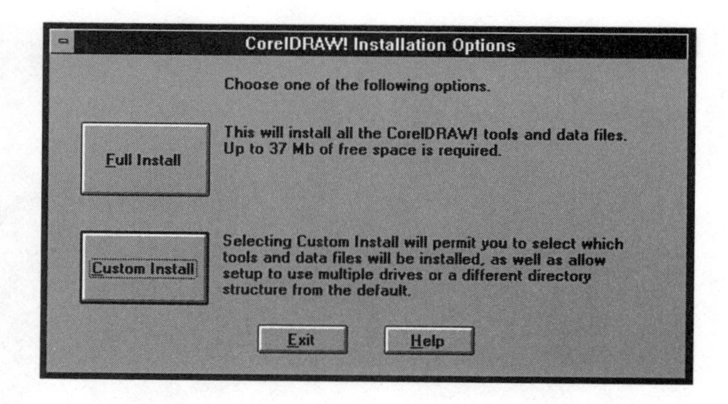

Selecting an installation type.

The serial numbers are not always in the front of the manual. Mine were loose in the box on a very small sheet of paper, and Corel shipped many boxes without the numbers anywhere in the box. You might have to look everywhere in the box for the serial numbers.

8. If you want to install all of the CorelDRAW! programs, choose the **Full Install** button. (You need at least 37 megs of free disk space to load CorelDRAW! completely.) If your hard disk doesn't have 37 free megs, or if you don't want to install all of the programs that come with CorelDRAW!, choose **Custom** Install. (You can also choose this installation type if you want to pick the type of import and export filters that are installed and to specify whether or not to install a program's sample and help files.)

After you choose an installation method, CorelDRAW! checks the available space on your hard disk and your directory configuration. When it finishes this check, the Destination Directory dialog box appears.

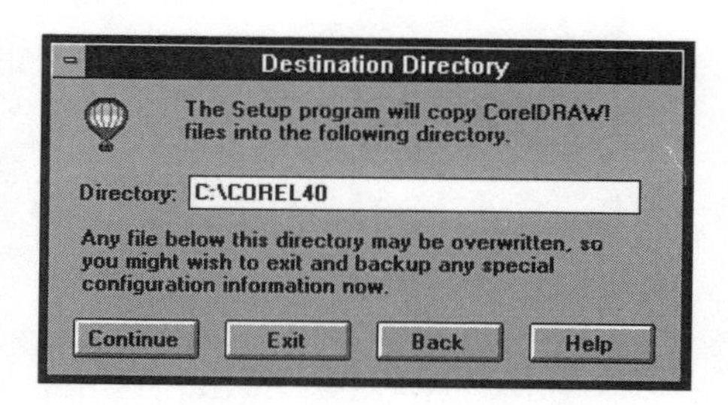

Unless you have something against the default directory, just click on Continue.

9. Click on **Continue** to accept the default directory. Or, if you want to install CorelDRAW! in a directory other than the one specified

in the Directory box, type a new directory path before clicking on Continue.

CorelDRAW! scans your drives and directories again. If you chose Custom Install, continue on to the next section for further instructions. If you chose Full Install, you're done with the preliminary setup! Now all you have to do is insert diskettes into your floppy drive as CorelDRAW! requests them. At the end of the process, CorelDRAW! will ask if you want to update your AUTOEXEC.BAT file. Go ahead and select **Update** to ensure that all the CorelDRAW! programs run correctly on your machine.

CorelDRAW! creates a Corel program group in the Program Manager. Inside the Corel program window, CorelDRAW! places program icons for each program you installed. To start a program, just double-click on the program icon. Feel free to drag the Corel window to a new position in the Program Manager or to resize the Corel window.

BY THE WAY

If at any time you need to exit or pause the install process, or back up to the previous window, do one of three things:

- To pause the install process so you can pick up where you left off later, click on the **minimize** button in the upper right corner of the Setup window.
- To shut down the setup process completely, click on the **Exit** button, which you can always find in the current dialog box.
- If you just want to go back to the previous window, click on the **Back** button. This button is not available in all of the dialog boxes.

For Custom Installers Only

More choice equals more dialog boxes. If you choose the Custom Install method, CorelDRAW! checks your drives and directories and then opens a dialog box that holds a list of all the programs you can install, and how much disk space each requires. It's all or nothing for CorelTRACE! and CorelMOSAIC!, but for the other programs, you can choose which program features to install (such as on-line help and sample files).

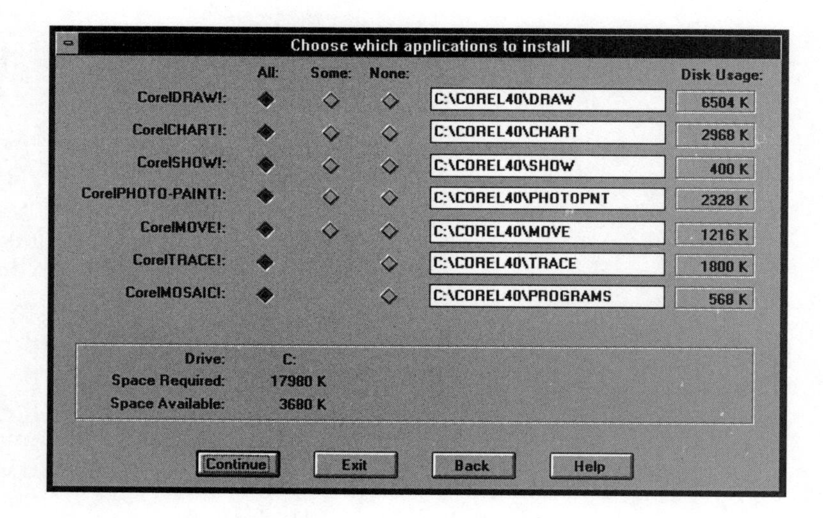

Choosing the applications you want to install.

Click on the **None** button for each program you don't want to install, and then click on **Continue**. If you do want to install only certain program features, click on the **Some** button for that program. An Options dialog box pops up on top of the current dialog box.

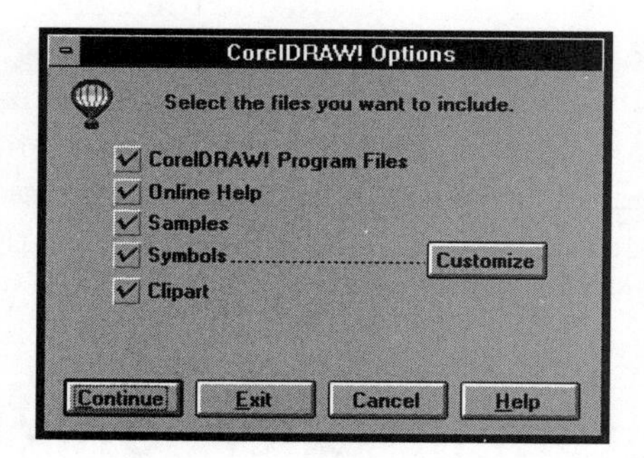

Choosing which program features to install.

By default, all program components are selected. Look through the list of files and decide what you want to leave out, if anything. For example, if you already have too much clip art than you know what to do with, you would deselect the **Clipart** option.

CorelDRAW! gives you an over abundant supply of symbols. If you don't want all of them installed, you can click on the **Customize** button and specify only certain categories of Symbols to install.

Since all of the files are selected, click on the files you don't want to install and click on **Continue**. The dialog box closes, and you are returned to the Choose Which Applications to Install dialog box. Click on **Continue**, and another dialog box greets you (don't worry, you're near the end of the process!). CorelDRAW! lets you decide which filters, fonts, and scanner drivers to install.

The window in which you select the programs and components of a program to install looks different in CorelDRAW! 5. A Customize button appears to the right of each program. So if you want to keep certain parts of a program from being installed, click on the Customize button.

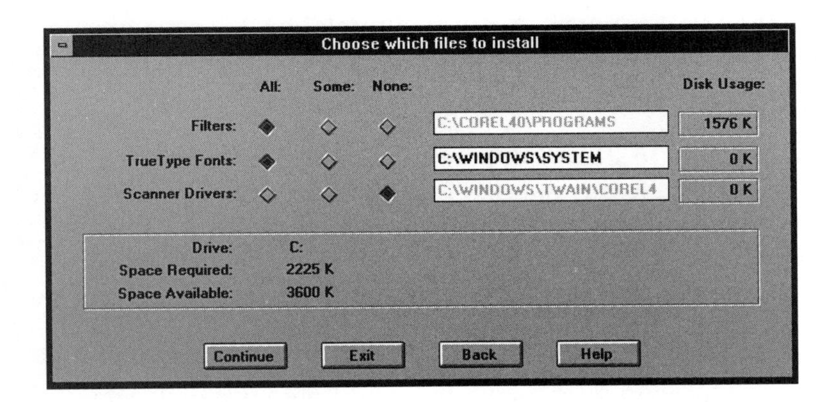

Selecting filters, fonts, and scanner drivers.

In the Choose Which Files to Install dialog box, click on **Some** if you want to specify which filters and fonts to install. Unless you have a scanner or will be using scanned images in CorelTRACE!, leave the Scanner Drivers option on **None**. If you select **Some** for the Filters option, the Filter Selection Window dialog box opens.

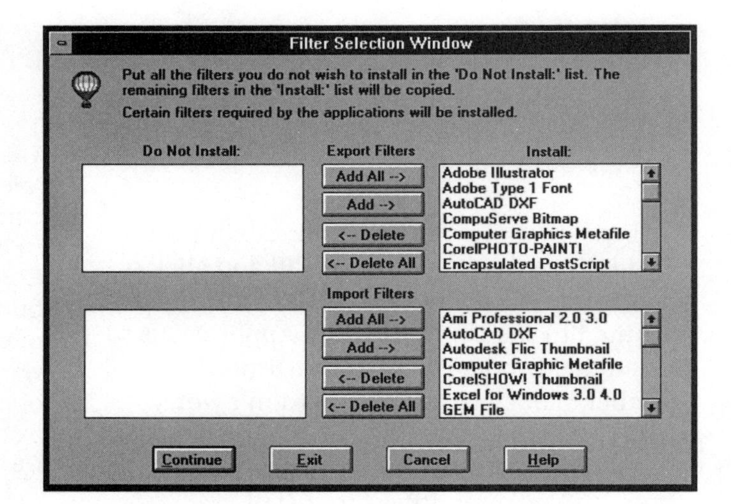

CorelDRAW! lets you pick which filters to install.

Go through the list of Export Filters in the Install box. For each filter you don't want to install, click on the filter and click on the **Delete** button.

Filter files enable you to share files between CorelDRAW! and other programs, such as Word, Excel, and Ami Pro. To be specific, *export filters* let you save work you've done in CorelDRAW! (or one of the accessory programs) for use in another program, and *import filters* let you open other program files in CorelDRAW! (or one of the accessory programs). Filters are even needed for sharing files between CorelDRAW! and CorelPHOTO-PAINT!. So don't be stingy in your selection of filters to install. Unless you're positive you have no need for a filter, you should install it. You never know, somewhere down the line a colleague may say, "Hey, I have just the graphics file you need," and then hand you a .GIF file. Unless you have the .GIF import filter installed, you won't be able to open the file in CorelDRAW!.

When you delete a filter, you're not actually deleting any files. The filter just moves over into the Do Not Install area. So you can always install a filter at a later time if you find you need it.

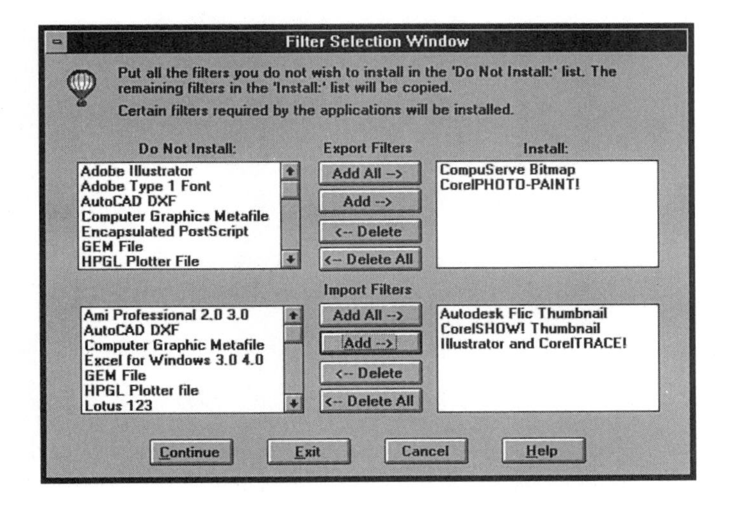

Selecting export and import filters to install.

Go through the list of Import Filters in the Install box. For each filter you don't want to install, click on the filter and click on the **Delete** button. When you're finished selecting filters, click on the **Continue** button.

And that's all folks! Well, not really. That's all the dialog boxes you have to go through for now. CorelDRAW! begins the installation process and asks you to insert one disk at a time. When all disks have been installed, DRAW asks you if you want to update your AUTOEXEC.BAT file to include a SHARE statement. You don't have to worry about the technicalities of this, just click on the **Update** button, and you're all set.

CorelDRAW! creates a Corel program group in the Program Manager. Inside the Corel program window, CorelDRAW! places program icons for each program you installed. To start a program, just double-click on the program icon.

What a waste it is to lose one's mind...

Speak Like a Geek: The Complete Archive

Active file An open file that you are currently working in. The cursor rests in the active file.

Actor A drawing that moves on the screen, consisting of a series of cels.

Animation A series of drawings strung together to give the illusion of movement.

Application A software program.

Artistic text Strings of text you enter using the Artistic text tool. You can apply special effects to artistic text.

Background layer The common full-page design used in all slides of a presentation.

Bézier mode A connect-the-dots approach to drawing curves. Bézier mode is a more accurate way to draw curves than Freehand mode is.

Bounding box A dotted line with nodes that appears when you select an envelope shape. You can change the shape of the envelope by moving the nodes on the bounding box.

Cell The intersection of a row and a column in a spreadsheet, a cell holds a piece of information, such as a number or title.

Cels The individual frames that make up an actor. Each cel is a single drawing in one stage of the movement.

CCapture A screen capturing program that can capture the current configuration of your screen.

Chart Shows the progress, cost, or value of items over time.

Chart styles Various formats that determine the layout of graphed data (for example, pie, bar, line, area, and so on). A chart style comes in several different types.

Chart types Variations of a chart style. There can be several different chart types associated with a chart style.

Child object A single object that is part of a grouped object.

Click To press the left mouse button once (unless you specified the right mouse button to be the primary button, in which case you would press the right mouse button).

Client A Windows application to which you can link or embed. Some CorelDRAW! applications are both clients and servers; others are only clients.

Clipboard A holding place for information you have copied in a Windows application. You can paste an item you have copied to the Clipboard in any other Windows application.

Clone A duplicate object linked to a master object. Most changes you make to the master object will automatically be applied to the clone.

Clone tool In CorelPHOTO-PAINT!, this tool copies one portion of an image to a new area in the image or to an entirely new picture, which can be helpful to cover obstructions.

Color Replacer tool In CorelPHOTO-PAINT!, this tool works together with the Color Selection roll-up (which opens automatically when you start CorelPHOTO-PAINT!) or with the Eyedropper tool. The Color Replacer tool removes an outline color and replaces it with a fill color.

Control points Points at the end of a dotted line that appear when you drag over a node. Control points determine the direction and shape of a segment.

Copy To copy something that you have selected to the Windows Clipboard.

Corel 4.0 group icon A miniature picture that represents the window holding all of Corel 4.0's program icons.

CorelCHART! A program that lets you create cool-looking charts, including 3D bar charts.

CorelDRAW! Both the package of graphics applications and the main drawing program in the package.

CorelMOSAIC! A visual file manager that comes with CorelDRAW!.

CorelMOVE! An animation program that lets you create and play back animations. This program does not come with CorelDRAW! 3.

CorelPHOTO-PAINT! A painting program that lets you add all sorts of neat effects to your images.

CorelQUERY! A database application that comes with CorelDRAW! 5. This program lets you extract information from a remote database.

CorelSHOW! A presentation program that lets you run screen shows and print presentation overheads.

CorelTRACE! A program that lets you convert scanned images for use in CorelDRAW!.

CorelVENTURA! A complete desktop publishing program that comes with CorelDRAW! 5.

Data A computer term for information; more technically, code.

Data Manager A spreadsheet program in which you enter all of your chart data, including titles, labels, and numbers.

Destination application The application into which you've placed a linked or embedded object.

Dialog boxes Windows that appear to request command options, information, or confirmation needed to carry out a particular task. A dialog box often appears as a result of the command you have just selected.

Disk A round magnetic storage unit.

Disk space The amount of space available on your computer's hard drive or on a floppy disk where data can be written.

Double-click To press the left mouse button twice quickly. If nothing happens, you didn't click fast enough. Try again, pressing twice in rapid succession.

Drag To press and hold down the mouse button as you move the mouse.

Editable view Draw's normal view that displays graphics as they will appear printed, with fill, colors, and attributes.

Ellipse An oblong circle.

Embed To insert an object from another application. Embedding an object allows you to use that application's tools to create something, from within a different application.

Embedded object An object created in a different application that is inserted into the present application. When you double-click on an embedded object, the application you used to create the object starts up.

Envelope editing style Determines how the envelope's shape moves when you drag its nodes and control points.

Envelope mode Determines the severity of the distortion created by an envelope.

Eraser tool In CorelPHOTO-PAINT!, this tool replaces portions of your picture with the background color. Drag this tool over the areas you want to replace. If you double-click this tool, it replaces the entire image with the background color.

Export To save a DRAW file in a different file format so that you can use the file in another application.

Extrusion lines Dotted lines that appear when you're extruding an object. These lines show you what the dimensions of the extruded object will be.

Eyedropper tool In CorelPHOTO-PAINT!, this tool picks up a color from another part of your picture, the same way a real eyedropper can suck up liquid.

File filter A code that can translate code formats from another program so that you can import files created in that program into DRAW.

File format The format in which an application stores its software code.

Fill A color or pattern inside a closed object.

Fitting text to a path Wrapping text around an object, curve, line, or DRAW symbol.

Flood Fill tool In CorelPHOTO-PAINT!, this tool fills an area with the outline or fill color currently selected in the Color Settings roll-up. Click the left mouse button to fill an area with the fill color, or the right mouse button to fill an area with the outline color.

Foreground layer The top layer of a slide on which you can place individual objects that sit on top of the background layer.

Fountain fill A fill that blends two colors or tints of colors.

Freehand mode Using the Pencil tool, to drag the mouse across the screen to draw an object the same way you would drag a pencil across a piece of paper.

Gradient Fill tool This tool fills an area with a wash of colors: the fill gradually changes from one color to another.

Grouped object Two or more objects that have been grouped together using the **A**rrange **G**roup command. With a few exceptions, DRAW treats a grouped object as if it were a single object, meaning you can move, transform, rotate, and size the group of objects as one object.

Highlighting box An invisible rectangle formed by the selection handles of a selected object.

Icon A tiny picture that represents a function, program, tool, group window—you name it! Clicking or double-clicking on an icon always does something wonderful, depending on which icon you fool with.

Idiot Anyone not reading this book (who uses CorelDRAW!).

Import To open a file in DRAW that was created in a different program.

Impressionist Clone tool In CorelPHOTO-PAINT!, this tool copies the color range of an area so that you can reproduce multitudinous dots of the same colors in the color range.

Intermediate shapes The same as steps, these are the objects between two blended objects.

Key combination A combination of keys you can use to execute a command without using the menu.

Lasso Selection tool Selects an irregular area. Drag this tool over the area you want to select so that you can modify the irregular shape within that area.

Legend Used in two-dimensional charts, a legend identifies which series items are represented by the colors of the bars or lines, depending on the chart style.

Library A CorelMOVE! file that contains actors, props, and sounds to be used in an animation.

Link To insert an object from another application so that whenever you change the original object in the original application, the linked object reflects those changes automatically.

Linked object An object that is connected to an object or file in another application. When you change the original object, the linked object is updated automatically.

Local Undo tool In CorelPHOTO-PAINT!, this tool returns specific portions of your picture to the appearance they had before the last command or tool you used. The Local Undo tool is not available once you use the **Edit Undo** command.

Magic Wand Selection tool Selects all areas in an image that are of the same hue. Click on a color to select all areas of that color.

Marquee box A dotted rectangle that appears as you drag the Selection tool over objects to select them.

Master An object that has been cloned. Most changes you make to the master are applied to the clone automatically.

Megabyte A unit which measures the storage capacity of a disk or the amount of memory. Also known as "meg," a megabyte is 1,048,576 bytes.

Menu bar Holds a program's menus.

Menu command A command that appears on a menu. You can activate menu commands using the mouse or keyboard.

Mouse A hand-held device that doesn't really look anything like a real mouse; its main purpose is to run a pointer around your computer screen whenever you drag it on your desktop. You can use a mouse to select icons or menu commands, draw objects, or do just about anything you want. Please purchase a computer mouse. If you attempt to use one you grabbed out of your backyard, you'll probably end up with a lot of scratches and bites!

Node An endpoint of an object. The number and position of nodes determine the shape of an object.

Paragraph text Blocks of text you enter using the Paragraph text tool. Paragraph text automatically wraps to the next line.

Parallel extrusion An evenly shaped object with a 3D appearance.

Paste To paste something that you have copied from the Windows Clipboard into the active file.

Path Essentially, the outline of any object, or a line or curve.

Perspective extrusion An unevenly shaped object with a 3D appearance.

Playback control buttons Buttons in CorelMOVE! that let you play or reverse an animation, and move to the last frame, first frame, next frame, or previous frame.

Point To move the mouse to a particular place on-screen. For example, to point to an object, move the mouse cursor so it rests on top of the object.

Pointillist Clone tool In CorelPHOTO-PAINT!, this tool copies the color range of an area so that you can reproduce brush strokes of the same colors in the color range.

Polygon Any multi-sided object that is not a square or rectangle.

Polygon Selection tool Does just what it says: selects a polygonal area. You click just as if you were drawing a polygon: click to start the selection, double-click to define each side, and click to end.

Presentation A collection of slides that can be printed, made into 35mm slides, used for overhead transparencies, or run in a screen show.

Program group A window in the Program Manager that holds program icons that you can use to start programs.

Program Manager The main Windows window, which serves as the Windows home base. You can start CorelDRAW! programs from the Program Manager.

Prop A drawing that remains stationary in the background as the actors move during an animation.

Roll-up A window that you can permanently display on the computer screen, which gives you direct access to loads of menu commands and functions.

Screen show A presentation that is played back on-screen. You can use CorelSHOW! to create screen shows.

Segment A line or curve between two nodes.

Selection handles Boxes that border a selected object. You'll use these boxes to size, transform, and skew objects.

Series Any data item that is being charted; series items are row labels.

Server A Windows application from which you can link or embed. Some CorelDRAW! applications are both clients and servers; others are only servers.

Slide A presentation page.

Sound What you hear when an animation is playing. Your computer system has to have sound capabilities (such as a SoundBlaster card and speakers) in order for you to use sounds in your animations.

Source application The application in which you created the original object.

Start and end objects The two original objects that you blended together.

Steps Intermediate shapes between two objects blended together with DRAW's Blend roll-up. You determine the number of steps in a blend.

Texture fill Predesigned fills CorelDRAW! provides that are mixtures of colors and gradients. Texture fills resemble designs such as water colors, gravel, clouds, satellite photography, recycled paper, minerals, and more.

Texture Fill tool Fills an area with a cool-looking texture.

Tile Fill tool This tool fills an area with a repeating pattern.

Toolbox A section of the screen that holds all of the drawing, painting, charting, moving, and showing tools available in the various CorelDRAW! programs.

Uniform fill A solid color fill, including white, black, and shades of gray.

Vanishing point An "x" that appears when you're extruding an object. By dragging the vanishing point, you can change the depth.

Windows The operating system that serves as the foundation for the CorelDRAW! programs. Many popular software programs run under Windows. You cannot run CorelDRAW! without Windows.

Windows Clipboard A holding place for information you have copied in a Windows application. You can paste an item you have copied to the Clipboard in any other Windows application.

Wireframe view Shows only the outlines of objects, making it somewhat easier and faster to edit your objects.

X-axis The bottom, outside edge of the chart frame; x-axis items are column heads.

Z-axis In 3D charts, the Z-axis title describes the type of measurement used to chart the data: for example, days, dollars, number of people (it depends on what you're measuring).

Index

Symbols

A

B

C

D

E

F

G

H

I

J–L

M

N

O

P

Q–R

S

T

U

V

W

X–Y–Z

Also Available!

The Complete Idiot's Guide to 1-2-3, New Edition
ISBN: 1-56761-404-3
Softbound, $14.95 USA

The Complete Idiot's Guide to 1-2-3 for Windows
ISBN: 1-56761-400-0
Softbound, $14.95 USA

The Complete Idiot's Guide to Ami Pro
ISBN: 1-56761-453-1
Softbound, $14.95 USA

The Complete Idiot's Guide to Buying & Upgrading PCs
ISBN: 1-56761-274-1
Softbound, $14.95 USA

The Complete Idiot's Guide to Computer Terms
ISBN: 1-56761-266-0
Softbound, $9.95 USA

The Complete Idiot's Guide to Excel
ISBN: 1-56761-318-7
Softbound, $14.95 USA

The Complete Idiot's Guide to Internet
ISBN: 1-56761-414-0
Softbound, $19.95 USA

The Complete Idiot's Guide to The Mac
ISBN: 1-56761-395-0
Softbound, $14.95 USA

The Complete Idiot's Guide to VCRs
ISBN: 1-56761-294-6
Softbound, $9.95 USA

The Complete Idiot's Guide to WordPerfect
ISBN: 1-56761-187-7
Softbound, $14.95 USA

The Complete Idiot's Guide to WordPerfect for Windows
ISBN: 1-56761-282-2
Softbound, $14.95 USA

The Complete Idiot's Guide to Word for Windows
ISBN: 1-56761-355-1
Softbound, $14.95 USA

The Complete Idiot's Guide to Works for Windows
ISBN: 1-56761-451-5
Softbound, $14.95 USA

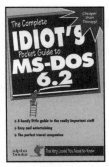

Who cares what you think? WE DO!

We take our customers' opinions very personally. After all, you're the reason we publish these books. If you're not happy, we're doing something wrong.

We'd appreciate it if you would take the time to drop us a note or fax us a fax. A real person—not a computer—reads every letter we get, and makes sure that your comments get relayed to the appropriate people.

Not sure what to say? Here are some details we'd like to know:

- Who you are (age, occupation, hobbies, etc.)
- Where you bought the book
- Why you picked this book instead of a different one
- What you liked best about the book
- What could have been done better
- Your overall opinion of the book
- What other topics you would purchase a book on

Mail, e-mail, or fax it to:

Faithe Wempen
Product Development Manager
Alpha Books
201 West 103rd Street
Indianapolis, IN 46290

FAX: (317) 581-4669
CIS: 75430,174

Special Offer!

Alpha Books needs people like you to give opinions about new and existing books. Product testers receive free books in exchange for providing their opinions about them. If you would like to be a product tester, please mention it in your letter, and make sure you include your full name, address, and daytime phone.